Texts and Monographs in Physics

Springer
Berlin
Heidelberg
New York
Hong Kong
London
Milan
Paris
Tokyo

H. J. Carmichael

Statistical Methods in Quantum Optics 1

Master Equations and Fokker–Planck Equations

With 28 Figures

 Springer

Professor Howard Carmichael

University of Auckland
Department of Physics
Private Bag 92019
Auckland, New Zealand

Library of Congress Cataloging-in-Publication Data

Carmichael, Howard, 1950– Statistical methods in quantum optics / H. J. Carmichael. p. cm. – (Texts and monographs in physics, 0172-5998) Includes bibliographical references and index.
1. Quantum optics–Statistical methods. 2. Quantum optics–Industrial applications. I. Title. II. Series. QC446.2.C374 1998 535–dc21 89-40873 CIP

First Edition 1999. Corrected Second Printing 2002

ISSN 0172-5998
ISBN 978-3-642-08133-0

Springer-Verlag Berlin Heidelberg New York
a member of BertelsmannSpringer Science+Business Media GmbH

http://www.springer.de

© Springer-Verlag Berlin Heidelberg 2010
Printed in Germany

Cover design: *design & production* GmbH, Heidelberg
Printed on acid-free paper

Dedication

To my mother
and to the memory of my father

"For even as He loves the arrow that flies, so He
loves also the bow that is stable."

Preface to the Second Corrected Printing

The material in the first volume of this book is standard and I have changed little in revising it for the second edition. My main task has been to correct the numerous errors that, having escaped detection in the original manuscript, were brought to my attention after the book came to print. My thanks to all those who have helped in this enterprise by informing me of an insidious misspelling, a mistake in an equation, or a mislabeled figure. A few other superficial changes have been made; I have reworked many of the figures and "prettied up" the typesetting in one or two spots.

There is just one change of real substance. I have chosen to replace the designation "quantum regression theorem," which has been standard in quantum optics circles for some three decades, with the more accurate "quantum regression formula." The replacement is perhaps not perfect, since the regression procedure introduced by Lax is expressed by different formulas on different occasions. In some cases the procedure runs very much parallel to Onsager's classical regression hypothesis – i.e., when a linearized treatment of fluctuations is carried out. In others it does not. The point to be made, however, is that the formula used, whatever its specific form, is never the expression of a "theorem"; it is the expression of a Markovian open system dynamics, reached from a microscopic model in quantum optics by way of an approximation, the same Markov dynamics that one finds defined, more formally, in the semigroup approach to open quantum systems. Physicists, all too often, remain unworried about semantic accuracy in a matter like this; the common designation is historical and no doubt harmless enough. On the other hand, for some reason, which was always difficult for me to understand, the quantum regression formula has attracted an undeserved level of suspicion throughout its 30 years of use; it seems not to be appreciated that the formula for multi time averages enjoys precisely the same grounding, Markov approximation and all, as the master equation itself – just as the master equation emerges in the Schrödinger picture, so, in the Heisenberg picture, emerges Lax's quantum regression. Given, then, what I perceive to be a background of misunderstanding, I think it wise to be as accurate and clear as possible. I have therefore avoided the word "theorem" in this second edition of Vol. 1, and also in Vol. 2. I have also added some commentary relating to this point in the section of Vol. 1 devoted to the quantum regression formula.

Auckland   *Howard Carmichael*
May 2002

Preface to the First Edition

As a graduate student working in quantum optics I encountered the question that might be taken as the theme of this book. The question definitely arose at that time though it was not yet very clearly defined; there was simply some deep irritation caused by the work I was doing, something quite fundamental I did not understand. Of course, so many things are not understood when one is a graduate student. However, my nagging question was not a technical issue, not merely a mathematical concept that was difficult to grasp. It was a sense that certain elementary notions that are accepted as starting points for work in quantum optics somehow had no fundamental foundation, no identifiable root. My inclination was to mine physics vertically, and here was a subject whose tunnels were dug horizontally. There were branches, certainly, going up and going down. Nonetheless, something major in the downwards direction was missing—at least in my understanding; no doubt others understood the connections downwards very well.

In retrospect I can identify the irritation. Quantum optics deals primarily with dynamics, quantum dynamics, and in doing so makes extensive use of words like "quantum fluctuations" and "quantum noise." The words seem harmless enough. Surely the ideas behind them are quite clear; after all, quantum mechanics is a statistical theory, and in its dynamical aspects it is therefore a theory of fluctuations. But there was my problem. Nothing in Schrödinger's equation fluctuates. What, then, *is* a quantum fluctuation?

In reply one might explore one of the horizontal tunnels. Statistical ideas became established in thermal physics during the early period of the quantum revolution. Although the central notions in this context are things like equilibrium ensembles, partition functions and the like, every graduate student is aware of the fluctuation aspect through the example of Brownian motion. Fluctuations are described using probability distributions, correlation functions, Fokker–Planck and Langevin equations, and mathematical devices such as these. In many instances the quantum analogs of these things are obvious. So, are quantum fluctuations simply thermal fluctuations that occur in the quantum realm? Well, once again, nothing fluctuates in Schrödinger's equation; yet the standard interpretation for the state solving this equation is statistical, and speaks of fluctuations, even when the most elementary system is described. Quantum fluctuations are therefore more fundamental than thermal fluctuations. They are a fundamental part of quantum theory—though apparently absent from its fundamental equation—and unlike thermal fluctuations, not comfortably accounted for by simply reflecting on the disorganized dynamics of a complex system.

I now appreciate more clearly where my question was headed: Yes it does head downwards, and it goes very deep. What is less clear is that there is a path in that direction understood by anyone very well. The direction is towards the foundations of quantum mechanics, and here one must face those notorious issues of interpretation that stimulate much confusion and contention but few definitive answers.

I must hasten to add that this book is not about the foundations of quantum mechanics—at least not in the formal sense; the subject is mentioned directly in only one chapter, near the end of Volume II. It is helpful to know, though, that this subject is the inevitable attractor to which four decades of development in quantum optics have been drawn. The book's real theme is quantum fluctuations, tackled for the most part at a pragmatic level. It is about the methods developed in quantum optics for analyzing quantum fluctuations in terms of a visualizable evolution over time. The qualifier "visualizable" is carried through as an informal connection to foundations. In view of it, I emphasize the Schrödinger and interaction pictures over the Heisenberg picture since in these pictures appropriate representations of the time-varying states (Glauber–Sudarshan or Wigner representations for example) can provide tangible access to something that fluctuates. Such mental props cannot be taken too literally, however, and the book is as much about their limitations as about their successes. I have written the book in a period when the demands for theoretical analyses of new experiments have required that the limitations be acknowledged and paid serious attention. The book meanders a bit in response to the proddings. Hopefully, though, there is always forwards momentum, towards methods of wider applicability and a more satisfying understanding of the foundations.

Quantum optics has a unique slant on quantum fluctuations, different from that of statistical physics with its emphasis on thermal equilibrium, and also differing from relativistic field theory where fluctuations refer either to virtual transitions—dressing stable objects—or little particle "explosions" (collisions) with a well-defined beginning and end. Quantum optics is concerned with matter interacting with electromagnetic waves at optical frequencies. At such frequencies, in terrestrial laboratories, it meets with quantum fluctuations that are real, and ongoing, and not inevitably buried in thermal noise; at least the latter has been the case since the invention of the laser; and it is the laser, overwhelmingly, that gives quantum optics its special perspective. The laser is basically a convenient source of coherence. Thought of simply, this is the coherence of a classical wave, but it is readily written into material systems where it must ultimately be seen as quantum coherence. The mix of coherence (waves) with the particle counting used to detect optical fields marks quantum optics for encounters with the difficult issues that arose around the ideas of Einstein and Bohr at the beginning of the century. Old issues are met with new clarity, but even more interesting are the entirely new dimensions. Seen as a quantum field, laser light is in a degenerate state, having a very large photon occupation number per mode. This property makes it easy to excite material systems far from thermal equilibrium, where simple perturbation theory is unable to account for the dynamics. In the classical limit one expects to encounter the gamut of non-linear phenomena: instability, bifurcation, multistability, chaos. One might ask where quantum fluctuations fit in the scheme of such things; no doubt as a minor perturbation in the approach to the classical limit. But in recent years the drive in optics towards precision and application has opened up the area of cavity QED. Here the electromagnetic field is confined within such a small volume that just one photon can supply the energy density needed to excite a system far from equilibrium. Under conditions like this, quantum fluctuations overwhelm the classical nonlinear dynamics. How, then, does the latter emerge from the fluctuations as the cavity QED limit is relaxed?

The book is divided into two volumes. This first volume deals with the statistical methods used in quantum optics up to the late 1970s. The material included here is based on a series of lectures I gave at the University of Texas at Austin during the fall semester of 1984. In this early period, methods for treating open systems in quantum optics were developed around two principal examples: the laser and resonance fluorescence. The two examples represent two defining themes for the subject, each identified with an innovation that extended the ways of thinking in some more established field. The laser required thinking in QED to be extended, beyond its focus on few-particle scattering to the treatment of many particle fields approaching the classical limit. The innovation was Glauber's coherence theory and the phase-space methods based on coherent states. The revival of the old topic of resonance fluorescence moved in the opposite direction. At first its concern was strong excitation—the nonperturbative limit which had been inaccessible to experiments before the laser was invented. Soon, however, a second theme developed. Contrasting with laser light and its approximation to a classical field, resonance fluorescence is manifestly a quantum field; its intensity fluctuations display features betraying their origin in *particle* scattering. The innovation here was in the theory of photoelectron counting—in the need to go beyond the semiclassical Mandel formula which holds only for statistical mixtures of coherent states. Thus, the study of resonance fluorescence began the preoccupation in quantum optics with the so-called nonclassical states of light.

Resonance fluorescence is treated in Chap. 2 and there are two chapters in this volume, Chaps. 7 and 8, on the theory of the laser. My aim with the example of resonance fluorescence is to illustrate the utility of the master equation and the quantum regression theorem for solving a significant problem, essentially exactly, with little more than some matrix algebra. Chapter 1 and the beginning of Chap. 2 fill in the background to the calculations. Here I provide derivations of the master equation and the quantum regression theorem. I think it important to emphasize that the quantum regression theorem is a derived result, equal in the firmness of its foundations to the master equation itself, and indeed a necessary adjunct to that equation if it is to be used to calculate anything other than the most trivial things (i.e. one-time operator averages).

Chapters 3–7 all lead up to a treatment of laser theory by the phase-space methods in Chap. 8. My purpose in Chap. 8 has been to carry through a systematic application of the phase-space methods to a nonequilibrium system of historical importance. Some readers will find the treatment overly detailed and be satisfied to simply skim the calculations. I would recommend the option, in fact, when the book is used as the basis for a course. In taking it, nothing need be lost with regard to the physics since the more useful results in laser theory are presented in Chap. 7 in a more accessible way. The earlier chapters have wide relevance in quantum optics. They deal with the properties of coherent states and the Glauber–Sudarshan P representation (Chap. 3), the Q and Wigner representations (Chap. 4), and the extension of these phase-space representations to two-state atoms (Chap. 6). Chapter 5 makes a short excursion to review those results from classical nonequilibrium statistical physics that are imported into quantum optics on the basis of the phase-space methods.

Volume I ends with Chap. 8 and the phase-space treatment of the laser. The treatment provides a rigorous basis for the standard visualization of amplitude and phase fluctuations in laser light. The visualization, however, is essentially classical, and the story of quantum fluctuations cannot be ended here. Being aware of the approximations used to derive the laser Fokker–Planck equation and having seen the example of resonance fluorescence, for which a similar simplification does not hold, it is clear that such classical visualizations cannot generally be sustained. Volume II will deal with the extension of the basic master equation approach to situations in which the naive phase-space visualization fails, where the quantum nature of the fluctuations has manifestations in the actual form of the evolution over time. Modern topics such as squeezing, the positive P representation, cavity QED, and quantum trajectory theory will be covered there.

I have sprinkled exercises throughout the book. In some cases they are included to excuse me from carrying through a calculation explicitly, or to repeat and generalize a calculation that has just been done. The exercises are integrated with the development of the subject matter and are intended, literally, as exercises, exercises for the practitioner, rather than an introduction to problems of topical interest. Their level varies. Some are quite difficult. The successful completion of the exercises will generally be aided by a detailed understanding of the calculations worked through in the book.

Numerous students and colleagues have read parts of this book as a manuscript and helped purge it of typographical errors or made other useful suggestions. I know I will not recall everyone, but I cannot overlook those whom I do remember. I am grateful for the interest and comments of Paul Alsing, Robert Ballagh, Young-Tak Chough, John Cooper, Rashed Haq, Wayne Itano, Jeff Kimble, Perry Rice, and Murray Wolinsky.

Eugene, Oregon   *Howard Carmichael*
August 1998

Table of Contents

Volume 2. Modern Topics

1. Dissipation in Quantum Mechanics: The Master Equation Approach

1.1 Introduction

This book deals with various quantum-statistical methods and their application to problems in quantum optics. The development of these methods arose out of the need to deal with dissipation in quantum optical systems. Thus, dissipation in quantized systems is a theme unifying the topics covered in the book. Two elementary systems provide the basic building blocks for a number of applications: the damped harmonic oscillator, which describes a single mode of the electromagnetic field in a lossy cavity (a cavity with imperfect mirrors), and the damped two-level atom. The need for a quantized treatment for the damped field mode arose originally in the context of the quantum theory of the maser and the laser. The damped two-level atom is, of course, of very general and fundamental interest, since it is just the problem of spontaneous emission. The book is structured around these two illustrative examples and their use in building quantum-theoretic treatments of resonance fluorescence and the single-mode laser. A second volume will extend the applications to the degenerate parametric oscillator and cavity quantum electrodynamics (cavity QED.). Discussion of the examples will guide the development of fundamental formalism. When we meet such things as master equations, phase-space representations, Fokker–Planck equations and stochastic differential equations, and the related methods of analysis, we will always have a specific application at hand with which to illustrate the formalism. Although formal methods will be introduced essentially from first principles, in places the treatment will necessarily be rather cursory. Ample references to the literature will hopefully offset any deficiencies.

Our objective in this book is to develop the background needed to gain access to issues of current research. The statistical methods we will cover were introduced over approximately two decades beginning in the early 1960's, stimulated by the invention of the laser. They are characterized by an emphasis on the two extremes of statistical physics – the single particle (resonance fluorescence) and very many particles (the single-mode laser). Where possible, they exploit analogies with the methods of classical statistical physics, though the incompatibility of a classical description with quantum mechanics is, in principle, always present. In the second volume we will enter into some of the modern research topics. The objective there will be to extend the methods discussed in this book, to move away from the one- and many-particle extremes and to face the quantum–classical incompatibility head on.

1.2 Inadequacy of an Ad Hoc Approach

In classical mechanics the essential features of dissipation, namely, the decay of oscillator amplitudes, particle velocities and energies, can be built into the theory by the simple addition of a velocity dependent force. For example, the harmonic oscillator, with Hamiltonian

$$H = \frac{p^2}{2m} + \tfrac{1}{2}m\omega^2 q^2 \tag{1.1}$$

and equations of motion

$$\dot{q} = p/m, \qquad \dot{p} = -m\omega^2 q, \tag{1.2}$$

becomes a damped harmonic oscillator with the addition of the force $-\gamma p$ to give

$$\dot{q} = p/m, \qquad \dot{p} = -\gamma p - m\omega^2 q, \tag{1.3}$$

or the familiar equation

$$\ddot{q} + \gamma\dot{q} + \omega^2 q = 0. \tag{1.4}$$

Can we simply transfer this approach to the quantized harmonic oscillator? For the quantized oscillator q and p become operators, $\hat{q}$ and $\hat{p}$, and (1.2) gives the Heisenberg equations of motion obtained from Hamiltonian (1.1) via the commutation relation

$$[\hat{q}, \hat{p}] = i\hbar. \tag{1.5}$$

After adding $-\gamma p$ to (1.3), the equations of motion remain linear; thus, the classical solution still holds when q and p become operators, and the expectation values of $\hat{q}$ and $\hat{p}$ will be damped in the same way as the classical variables. We seem to be in good shape. Consider, however, the evolution of the commutator $[\hat{q}, \hat{p}]$. From (1.3)

$$\frac{d}{dt}[\hat{q}, \hat{p}] = \dot{\hat{q}}\hat{p} + \hat{q}\dot{\hat{p}} - \dot{\hat{p}}\hat{q} - \hat{p}\dot{\hat{q}}$$
$$= -\gamma[\hat{q}, \hat{p}],$$

and

$$[\hat{q}(t), \hat{p}(t)] = e^{-\gamma t}[\hat{q}(0), \hat{p}(0)] = e^{-\gamma t}i\hbar. \tag{1.6}$$

As a consequence of this decay of the commutator the Heisenberg uncertainty also decays; the Heisenberg uncertainty relation becomes

$$\Delta q \Delta p \geq \tfrac{1}{2}\hbar e^{-\gamma t}. \tag{1.7}$$

In the face of this difficulty there have been various attempts to consistently incorporate dissipation into quantum mechanics. Some approaches based on novel quantization procedures remain controversial. We will not review these issues here. Of course, in many of the traditional domains of quantum mechanics dissipation plays no role: in the analysis of atomic structure, or the calculation of harmonic oscillator eigenstates and the like. The situation is quite different, though, in quantum optics. For example, the phenomenon of laser action, which gave birth to this field, takes place in a *lossy* cavity. In fact, applications in quantum optics have played a central role in developing methods to treat quantum-mechanical dissipation. We follow the widely accepted approach pioneered by Senitzky [1.1] for describing lossy maser cavities. Some discussion of alternative points of view can be found in papers by Ray [1.2] and Caldeira and Leggett [1.3], and references therein.

1.3 System Plus Reservoir Approach

The system plus reservoir approach begins from a microscopic view of the mechanism underlying dissipation. Although the procedure leading to (1.3) and (1.4) is often adequate in classical mechanics, even there it provides an incomplete description. In particular, equations (1.2) are time-reversal invariant, while in (1.3) this symmetry has been broken. If we want to understand the origin of this irreversibility we must begin by recognizing that the oscillator is damped through interactions with a large and complex system – its environment. This recognition also leads us to the fundamental relationship between dissipation and fluctuations. If the environment is some large system in thermal equilibrium, it will exert a fluctuating force $F(t)$ on an oscillator coupled to it, in addition to soaking up the oscillator's energy. Equation (1.4) must generally be replaced by a stochastic equation

$$\ddot{q} + \gamma \dot{q} + \omega^2 q = F(t)/m. \tag{1.8}$$

In many situations the added noise source cannot be overlooked – in electrical circuits, for example.

We observe that damping takes place through the coupling of the damped system to its environment. Is there anything in this observation to suggest a resolution of our problem with commutators? Well, the interaction between systems mixes their operators in a way which certainly does play a role in preserving commutators in time. Consider resonant harmonic oscillators coupled in the rotating-wave approximation. The Hamiltonian is

$$H = \hbar \omega a^\dagger a + \hbar \omega b^\dagger b + \hbar \kappa (a^\dagger b + a b^\dagger), \tag{1.9}$$

where ω is the frequency of the oscillators, κ is a coupling constant, $a^\dagger$ and $b^\dagger$ are creation operators, and a and b are the corresponding annihilation operators, satisfying commutation relations

$$[a, a^\dagger] = 1, \qquad [b, b^\dagger] = 1. \tag{1.10}$$

Note 1.1 To understand the origin of the Hamiltonian (1.9) first note that the free oscillator Hamiltonian (1.1) becomes

$$H = \hbar\omega(a^\dagger a + \tfrac{1}{2}) \tag{1.11}$$

where $\frac{1}{2}\hbar\omega$ is the zero-point energy, under the transformation

$$a \equiv \frac{1}{\sqrt{2\hbar m\omega}}(m\omega\hat{q} + i\hat{p}), \tag{1.12a}$$

$$a^\dagger \equiv \frac{1}{\sqrt{2\hbar m\omega}}(m\omega\hat{q} - i\hat{p}). \tag{1.12b}$$

Then (1.6) becomes

$$[a, a^\dagger] = e^{-\gamma t}. \tag{1.13}$$

In the *rotating-wave approximation* an interaction energy proportional to $\hat{q}_a\hat{q}_b$ gives the interaction Hamiltonian $\hbar\kappa(a^\dagger b + ab^\dagger)$ after the highly oscillatory terms (energy nonconserving terms) ab and $a^\dagger b^\dagger$ are neglected.

The solutions to the Heisenberg equations of motion following from (1.9) are

$$a(t) = e^{-i\omega t}[a(0)\cos\kappa t - ib(0)\sin\kappa t], \tag{1.14a}$$
$$b(t) = e^{-i\omega t}[b(0)\cos\kappa t - ia(0)\sin\kappa t]. \tag{1.14b}$$

Then

$$[a(t), a^\dagger(t)] = [a(0), a^\dagger(0)]\cos^2\kappa t + [b(0), b^\dagger(0)]\sin^2\kappa t = 1. \tag{1.15}$$

We see that the commutator for $a(t)$ and $a^\dagger(t)$ is preserved in time only by the presence of the operator $b(0)$ mixed into the solution for $a(t)$. Taking the environmental interaction into account in the treatment of dissipation, we might anticipate a similar mixing of environmental operators into the operators of the damped system in such a way as to preserve commutation relations. This is precisely what Senitzky found [1.1]. The fluctuating force in (1.8) becomes an operator in Senitzky's theory. Contributions from this environmental operator in the solutions for $\hat{q}(t)$ and $\hat{p}(t)$ introduce thermal fluctuations, and also preserve the commutation relations.

The master equation method we now discuss is essentially a Schrödinger picture version of Senitzky's theory. It is somewhat less transparent on this point about preserving commutation relations, so it is valuable to study Senitzky's calculation in the Heisenberg picture as well as the following. In both

the philosophy is to model environmental interactions by coupling the undamped system S to a reservoir R, beginning with a Hamiltonian in the
general form

$$H = H_S + H_R + H_{SR}, \tag{1.16}$$

where H_S and H_R are Hamiltonians for S and R, respectively, and H_{SR} is
an interaction Hamiltonian. The reservoir is only of indirect interest, and
its properties need only be specified in very general terms; for example, by
a temperature and an energy density of states. For illustrative purposes we
will give H_R and H_{SR} an explicit form once we get a little further into the
calculation.

The derivation given here follows the treatments by Louisell [1.4] and
Haken [1.5] fairly closely. There are some minor differences in the way approximations are introduced, and no attempt is made to follow either author's
notation. A rather different and more specialized approach is taken by Sargent, Scully and Lamb [1.6]. These authors get away without having to deal
with the complicated frequency and time integrals we will meet in our calculation. It is a useful exercise to study their calculation and try to find
where they introduce the physical assumptions we will use to deal with these
integrals. The physics must, of course, be the same.

We are seeking information about the system S without requiring detailed
information about the composite system $S \otimes R$. We will let $\chi(t)$ be the density
operator for $S \otimes R$ and define the reduced density operator $\rho(t)$ by

$$\rho(t) \equiv \mathrm{tr}_R[\chi(t)], \tag{1.17}$$

where the trace is taken over the reservoir states. Clearly, if $\hat{O}$ is an operator
in the Hilbert space of S we can calculate its average in the Schrödinger
picture if we have knowledge of $\rho(t)$ alone, and not of the full $\chi(t)$:

$$\langle \hat{O} \rangle = \mathrm{tr}_{S \otimes R}[\hat{O}\chi(t)] = \mathrm{tr}_S\{\hat{O}\mathrm{tr}_R[\chi(t)]\} = \mathrm{tr}_S[\hat{O}\rho(t)]. \tag{1.18}$$

Our objective is to obtain an equation for $\rho(t)$ with the properties of R
entering only as parameters.

1.3.1 The Schrödinger Equation in Integro-Differential Form

The Schrödinger equation for χ reads

$$\dot{\chi} = \frac{1}{i\hbar}[H, \chi], \tag{1.19}$$

where H is given by (1.16). We transform (1.19) into the interaction picture,
separating the rapid motion generated by $H_S + H_R$ from the slow motion
generated by the interaction H_{SR}. Defining

$$\tilde{\chi}(t) \equiv e^{(i/\hbar)(H_S+H_R)t}\chi(t)e^{-(i/\hbar)(H_S+H_R)t}, \tag{1.20}$$

from (1.16) and (1.19), we obtain

$$\dot{\tilde{\chi}} = \frac{i}{\hbar}(H_S + H_R)\tilde{\chi} - \frac{i}{\hbar}\tilde{\chi}(H_S + H_R) + e^{(i/\hbar)(H_S+H_R)t}\dot{\chi}e^{-(i/\hbar)(H_S+H_R)t}$$

$$= \frac{1}{i\hbar}[\tilde{H}_{SR}(t), \tilde{\chi}], \tag{1.21}$$

where $\tilde{H}_{SR}(t)$ is explicitly time-dependent:

$$\tilde{H}_{SR}(t) \equiv e^{(i/\hbar)(H_S+H_R)t}H_{SR}e^{-(i/\hbar)(H_S+H_R)t}. \tag{1.22}$$

We now integrate (1.21) formally to give

$$\tilde{\chi}(t) = \chi(0) + \frac{1}{i\hbar}\int_0^t dt'\,[\tilde{H}_{SR}(t'), \tilde{\chi}(t')], \tag{1.23}$$

and substitute for $\tilde{\chi}(t)$ inside the commutator in (1.21):

$$\dot{\tilde{\chi}} = \frac{1}{i\hbar}[\tilde{H}_{SR}(t), \chi(0)] - \frac{1}{\hbar^2}\int_0^t dt'\,[\tilde{H}_{SR}(t), [\tilde{H}_{SR}(t'), \tilde{\chi}(t')]]. \tag{1.24}$$

This equation is exact. Equation (1.19) has simply been cast into a convenient form which helps us identify reasonable approximations.

1.3.2 Born and Markov Approximations

We will assume that the interaction is turned on at $t = 0$ and that no correlations exist between S and R at this initial time. Then $\chi(0) = \tilde{\chi}(0)$ factorizes as

$$\chi(0) = \rho(0)R_0, \tag{1.25}$$

where R_0 is an initial reservoir density operator. Then, noting that

$$\text{tr}_R[\tilde{\chi}(t)] = e^{(i/\hbar)H_S t}\rho(t)e^{-(i/\hbar)H_S t} \equiv \tilde{\rho}(t), \tag{1.26}$$

after tracing over the reservoir, (1.24) gives the *master equation*

$$\dot{\tilde{\rho}} = -\frac{1}{\hbar^2}\int_0^t dt'\,\text{tr}_R\{[\tilde{H}_{SR}(t), [\tilde{H}_{SR}(t'), \tilde{\chi}(t')]]\}, \tag{1.27}$$

where, for simplicity, we have eliminated the term $(1/i\hbar)\text{tr}_R\{[\tilde{H}_{SR}(t), \chi(0)]\}$ with the assumption

$$\text{tr}_R[\tilde{H}_{SR}(t)R_0] = 0. \tag{1.28}$$

This is guaranteed if the reservoir operators coupling to S have zero mean in the state R_0, a condition which can always be arranged by simply including $\text{tr}_R(H_{SR}R_0)$ in the system Hamiltonian (see Sect. 2.2.4 and Note 8.8).

We have stated that $\tilde{\chi}$ factorizes at $t = 0$. At later times correlations between S and R will arise due to the coupling between the system and the reservoir. We have assumed, however, that this coupling is very weak, and at all times $\chi(t)$ should only show deviations of order H_{SR} from an uncorrelated state. Furthermore, R is a large system whose state should be virtually unaffected by its coupling to S (of course, we expect the state of S to be significantly affected by R – we want it to be damped). We therefore write

$$\tilde{\chi}(t) = \tilde{\rho}(t)R_0 + O(H_{SR}).$$ (1.29)

Now we can make our first major approximation, a *Born approximation*. Neglecting terms higher than second order in H_{SR}, we write (1.27) as

$$\dot{\tilde{\rho}} = -\frac{1}{\hbar^2} \int_0^t dt'\, \text{tr}_R\{[\tilde{H}_{SR}(t), [\tilde{H}_{SR}(t'), \tilde{\rho}(t')R_0]]\}.$$ (1.30)

A detailed discussion of this approximation can be found in the work of Haake [1.7, 1.8].

Equation (1.30) is still a complicated equation. In particular, it is not Markovian since the future evolution of $\tilde{\rho}(t)$ depends on its past history through the integration over $\tilde{\rho}(t')$ (the future behavior of a Markovian system depends only on its present state). Our second major approximation, the *Markov approximation*, replaces $\tilde{\rho}(t')$ by $\tilde{\rho}(t)$ to obtain a *master equation in the Born–Markov approximation*:

$$\dot{\tilde{\rho}} = -\frac{1}{\hbar^2} \int_0^t dt'\, \text{tr}_R\{[\tilde{H}_{SR}(t), [\tilde{H}_{SR}(t'), \tilde{\rho}(t)R_0]]\}.$$ (1.31)

1.3.3 The Markov Approximation and Reservoir Correlations

Markovian behavior seems reasonable on physical grounds. Potentially, S can depend on its past history because its earlier states become imprinted as changes in the reservoir state through the interaction H_{SR}; earlier states are then reflected back on the future evolution of S as it interacts with the changed reservoir. If, however, the reservoir is a large system maintained in thermal equilibrium, we do not expect it to preserve the minor changes brought by its interaction with S for very long; not for long enough to significantly affect the future evolution of S. It becomes a question of reservoir correlation time versus the time scale for significant change in S. By studying the integrand of (1.30) with this view in mind we can make the underlying assumption of the Markov approximation more explicit.

Let us make our model a little more specific by writing

$$H_{SR} = \hbar \sum_i s_i \Gamma_i,$$ (1.32)

where the s_i are operators in the Hilbert space of S and the Γ_i are reservoir operators, operators in the Hilbert space of R. Then

$$\tilde{H}_{SR}(t) = \hbar \sum_i e^{(i/\hbar)(H_S+H_R)t} s_i \Gamma_i e^{-(i/\hbar)(H_S+H_R)t}$$

$$= \hbar \sum_i \left(e^{(i/\hbar)H_S t} s_i e^{-(i/\hbar)H_S t} \right) \left(e^{(i/\hbar)H_R t} \Gamma_i e^{-(i/\hbar)H_R t} \right)$$

$$= \hbar \sum_i \tilde{s}_i(t) \tilde{\Gamma}_i(t). \tag{1.33}$$

The master equation in the Born approximation [Eq. (1.30)] is now

$$\dot{\tilde{\rho}} = -\sum_{i,j} \int_0^t dt' \, \mathrm{tr}_R \big\{ [\tilde{s}_i(t)\tilde{\Gamma}_i(t), [\tilde{s}_j(t')\tilde{\Gamma}_j(t'), \tilde{\rho}(t')R_0]] \big\}$$

$$= -\sum_{i,j} \int_0^t dt' \Big\{ \tilde{s}_i(t)\tilde{s}_j(t')\tilde{\rho}(t') \, \mathrm{tr}_R[\tilde{\Gamma}_i(t)\tilde{\Gamma}_j(t')R_0]$$

$$- \tilde{s}_i(t)\tilde{\rho}(t')\tilde{s}_j(t') \mathrm{tr}_R[\tilde{\Gamma}_i(t)R_0\tilde{\Gamma}_j(t')] - \tilde{s}_j(t')\tilde{\rho}(t')\tilde{s}_i(t)$$

$$\times \mathrm{tr}_R[\tilde{\Gamma}_j(t')R_0\tilde{\Gamma}_i(t)] + \tilde{\rho}(t')\tilde{s}_j(t')\tilde{s}_i(t) \, \mathrm{tr}_R[R_0\tilde{\Gamma}_j(t')\tilde{\Gamma}_i(t)] \Big\}$$

$$= -\sum_{i,j} \int_0^t dt' \Big\{ [\tilde{s}_i(t)\tilde{s}_j(t')\tilde{\rho}(t') - \tilde{s}_j(t')\tilde{\rho}(t')\tilde{s}_i(t)]\langle \tilde{\Gamma}_i(t)\tilde{\Gamma}_j(t')\rangle_R$$

$$+ [\tilde{\rho}(t')\tilde{s}_j(t')\tilde{s}_i(t) - \tilde{s}_i(t)\tilde{\rho}(t')\tilde{s}_j(t')]\langle \tilde{\Gamma}_j(t')\tilde{\Gamma}_i(t)\rangle_R \Big\}, \tag{1.34}$$

where we have used the cyclic property of the trace – $\mathrm{tr}(\hat{A}\hat{B}\hat{C}) = \mathrm{tr}(\hat{C}\hat{A}\hat{B}) = \mathrm{tr}(\hat{B}\hat{C}\hat{A})$ – and write

$$\langle \tilde{\Gamma}_i(t)\tilde{\Gamma}_j(t')\rangle_R = \mathrm{tr}_R[R_0\tilde{\Gamma}_i(t)\tilde{\Gamma}_j(t')], \tag{1.35a}$$

$$\langle \tilde{\Gamma}_j(t')\tilde{\Gamma}_i(t)\rangle_R = \mathrm{tr}_R[R_0\tilde{\Gamma}_j(t')\tilde{\Gamma}_i(t)]. \tag{1.35b}$$

The properties of the reservoir enter (1.34) through the two correlation functions (1.35a) and (1.35b). We can justify the replacement of $\tilde{\rho}(t')$ by $\tilde{\rho}(t)$ if these correlation functions decay very rapidly on the timescale on which $\tilde{\rho}(t)$ varies. Ideally, we might take

$$\langle \tilde{\Gamma}_i(t)\tilde{\Gamma}_j(t')\rangle_R \propto \delta(t-t'). \tag{1.36}$$

The Markov approximation then relies, as suggested, on the existence of two widely separated time scales: a slow time scale for the dynamics of the system S, and a fast time scale characterizing the decay of reservoir correlation functions. Further discussion of this point is given by Schieve and Middleton [1.9]. We will look explicitly at reservoir correlation functions and the separation of time scales in our first example.

1.4 The Damped Harmonic Oscillator

1.4.1 Master Equation for the Damped Harmonic Oscillator

We now adopt an explicit model. For the Hamiltonian of the composite system $S \otimes R$ we write

$$H_S \equiv \hbar \omega_0 a^\dagger a, \tag{1.37a}$$

$$H_R \equiv \sum_j \hbar \omega_j r_j{}^\dagger r_j, \tag{1.37b}$$

$$H_{SR} \equiv \sum_j \hbar (\kappa_j^* a r_j{}^\dagger + \kappa_j a^\dagger r_j) = \hbar (a \Gamma^\dagger + a^\dagger \Gamma). \tag{1.37c}$$

The system S is a harmonic oscillator with frequency ω_0 and creation and annihilation operators $a^\dagger$ and a, respectively; the reservoir R is modeled as a collection of harmonic oscillators with frequencies ω_j, and corresponding creation and annihilation operators $r_j{}^\dagger$ and r_j, respectively; the oscillator a couples to the jth reservoir oscillator via a coupling constant κ_j in the rotating-wave approximation. We take the reservoir to be in thermal equilibrium at temperature T, with density operator

$$R_0 = \prod_j e^{-\hbar \omega_j r_j{}^\dagger r_j / k_B T} \left(1 - e^{-\hbar \omega_j / k_B T} \right), \tag{1.38}$$

where k_B is Boltzmann's constant. It is not necessary to be so specific about the reservoir model. Haken [1.5], for example, keeps his discussion quite general. Aside, however, from its pedagogical clarity, the oscillator model is physically reasonable in many circumstances. The reservoir oscillators might be the many modes of the vacuum radiation field into which an optical cavity mode decays through partially transmitting mirrors, or into which an excited atom decays via spontaneous emission; alternatively, they might represent phonon modes in a solid.

The identification with (1.34) is made by setting

$$s_1 = a, \qquad s_2 = a^\dagger, \tag{1.39a}$$

$$\Gamma_1 = \Gamma^\dagger \equiv \sum_j \kappa_j^* r_j{}^\dagger, \qquad \Gamma_2 = \Gamma \equiv \sum_j \kappa_j r_j, \tag{1.39b}$$

and then from (1.33) and (1.37), the operators in the interaction picture are

$$\tilde{s}_1(t) = e^{i \omega_0 a^\dagger a t} a e^{-i \omega_0 a^\dagger a t} = a e^{-i \omega_0 t}, \tag{1.40a}$$

$$\tilde{s}_2(t) = e^{i \omega_0 a^\dagger a t} a^\dagger e^{-i \omega_0 a^\dagger a t} = a^\dagger e^{i \omega_0 t}, \tag{1.40b}$$

and

$$\tilde{\Gamma}_1(t) = \tilde{\Gamma}^\dagger(t) = \exp\left(i\sum_n \omega_n r_n{}^\dagger r_n t\right)\sum_j \kappa_j^* r_j{}^\dagger \exp\left(-i\sum_m \omega_m r_m{}^\dagger r_m t\right)$$

$$= \sum_j \kappa_j^* r_j{}^\dagger e^{i\omega_j t}, \tag{1.41a}$$

$$\tilde{\Gamma}_2(t) = \tilde{\Gamma}(t) = \exp\left(i\sum_n \omega_n r_n{}^\dagger r_n t\right)\sum_j \kappa_j r_j \exp\left(-i\sum_m \omega_m r_m{}^\dagger r_m t\right)$$

$$= \sum_j \kappa_j r_j e^{-i\omega_j t}, \tag{1.41b}$$

where in (1.41) we use the fact that operators for different reservoir oscillators commute. To show, for example, that $e^{i\omega_0 a^\dagger a t} a e^{-i\omega_0 a^\dagger a t} = a e^{-i\omega_0 t}$, observe that the left hand side is just the formal solution to the Heisenberg equation of motion $\dot{a} = -i\omega_0[a, a^\dagger a] = -i\omega_0 a$. Note that, from (1.38) and (1.41), $\langle \tilde{\Gamma}_1(t)\rangle_R = \langle \tilde{\Gamma}_2(t)\rangle_R = 0$, as required by the assumption (1.28).

Now, since the summation in (1.34) runs over $i = 1, 2$ and $j = 1, 2$, the integrand involves sixteen terms. We write

$$\dot{\tilde{\rho}} = -\int_0^t dt'\left\{\left[aa\tilde{\rho}(t') - a\tilde{\rho}(t')a\right]e^{-i\omega_0(t+t')}\langle\tilde{\Gamma}^\dagger(t)\tilde{\Gamma}^\dagger(t')\rangle_R + \text{h.c.}\right.$$

$$+ \left[a^\dagger a^\dagger \tilde{\rho}(t') - a^\dagger \tilde{\rho}(t')a^\dagger\right]e^{i\omega_0(t+t')}\langle\tilde{\Gamma}(t)\tilde{\Gamma}(t')\rangle_R + \text{h.c.}$$

$$+ \left[aa^\dagger \tilde{\rho}(t') - a^\dagger \tilde{\rho}(t')a\right]e^{-i\omega_0(t-t')}\langle\tilde{\Gamma}^\dagger(t)\tilde{\Gamma}(t')\rangle_R + \text{h.c.}$$

$$\left.+ \left[a^\dagger a\tilde{\rho}(t') - a\tilde{\rho}(t')a^\dagger\right]e^{i\omega_0(t-t')}\langle\tilde{\Gamma}(t)\tilde{\Gamma}^\dagger(t')\rangle_R + \text{h.c.}\right\}, \tag{1.42}$$

where the reservoir correlation functions are explicitly:

$$\langle\tilde{\Gamma}^\dagger(t)\tilde{\Gamma}^\dagger(t')\rangle_R = \sum_{j,k} \kappa_j^* \kappa_k^* e^{i\omega_j t} e^{i\omega_k t'} \text{tr}_R(R_0 r_j{}^\dagger r_k{}^\dagger) = 0, \tag{1.43}$$

$$\langle\tilde{\Gamma}(t)\tilde{\Gamma}(t')\rangle_R = \sum_{j,k} \kappa_j \kappa_k e^{-i\omega_j t} e^{-i\omega_k t'} \text{tr}_R(R_0 r_j r_k) = 0, \tag{1.44}$$

$$\langle\tilde{\Gamma}^\dagger(t)\tilde{\Gamma}(t')\rangle_R = \sum_{j,k} \kappa_j^* \kappa_k e^{i\omega_j t} e^{-i\omega_k t'} \text{tr}_R(R_0 r_j{}^\dagger r_k)$$

$$= \sum_j |\kappa_j|^2 e^{i\omega_j(t-t')}\bar{n}(\omega_j, T), \tag{1.45}$$

$$\langle\tilde{\Gamma}(t)\tilde{\Gamma}^\dagger(t')\rangle_R = \sum_{j,k} \kappa_j \kappa_k^* e^{-i\omega_j t} e^{i\omega_k t'} \text{tr}_R(R_0 r_j r_k{}^\dagger)$$

$$= \sum_j |\kappa_j|^2 e^{-i\omega_j(t-t')}[\bar{n}(\omega_j, T) + 1], \tag{1.46}$$

with

$$\bar{n}(\omega_j, T) = \mathrm{tr}_R(R_0 r_j{}^\dagger r_j) = \frac{e^{-\hbar\omega_j/k_BT}}{1 - e^{-\hbar\omega_j/k_BT}}. \tag{1.47}$$

The correlation functions (1.43)–(1.46) follow quite readily by evaluating the trace using the multimode Fock states as a basis. $\bar{n}(\omega_j, T)$ is the mean photon number for an oscillator with frequency ω_j in thermal equilibrium at temperature T.

The nonvanishing reservoir correlation functions (1.45) and (1.46) involve a summation over the reservoir oscillators. We change this summation to an integration by introducing a density of states $g(\omega)$ such that $g(\omega)d\omega$ gives the number of oscillators with frequencies in the interval ω to $\omega + d\omega$. Making the change of variable

$$\tau = t - t', \tag{1.48}$$

(1.42) can then be restated as

$$\dot{\tilde{\rho}} = -\int_0^t d\tau \left\{ \left[aa^\dagger \tilde{\rho}(t - \tau) - a^\dagger \tilde{\rho}(t - \tau)a \right] e^{-i\omega_0\tau} \langle \tilde{\Gamma}^\dagger(t)\tilde{\Gamma}(t - \tau)\rangle_R + \mathrm{h.c.} \right.$$
$$\left. + \left[a^\dagger a\tilde{\rho}(t - \tau) - a\tilde{\rho}(t - \tau)a^\dagger \right] e^{i\omega_0\tau} \langle \tilde{\Gamma}(t)\tilde{\Gamma}^\dagger(t - \tau)\rangle_R + \mathrm{h.c.} \right\}, \tag{1.49}$$

where the nonzero reservoir correlation functions are

$$\langle \tilde{\Gamma}^\dagger(t)\tilde{\Gamma}(t - \tau)\rangle_R = \int_0^\infty d\omega\, e^{i\omega\tau} g(\omega)|\kappa(\omega)|^2 \bar{n}(\omega, T), \tag{1.50}$$

$$\langle \tilde{\Gamma}(t)\tilde{\Gamma}^\dagger(t - \tau)\rangle_R = \int_0^\infty d\omega\, e^{-i\omega\tau} g(\omega)|\kappa(\omega)|^2 [\bar{n}(\omega, T) + 1], \tag{1.51}$$

with

$$\bar{n}(\omega, T) = \frac{e^{-\hbar\omega/k_BT}}{1 - e^{-\hbar\omega/k_BT}}. \tag{1.52}$$

We can now argue more specifically about the Markov approximation. Are (1.50) and (1.51) approximately proportional to $\delta(\tau)$? We can certainly see that for τ "large enough" the oscillating exponential will average the "slowly varying" functions $g(\omega)$, $|\kappa(\omega)|^2$, and $\bar{n}(\omega, T)$ essentially to zero. However, how large is large enough? Can we get some idea of the width of these correlation functions? Let us look at (1.50), taking $g(\omega)|\kappa(\omega)|^2 = C\omega$, with C a constant. This correlation function may be evaluated in terms of the trigamma function [1.10]:

$$\langle \tilde{\Gamma}^\dagger(t)\tilde{\Gamma}(t - \tau)\rangle_R = C\int_0^\infty d\omega\, e^{i\omega\tau} \frac{\omega e^{-\hbar\omega/k_BT}}{1 - e^{-\hbar\omega/k_BT}}$$
$$= Ct_R^{-2}\int_0^\infty dx\, \frac{xe^{-(1-i\tau/t_R)x}}{1 - e^{-x}}$$
$$= Ct_R^{-2}\psi'(1 - i\tau/t_R), \tag{1.53}$$

where we have defined the reservoir correlation time $t_R = \hbar/k_B T$. A simple approximation gives some insight into the behavior of the trigamma function. Set

$$\frac{\omega e^{-\hbar\omega/k_B T}}{1 - e^{-\hbar\omega/k_B T}} \approx \frac{k_B T}{\hbar} e^{-\hbar\omega/k_B T}; \tag{1.54}$$

then

$$\langle \tilde{\Gamma}^\dagger(t)\tilde{\Gamma}(t-\tau)\rangle_R \approx C\frac{k_B T}{\hbar} \int_0^\infty d\omega\, e^{i\omega\tau} e^{-\hbar\omega/k_B T}$$

$$\approx C t_R^{-2} \frac{1 + i\tau/t_R}{1 + (\tau/t_R)^2}. \tag{1.55}$$

The approximation is accurate for low frequencies, but is not so good for $\omega \sim k_B T/\hbar = t_R^{-1}$; here the error is $\sim 40\%$. It is adequate, nevertheless, to give us a feel for the qualitative behavior of the reservoir correlation function. Actually, the exact result for the *real* part of the correlation function can be computed with little effort using the formula [1.10]

$$\mathrm{Re}[\psi'(1 - i\tau/t_R)] = \tfrac{1}{2}\pi^2\left[1 - \coth^2(\pi\tau/t_R)\right] + \tfrac{1}{2}(\tau/t_R)^{-2}. \tag{1.56}$$

The exact result is plotted together with the real part of (1.55) for comparison in Fig. 1.1(a).

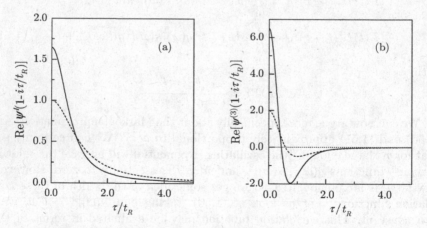

Fig. 1.1 (a) Real part of the reservoir correlation function for $g(\omega)|\kappa(\omega)|^2 = C\omega$ plotted from (1.56) (solid line) and (1.55) (dashed line). (b) Real part of the reservoir correlation function for $g(\omega)|\kappa(\omega)|^2 = C\omega^3$ plotted from (1.61) (solid line) and (1.60) (dashed line).

Equation (1.55) indicates a correlation function peaked about $\tau = 0$ with a width $t_R = \hbar/k_B T$. In (1.49) the reservoir correlation functions are integrated against two time-dependent terms: $\tilde{\rho}(t-\tau)$ and $e^{\pm i\omega_0\tau}$. Now at room temperature $\hbar/k_B T \approx 0.25 \times 10^{-13}$s. If the oscillator a represents an optical cavity mode, we expect $\tilde{\rho}(t-\tau)$ to vary on the time scale of a typical cavity decay time, $t_S \sim 10^{-8}$s, and if ω_0 is an optical frequency, $e^{\pm i\omega_0\tau}$ oscillates on a time scale $t_0 \sim 10^{-15}$s. Then since $t_S/t_R \sim 10^5$, it seems we can justify the Markov approximation and replace $\tilde{\rho}(t-\tau)$ by $\tilde{\rho}(t)$. But, with $t_0/t_R \sim 10^{-2}$, we cannot set $\tau = 0$ in the terms $e^{\pm i\omega_0\tau}$. Rather, integrating the reservoir correlation functions against these oscillating terms will extract their ω_0 frequency components, just as in a Fourier transform.

After taking a closer look we might worry a little about the imaginary part of (1.55). This has a long tail which decays as $(\tau/t_R)^{-1}$; the integral of this tail is logarithmically divergent; far out in the tail the replacement of $\tilde{\rho}(t-\tau)$ by $\tilde{\rho}(t)$ will not be justified. It is, however, the ω_0 frequency component of the product $\tilde{\rho}(t-\tau)\langle\tilde{\Gamma}^\dagger(t)\tilde{\Gamma}(t-\tau)\rangle_R$ that survives the integral in (1.49), and, with $t_0 < t_R << t_S$, this frequency component is contributed by the short-time behavior of (1.55), where the replacement of $\tilde{\rho}(t-\tau)$ by $\tilde{\rho}(t)$ is justified.

In fact, the divergent tail is a consequence of the form we have chosen for $g(\omega)\kappa^2(\omega)$. More generally, if we take $g(\omega)|\kappa(\omega)|^2 = C\omega^n$, with n a positive integer,

$$\langle\tilde{\Gamma}^\dagger(t)\tilde{\Gamma}(t-\tau)\rangle_R = (-i)^{n-1}\frac{d^{n-1}}{d\tau^{n-1}}\left[Ct_R^{-2}\psi'(1-i\tau/t_R)\right]$$

$$= Ct_R^{-(n+1)}(-1)^{n-1}\psi^{(n)}(1-i\tau/t_R), \quad (1.57)$$

where the $\psi^{(n)}$ are the polygamma functions [2.10]. In the approximation (1.54)

$$\langle\tilde{\Gamma}^\dagger(t)\tilde{\Gamma}(t-\tau)\rangle_R = (-i)^{n-1}\frac{d^{n-1}}{d\tau^{n-1}}\left[Ct_R^{-2}\frac{1+i\tau/t_R}{1+(\tau/t_R)^2}\right]$$

$$= Ct_R^{-(n+1)}(-i)^{n-1}\frac{d^{n-1}}{d(\tau/t_R)^{n-1}}\frac{1+i\tau/t_R}{1+(\tau/t_R)^2}. \quad (1.58)$$

For $\tau/t_R >> 1$ the asymptotic form of the polygamma function gives

$$\langle\tilde{\Gamma}^\dagger(t)\tilde{\Gamma}(t-\tau)\rangle_R \sim -Ct_R^{-(n+1)}\left[i^{n+1}(n-1)!\right]\left[\tfrac{1}{2}n(\tau/t_R)^{-(n+1)}\right.$$

$$\left. -i(\tau/t_R)^{-n}\right], \quad (1.59)$$

which has no $(\tau/t_R)^{-1}$ tail for $n > 1$.

The case $n = 3$ is of special interest since this corresponds to the form of $g(\omega)|\kappa(\omega)|^2$ that we will meet when we apply our theory to the damped two-level atom (Sect. 2.2). The approximate result (1.58) gives

$$\langle\tilde{\Gamma}^\dagger(t)\tilde{\Gamma}(t-\tau)\rangle_R \approx Ct_R^{-4}\frac{2\left[1-3(\tau/t_R)^2\right]+i2(\tau/t_R)\left[3-(\tau/t_R)^2\right]}{\left[1+(\tau/t_R)^2\right]^3}. \quad (1.60)$$

For comparison with the real part of this result, the real part of the exact correlation function can be computed from (1.57) using the formula

$$\text{Re}\left[\psi^{(3)}(1 - i\tau/t_R)\right] = \pi^4\left[1 - \coth^2(\pi\tau/t_R)\right]\left[1 - 3\coth^2(\pi\tau/t_R)\right]$$
$$- 3(\tau/t_R)^{-4}. \tag{1.61}$$

This formula is obtained by taking two derivatives of (1.56). The exact and approximate results for the real part of the correlation function are plotted in Fig. 1.1(b). Again the correlation function is peaked around $\tau = 0$ with a width $\sim t_R$. The approximate correlation function (1.60) explicitly shows the $(\tau/t_R)^{-4}$ and $(\tau/t_R)^{-3}$ dependence for the real and imaginary parts, respectively, in the large τ limit, as given by (1.59).

Exercise 1.1 Consider the correlation function (1.51). The second term inside the square bracket comes from quantum (vacuum) fluctuations. It arose from our use of the boson commutation relation in the derivation of (1.46). What contribution does this term make to the correlation function?

Continuing our derivation now from (1.49), it is actually more straightforward to evaluate the time integral first, without performing the frequency integrals to obtain an explicit form for the reservoir correlation functions. This is possible now we are satisfied that the τ integration is dominated by times that are much shorter than the time scale for the evolution of $\tilde{\rho}$. With $\tilde{\rho}(t - \tau)$ replaced by $\tilde{\rho}(t)$ (Markov approximation), (1.49) becomes

$$\dot{\tilde{\rho}} = \alpha(a\tilde{\rho}a^\dagger - a^\dagger a\tilde{\rho}) + \beta(a\tilde{\rho}a^\dagger + a^\dagger\tilde{\rho}a - a^\dagger a\tilde{\rho} - \tilde{\rho}aa^\dagger) + \text{h.c.}, \tag{1.62}$$

with

$$\alpha \equiv \int_0^t d\tau \int_0^\infty d\omega\, e^{-i(\omega-\omega_0)\tau} g(\omega)|\kappa(\omega)|^2, \tag{1.63}$$

$$\beta \equiv \int_0^t d\tau \int_0^\infty d\omega\, e^{-i(\omega-\omega_0)\tau} g(\omega)|\kappa(\omega)|^2\bar{n}(\omega, T). \tag{1.64}$$

Then, since t is of the order of t_S and the τ integration is dominated by much shorter times $\sim t_R$, we can extend the τ integration to infinity and evaluate α and β using

$$\lim_{t\to\infty}\int_0^t d\tau\, e^{-i(\omega-\omega_0)\tau} = \pi\delta(\omega - \omega_0) + i\frac{P}{\omega_0 - \omega}, \tag{1.65}$$

where P indicates the Cauchy principal value. We find

$$\alpha = \pi g(\omega_0)|\kappa(\omega_0)|^2 + i\Delta, \tag{1.66}$$
$$\beta = \pi g(\omega_0)|\kappa(\omega_0)|^2\bar{n}(\omega_0) + i\Delta', \tag{1.67}$$

with

$$\Delta \equiv P \int_0^\infty d\omega \, \frac{g(\omega)|\kappa(\omega)|^2}{\omega_0 - \omega}, \tag{1.68}$$

$$\Delta' \equiv P \int_0^\infty d\omega \, \frac{g(\omega)|\kappa(\omega)|^2}{\omega_0 - \omega} \bar{n}(\omega, T). \tag{1.69}$$

Note 1.2 To obtain (1.65), we have

$$\int_0^t d\tau \, e^{-i(\omega - \omega_0)\tau} = \frac{\sin(\omega - \omega_0)t}{\omega - \omega_0} - i\frac{1 - \cos(\omega - \omega_0)t}{\omega - \omega_0}.$$

The limit as t tends to infinity is defined anticipating the role of the right-hand side inside an integration over ω, thus:

$$\lim_{t \to \infty} \int_{-\infty}^\infty d\omega f(\omega) \frac{\sin(\omega - \omega_0)t}{\omega - \omega_0} = f(\omega_0) \lim_{t \to \infty} \int_{-\infty}^\infty d\omega \, \frac{\sin(\omega - \omega_0)t}{\omega - \omega_0}$$

$$= \pi f(\omega_0)$$

$$= \int_{-\infty}^\infty d\omega \, \pi\delta(\omega - \omega_0)f(\omega);$$

also

$$\lim_{t \to \infty} \int_{-\infty}^\infty d\omega f(\omega) \frac{1 - \cos(\omega - \omega_0)t}{\omega - \omega_0}$$

$$= \int_{-\infty}^\infty d\omega \, \frac{f(\omega)}{\omega - \omega_0} - \lim_{t \to \infty} \int_{-\infty}^\infty d\omega \, \frac{f(\omega)\cos(\omega - \omega_0)t}{\omega - \omega_0}$$

$$= P \int_{-\infty}^\infty d\omega \, \frac{f(\omega)}{\omega - \omega_0},$$

where the term

$$\lim_{t \to \infty} \int_{-\infty}^\infty d\omega \, \frac{f(\omega)\cos(\omega - \omega_0)t}{\omega - \omega_0}$$

subtracts the singularity at $\omega = \omega_0$ to give the principal value integral [1.11].

We finally have our master equation for the damped harmonic oscillator. After defining

$$\gamma \equiv 2\pi g(\omega_0)|\kappa(\omega_0)|^2, \tag{1.70a}$$

$$\bar{n} \equiv \bar{n}(\omega_0, T), \tag{1.70b}$$

from (1.62), (1.66), and (1.67), we obtain

$$\dot{\tilde{\rho}} = -i\Delta[a^\dagger a, \tilde{\rho}] + \frac{\gamma}{2}(2a\tilde{\rho}a^\dagger - a^\dagger a\tilde{\rho} - \tilde{\rho}a^\dagger a)$$

$$+ \gamma\bar{n}(a\tilde{\rho}a^\dagger + a^\dagger\tilde{\rho}a - a^\dagger a\tilde{\rho} - \tilde{\rho}aa^\dagger). \tag{1.71}$$

Here $\tilde{\rho}$ is still in the interaction picture. To transform back to the Schrödinger picture we use (1.26) to obtain

$$\dot{\rho} = \frac{1}{i\hbar}[H_S, \rho] + e^{-(i/\hbar)H_S t}\dot{\tilde{\rho}}e^{(i/\hbar)H_S t}. \tag{1.72}$$

With $H_S = \hbar\omega_0 a^\dagger a$, we substitute for $\dot{\tilde{\rho}}$ and use (1.26) and (1.40) to write, for example,

$$e^{-i\omega_0 a^\dagger a t} a \tilde{\rho} a^\dagger e^{i\omega_0 a^\dagger a t} = e^{-i\omega_0 a^\dagger a t} a \left(e^{i\omega_0 a^\dagger a t} \rho e^{-i\omega_0 a^\dagger a t}\right) a^\dagger e^{i\omega_0 a^\dagger a t}$$

$$= \left(e^{-i\omega_0 a^\dagger a t} a e^{i\omega_0 a^\dagger a t}\right) \rho \left(e^{-i\omega_0 a^\dagger a t} a^\dagger e^{i\omega_0 a^\dagger a t}\right)$$

$$= a\rho a^\dagger.$$

Each term can be treated similarly. We arrive at the *master equation for the damped harmonic oscillator*

$$\dot{\rho} = -i\omega_0'[a^\dagger a, \rho] + \frac{\gamma}{2}(2a\rho a^\dagger - a^\dagger a\rho - \rho a^\dagger a)$$

$$+ \gamma\bar{n}(a\rho a^\dagger + a^\dagger\rho a - a^\dagger a\rho - \rho a a^\dagger), \tag{1.73}$$

where

$$\omega_0' \equiv \omega_0 + \Delta. \tag{1.74}$$

Note 1.3 An alternate, more compact, writing of the master equation (1.73) may be given in the form

$$\dot{\rho} = -i\omega_0'[a^\dagger a, \rho] + \frac{\gamma}{2}([a, \rho a^\dagger] + [a\rho, a^\dagger])$$

$$+ \frac{\gamma}{2}\bar{n}([a\rho, a^\dagger] + [a^\dagger, \rho a]). \tag{1.75}$$

In both this form and (1.73) the damping terms are grouped according to whether they are proportional to $\bar{n}$ or not. This is a natural grouping from the point of view of the phase-space representations commonly used in quantum optics, which we meet in Chaps. 3 and 4 [see (3.47), for example, where the terms proportional and not proportional to $\bar{n}$ have distinct physical interpretations]. Nowadays it is more usual to group the terms so that the Lindblad form of the master equation is explicit [1.12], writing

$$\dot{\rho} = -i\omega_0'[a^\dagger a, \rho] + \frac{\gamma}{2}(\bar{n} + 1)(2a\rho a^\dagger - a^\dagger a\rho - \rho a^\dagger a)$$

$$+ \frac{\gamma}{2}\bar{n}(2a^\dagger\rho a - aa^\dagger\rho - \rho aa^\dagger). \tag{1.76}$$

Here the physical interpretation follows from the rate equations satisfied by the probabilities $p_n \equiv \langle n|\rho|n\rangle$ for the oscillator to be found in its nth energy eigenstate:

$$\dot{p}_n = \gamma(\bar{n} + 1)(n + 1)p_{n+1} - \gamma\bar{n}np_n$$

$$+ \gamma\bar{n}np_{n-1} - \gamma\bar{n}(n + 1)p_n. \tag{1.77}$$

The terms on the right-hand side of (1.77) describe transition rates into and out of the nth energy level (see Fig. 7.4) and originate, respectively, in the terms proportional to $2a\rho a^\dagger$, $-(a^\dagger a\rho + \rho a^\dagger a)$, $2a^\dagger \rho a$, and $-(aa^\dagger \rho + \rho aa^\dagger)$ in (1.76) [also see the discussion below (2.27) and (2.36d)].

Note 1.4 There is a large literature on the treatment of dissipative quantum systems using semigroups, from which the work of Lindblad on the form of the generator for physical semigroup dynamics [1.12] is a result of particular relevance to quantum optics; thus, the master equations we met in this book are all of Lindblad form. The foundational work of Davies [1.13] has also been influential in quantum optics, particularly in relation to the theory of photon counting [1.14]. We will have more to say about this topic when we discuss quantum trajectories in Volume 2 (Chaps. 15 and 16). More generally, the orientation in the literature on semigroups is towards the proof of rigorous mathematical results and hence the connections to quantum optics applications are somewhat indirect.

1.4.2 Some Limitations

Equation (1.73) is one of the central equations for future applications. Before proceeding we should note its limitations as a general equation for the damped harmonic oscillator.

First, it is derived in the rotating-wave approximation (R.W.A.). We expect this to be a good approximation for oscillators at optical frequencies [1.15], but for low frequency oscillators (strong damping, where the decay time approaches the oscillator period) we would not expect the R.W.A. to work well. In fact, even at optical frequencies the R.W.A. brings one notable inaccuracy. The frequency shift Δ in (1.74) is small, and generally neglected. However, in the example of the damped two-level atom this is the Lamb shift, and it is therefore of fundamental importance. Of course, an accurate calculation of the Lamb shift must include many things that we do not discuss – for example, relativistic effects. Nevertheless, it is as well to know that the (two-level) nonrelativistic contribution to the Lamb shift is not obtained correctly when the master equation is derived using the rotating-wave approximation. A derivation that does not use the R.W.A. is quite straightforward and proceeds along the same lines as the calculation in Sect. 1.4.1. The details are given by Agarwal [1.16, 1.17], who, in Ref. [1.17] in particular, discusses the question of the frequency shift.

Secondly, (1.73) is not valid at low temperatures. At sufficiently low temperatures the reservoir correlation functions can no longer be treated as δ-functions. There is quite an active interest in this low temperature regime. Discussions can be found in recent papers by Caldeira and Leggett [1.3], Lindenberg and West [1.18], and Grabert et al. [1.19].

1.4.3 Expectation Values and Commutation Relations

Let us make some simple checks to see if (1.73) predicts the behavior we expect from a damped harmonic oscillator. Since we have formulated our theory in the Schrödinger picture, we cannot obtain solutions for the operators themselves, but only for their expectation values. For example, if we multiply (1.73) on the left by a and take the trace (over the system S) we obtain an equation for $\langle a \rangle = \mathrm{tr}(a\rho)$:

$$
\begin{aligned}
\langle \dot{a} \rangle = &- i\omega_0 \,\mathrm{tr}(aa^\dagger a\rho - a\rho a^\dagger a) + \frac{\gamma}{2}\,\mathrm{tr}(2a^2\rho a^\dagger - aa^\dagger a\rho - a\rho a^\dagger a) \\
&+ \gamma\bar{n}\,\mathrm{tr}(a^2\rho a^\dagger + aa^\dagger\rho a - aa^\dagger a\rho - a\rho aa^\dagger) \\
= &- i\omega_0\,\mathrm{tr}\big[(aa^\dagger - a^\dagger a)a\rho\big] + \frac{\gamma}{2}\,\mathrm{tr}\big[(a^\dagger a - aa^\dagger)a\rho\big] \\
&+ \gamma\bar{n}\,\mathrm{tr}\big[(a^\dagger a - aa^\dagger)a\rho + a(aa^\dagger - a^\dagger a)\rho\big] \\
= &-\left(\frac{\gamma}{2} + i\omega_0\right)\langle a \rangle,
\end{aligned}
\tag{1.78}
$$

where we have used the cyclic property of the trace and the boson commutation relation (1.10). From now on we assume that the frequency shift Δ is included in the resonance frequency of the oscillator and do not distinguish ω_0' from ω_0. Equation (1.78) correctly describes the damped mean oscillator amplitude.

As a second example consider $\langle \hat{n} \rangle = \langle a^\dagger a \rangle$:

$$
\begin{aligned}
\langle \dot{\hat{n}} \rangle = &- i\omega_0\,\mathrm{tr}(a^\dagger aa^\dagger a\rho - a^\dagger a\rho a^\dagger a) + \frac{\gamma}{2}\,\mathrm{tr}(2a^\dagger a^2\rho a^\dagger - a^\dagger aa^\dagger a\rho \\
&- a^\dagger a\rho a^\dagger a) + \gamma\bar{n}\,\mathrm{tr}(a^\dagger a^2\rho a^\dagger + a^\dagger aa^\dagger\rho a - a^\dagger aa^\dagger a\rho - a^\dagger a\rho aa^\dagger) \\
= &\,\gamma\,\mathrm{tr}\big[a^{\dagger 2}a^2\rho - (a^\dagger a)^2\rho\big] \\
&+ \gamma\bar{n}\,\mathrm{tr}\big[a^{\dagger 2}a^2\rho + (aa^\dagger)^2\rho - (a^\dagger a)^2\rho - aa^{\dagger 2}a\rho\big] \\
= &- \gamma(\langle \hat{n} \rangle - \bar{n}),
\end{aligned}
\tag{1.79}
$$

with the solution

$$
\langle \hat{n}(t) \rangle = \langle \hat{n}(0) \rangle e^{-\gamma t} + \bar{n}(1 - e^{-\gamma t}).
\tag{1.80}
$$

Notice how thermal fluctuations are fed into the oscillator from the reservoir; the mean energy does not decay to zero but to the mean energy for an oscillator with frequency ω_0 in thermal equilibrium at temperature T.

Exercise 1.2 Show that the thermal equilibrium density operator

$$
\rho_{\mathrm{eq}} = \frac{e^{-H_S/k_B T}}{\mathrm{tr}\left(e^{-H_S/k_B T}\right)} = \frac{e^{-\hbar\omega_0 a^\dagger a/k_B T}}{1 - e^{-\hbar\omega_0/k_B T}}
$$

satisfies (1.73) in the steady state.

As a final observation we note that the boson commutation relation is preserved in time – at least in the mean, which is all we can say in the Schrödinger picture. Using the initial time commutator we find

$$\langle[a,a^\dagger](t)\rangle = \mathrm{tr}\{[a,a^\dagger]\rho(t)\} = \mathrm{tr}\{\rho(t)\} = 1;$$

it is readily shown that (1.73) preserves the trace of the density operator.

1.5 Two-Time Averages and the Quantum Regression Formula

We have developed a formalism which allows us, in principle, to solve for the density operator (reduced density operator) for a system interacting with a reservoir. From this density operator we can obtain time-dependent expectation values for any operator acting in the Hilbert space of the system S. What, however, about products of operators evaluated at two different times? Of particular interest, for example, will be the first-order and second-order correlation functions of the electromagnetic field. For a single mode these are given by

$$G^{(1)}(t,t+\tau) \propto \langle a^\dagger(t)a(t+\tau)\rangle,$$
$$G^{(2)}(t,t+\tau) \propto \langle a^\dagger(t)a^\dagger(t+\tau)a(t+\tau)a(t)\rangle.$$

The first-order correlation function is required for calculating the spectrum of the field. The second-order correlation function gives information about the photon statistics and describes photon bunching and antibunching.

Note 1.4 It may seem a strange talking about the spectrum of a single mode field since we normally associate a single mode with a single frequency. Here we are dealing, however, with what should more correctly be called a quasimode – a mode defined in a *lossy* optical cavity, which therefore has a finite linewidth.

Clearly, averages involving two times cannot be calculated directly from the master equation – at least, not without a little extra thought. We need to return to the microscopic picture of system plus reservoir. At this level two-time averages are defined in the usual way in the Heisenberg representation. Our objective, then, is to derive a formula which allows us to calculate these averages at the macroscopic level using the master equation for the reduced density operator alone; thus, in some approximate way we wish to carry out the trace over reservoir variables explicitly, as we did in deriving the master equation itself. The result we obtain is the so-called quantum regression theorem and is attributed to Lax [1.20, 1.21]. The particular designation, as a theorem, seems to begin with Mollow's classic paper on the spectrum of resonance fluorescence, where it appears in a footnote [1.22]. In a way the designation is unfortunate, since the so-called "theorem" is not a theorem at all, but a formula for two-time, or, more generally, multi-time averages, which follows from the Heisenberg equations of motion for $S \otimes R$ *under the Born-Markov approximation*. Lax did not speak of a theorem in his original paper [1.20], and there he makes it clear that a Markov assumption is used to arrive at his principal result, that "*even in the nonequilibrium case the regression of fluctuations obeys the macroscopic equations.*" The focus on "the nonequilibrium case" contrasts the case of thermal equilibrium, for which Onsager was the first to suggest that the regression of fluctuations obeys the macroscopic equations of motion; Onsager used this hypothesis to arrive at his famous reciprocity relations [1.23]. We are certainly not in the business of proving theorems, and since an informal use of a word like "theorem" is hardly appropriate, we will drop the "theorem" and speak of the quantum regression formula – a formula that, as we will see, may take on different forms for different occasions.

Note 1.5 Those interested in theorems might look at the literature on the semigroup approach to open quantum systems which develops its mathematics in a rigorous manner [1.12, 1.14]. That is not to suggest that the quantum regression formula of quantum optics is a theorem there either; it is, rather, a straightforward expression of the axiomatic definition of a Markovian open system dynamics. Our derivation of it, below, merely connects a microscopic model for $S \otimes R$ to the mathematics of semigroups by introducing the Markov property through an approximation. Interestingly, the rigorous mathematical approach concerns itself with a question moving in the opposite direction: given a semigroup evolution for a system S, can this dynamics be rigorously embedded in some unitary evolution for a larger system $S \otimes R$? The answer turns out to be in the affirmative; the embedding is executed by what the mathematicians call a dilation; there is no assertion, however, that the $S \otimes R$ so obtained is precisely that which a physicist would adopt as a fundamental model for the microscopic world.

Note 1.6 In a recent series of papers, Ford and O'Connell have argued that "There is No Quantum Regression Theorem" [1.24]. These authors consider the case of a harmonic oscillator coupled to a reservoir of harmonic oscillators in thermal equilibrium. They identify weak coupling and resonance assumptions – used in quantum optics – which allow the frequency-dependent energy of the reservoir oscillators to be replaced by a constant; thus, they correctly recognize that quantum optics assumes a (locally) flat reservoir spectrum where the reservoir spectrum is strictly not flat. Ford and O'Connell then note that it is precisely the non-flatness of the quantum mechanical reservoir spectrum that reveals itself, at low temperatures, in corrections to the regression of fluctuations given by the classical Onsager hypothesis (where the spectrum is kT per reservoir oscillator, independent of frequency – the equipartition result). The appearance, and the authors' suggestion from all of this is that having eliminated the quantum corrections, the approximations of quantum optics yield a regression formula that simply implements the classical Onsager hypothesis; nothing has been added to Onsager to justify the designation of a separate *quantum* regression formula. Their conclusion aside, Ford and O'Connell make a valid and fundamental point; we might recall, after all, that it is the same non-flat spectrum that underlies Einstein's quantum mechanical explanation for the low temperature behavior of specific heats [1.25]. What then might be said to put Ford and O'Connell's observations in perspective? First, there can be no quibbling about the approximations. They are indeed used in quantum optics in order to arrive at a Markovian description [see the discussion following Eq. (1.52)]; there is also the rotating-wave approximation, which fails to give the correct frequency shifts (Sect. 1.4.2) [1.26, 1.27]. The approximations themselves are not the main point, though, since no one disputes that they are both appropriate and extremely accurate in quantum optical applications. Lax demonstrates this explicitly in his response to Ford and O'Connell [1.28]. The central question is this: what is one left with after making the approximations? Does the quantum regression formula take us beyond the classical Onsager hypothesis? For Lax, the main thing was to establish the validity of a procedure for calculating two-time correlation functions in "the nonequilibrium case" [1.28]. Onsager's hypothesis concerned fluctuations about equilibrium. This is also the focus of Ford and O'Connell; their quantum corrections are typical of the sort of thing found in equilibrium statistical physics. Quantum optics looks in a different direction, away from equilibrium, to the strong resonant, or near resonant, interactions made accessible by coherent light sources. Coherence is very much the name of the game; it is established away from thermal equilibrium and generally has quantum mechanical consequences; the quantum regression formula carries through those aspects of quantum mechanics dealing with things like coherence, probability amplitudes, entanglement and so on. This is clear from the more formal versions of the formula [Eqs. (1.97) and (1.98)] which are manifestly quantum mechanical expressions, respecting operator order and employing a quantum mechanical propagator. For those cases where a linearized treatment of fluctuations may be made, there is, it is true, a version of the formula much closer to Onsager, where one has a linear set of mean-value equations analogous to macroscopic transport equations [Eqs. (1.107) and (1.108)]. Even here, though, quantum mechanical features can turn up, such as antibunched or squeezed fluctuations. It is important to note, furthermore, that the quantum regression formula is not limited to the small fluctuation regime. It can treat large fluctuations, where a classical evolution would necessarily be nonlinear. Of course quantum mechanics is linear, even in this case. There is a quantum mechanical substitute for nonlinearity, however, expressed through multiphoton processes, which prohibit a reduction of the dynamics to a simple set of transport-like equations recognizable from a classical or semi-classical treatment.

1.5.1 Formal Results

We will not follow Lax in detail, but our method is fundamentally the same as his. Recall our microscopic formulation of system S coupled to reservoir R. The Hamiltonian for the composite system $S \otimes R$ takes the form given in (1.16). The density operator is designated $\chi(t)$ and satisfies Schrödinger's equation (1.19). Our derivation of the master equation has given us an equation for the reduced density operator (1.17), which we will now write formally as

$$\dot{\rho} = \mathcal{L}\rho; \tag{1.81}$$

$\mathcal{L}$ is a generalized Liouvillian, a "superoperator" in the language of the Brussells-Austin group [1.29]; $\mathcal{L}$ operates on operators rather than on states. For the damped harmonic oscillator, from (1.73), the action of $\mathcal{L}$ on an arbitrary operator $\hat{O}$ is defined by the equation

$$\mathcal{L}\hat{O} \equiv -i\omega_0[a^\dagger a, \hat{O}] + \frac{\gamma}{2}(2a\hat{O}a^\dagger - a^\dagger a\hat{O} - \hat{O}a^\dagger a)$$
$$+ \gamma\bar{n}(a\hat{O}a^\dagger + a^\dagger\hat{O}a - a^\dagger a\hat{O} - \hat{O}aa^\dagger). \tag{1.82}$$

Within the microscopic formalism multi-time averages are straightforwardly defined in the Heisenberg picture. In particular, the average of a product of operators evaluated at two different times is given by

$$\langle \hat{O}_1(t)\hat{O}_2(t') \rangle = \mathrm{tr}_{S \otimes R}[\chi(0)\hat{O}_1(t)\hat{O}_2(t')], \tag{1.83}$$

where $\hat{O}_1$ and $\hat{O}_2$ are any two system operators. These operators satisfy the Heisenberg equations of motion

$$\dot{\hat{O}}_1 = \frac{1}{i\hbar}[\hat{O}_1, H], \tag{1.84a}$$

$$\dot{\hat{O}}_2 = \frac{1}{i\hbar}[\hat{O}_2, H], \tag{1.84b}$$

with the formal solutions

$$\hat{O}_1(t) = e^{(i/\hbar)Ht}\hat{O}_1(0)e^{-(i/\hbar)Ht}, \tag{1.85a}$$

$$\hat{O}_2(t') = e^{(i/\hbar)Ht'}\hat{O}_2(0)e^{-(i/\hbar)Ht'}. \tag{1.85b}$$

From (1.19), the formal solution for χ gives

$$\chi(0) = e^{(i/\hbar)Ht}\chi(t)e^{-(i/\hbar)Ht}. \tag{1.86}$$

We substitute these formal solutions into (1.83) and use the cyclic property of the trace to obtain

$$\langle \hat{O}_1(t)\hat{O}_2(t') \rangle = \mathrm{tr}_{S\otimes R}\left[e^{(i/\hbar)Ht}\chi(t)\hat{O}_1(0)e^{(i/\hbar)H(t'-t)}\hat{O}_2(0)e^{-(i/\hbar)Ht'} \right]$$

$$= \mathrm{tr}_{S\otimes R}\left[\hat{O}_2(0)e^{-(i/\hbar)H(t'-t)}\chi(t)\hat{O}_1(0)e^{(i/\hbar)H(t'-t)} \right]$$

$$= \mathrm{tr}_S\left\{ \hat{O}_2(0)\mathrm{tr}_R\left[e^{-(i/\hbar)H(t'-t)}\chi(t)\hat{O}_1(0)e^{(i/\hbar)H(t'-t)} \right] \right\}. \tag{1.87}$$

In the final step we have used the fact that $\hat{O}_2$ is an operator in the Hilbert space of S alone.

We now specialize to the case $t' \geq t$ and define

$$\tau \equiv t' - t, \tag{1.88}$$

$$\chi_{\hat{O}_1}(\tau) \equiv e^{-(i/\hbar)H\tau}\chi(t)\hat{O}_1(0)e^{(i/\hbar)H\tau}. \tag{1.89}$$

Clearly, $\chi_{\hat{O}_1}$ satisfies the equation

$$\frac{d\chi_{\hat{O}_1}}{d\tau} = \frac{1}{i\hbar}\left[H, \chi_{\hat{O}_1} \right] \tag{1.90}$$

with

$$\chi_{\hat{O}_1}(0) = \chi(t)\hat{O}_1(0). \tag{1.91}$$

If we are to eliminate explicit reference to the reservoir in (1.87), we need to evaluate the reservoir trace over $\chi_{\hat{O}_1}(\tau)$ to obtain the reduced operator

$$\rho_{\hat{O}_1}(\tau) \equiv \mathrm{tr}_R\left[\chi_{\hat{O}_1}(\tau) \right], \tag{1.92}$$

where

$$\rho_{\hat{O}_1}(0) = \mathrm{tr}_R[\chi(t)\hat{O}_1(0)] = \mathrm{tr}_R[\chi(t)]\hat{O}_1(0) = \rho(t)\hat{O}_1(0); \tag{1.93}$$

notice that $\rho_{\hat{O}_1}(\tau)$ is just the term $\mathrm{tr}_R[\cdots]$ inside the curly brackets in (1.87). If we then assume that $\chi(t)$ factorizes as $\rho(t)R_0$, in the spirit of (1.29), from (1.91) and (1.93) we can write

$$\chi_{\hat{O}_1}(0) = R_0[\rho(t)\hat{O}_1(0)] = R_0\,\rho_{\hat{O}_1}(0). \tag{1.94}$$

Equations (1.90), (1.92), and (1.94) are now equivalent to (1.19), (1.17), and (1.25) – namely, to the starting equations in our derivation of the master equation. We can find an equation for $\rho_{\hat{O}_1}(\tau)$ in the Born–Markov approximation following a completely analogous course to that followed in Sects. 1.3 and 1.4. Since (1.19) and (1.90) contain the same Hamiltonian H, using the formal notation of (1.81), we arrive at the equation

$$\frac{d\rho_{\hat{O}_1}}{d\tau} = \mathcal{L}\rho_{\hat{O}_1}, \tag{1.95}$$

with solution

$$\rho_{\hat{O}_1}(\tau) = e^{\mathcal{L}\tau}\left[\rho_{\hat{O}_1}(0)\right] = e^{\mathcal{L}\tau}[\rho(t)\hat{O}_1(0)]. \tag{1.96}$$

When we substitute for $\rho_{\hat{O}_1}(\tau)$ in (1.87), we have ($\tau \geq 0$)

$$\langle \hat{O}_1(t)\hat{O}_2(t+\tau)\rangle = \text{tr}_S\{\hat{O}_2(0)e^{\mathcal{L}\tau}[\rho(t)\hat{O}_1(0)]\}. \tag{1.97}$$

Exercise 1.3 Follow the same procedure to obtain ($\tau \geq 0$)

$$\langle \hat{O}_1(t+\tau)\hat{O}_2(t)\rangle = \text{tr}_S\{\hat{O}_1(0)e^{\mathcal{L}\tau}[\hat{O}_2(0)\rho(t)]\}. \tag{1.98}$$

Equations (1.97) and (1.98) give formal statements of the *quantum regression formula* for two-time averages. To calculate a correlation function $\langle \hat{O}_1(t)\hat{O}_2(t')\hat{O}_3(t)\rangle$ we cannot use (1.97) and (1.98) because noncommuting operators do not allow the reordering necessary to bring $\hat{O}_1(t)$ next to $\hat{O}_3(t)$. We may, however, generalize the approach taken above. Specifically, we have

$$\langle \hat{O}_1(t)\hat{O}_2(t')\hat{O}_3(t)\rangle$$
$$= \text{tr}_{S\otimes R}\left[e^{(i/\hbar)Ht}\chi(t)\hat{O}_1(0)e^{(i/\hbar)H(t'-t)}\hat{O}_2(0)e^{-(i/\hbar)H(t'-t)}\right.$$
$$\left.\times\hat{O}_3(0)e^{-(i/\hbar)Ht}\right]$$
$$= \text{tr}_{S\otimes R}\left[\hat{O}_2(0)e^{-(i/\hbar)H(t'-t)}\hat{O}_3(0)\chi(t)\hat{O}_1(0)e^{(i/\hbar)H(t'-t)}\right]$$
$$= \text{tr}_S\left\{\hat{O}_2(0)\text{tr}_R\left[e^{-(i/\hbar)H(t'-t)}\hat{O}_3(0)\chi(t)\hat{O}_1(0)e^{(i/\hbar)H(t'-t)}\right]\right\}. \tag{1.99}$$

Defining

$$\chi_{\hat{O}_3\hat{O}_1}(\tau) \equiv e^{-(i/\hbar)H\tau}\hat{O}_3(0)\chi(t)\hat{O}_1(0)e^{(i/\hbar)H\tau} \tag{1.100}$$

and

$$\rho_{\hat{O}_3\hat{O}_1}(\tau) \equiv \text{tr}_R\left[\chi_{\hat{O}_3\hat{O}_1}(\tau)\right] \tag{1.101}$$

as analogs of (1.89) and (1.92), we can proceed as before to the result ($\tau \geq 0$)

$$\langle \hat{O}_1(t)\hat{O}_2(t+\tau)\hat{O}_3(t)\rangle = \text{tr}_S\{\hat{O}_2(0)e^{\mathcal{L}\tau}[\hat{O}_3(0)\rho(t)\hat{O}_1(0)]\}. \tag{1.102}$$

Equations (1.97) and (1.98) are, in fact, just special cases of (1.102) with either $\hat{O}_1(t)$ or $\hat{O}_3(t)$ set equal to the unit operator.

1.5.2 Quantum Regression for a Complete Set of Operators

It is possible to work directly with the rather formal expressions derived above. The formal expressions can also be reduced, however, to a more familiar form [1.20], which is often more convenient for doing calculations. Essentially, we will find that the equations of motion for expectation values of system operators (one-time averages) are also the equations of motion for correlation functions (two-time averages).

We begin by assuming that there exists a complete set of system operators $\hat{A}_\mu$, $\mu = 1, 2, \ldots$, in the following sense: that for an arbitrary operator $\hat{O}$, and for each $\hat{A}_\mu$,

$$\mathrm{tr}_S[\hat{A}_\mu(\mathcal{L}\hat{O})] = \sum_\lambda M_{\mu\lambda}\mathrm{tr}_S(\hat{A}_\lambda\hat{O}), \qquad (1.103)$$

where the $M_{\mu\lambda}$ are constants. In particular, from this it follows that

$$\begin{aligned}
\langle \dot{\hat{A}}_\mu \rangle = \mathrm{tr}_S(\hat{A}_\mu\dot{\rho}) &= \mathrm{tr}_S[\hat{A}_\mu(\mathcal{L}\rho)] \\
&= \sum_\lambda M_{\mu\lambda}\mathrm{tr}_S(\hat{A}_\lambda\rho) \\
&= \sum_\lambda M_{\mu\lambda}\langle \hat{A}_\lambda \rangle.
\end{aligned} \qquad (1.104)$$

Thus, expectation values $\langle \hat{A}_\mu \rangle$, $\mu = 1, 2, \ldots$, obey a coupled set of linear equations with the evolution matrix $\boldsymbol{M}$ defined by the $M_{\mu\lambda}$ that appear in (1.103). In vector notation,

$$\langle \dot{\hat{\boldsymbol{A}}} \rangle = \boldsymbol{M}\langle \hat{\boldsymbol{A}} \rangle, \qquad (1.105)$$

where $\hat{\boldsymbol{A}}$ is the column vector of operators $\hat{A}_\mu$, $\mu = 1, 2, \ldots$. Now, using (1.97) and (1.103) ($\tau \geq 0$):

$$\begin{aligned}
\frac{d}{d\tau}\langle \hat{O}_1(t)\hat{A}_\mu(t+\tau) \rangle &= \mathrm{tr}_S\{\hat{A}_\mu(0)(\mathcal{L}e^{\mathcal{L}\tau}[\rho(t)\hat{O}_1(0)])\} \\
&= \sum_\lambda M_{\mu\lambda}\mathrm{tr}_S\{\hat{A}_\lambda(0)e^{\mathcal{L}\tau}[\rho(t)\hat{O}_1(0)]\} \\
&= \sum_\lambda M_{\mu\lambda}\langle \hat{O}_1(t)\hat{A}_\lambda(t+\tau) \rangle,
\end{aligned} \qquad (1.106)$$

or,

$$\frac{d}{d\tau}\langle \hat{O}_1(t)\hat{\boldsymbol{A}}(t+\tau) \rangle = \boldsymbol{M}\langle \hat{O}_1(t)\hat{\boldsymbol{A}}(t+\tau) \rangle, \qquad (1.107)$$

where $\hat{O}_1$ can be any system operator, not necessarily one of the $\hat{A}_\mu$. This result is just what would be obtained by removing the angular brackets from (1.105) (written with $t \to t+\tau$, and $\cdot \equiv d/dt \to d/d\tau$), multiplying on the left by $\hat{O}_1(t)$, and then replacing the angular brackets. Hence, for each operator $\hat{O}_1$, the set of correlation functions $\langle \hat{O}_1(t)\hat{A}_\mu(t+\tau) \rangle$, $\mu = 1, 2, \ldots$, with $\tau \geq 0$, satisfies the same equations (as functions of τ) as do the averages $\langle \hat{A}_\mu(t+\tau) \rangle$. This is perhaps the more familiar statement of the content of the *quantum regression formula*.

Exercise 1.4 For $\tau \geq 0$ show that

$$\frac{d}{d\tau} \langle \hat{A}(t+\tau)\hat{O}_2(t) \rangle = \boldsymbol{M} \langle \hat{A}(t+\tau)\hat{O}_2(t) \rangle. \tag{1.108}$$

Thus, we can also multiply (1.105) on the right by $\hat{O}_2(t)$, inside the average. Also show that

$$\frac{d}{d\tau} \langle \hat{O}_1(t)\hat{A}(t+\tau)\hat{O}_2(t) \rangle = \boldsymbol{M} \langle \hat{O}_1(t)\hat{A}(t+\tau)\hat{O}_2(t) \rangle. \tag{1.109}$$

It may appear that this form of the quantum regression formula is quite restricted, since its derivation relies on the existence of a set of operators $\hat{A}_\mu$, $\mu = 1, 2, \ldots$, for which (1.103) holds. We can show that this is always so, however, if a discrete basis $|n\rangle$, $n = 1, 2, \ldots$, exists; although, in general, the complete set of operators may be very large. Consider the operators

$$\hat{A}_\mu = \hat{A}_{nm} \equiv |n\rangle\langle m|. \tag{1.110}$$

Then

$$\begin{aligned}
\mathrm{tr}_S[\hat{A}_{nm}(\mathcal{L}\hat{O})] &= \mathrm{tr}_S[|n\rangle\langle m|(\mathcal{L}\hat{O})] \\
&= \langle m|(\mathcal{L}\hat{O})|n\rangle \\
&= \langle m| \left(\mathcal{L} \sum_{n',m'} |n'\rangle\langle m'| \langle n'|\hat{O}|m'\rangle \right) |n\rangle \\
&= \sum_{n',m'} \langle m|(\mathcal{L}|n'\rangle\langle m'|)|n\rangle \langle n'|\hat{O}|m'\rangle \\
&= \sum_{n',m'} \langle m|(\mathcal{L}|n'\rangle\langle m'|)|n\rangle \, \mathrm{tr}_S(|m'\rangle\langle n'|\hat{O}) \\
&= \sum_{n',m'} M_{nm;n'm'} \, \mathrm{tr}_S(\hat{A}_{n'm'}\hat{O}), \tag{1.111}
\end{aligned}$$

with

$$M_{nm;n'm'} \equiv \langle m| \left(\mathcal{L}|m'\rangle\langle n'| \right) |n\rangle. \tag{1.112}$$

In the last step we have interchanged the indices n' and m'. Equation (1.111) gives an expansion in the form of (1.103). The complete set of operators includes all the outer products $|n\rangle\langle m|$, $n = 1, 2, \ldots$, $m = 1, 2, \ldots$; this may be a small number of operators, a large but finite number of operators, or a double infinity of operators in the case of the Fock state basis.

1.5.3 Correlation Functions for the Damped Harmonic Oscillator

We will conclude our discussion of two-time averages with two simple examples based on the equations for expectation values for the damped harmonic oscillator [Eqs. (1.78) and (1.79)]. We first calculate the first-order correlation function $\langle a^\dagger(t)a(t+\tau)\rangle$. Equation (1.78) gives the equation of motion for the mean oscillator amplitude:

$$\langle \dot{a}\rangle = -\left(\frac{\gamma}{2}+i\omega_0\right)\langle a\rangle. \tag{1.113}$$

Then, with $\hat{A}_1 = a$ and $\hat{O}_1 = a^\dagger$, from (1.105) and (1.107), we may write

$$\frac{d}{d\tau}\langle a^\dagger(t)a(t+\tau)\rangle = -\left(\frac{\gamma}{2}+i\omega_0\right)\langle a^\dagger(t)a(t+\tau)\rangle. \tag{1.114}$$

Thus,

$$\langle a^\dagger(t)a(t+\tau)\rangle = \langle a^\dagger(t)a(t)\rangle e^{-(\gamma/2+i\omega_0)\tau}$$
$$= \left[\langle \hat{n}(0)\rangle e^{-\gamma t}+\bar{n}(1-e^{-\gamma t})\right]e^{-(\gamma/2+i\omega_0)\tau}, \tag{1.115}$$

where the last line follows from (1.80). If the oscillator describes a lossy cavity mode, in the long-time limit the Fourier transform of the first-order correlation function

$$\langle a^\dagger(0)a(\tau)\rangle_{\mathrm{ss}} \equiv \lim_{t\to\infty}\langle a^\dagger(t)a(t+\tau)\rangle = \bar{n}e^{-(\gamma/2+i\omega_0)\tau} \tag{1.116}$$

gives the spectrum of the light at the cavity output. This is clearly a Lorentzian with width γ (full-width at half-maximum).

Note 1.7 This statement about the spectrum of the light at the cavity output is not strictly correct for the lossy cavity model as we have described it. The reason is that we have taken the environment outside the cavity to be in thermal equilibrium at temperature T (it is the environment that is modeled by the reservoir). Given this, the light detected in the cavity output will be a sum of transmitted light – light that passes from inside the cavity, through the cavity output mirror, into the environment – and thermal radiation reflected from the outside of the output mirror. Calculating the spectrum at the cavity output for this situation is more involved (Sect. 7.3.4). Physically, however, the result is clear; the spectrum must be a blackbody spectrum. The Lorentzian spectrum obtained from (1.116) would be observed, as filtered thermal radiation, for a cavity coupled to two reservoirs, one at temperature T and the other at zero temperature. If the bandwidth for coupling to the reservoir at temperature T is much larger than for coupling to the zero temperature reservoir, the master equation (1.73) is basically unchanged. Light emitted into the zero temperature reservoir then shows the Lorentzian spectrum obtained from the Fourier transform of (1.116).

For a second example we calculate the second-order correlation function $\langle a^\dagger(t)a^\dagger(t+\tau)a(t+\tau)a(t)\rangle = \langle a^\dagger(t)\hat{n}(t+\tau)a(t)\rangle$. Writing (1.79) in the form

$$\frac{d}{dt}\begin{pmatrix}\langle\hat{n}\rangle\\\bar{n}\end{pmatrix} = \begin{pmatrix}-\gamma & \gamma\\0 & 0\end{pmatrix}\begin{pmatrix}\langle\hat{n}\rangle\\\bar{n}\end{pmatrix}, \tag{1.117}$$

we set $\hat{A}_1 = \hat{n} = a^\dagger a$ and $\hat{A}_2 = \bar{n}$ (a constant). Then, from (1.105) and (1.109), with $\hat{O}_1 = a^\dagger$ and $\hat{O}_2 = a$,

$$\frac{d}{d\tau}\begin{pmatrix}\langle a^\dagger(t)\hat{n}(t+\tau)a(t)\rangle\\\bar{n}\langle\hat{n}(t)\rangle\end{pmatrix} = \begin{pmatrix}-\gamma & \gamma\\0 & 0\end{pmatrix}\begin{pmatrix}\langle a^\dagger(t)\hat{n}(t+\tau)a(t)\rangle\\\bar{n}\langle\hat{n}(t)\rangle\end{pmatrix}. \tag{1.118}$$

Thus,

$$\langle a^\dagger(t)\hat{n}(t+\tau)a(t)\rangle = \langle a^\dagger(t)\hat{n}(t)a(t)\rangle e^{-\gamma\tau} + \bar{n}\langle\hat{n}(t)\rangle(1-e^{-\gamma\tau}). \tag{1.119}$$

We obtained an expression for $\langle\hat{n}(t)\rangle$ in (1.80). The calculation of $\langle a^\dagger(t)\hat{n}(t)a(t)\rangle$ is left as an exercise:

Exercise 1.5 Derive an equation of motion for the expectation value $\langle a^\dagger(t)\hat{n}(t)a(t)\rangle = \langle a^{\dagger 2}(t)a^2(t)\rangle$ from the master equation (1.73) and show that

$$\langle a^\dagger(t)\hat{n}(t)a(t)\rangle = \left[\langle\hat{n}^2(0)\rangle - \langle\hat{n}(0)\rangle\right]e^{-2\gamma t} + 2\bar{n}(1-e^{-\gamma t})$$
$$\times\left[2\langle\hat{n}(0)\rangle e^{-\gamma t} + \bar{n}(1-e^{-\gamma t})\right]. \tag{1.120}$$

Now, substituting from (1.80) and (1.120) into (1.119),

$$\langle a^\dagger(t)a^\dagger(t+\tau)a(t+\tau)a(t)\rangle$$
$$= \left\{\left[\langle\hat{n}^2(0)\rangle - \langle\hat{n}(0)\rangle\right]e^{-2\gamma t} + 2\bar{n}(1-e^{-\gamma t})\left[2\langle\hat{n}(0)\rangle e^{-\gamma t}\right.\right.$$
$$\left.\left.+\bar{n}(1-e^{-\gamma t})\right]\right\}e^{-\gamma\tau} + \bar{n}\left[\langle\hat{n}(0)\rangle e^{-\gamma t} + \bar{n}(1-e^{-\gamma t})\right](1-e^{-\gamma\tau}). \tag{1.121}$$

In the long-time limit, the second-order correlation function is

$$\langle a^\dagger(0)a^\dagger(\tau)a(\tau)a(0)\rangle_{\text{ss}} \equiv \lim_{t\to\infty}\langle a^\dagger(t)a^\dagger(t+\tau)a(t+\tau)a(t)\rangle$$
$$= \bar{n}^2(1+e^{-\gamma\tau}). \tag{1.122}$$

This expression describes the well-known Hanbury-Brown-Twiss effect, or photon bunching, for thermal light [1.30]; at zero delay the correlation function has twice the value it has for long delays ($\gamma\tau \gg 1$).

Note 1.8 The correlation time, $1/\gamma$, in (1.122) holds for filtered thermal light in accord with the comments in Note 1.7.

2. Two-Level Atoms
and Spontaneous Emission

The damped harmonic oscillator provides our elementary description for the electromagnetic field in a lossy cavity. The damped two-level atom will provide our elementary description for the matter with which this field interacts. In an atomic vapor, loss of energy from an excited atom may take place via spontaneous emission or inelastic collisions. Elastic collisions can also play an important damping role; although, of course, they do not carry away energy; elastic collisions interrupt the phase of induced electronic oscillations and in this way damp the atomic polarization. We will first restrict our treatment to the case of purely radiative damping, assuming conditions in which collisions are unimportant. Such conditions are achieved, for example, in atomic beams. Later we will derive the terms that must be added to the master equation to describe additional phase destroying processes such as elastic collisions.

We consider an atom with two states, designated $|1\rangle$ and $|2\rangle$, having energies E_1 and E_2 with $E_1 < E_2$. Radiative transitions between $|1\rangle$ and $|2\rangle$ are allowed in the dipole approximation. Our objective is to describe energy dissipation and polarization damping through the coupling of the $|1\rangle \rightarrow |2\rangle$ transition to the many modes of the vacuum radiation field (a reservoir of harmonic oscillators). For simplicity we assume that there are no transitions between $|1\rangle$ and $|2\rangle$ and any other states of the atom. The extension to multilevel atoms can be found in Louisell [2.1] and Haken [2.2]. A treatment for just two levels which corresponds closely to our own is given in Sargent, Scully and Lamb [2.3].

2.1 Two-Level Atom as a Pseudo-Spin System

A two-state system can be described in terms of the Pauli spin operators. We will be using this description extensively and we therefore begin by briefly reviewing the relationship between these operators and quantities of physical interest, such as the atomic inversion and polarization. A more complete coverage of this subject is given by Allen and Eberly [2.4].

If we have a representation in terms of a complete set of states $|n\rangle, n = 1, 2, \ldots$, any operator $\hat{O}$ can be expanded as

$$\hat{O} = \sum_{n,m} \langle n|\hat{O}|m\rangle |n\rangle \langle m|. \tag{2.1}$$

This follows after multiplying on the left and right by the identity operator $\hat{I} = \sum_n |n\rangle\langle n|$. The $\langle n|\hat{O}|m\rangle$ define the matrix representation of $\hat{O}$ with respect to the basis $|n\rangle$. If we adopt the energy eigenstates $|1\rangle$ and $|2\rangle$ as a basis for our two-level atom, the unperturbed atomic Hamiltonian H_A can then be written in the form

$$\begin{aligned} H_A &= E_1|1\rangle\langle 1| + E_2|2\rangle\langle 2| \\ &= \tfrac{1}{2}(E_1 + E_2)\hat{I} + \tfrac{1}{2}(E_2 - E_1)\sigma_z, \end{aligned} \tag{2.2}$$

where

$$\sigma_z \equiv |2\rangle\langle 2| - |1\rangle\langle 1|. \tag{2.3}$$

The first term in (2.2) is a constant which may be eliminated by referring the atomic energies to the middle of the atomic transition, as in Fig. 2.1. We then write

$$H_A = \tfrac{1}{2}\hbar\omega_A\sigma_z, \qquad \omega_A \equiv (E_2 - E_1)/\hbar. \tag{2.4}$$

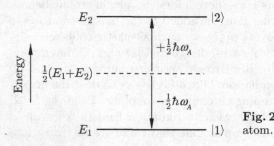

Fig. 2.1 Energy levels for a two-level atom.

Consider now the dipole moment operator $e\hat{\boldsymbol{q}}$, where e is the electronic charge and $\hat{\boldsymbol{q}}$ is the coordinate operator for the bound electron:

$$\begin{aligned} e\hat{\boldsymbol{q}} &= e \sum_{n,m=1}^{2} \langle n|\hat{\boldsymbol{q}}|m\rangle |n\rangle\langle m| \\ &= e\big(\langle 1|\hat{\boldsymbol{q}}|2\rangle |1\rangle\langle 2| + \langle 2|\hat{\boldsymbol{q}}|1\rangle |2\rangle\langle 1|\big) \\ &= \boldsymbol{d}_{12}\sigma_- + \boldsymbol{d}_{21}\sigma_+, \end{aligned} \tag{2.5}$$

where we have set $\langle 1|\hat{\boldsymbol{q}}|1\rangle = \langle 2|\hat{\boldsymbol{q}}|2\rangle = 0$, assuming atomic states whose symmetry guarantees zero permanent dipole moment, and we have introduced the *atomic dipole matrix elements*

$$\boldsymbol{d}_{12} \equiv e\langle 1|\hat{\boldsymbol{q}}|2\rangle, \qquad \boldsymbol{d}_{21} = (\boldsymbol{d}_{12})^*, \tag{2.6}$$

and *atomic lowering and raising operators*

$$\sigma_- \equiv |1\rangle\langle 2|, \qquad \sigma_+ \equiv |2\rangle\langle 1|. \tag{2.7}$$

The matrix representations for the operators introduced in (2.3) and (2.7) are

$$\sigma_z = \begin{pmatrix} 1 & 0 \\ 0 & -1 \end{pmatrix}, \qquad \sigma_- = \begin{pmatrix} 0 & 0 \\ 1 & 0 \end{pmatrix}, \qquad \sigma_+ = \begin{pmatrix} 0 & 1 \\ 0 & 0 \end{pmatrix}. \tag{2.8}$$

By writing

$$\sigma_\pm = \tfrac{1}{2}(\sigma_x \pm i\sigma_y), \tag{2.9}$$

with

$$\sigma_x = \begin{pmatrix} 0 & 1 \\ 1 & 0 \end{pmatrix}, \qquad \sigma_y = \begin{pmatrix} 0 & -i \\ i & 0 \end{pmatrix}, \tag{2.10}$$

we see that σ_x, σ_y, and σ_z are the *Pauli spin matrices* introduced initially in the context of magnetic transitions in spin-$\tfrac{1}{2}$ systems [2.5]. When applied to two-level atoms σ_z, σ_-, and σ_+ are referred to as *pseudo-spin operators*, since, in this context the two levels are not associated with the states of a real spin.

Exercise 2.1 From the relationships above, deduce the following:

1. the commutation relations

$$[\sigma_+, \sigma_-] = \sigma_z, \qquad [\sigma_\pm, \sigma_z] = \mp 2\sigma_\pm; \tag{2.11}$$

2. the action on atomic states:

$$\sigma_z|1\rangle = -|1\rangle, \qquad \sigma_z|2\rangle = |2\rangle, \tag{2.12a}$$
$$\sigma_-|1\rangle = 0, \qquad \sigma_-|2\rangle = |1\rangle, \tag{2.12b}$$
$$\sigma_+|1\rangle = |2\rangle, \qquad \sigma_+|2\rangle = 0. \tag{2.12c}$$

From (2.12b) and (2.12c) the designation of σ_- and σ_+ as atomic lowering and raising operators is clear.

We will formulate our description of two-level atoms in terms of the operators σ_z, σ_-, and σ_+. For an atomic state specified by a density operator ρ, expectation values of σ_z, σ_-, and σ_+ are just the matrix elements of the density operator, and give the *population difference*

$$\langle \sigma_z \rangle = \mathrm{tr}(\rho\sigma_z) = \langle 2|\rho|2\rangle - \langle 1|\rho|1\rangle = \rho_{22} - \rho_{11}, \tag{2.13}$$

and the mean *atomic polarization*

$$\begin{aligned} \langle e\hat{q} \rangle &= \boldsymbol{d}_{12}\mathrm{tr}(\rho\sigma_-) + \boldsymbol{d}_{21}\mathrm{tr}(\rho\sigma_+) \\ &= \boldsymbol{d}_{12}\langle 2|\rho|1\rangle + \boldsymbol{d}_{21}\langle 1|\rho|2\rangle \\ &= \boldsymbol{d}_{12}\,\rho_{21} + \boldsymbol{d}_{21}\,\rho_{12}. \end{aligned} \tag{2.14}$$

2.2 Spontaneous Emission
in the Master Equation Approach

2.2.1 Master Equation for a Radiatively Damped Two-Level Atom

We consider an atom that is radiatively damped by its interaction with the many modes of the radiation field taken in thermal equilibrium at temperature T. This field acts as a reservoir of harmonic oscillators. Within the general formula for a system S interacting with a reservoir R, the Hamiltonian (1.16) is given in the rotating-wave and dipole approximations by [2.6, 2.7]

$$H_S \equiv \tfrac{1}{2}\hbar\omega_A\sigma_z, \tag{2.15a}$$

$$H_R \equiv \sum_{k,\lambda} \hbar\omega_k r^{\dagger}_{k,\lambda} r_{k,\lambda}, \tag{2.15b}$$

$$H_{SR} \equiv \sum_{k,\lambda} \hbar\big(\kappa^*_{k,\lambda} r^{\dagger}_{k,\lambda}\sigma_- + \kappa_{k,\lambda} r_{k,\lambda}\sigma_+\big), \tag{2.15c}$$

with

$$\kappa_{k,\lambda} \equiv -ie^{ik\cdot r_A}\sqrt{\frac{\omega_k}{2\hbar\epsilon_0 V}}\,\hat{e}_{k,\lambda}\cdot d_{21}. \tag{2.16}$$

The summation extends over reservoir oscillators (modes of the electromagnetic field) with wavevectors k and polarization states λ, and corresponding frequencies ω_k and unit polarization vectors $\hat{e}_{k,\lambda}$. The atom is positioned at r_A, and V is the quantization volume. $\kappa_{k,\lambda}$ is the *dipole coupling constant* for the electromagnetic field mode with wavevector k and polarization λ. The general formalism from Sect. 1.3 now takes us directly to (1.34), where from (1.32) and (2.15) we must make the identification:

$$s_1 = \sigma_-, \qquad s_2 = \sigma_+, \tag{2.17a}$$

$$\Gamma_1 = \Gamma^{\dagger} \equiv \sum_{k,\lambda} \kappa^*_{k,\lambda} r^{\dagger}_{k,\lambda}, \qquad \Gamma_2 = \Gamma \equiv \sum_{k,\lambda} \kappa_{k,\lambda} r_{k,\lambda}. \tag{2.17b}$$

In the interaction picture,

$$\tilde{\Gamma}_1(t) = \tilde{\Gamma}^{\dagger}(t) = \sum_{k,\lambda} \kappa^*_{k,\lambda} r^{\dagger}_{k,\lambda} e^{i\omega_k t}, \tag{2.18a}$$

$$\tilde{\Gamma}_2(t) = \tilde{\Gamma}(t) = \sum_{k,\lambda} \kappa_{k,\lambda} r_{k,\lambda} e^{-i\omega_k t}, \tag{2.18b}$$

and

$$\tilde{s}_1(t) = e^{i(\omega_A\sigma_z/2)t}\sigma_- e^{-i(\omega_A\sigma_z/2)t} = \sigma_- e^{-i\omega_A t}, \tag{2.19a}$$

$$\tilde{s}_2(t) = e^{i(\omega_A\sigma_z/2)t}\sigma_+ e^{-i(\omega_A\sigma_z/2)t} = \sigma_+ e^{i\omega_A t}. \tag{2.19b}$$

Note 2.1 To obtain (2.19), consider the Heisenberg equation of motion

$$\dot{\tilde{s}}_1 = i\tfrac{1}{2}\omega_A e^{i(\omega_A \sigma_z/2)t}(\sigma_z \sigma_- - \sigma_- \sigma_z)e^{-i(\omega_A \sigma_z/2)t}$$
$$= -i\omega_A \tilde{s}_1.$$

This is trivially solved to give

$$\tilde{s}_1(t) = \tilde{s}_1(0)e^{-i\omega_A t} = \sigma_- e^{-i\omega_A t}.$$

Aside from the obvious notational differences, (2.18) and (2.19) are the same as (1.41) and (1.40), respectively, with the substitution $a \to \sigma_-$, $a^\dagger \to \sigma_+$. The derivation of the master equation for a two-level atom then follows in complete analogy to the derivation of the master equation for the harmonic oscillator, aside from two minor differences: (1) The explicit evaluation of the summation over reservoir oscillators now involves a summation over wavevector directions and polarization states. (2) The commutation relations used to reduce the master equation to its simplest form are different. Neither of these steps are taken in passing from (1.34) to (1.62), or in evaluating the time integrals using (1.65). We can therefore simply make the substitution $a \to \sigma_-$, $a^\dagger \to \sigma_+$ in (1.62) to write

$$\dot{\tilde{\rho}} = \left[\frac{\gamma}{2}(\bar{n}+1) + i(\Delta' + \Delta)\right](\sigma_- \tilde{\rho}\sigma_+ - \sigma_+\sigma_-\tilde{\rho})$$
$$+ \left(\frac{\gamma}{2}\bar{n} + i\Delta'\right)(\sigma_+ \tilde{\rho}\sigma_- - \tilde{\rho}\sigma_-\sigma_+) + \text{h.c.}, \qquad (2.20)$$

with $\bar{n} \equiv \bar{n}(\omega_A, T)$ and

$$\gamma \equiv 2\pi \sum_\lambda \int d^3k \, g(\boldsymbol{k})|\kappa(\boldsymbol{k},\lambda)|^2 \delta(kc - \omega_A), \qquad (2.21)$$

$$\Delta \equiv \sum_\lambda P \int d^3k \, \frac{g(\boldsymbol{k})|\kappa(\boldsymbol{k},\lambda)|^2}{\omega_A - kc}, \qquad (2.22)$$

$$\Delta' \equiv \sum_\lambda P \int d^3k \, \frac{g(\boldsymbol{k})|\kappa(\boldsymbol{k},\lambda)|^2}{\omega_A - kc}\bar{n}(kc, T). \qquad (2.23)$$

We have grouped the terms slightly differently in (2.20), but the correspondence to (1.62) is clear when we note that, there, $\alpha = \gamma/2 + i\Delta$ and $\beta = (\gamma/2)\bar{n} + i\Delta'$. Equation (2.20) gives

$$\dot{\tilde{\rho}} = \frac{\gamma}{2}(\bar{n}+1)(2\sigma_-\tilde{\rho}\sigma_+ - \sigma_+\sigma_-\tilde{\rho} - \tilde{\rho}\sigma_+\sigma_-) - i(\Delta'+\Delta)[\sigma_+\sigma_-,\tilde{\rho}]$$
$$+ \frac{\gamma}{2}\bar{n}(2\sigma_+\tilde{\rho}\sigma_- - \sigma_-\sigma_+\tilde{\rho} - \tilde{\rho}\sigma_-\sigma_+) + i\Delta'[\sigma_-\sigma_+,\tilde{\rho}]$$
$$= -i\frac{1}{2}(2\Delta' + \Delta)[\sigma_z,\tilde{\rho}] + \frac{\gamma}{2}(\bar{n}+1)(2\sigma_-\tilde{\rho}\sigma_+ - \sigma_+\sigma_-\tilde{\rho} - \tilde{\rho}\sigma_+\sigma_-)$$
$$+ \frac{\gamma}{2}\bar{n}(2\sigma_+\tilde{\rho}\sigma_- - \sigma_-\sigma_+\tilde{\rho} - \tilde{\rho}\sigma_-\sigma_+), \qquad (2.24)$$

where we have used

$$\sigma_+\sigma_- = |2\rangle\langle 1|1\rangle\langle 2| = |2\rangle\langle 2| = \tfrac{1}{2}(1+\sigma_z), \qquad (2.25a)$$

$$\sigma_-\sigma_+ = |1\rangle\langle 2|2\rangle\langle 1| = |1\rangle\langle 1| = \tfrac{1}{2}(1-\sigma_z). \qquad (2.25b)$$

Finally, transforming back to the Schrödinger picture using (1.72), we obtain the *master equation for a radiatively damped two-level atom*:

$$\dot\rho = -i\tfrac{1}{2}\omega'_A[\sigma_z,\rho] + \frac{\gamma}{2}(\bar n+1)(2\sigma_-\rho\sigma_+ - \sigma_+\sigma_-\rho - \rho\sigma_+\sigma_-)$$

$$+ \frac{\gamma}{2}\bar n(2\sigma_+\rho\sigma_- - \sigma_-\sigma_+\rho - \rho\sigma_-\sigma_+), \qquad (2.26)$$

with

$$\omega'_A \equiv \omega_A + 2\Delta' + \Delta. \qquad (2.27)$$

The symmetric grouping of terms we have adopted identifies a transition rate from $|2\rangle \to |1\rangle$, described by the term proportional to $(\gamma/2)(\bar n+1)$, and a transition rate from $|1\rangle \to |2\rangle$, described by the term proportional to $(\gamma/2)\bar n$. The former contains a rate for spontaneous transitions, independent of $\bar n$, and a rate for stimulated transitions induced by thermal photons, proportional to $\bar n$; the latter gives a rate for absorptive transitions which take thermal photons from the equilibrium electromagnetic field. We will have more to say about this point later. Notice that the Lamb shift given by $\omega'_A - \omega_A$ includes a temperature-dependent contribution $2\Delta'$ which did not appear for the harmonic oscillator. Its appearance here is a consequence of the commutator $[\sigma_-,\sigma_+] = -\sigma_z$, in place of the corresponding $[a,a^\dagger] = 1$ for the harmonic oscillator. From (2.22), (2.23), and (1.52)

$$2\Delta' + \Delta = \sum_\lambda P\int d^3k\, \frac{g(\boldsymbol{k})|\kappa(\boldsymbol{k},\lambda)|^2}{\omega_A - kc}[1 + 2\bar n(kc,T)]$$

$$= \sum_\lambda P\int d^3k\, \frac{g(\boldsymbol{k})|\kappa(\boldsymbol{k},T)|^2}{\omega_A - kc}\coth\left(\frac{\hbar kc}{2k_BT}\right), \qquad (2.28)$$

where k_B is Boltzmann's constant. The temperature independent term in the square bracket gives the normal *Lamb shift*, while the term proportional to $2\bar n$ gives the frequency shift induced via the ac Stark effect by the thermal reservoir field. We will discuss the ac Stark effect later in this chapter. It is only quite recently that attention has been paid to this temperature-dependent frequency shift, following the work of Gallagher and Cook [2.8]. A thorough discussion for real atoms is given by Farley and Wing [2.9]. Beautiful experiments by Hollberg and Hall using highly stabilized lasers have measured the temperature-dependent shift in Rydberg atoms [2.10].

Note 2.2 Recall from Sect. 1.4.2 that the rotating-wave approximation does not give the correct nonrelativistic result for the Lamb shift [2.11]. Actually, $(\omega_A - kc)^{-1}$ should read $(\omega_A - kc)^{-1} + (\omega_A + kc)^{-1}$ in (2.28) (Exercise 2.2).

2.2.2 The Einstein A Coefficient

If we have a correct description of spontaneous emission we must expect the damping constant γ appearing in (2.26) to give us the correct result for the Einstein A coefficient. We can check this by performing the integration over wavevectors and the polarization summation in (2.21).

Adopting spherical coordinates in k-space, the density of states for each polarization state λ is given by [2.12]

$$g(\boldsymbol{k})d^3k = \frac{\omega^2 V}{8\pi^3 c^3} d\omega \sin\theta d\theta d\phi. \tag{2.29}$$

Substituting from (2.29) and (2.16) into (2.21),

$$\gamma = 2\pi \sum_\lambda \int_0^\infty d\omega \int_0^\pi \sin\theta d\theta \int_0^{2\pi} d\phi \, \frac{\omega^2 V}{8\pi^3 c^3} \frac{\omega}{2\hbar\epsilon_0 V} (\hat{e}_{\boldsymbol{k},\lambda} \cdot \boldsymbol{d}_{12})^2 \delta(\omega - \omega_A)$$

$$= \frac{\omega_A^3}{8\pi^2 \epsilon_0 \hbar c^3} \sum_\lambda \int_0^\pi \sin\theta d\theta \int_0^{2\pi} d\phi \, (\hat{e}_{\boldsymbol{k},\lambda} \cdot \boldsymbol{d}_{12})^2. \tag{2.30}$$

Now, for each $\boldsymbol{k}$ we can choose polarization states λ_1 and λ_2 so that the first polarization state gives $\hat{e}_{\boldsymbol{k},\lambda_1} \cdot \boldsymbol{d}_{12} = 0$ (taking $\boldsymbol{d}_{12}$ real for simplicity). This is achieved with the geometry illustrated in Fig. 2.2. Then, for the second polarization state, we find

$$(\hat{e}_{\boldsymbol{k},\lambda_2} \cdot \boldsymbol{d}_{12})^2 = d_{12}^2(1 - \cos^2\alpha) = d_{12}^2\left[1 - (\hat{d}_{12} \cdot \hat{k})^2\right], \tag{2.31}$$

where $\hat{d}_{12}$ and $\hat{k}$ are unit vectors in the directions of $\boldsymbol{d}_{12}$ and $\boldsymbol{k}$, respectively. The angular integrals are now easily performed if we choose the k_z-axis to correspond to the $\hat{d}_{12}$ direction. We have

$$\int_0^\pi \sin\theta d\theta \int_0^{2\pi} d\phi \, (\hat{e}_{\boldsymbol{k},\lambda} \cdot \boldsymbol{d}_{12})^2 = d_{12}^2 \int_0^{2\pi} d\phi \int_0^\pi d\theta \sin\theta (1 - \cos^2\theta)$$

$$= \frac{8\pi}{3} d_{12}^2. \tag{2.32}$$

From (2.30) and (2.32)

$$\gamma = \frac{1}{4\pi\epsilon_0} \frac{4\omega_A^3 d_{12}^2}{3\hbar c^3}. \tag{2.33}$$

This is the correct result for the *Einstein A coefficient*, as obtained from the Wigner-Weisskopf theory of natural linewidth [2.13, 2.14].

Exercise 2.2 After replacing $(\omega_A - kc)^{-1}$ by $(\omega_A - kc)^{-1} + (\omega_A + kc)^{-1}$ in (2.23), show that this equation gives the formula for the *temperature-dependent shift* derived in Ref. [2.9]:

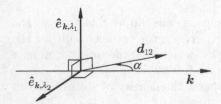

Fig. 2.2 Polarization states used in the evaluation of (2.30).

$$2\Delta' = \frac{1}{4\pi\epsilon_0} \frac{4d_{12}^2}{3\hbar\pi c^3} \, P \int_0^\infty d\omega \, \omega^3 \left(\frac{1}{\omega_A - \omega} + \frac{1}{\omega_A + \omega} \right) \frac{1}{e^{\hbar\omega/k_B T} - 1}. \quad (2.34)$$

The corresponding formula for the *Lamb shift* is

$$\Delta = \frac{1}{4\pi\epsilon_0} \frac{2d_{12}^2}{3\hbar\pi c^3} \, P \int_0^\infty d\omega \, \omega^3 \left(\frac{1}{\omega_A - \omega} + \frac{1}{\omega_A + \omega} \right). \quad (2.35)$$

2.2.3 Matrix Element Equations, Correlation Functions, and Spontaneous Emission Spectrum

We mentioned earlier that $\langle\sigma_z\rangle$, $\langle\sigma_-\rangle$, and $\langle\sigma_+\rangle$ are simply related to the matrix elements of ρ. We can derive equations of motion for these expectation values from (2.26) as we did for the harmonic oscillator, or, alternatively, we can simply take the matrix elements of (2.26) directly. Following the second approach, we use (2.12) to find

$$\begin{aligned}
\dot{\rho}_{22} = &- i\tfrac{1}{2}\omega_A \langle 2|(\sigma_z\rho - \rho\sigma_z)|2\rangle \\
&+ \frac{\gamma}{2}(\bar{n}+1)\langle 2|(2\sigma_-\rho\sigma_+ - \sigma_+\sigma_-\rho - \rho\sigma_+\sigma_-)|2\rangle \\
&+ \frac{\gamma}{2}\bar{n}\langle 2|(2\sigma_+\rho\sigma_- - \sigma_-\sigma_+\rho - \rho\sigma_-\sigma_+)|2\rangle \\
= &- \gamma(\bar{n}+1)\rho_{22} + \gamma\bar{n}\rho_{11}, \quad (2.36a)
\end{aligned}$$

and, similarly:

$$\dot{\rho}_{11} = -\gamma\bar{n}\rho_{11} + \gamma(\bar{n}+1)\rho_{22}, \quad (2.36b)$$

$$\dot{\rho}_{21} = -\left[\frac{\gamma}{2}(2\bar{n}+1) + i\omega_A\right]\rho_{21}, \quad (2.36c)$$

$$\dot{\rho}_{12} = -\left[\frac{\gamma}{2}(2\bar{n}+1) - i\omega_A\right]\rho_{12}. \quad (2.36d)$$

We have dropped the distinction between ω_A and ω_A'. Equations (2.36a) and (2.36b) clearly illustrate our interpretation of the two terms – proportional to $(\gamma/2)(\bar{n}+1)$ and $(\gamma/2)\bar{n}$ – in the master equation; the former describes $|2\rangle \rightarrow |1\rangle$ transitions at a rate $\gamma(\bar{n}+1)$, and the latter describes $|1\rangle \rightarrow |2\rangle$ transitions at a rate $\gamma\bar{n}$. Of course, probability leaves and enters the two states in such a way that the total probability is preserved – $\dot{\rho}_{11} + \dot{\rho}_{22} = 0$. Equations (2.36a) and (2.36b) are in fact just the rate equations of Einstein A and B theory.

Exercise 2.3 Show that in the steady state the balance between upwards and downwards transitions leads to a thermal equilibrium distribution between the states $|1\rangle$ and $|2\rangle$.

Using the relations $\langle\sigma_z\rangle = \rho_{22} - \rho_{11}$, $\langle\sigma_-\rangle = \rho_{21}$, $\langle\sigma_+\rangle = \rho_{12}$, and $\rho_{11} + \rho_{22} = 1$, the matrix element equations can be written in the alternative form:

$$\langle\dot{\sigma}_z\rangle = -\gamma\big[\langle\sigma_z\rangle(2\bar{n}+1)+1\big], \qquad (2.37a)$$

$$\langle\dot{\sigma}_-\rangle = -\Big[\frac{\gamma}{2}(2\bar{n}+1)+i\omega_A\Big]\langle\sigma_-\rangle, \qquad (2.37b)$$

$$\langle\dot{\sigma}_+\rangle = -\Big[\frac{\gamma}{2}(2\bar{n}+1)-i\omega_A\Big]\langle\sigma_+\rangle. \qquad (2.37c)$$

These provide us with a simple illustration of the use of the quantum regression formula (Sect. 1.5). At optical frequencies and normal laboratory temperatures $\bar{n}$ is negligible, and for simplicity we drop it here. Then, using (2.25a), we may write the mean-value equations in vector form:

$$\langle\dot{s}\rangle = M\langle s\rangle, \qquad (2.38)$$

with

$$s \equiv \begin{pmatrix} \sigma_- \\ \sigma_+ \\ \sigma_+\sigma_- \end{pmatrix}, \qquad (2.39)$$

$$M \equiv \mathrm{diag}\Big[-\Big(\frac{\gamma}{2}+i\omega_A\Big), -\Big(\frac{\gamma}{2}-i\omega_A\Big), -\gamma\Big]. \qquad (2.40)$$

For $\tau \geq 0$, equations for nine correlation functions are obtained from (1.107):

$$\frac{d}{d\tau}\langle\sigma_-(t)s(t+\tau)\rangle = M\langle\sigma_-(t)s(t+\tau)\rangle, \qquad (2.41a)$$

$$\frac{d}{d\tau}\langle\sigma_+(t)s(t+\tau)\rangle = M\langle\sigma_+(t)s(t+\tau)\rangle, \qquad (2.41b)$$

$$\frac{d}{d\tau}\langle\sigma_+(t)\sigma_-(t)s(t+\tau)\rangle = M\langle\sigma_+(t)\sigma_-(t)s(t+\tau)\rangle. \qquad (2.41c)$$

Equations for a further nine correlation functions with reverse time order are obtained from (1.108); alternatively, this second set of correlation functions can be derived from the first, using

$$\langle\hat{A}(t+\tau)\hat{A}_\nu(t)\rangle = \langle\hat{A}_\nu^\dagger(t)\hat{A}^\dagger(t+\tau)\rangle^*. \qquad (2.42)$$

Equation (1.109) defines a further twenty-seven correlation functions.

Let us consider an atom prepared initially in its excited state. For this initial condition $\langle\sigma_-\rangle = \langle\sigma_+\rangle = 0$, $\langle\sigma_+\sigma_-\rangle = \rho_{22} = 1$, and the solution to (2.38) is

$$\langle s \rangle = \begin{pmatrix} 0 \\ 0 \\ e^{-\gamma t} \end{pmatrix}. \tag{2.43}$$

Initial conditions for (2.41a)–(2.41c) are then, respectively,

$$\langle \sigma_-(t)s(t) \rangle = \begin{pmatrix} 0 \\ 1 - e^{-\gamma t} \\ 0 \end{pmatrix}, \tag{2.44a}$$

$$\langle \sigma_+(t)s(t) \rangle = \begin{pmatrix} e^{-\gamma t} \\ 0 \\ 0 \end{pmatrix}, \tag{2.44b}$$

$$\langle \sigma_+(t)\sigma_-(t)s(t) \rangle = \begin{pmatrix} 0 \\ 0 \\ e^{-\gamma t} \end{pmatrix}, \tag{2.44c}$$

where we have used (2.25), together with the following:

$$\sigma_+^2 = |2\rangle\langle 1|2\rangle\langle 1| = 0, \tag{2.45a}$$

$$\sigma_-^2 = |1\rangle\langle 2|1\rangle\langle 2| = 0, \tag{2.45b}$$

$$\sigma_+\sigma_-\sigma_+ = |2\rangle\langle 1|1\rangle\langle 2|2\rangle\langle 1| = |2\rangle\langle 1| = \sigma_+, \tag{2.45c}$$

$$\sigma_-\sigma_+\sigma_- = |1\rangle\langle 2|2\rangle\langle 1|1\rangle\langle 2| = |1\rangle\langle 2| = \sigma_-. \tag{2.45d}$$

The nonzero correlation functions obtained from (2.41) with initial conditions (2.44) are ($\tau \geq 0$)

$$\langle \sigma_-(t)\sigma_+(t+\tau) \rangle = e^{i\omega_A\tau}e^{-(\gamma/2)\tau}\left(1 - e^{-\gamma t}\right), \tag{2.46}$$

$$\langle \sigma_+(t)\sigma_-(t+\tau) \rangle = e^{-i\omega_A\tau}e^{-(\gamma/2)\tau}e^{-\gamma t}, \tag{2.47}$$

$$\langle \sigma_+(t)\sigma_-(t)\sigma_+(t+\tau)\sigma_-(t+\tau) \rangle = e^{-\gamma\tau}e^{-\gamma t}. \tag{2.48}$$

Equation (2.47) provides the result for the *spontaneous emission spectrum*. For an ideal detector, the probability of detecting a photon of frequency ω during the interval $t = 0$ to $t = T$ is given by [2.15]

$$P(\omega) \propto \int_0^T dt \int_0^T dt'\, e^{-i\omega(t-t')}\langle\sigma_+(t)\sigma_-(t')\rangle. \tag{2.49}$$

We will see how the field at the detector is related to the atomic operators σ_- and σ_+ shortly (Sect. 2.3.1); clearly, such a relationship is needed to write (2.49). Using (2.47) and

$$\langle\sigma_+(t+\tau)\sigma_-(t)\rangle = \langle\sigma_+(t)\sigma_-(t+\tau)\rangle^*, \tag{2.50}$$

we find, for all t and t',

$$\langle \sigma_+(t)\sigma_-(t') \rangle = e^{i\omega_A(t-t')}e^{-(\gamma/2)(t+t')}. \tag{2.51}$$

Then,

$$P(\omega) \propto \int_0^T dt\, e^{-[(\gamma/2)+i(\omega-\omega_A)]t} \int_0^T dt'\, e^{-[(\gamma/2)-i(\omega-\omega_A)]t'}$$

$$\propto \frac{1 - e^{-(\gamma/2)T}e^{-i(\omega-\omega_A)T}}{\gamma/2 + i(\omega - \omega_A)} \frac{1 - e^{-(\gamma/2)T}e^{i(\omega-\omega_A)T}}{\gamma/2 - i(\omega - \omega_A)}. \tag{2.52}$$

For long times, $T \gg 1/\gamma$, this gives the Lorentzian lineshape

$$P(\omega) \propto \frac{1}{(\gamma/2)^2 + (\omega - \omega_A)^2}. \tag{2.53}$$

2.2.4 Phase Destroying Processes

The interaction with the many mode electromagnetic field that gives rise to spontaneous emission causes both energy loss from the atom and damping of the atomic polarization. Polarization damping is described by the loss terms proportional to $(\gamma/2)(2\bar{n} + 1)$ in (2.36c) and (2.36d). This damping results from a randomization of the phases of the atomic wavefunctions by thermal and vacuum fluctuations in the electromagnetic field, causing the overlap of the upper and lower state wavefunctions to decay in time. It is often necessary to account for additional dephasing interactions; these might arise from elastic collisions in an atomic vapor, or elastic phonon scattering in a solid. What terms must we add to the master equation (2.26) to describe such processes?

A phenomenological model describing atomic dephasing can be obtained by adding two further reservoir interactions to the Hamiltonian (2.15). We add

$$H_{\text{dephase}} \equiv H_{R_1} + H_{R_2} + H_{SR_1} + H_{SR_2}, \tag{2.54}$$

with

$$H_{R_1} + H_{R_2} \equiv \sum_j \hbar\omega_{1j}\, r_{1j}^\dagger r_{1j} + \sum_j \hbar\omega_{2j}\, r_{2j}^\dagger r_{2j}, \tag{2.55a}$$

$$H_{SR_1} + H_{SR_2} \equiv \sum_{j,k} \hbar\kappa_{1jk}\, r_{1j}^\dagger r_{1k}\, \sigma_-\sigma_+ + \sum_{j,k} \hbar\kappa_{2jk}\, r_{2j}^\dagger r_{2k}\, \sigma_+\sigma_-. \tag{2.55b}$$

The complete reservoir seen by the atom is now composed of three subsystems: $R = R_{12} \otimes R_1 \otimes R_2$, where R_{12} is the reservoir defined by (2.15b). These reservoir subsystems are assumed to be statistically independent, with the density operator R_0 given by the product of three thermal equilibrium operators in the form of (1.38). The interactions H_{SR_1} and H_{SR_2} describe the scattering of quanta from the atom while it is in states $|1\rangle$ and $|2\rangle$, respectively; they sum over virtual processes that scatter quanta with energies $\hbar\omega_{1k}$ and $\hbar\omega_{2k}$ into quanta with energies $\hbar\omega_{1j}$ and $\hbar\omega_{2j}$ while leaving the state of the atom unchanged.

The terms that are added to the master equation by these new reservoir interactions follow in a rather straightforward manner from the general form (1.34) for the master equation in the Born approximation. In addition to the reservoir operators $\tilde{\Gamma}_1(t)$ and $\tilde{\Gamma}_2(t)$ that are defined by the interaction with R_{12} [Eqs. (2.18)], we must introduce operators $\tilde{\Gamma}_3(t)$ and $\tilde{\Gamma}_4(t)$ to account for the interactions with R_1 and R_2. First, however, we have to take care of a problem, one which was not met in deriving master equations for the damped harmonic oscillator and the radiatively damped atom. Equation (1.34) was obtained using the assumption (1.28) that all reservoir operators coupling to the system S have zero mean in the state R_0. This is not true for the reservoir operators coupling to $\sigma_-\sigma_+$ and $\sigma_+\sigma_-$ in (2.55b); terms with $j = k$ in the summation over reservoir modes have nonzero averages proportional to mean thermal occupation numbers. To overcome this difficulty the interaction between S and the mean reservoir "field" can be included in H_S rather than H_{SR}. With the use of (2.25), in place of (2.55a) and (2.55b) we may write

$$H_S \equiv \tfrac{1}{2}\hbar(\omega_A + \delta_p)\sigma_z, \tag{2.56}$$

and

$$
\begin{aligned}
& H_{SR_1} + H_{SR_2} \\
& \equiv \sum_{j,k} \hbar\kappa_{1jk}(r_{1j}^\dagger r_{1k} - \delta_{jk}\bar{n}_{1j})\sigma_-\sigma_+ + \sum_{j,k} \hbar\kappa_{2jk}(r_{2j}^\dagger r_{2k} - \delta_{jk}\bar{n}_{2j})\sigma_+\sigma_-,
\end{aligned}
\tag{2.57}
$$

with the frequency shift δ_p given by

$$
\begin{aligned}
\delta_p &= \sum_j (\kappa_{2jj}\bar{n}_{2j} - \kappa_{1jj}\bar{n}_{1j}) \\
&= \int_0^\infty d\omega\,[g_2(\omega)\kappa_2(\omega,\omega) - g_1(\omega)\kappa_1(\omega,\omega)]\bar{n}(\omega,T). \tag{2.58}
\end{aligned}
$$

$\bar{n}_{1j} \equiv \bar{n}(\omega_{1j}, T)$ and $\bar{n}_{2j} \equiv \bar{n}(\omega_{2j}, T)$ are mean occupation numbers for reservoir modes with frequencies ω_{1j} and ω_{2j}, respectively, and in (2.58) the summation over reservoir modes has been converted to an integration by introducing the densities of states $g_1(\omega)$ and $g_2(\omega)$. The sum of (2.56) and (2.57) gives the same Hamiltonian as the sum of (2.55a) and (2.55b); but now the reservoir operators that appear in H_{SR_1} and H_{SR_2} have zero mean.

We may now proceed directly from (1.34). After transforming to the interaction picture, the interaction Hamiltonian (2.57) is written in the form (1.33) with

$$\tilde{s}_3(t) = \sigma_-\sigma_+, \tag{2.59a}$$

$$\tilde{s}_4(t) = \sigma_+\sigma_-, \tag{2.59b}$$

and

$$\tilde{\Gamma}_3(t) = \sum_{j,k} \kappa_{1jk} \left(r_{1j}^\dagger r_{1k}\, e^{i(\omega_{1j}-\omega_{1k})t} - \delta_{jk}\bar{n}_{1j} \right), \tag{2.60a}$$

$$\tilde{\Gamma}_4(t) = \sum_{j,k} \kappa_{2jk} \left(r_{2j}^\dagger r_{2k}\, e^{i(\omega_{2j}-\omega_{2k})t} - \delta_{jk}\bar{n}_{2j} \right). \tag{2.60b}$$

These are to be substituted – together with $\tilde{s}_1(t)$, $\tilde{s}_2(t)$, $\tilde{\Gamma}_1(t)$, and $\tilde{\Gamma}_2(t)$ from (2.18) and (2.19) – into (1.34). Since the reservoir subsystems are statistically independent and all reservoir operators have zero mean, all of the cross terms involving correlation functions for products of operators from different reservoir subsystems will vanish. Thus, the spontaneous emission terms arising from the interaction with $\tilde{\Gamma}_1$ and $\tilde{\Gamma}_2$ are obtained exactly as in Sect. 2.2.1. The additional terms from the interaction with $\tilde{\Gamma}_3$ and $\tilde{\Gamma}_4$ take the form

$$\begin{aligned}
\left(\dot{\tilde{\rho}} \right)_{\text{dephase}} = &-\int_0^t dt' [\sigma_-\sigma_+\sigma_-\sigma_+\tilde{\rho}(t') - \sigma_-\sigma_+\tilde{\rho}(t')\sigma_-\sigma_+]\langle\tilde{\Gamma}_3(t)\tilde{\Gamma}_3(t')\rangle_{R_1} \\
&+[\tilde{\rho}(t')\sigma_-\sigma_+\sigma_-\sigma_+ - \sigma_-\sigma_+\tilde{\rho}(t')\sigma_-\sigma_+]\langle\tilde{\Gamma}_3(t')\tilde{\Gamma}_3(t)\rangle_{R_1} \\
&+[\sigma_+\sigma_-\sigma_+\sigma_-\tilde{\rho}(t') - \sigma_+\sigma_-\tilde{\rho}(t')\sigma_+\sigma_-]\langle\tilde{\Gamma}_4(t)\tilde{\Gamma}_4(t')\rangle_{R_2} \\
&+[\tilde{\rho}(t')\sigma_+\sigma_-\sigma_+\sigma_- - \sigma_+\sigma_-\tilde{\rho}(t')\sigma_+\sigma_-]\langle\tilde{\Gamma}_4(t')\tilde{\Gamma}_4(t)\rangle_{R_2}.
\end{aligned} \tag{2.61}$$

We will evaluate the first of the reservoir correlation functions appearing in (2.61); the others follow in a similar form. From (2.59a),

$$\begin{aligned}
&\langle\tilde{\Gamma}_3(t)\tilde{\Gamma}_3(t')\rangle_{R_1} \\
&= \text{tr}\Big[R_{10} \sum_{j,k}\sum_{j',k'} \kappa_{1jk}\,\kappa_{1j'k'} \left(r_{1j}^\dagger r_{1k}e^{i(\omega_{1j}-\omega_{1k})t} - \delta_{jk}\bar{n}_{1j} \right) \\
&\quad \times \left(r_{1j'}^\dagger r_{1k'}e^{i(\omega_{1j'}-\omega_{1k'})t'} - \delta_{j'k'}\bar{n}_{1j'} \right)\Big] \\
&= \text{tr}\Big[R_{10}\Big(\sum_{j,k}\sum_{j',k'} \kappa_{1jk}\,\kappa_{1j'k'}\, r_{1j}^\dagger r_{1k}\, r_{1j'}^\dagger r_{1k'}\, e^{i(\omega_{1j}-\omega_{1k})t}e^{i(\omega_{1j'}-\omega_{1k'})t'} \\
&\quad - \sum_j\sum_{j',k'} \kappa_{1jj}\,\kappa_{1j'k'}\, \bar{n}_{1j}\, r_{1j'}^\dagger r_{1k'}\, e^{i(\omega_{1j'}-\omega_{1k'})t'} \\
&\quad - \sum_{j,k}\sum_{j'} \kappa_{1jk}\,\kappa_{1j'j'}\, r_{1j}^\dagger r_{1k}\, \bar{n}_{1j'}\, e^{i(\omega_{1j}-\omega_{1k})t} \Big)\Big] \\
&\quad + \sum_j\sum_{j'} \kappa_{1jj}\,\kappa_{1j'j'}\, \bar{n}_{1j}\, \bar{n}_{1j'},
\end{aligned}$$

where $R_{\tilde{1}0}$ is the thermal equilibrium density operator [Eq. (1.38)] for the reservoir subsystem R_1. The nonvanishing contributions to the trace are now obtained as follows: the first double sum contributes for $j = k \neq j' = k'$, for $j = k' \neq k = j'$, and for $j = k = j' = k'$; the second double sum contributes for $j' = k'$; and the third double sum for $j = k$. The correlation function becomes

$$\langle \tilde{\Gamma}_3(t)\tilde{\Gamma}_3(t')\rangle_{R_1}$$
$$= \sum_{\substack{j,j' \\ j \neq j'}} \kappa_{1jj}\,\kappa_{1j'j'}\,\bar{n}_{1j}\,\bar{n}_{1j'} + \sum_{\substack{j,j' \\ j \neq j'}} \kappa_{1jj'}\,\kappa_{1j'j}\,\bar{n}_{1j}(\bar{n}_{1j'}+1)e^{i(\omega_{1j}-\omega_{1j'})(t-t')}$$
$$+ \sum_j \kappa_{1jj}^2\overline{n_{1j}^2} - 2\sum_{j,j'} \kappa_{1jj}\,\kappa_{1j'j'}\,\bar{n}_{1j}\,\bar{n}_{1j'} + \sum_{j,j'} \kappa_{1jj}\,\kappa_{1j'j'}\,\bar{n}_{1j}\,\bar{n}_{1j'},$$

where the first three terms come from the first double sum, and the fourth term comes from the second and third double sums. Noting that $\overline{n_{1j}^2} = \bar{n}_{1j}^2 + \bar{n}_{1j}(\bar{n}_{1j}+1)$, we see that the sums for $j \neq j'$ are completed for all j and j' by the third term in this expression; setting $\kappa_{1jj'}\,\kappa_{1j'j} = |\kappa_{1jj'}|^2$ – required for (2.55b) to be Hermitian – we arrive at the result

$$\langle \tilde{\Gamma}_3(t)\tilde{\Gamma}_3(t')\rangle_{R_1} = \sum_{j,j'} |\kappa_{1jj'}|^2\bar{n}_{1j}(\bar{n}_{1j'}+1)e^{i(\omega_{1j}-\omega_{1j'})(t-t')}. \qquad (2.62a)$$

Similar expressions follow for the other reservoir correlation functions:

$$\langle \tilde{\Gamma}_4(t)\tilde{\Gamma}_4(t')\rangle_{R_2} = \sum_{j,j'} |\kappa_{2jj'}|^2\bar{n}_{2j}(\bar{n}_{2j'}+1)e^{i(\omega_{2j}-\omega_{2j'})(t-t')}, \qquad (2.62b)$$

and

$$\langle \tilde{\Gamma}_3(t')\tilde{\Gamma}_3(t)\rangle_{R_1} = \left(\langle \tilde{\Gamma}_3(t)\tilde{\Gamma}_3(t')\rangle_{R_1}\right)^*, \qquad (2.62c)$$

$$\langle \tilde{\Gamma}_4(t')\tilde{\Gamma}_4(t)\rangle_{R_2} = \left(\langle \tilde{\Gamma}_4(t)\tilde{\Gamma}_4(t')\rangle_{R_2}\right)^*. \qquad (2.62d)$$

If reservoir correlation times are very short compared to the timescale for the system dynamics, the time integral in (2.61) can be treated in the same fashion as in Sect. 1.4.1. After simplifying the operator products using (2.25), (2.61) then gives

$$\left(\dot{\tilde{\rho}}\right)_{\text{dephase}} = -i\frac{1}{2}\Delta_p[\sigma_z, \tilde{\rho}] + \frac{\gamma_p}{2}(\sigma_z\tilde{\rho}\sigma_z - \tilde{\rho}), \qquad (2.63)$$

with

$$\gamma_p \equiv \pi \int_0^\infty d\omega \left[g_2(\omega)^2 |\kappa_2(\omega,\omega)|^2 + g_1(\omega)|\kappa_1(\omega,\omega)|^2 \right]$$
$$\times \bar{n}(\omega,T)[\bar{n}(\omega,T)+1], \tag{2.64}$$

$$\Delta_p \equiv P \int_0^\infty d\omega \int_0^\infty d\omega' \frac{g_2(\omega)g_2(\omega')|\kappa_2(\omega,\omega')|^2 - g_1(\omega)g_1(\omega')|\kappa_1(\omega,\omega')|^2}{\omega - \omega'}$$
$$\times \bar{n}(\omega,T). \tag{2.65}$$

We add (2.63) to the terms describing radiative damping given by (2.24), and transform back to the Schrödinger picture using (1.72) and (2.56) to obtain the *master equation for a radiatively damped two-level atom with nonradiative dephasing*:

$$\dot{\rho} = -i\tfrac{1}{2}\omega_A'[\sigma_z, \rho] + \frac{\gamma}{2}(\bar{n}+1)(2\sigma_-\rho\sigma_+ - \sigma_+\sigma_-\rho - \rho\sigma_+\sigma_-)$$
$$+ \frac{\gamma}{2}\bar{n}(2\sigma_+\rho\sigma_- - \sigma_-\sigma_+\rho - \rho\sigma_-\sigma_+) + \frac{\gamma_p}{2}(\sigma_z\rho\sigma_z - \rho), \tag{2.66}$$

where the shifted atomic frequency is now

$$\omega_A' \equiv \omega_A + 2\Delta' + \Delta + \delta_p + \Delta_p, \tag{2.67}$$

with $2\Delta' + \Delta$, δ_p, and Δ_p given by (2.28), (2.58), and (2.65).

2.3 Resonance Fluorescence

The theory of resonance fluorescence provides a good illustration of the methods we have learned so far, and a simple situation in which to introduce some of the subtleties that arise in the treatment of damping for interacting atoms and fields. We are concerned here with a two-level atom irradiated by a strong monochromatic laser beam tuned to the atomic transition. Photons may be absorbed from this beam and emitted to the many modes of the vacuum electromagnetic field as fluorescent scattering. This scattering process is mediated by the reservoir interaction (2.15c) underlying our treatment of spontaneous emission.

The phenomenon of fluorescence has fascinated physicists for over a century [2.16, 2.17]. A simple classical picture can be given in terms of the Lorentz oscillator model which underlies the classical theory of dispersion [2.18–2.20]. In this picture, a harmonic electron oscillator is set into forced oscillation by the incident light and reradiates as a dipole source according to the laws of classical electrodynamics. Of course, in the absence of damping the amplitude of a resonantly forced oscillator grows without bound; to avoid this divergence some account of atomic damping must be given. In the classical theory this is achieved with the introduction of a velocity-dependent force derived from radiation reaction. The damping constant introduced in this way ensures that the energy appearing in the reradiated field is matched by energy loss from the oscillator. This classical theory does pretty well at weak excitation. In particular, the relationship between the fluorescence spectrum and the spectrum of the excitation is correctly obtained; single-frequency excitation produces a forced response of the electron oscillator and a reradiated field with the same frequency. A hastily drawn conclusion for a two-level atom might expect the fluorescence spectrum to show the natural linewidth [Eq. (2.53)]. This would follow if the atomic dynamics proceeded by independent absorption and spontaneous emission events. However, this is an incorrect view of the scattering process. A perturbative treatment of the quantum-mechanical problem is adequate to show that at weak intensities the classical result is correct [2.21]. We must view the scattering as an essentially coherent process, passing energy from the incident beam to the scattered field without lingering en route in the excited state [2.22].

Of course, a two-level atom is not a harmonic oscillator, and the classical theory fails at sufficiently high laser intensities – in fact, it fails even at weak intensities if we look more carefully at the statistics of the scattered photons. As we will see, a two-level atom responds nonlinearly to increasing intensity; also, while a harmonic oscillator can be excited ever higher up its ladder of Fock states, a two-level atom can only store a single quantum of energy. From a quantum treatment we will find the following: With increasing incident intensity, the fluorescence spectrum picks up an incoherent component having the natural linewidth. This incoherent spectrum splits into a three-peaked structure and eventually accounts for nearly all of the scattered intensity. This behavior was first predicted by Mollow [2.23] and has been observed in a number of experiments [2.24–2.26]. The incoherent spectral component arises from quantum fluctuations around the nonequilibrium steady state established by the balance between excitation and emission processes. These quantum fluctuations are inherent in the probabilistic character of quantum dynamics, and are not introduced by any external stochastic agent.

Quantum mechanics makes its mark even at weak laser intensities if we ask the right question. We will find that there is zero probability of detecting two scattered photons emitted at the same time, independent of the incident intensity. This photon "antibunching" is a consequence of the fact that the atom can store just a single quantum of energy, and, after emitting this quantum, cannot produce a second until it is reexcited. It is the inverse of the photon "bunching" associated with the famous Hanbury-Brown-Twiss effect (Sect. 1.5.3) – there the probability for detecting two simultaneous photons is twice that expected for random photon arrivals [2.27]. Photon antibunching cannot be treated using a classical statistical description for the scattered field, and has therefore received special attention as a phenomenon requiring the quantized electromagnetic field [2.28–2.30]. The earliest reference to the vanishing probability for simultaneous photon detection in resonance fluorescence is contained in the work of Mollow [2.31]. Carmichael and Walls [2.32] calculated the second-order correlation function for the scattered light, explicitly demonstrating antibunching in contrast to the bunching of Hanbury-Brown and Twiss. Shortly thereafter photon antibunching was observed by Kimble et al. [2.33] in the fluorescence from a dilute sodium atomic beam.

We will obtain the fluorescence spectrum and a description of photon antibunching using the master equation methods we have developed. This is not the only approach to these problems and an extensive literature is available on this subject. A good review with complete references is given by Cresser et al. [2.34].

2.3.1 The Scattered Field

The incident laser mode is in a highly excited state that is essentially unaffected by its interaction with the single atom. We can treat this field as a classical driving force. Then the Hamiltonian for the resonantly driven two-level atom interacting with the many modes of the electromagnetic field separates into system and reservoir terms, as in (1.16), with

$$H_S \equiv \tfrac{1}{2}\hbar\omega_A\sigma_z - dE\big(e^{-i\omega_A t}\sigma_+ + e^{i\omega_A t}\sigma_-\big), \tag{2.68a}$$

$$H_R \equiv \sum_{k,\lambda} \hbar\omega_k r^\dagger_{k,\lambda}r_{k,\lambda}, \tag{2.68b}$$

$$H_{SR} \equiv \sum_{k,\lambda} \hbar\big(\kappa^*_{k,\lambda}r^\dagger_{k,\lambda}\sigma_- + \kappa_{k,\lambda}r_{k,\lambda}\sigma_+\big); \tag{2.68c}$$

both interactions are written in the dipole and rotating-wave approximations. The laser field at the site of the atom is

$$E(t) \equiv \hat{e}2E\cos(\omega_A t + \phi), \tag{2.69}$$

where $\hat{e}$ is a unit polarization vector, E is a real amplitude, and the phase ϕ is chosen so that $d \equiv \hat{e}\cdot d_{12}e^{i\phi}$ is also real.

The master equation approach focuses on the dynamics of the atom. We are ultimately interested, however, in the properties of the fluorescence. The scattered field is given in terms of the reservoir operators – in the Heisenberg picture

$$\hat{E}(r,t) = \hat{E}^{(+)}(r,t) + \hat{E}^{(-)}(r,t), \tag{2.70a}$$

with

$$\hat{E}^{(+)}(r,t) = i\sum_{k,\lambda} \sqrt{\frac{\hbar\omega_k}{2\epsilon_0 V}}\,\hat{e}_{k,\lambda}r_{k,\lambda}(t)e^{ik\cdot r}, \tag{2.70b}$$

$$\hat{E}^{(-)}(r,t) = \hat{E}^{(+)}(r,t)^\dagger. \tag{2.70c}$$

We will need the correlation functions

$$G^{(1)}(t, t+\tau) \equiv \langle \hat{E}^{(-)}(t)\hat{E}^{(+)}(t+\tau)\rangle, \qquad (2.71)$$

and

$$G^{(2)}(t, t+\tau) \equiv \langle \hat{E}^{(-)}(t)\hat{E}^{(-)}(t+\tau)\hat{E}^{(+)}(t+\tau)\hat{E}^{(+)}(t)\rangle, \qquad (2.72)$$

where the field operators are evaluated at the position of an idealized point-like detector. Since we trace over the reservoir variables in deriving the master equation for S, our first task is to relate the scattered field to atomic source operators, so that (2.71) and (2.72) can be expressed in terms of operators of the system S.

We begin with the Heisenberg equations of motion for the electromagnetic field modes:

$$\dot{r}_{k,\lambda} = -i\omega_k r_{k,\lambda} - i\kappa_{k,\lambda}^* \sigma_-. \qquad (2.73)$$

Writing

$$r_{k,\lambda} = \tilde{r}_{k,\lambda} e^{-i\omega_k t}, \qquad (2.74a)$$

$$\sigma_- = \tilde{\sigma}_- e^{-i\omega_A t}, \qquad (2.74b)$$

and integrating (2.73) formally, gives

$$\tilde{r}_{k,\lambda}(t) = r_{k,\lambda}(0) - i\kappa_{k,\lambda}^* \int_0^t dt'\, \tilde{\sigma}_-(t')e^{i(\omega_k - \omega_A)t'}. \qquad (2.75)$$

The separation of the rapidly oscillating term in (2.74b) is motivated by the solution to the Heisenberg equations for the free atom [Eqs. (2.19)]. Now, substituting $r_{k,\lambda}(t)$ into (2.70a), and introducing the explicit form of the coupling constant from (2.16), the field operator becomes

$$\hat{E}^{(+)}(r, t) = \hat{E}_f^{(+)}(r, t) + \hat{E}_s^{(+)}(r, t), \qquad (2.76)$$

with

$$\hat{E}_f^{(+)}(r, t) = i\sum_{k,\lambda} \sqrt{\frac{\hbar\omega_k}{2\epsilon_0 V}}\, \hat{e}_{k,\lambda} r_{k,\lambda}(0) e^{-i(\omega_k t - k\cdot r)}, \qquad (2.77)$$

and

$$\hat{E}_s^{(+)}(r, t) = i\frac{1}{2\epsilon_0 V} e^{-i\omega_A t} \sum_{k,\lambda} \omega_k \hat{e}_{k,\lambda}(\hat{e}_{k,\lambda}\cdot d_{12}) e^{ik\cdot(r-r_A)}$$

$$\times \int_0^t dt'\, \tilde{\sigma}_-(t')e^{i(\omega_k - \omega_A)(t'-t)}. \qquad (2.78)$$

Here $\hat{E}_f^{(+)}(r, t)$ describes the free evolution of the electromagnetic field, in the absence of the atomic scatterer; $\hat{E}_s^{(+)}(r, t)$ is the source field radiated by the atom. It remains to perform the summation and integration in (2.78).

The summation over k is performed by introducing the density of states (2.29) and converting the sum into an integration:

$$\hat{E}_s^{(+)}(r,t) = i\frac{1}{16\pi^3\epsilon_0 c^3}e^{-i\omega_A t}\sum_\lambda \int_0^\infty d\omega \int_0^\pi \sin\theta\, d\theta \int_0^{2\pi} d\phi$$

$$\times\, \omega^3 \hat{e}_{k,\lambda}(\hat{e}_{k,\lambda}\cdot d_{12})e^{i(\omega r/c)\cos\theta}\int_0^t dt'\,\tilde\sigma_-(t')e^{i(\omega-\omega_A)(t'-t)};$$

$$(2.79)$$

we have chosen a geometry with the origin in r-space at the site of the atom and the k_z-axis in the direction of r. One polarization state may be chosen perpendicular to both k and d_{12}, as in Fig. 2.2, and for the second we can write

$$\hat{e}_{k,\lambda_2}(\hat{e}_{k,\lambda_2}\cdot d_{12}) = -\hat{e}_{k,\lambda_2}d_{12}\sin\alpha = -(d_{12}\times\hat{k})\times\hat{k}, \qquad (2.80)$$

where $\hat{k}$ is a unit vector in the direction of k. Setting

$$\hat{k} = \hat{r}\cos\theta + \hat{k}_x\sin\theta\cos\phi + \hat{k}_y\sin\theta\sin\phi, \qquad (2.81)$$

where $\hat{k}_x$, $\hat{k}_y$, and $\hat{r}\equiv r/r$ are unit vectors along the Cartesian axes in k-space, the angular integrals are then readily evaluated to give

$$\hat{E}_s^{(+)}(r,t) = \frac{1}{8\pi^2\epsilon_0 c^2 r}(d_{12}\times\hat{r})\times\hat{r}\int_0^\infty d\omega\,\omega^2\left[e^{-i\omega_A(t+r/c)}\right.$$

$$\times\int_0^t dt'\,\tilde\sigma_-(t')e^{i(\omega-\omega_A)(t'-t-r/c)} - e^{-i\omega_A(t-r/c)}$$

$$\left.\times\int_0^t dt'\,\tilde\sigma_-(t')e^{i(\omega-\omega_A)(t'-t+r/c)}\right]. \qquad (2.82)$$

Now, since the transformation (2.74b) removes the rapid oscillation at the atomic resonance frequency, $\tilde\sigma_-$ is expected to vary slowly in comparison with the optical period – on a time scale characterized by $\gamma^{-1}\sim 10^{-8}$s (for optical frequencies), compared with $\omega_A^{-1}\sim 10^{-15}$s. Thus, for frequencies outside the range $-100\gamma \le \omega - \omega_A \le 100\gamma$, say, the time integrals in (2.82) average to zero. This means that over the important range of the frequency integral $\omega^2 \approx \omega_A^2 + 2(\omega-\omega_A)\omega_A$ varies by less than 0.01% from $\omega^2 = \omega_A^2$. We therefore replace ω^2 by ω_A^2 and extend the frequency integral to $-\infty$. We then find

$$\hat{E}_s^{(+)}(r,t)$$

$$= \frac{\omega_A^2}{4\pi\epsilon_0 c^2 r}(d_{12}\times\hat{r})\times\hat{r}\left[e^{-i\omega_A(t+r/c)}\int_0^t dt'\,\tilde\sigma_-(t')\delta(t'-t-r/c)\right.$$

$$\left.- e^{-i\omega_A(t-r/c)}\int_0^t dt'\,\tilde\sigma_-(t')\delta(t'-t+r/c)\right]$$

$$= -\frac{\omega_A^2}{4\pi\epsilon_0 c^2 r}(d_{12}\times\hat{r})\times\hat{r}\,\sigma_-(t-r/c). \qquad (2.83)$$

This is precisely the familiar result for classical dipole radiation with the dipole moment *operator* $d_{12}\sigma_-$ in place of the classical dipole moment.

Since thermal effects are negligible at optical frequencies ($\hbar\omega_A \gg k_B T$), we will take the reservoir state to correspond to the vacuum electromagnetic field – the thermal equilibrium state at $T = 0$. Then, the free field (2.77) makes no contribution to normal-ordered correlation functions such as (2.71) and (2.72); thus, from (2.83) we may now write

$$G^{(1)}(t + r/c, t + r/c + \tau) = f(r)\langle\sigma_+(t)\sigma_-(t+\tau)\rangle, \qquad (2.84)$$

and

$$G^{(2)}(t + r/c, t + r/c + \tau) = f(r)^2\langle\sigma_+(t)\sigma_+(t+\tau)\sigma_-(t+\tau)\sigma_-(t)\rangle. \quad (2.85)$$

$f(r)$ is the geometrical factor

$$f(r) \equiv \left(\frac{\omega_A^2 d_{12}}{4\pi\epsilon_0 c^2}\right)^2 \frac{\sin^2\theta}{r^2}, \qquad (2.86)$$

where θ is the angle between d_{12} and r. (Recall that r measures positions with respect to an origin at the location of the atom.)

2.3.2 Master Equation for a Two-Level Atom Driven by a Classical Field

In deriving the master equation for resonance fluorescence we may go directly to (1.34), with s_1, s_2, Γ_1, and Γ_2 identified as in (2.17). We meet only one minor difference from our treatment of spontaneous emission in proceeding from this equation to the final result: The reservoir operators in the interaction picture are again given by (2.18); but the system operators $\tilde{s}_1$ and $\tilde{s}_2$ are now given by

$$\tilde{s}_1(t) = \sigma_-(t) \equiv \exp\left[(i/\hbar)\int_0^t dt' H_S(t')\right]\sigma_- \exp\left[-(i/\hbar)\int_0^t dt' H_S(t')\right],$$
$$(2.87a)$$

$$\tilde{s}_2(t) = \sigma_+(t) \equiv \exp\left[(i/\hbar)\int_0^t dt' H_S(t')\right]\sigma_+ \exp\left[-(i/\hbar)\int_0^t dt' H_S(t')\right],$$
$$(2.87b)$$

where H_S includes the interaction with the laser. What effect does this interaction have on the atomic damping? It will turn out, in fact, that any changes in the treatment of the damping are negligible under normal conditions. However, let us spend some time discussing this question anyway so that we have an idea of the approximation involved. The same approximation is made, often without mention, in laser theory and in cavity QED.

Equations (2.87) are just the formal solutions to the Heisenberg equations of motion for the atom-field interaction described by Hamiltonian (2.68a). These equations are given by

$$\dot{\sigma}_- = \frac{1}{i\hbar}[\sigma_-, H_S]$$

$$= -i\tfrac{1}{2}\omega_A[\sigma_-, \sigma_z] + i\left(\frac{d}{\hbar}E\right)e^{-i\omega_A t}[\sigma_-, \sigma_+]$$

$$= -i\omega_A\sigma_- - i\left(\frac{d}{\hbar}E\right)e^{-i\omega_A t}\sigma_z, \qquad (2.88a)$$

and, similarly,

$$\dot{\sigma}_+ = i\omega_A\sigma_+ + i\left(\frac{d}{\hbar}E\right)e^{i\omega_A t}\sigma_z, \qquad (2.88b)$$

$$\dot{\sigma}_z = 2i\left(\frac{d}{\hbar}E\right)e^{-i\omega_A t}\sigma_+ - 2i\left(\frac{d}{\hbar}E\right)e^{i\omega_A t}\sigma_-. \qquad (2.88c)$$

Defining

$$\tilde{\sigma}_x \equiv \sigma_+ e^{-i\omega_A t} + \sigma_- e^{i\omega_A t}, \qquad (2.89a)$$

$$i\tilde{\sigma}_y \equiv \sigma_+ e^{-i\omega_A t} - \sigma_- e^{i\omega_A t}, \qquad (2.89b)$$

(2.88a)–(2.88c) become

$$\dot{\tilde{\sigma}}_x = 0, \qquad (2.90a)$$

$$\dot{\tilde{\sigma}}_y = \Omega\sigma_z, \qquad (2.90b)$$

$$\dot{\sigma}_z = -\Omega\tilde{\sigma}_y, \qquad (2.90c)$$

where

$$\Omega \equiv 2\left(\frac{d}{\hbar}E\right). \qquad (2.91)$$

In particular, from (2.90b) and (2.90c),

$$\ddot{\sigma}_z = -\Omega^2\sigma_z. \qquad (2.92)$$

Then, for an atom initially in its lower state $[\langle\tilde{\sigma}_y(0)\rangle = 0, \langle\sigma_z(0)\rangle = -1]$,

$$\langle\sigma_z(t)\rangle = -\cos\Omega t. \qquad (2.93)$$

Ω is the *Rabi frequency* [2.30]; the frequency at which the atom periodically cycles between its lower and upper states, following absorption from the laser field with stimulated emission, then again absorption, and so on.

The general solution to (2.90) is

$$\tilde{s}_1(t) = \sigma_-(t) = e^{-i\omega_A t}\left[\sigma_- + \tfrac{1}{2}(1 - \cos\Omega t)(\sigma_+ - \sigma_-) - \tfrac{1}{2}i(\sin\Omega t)\sigma_z\right],$$
$$(2.94a)$$

$$\tilde{s}_2(t) = \sigma_+(t) = e^{i\omega_A t}\left[\sigma_+ - \tfrac{1}{2}(1 - \cos\Omega t)(\sigma_+ - \sigma_-) + \tfrac{1}{2}i(\sin\Omega t)\sigma_z\right],$$
$$(2.94b)$$

where $\sigma_+ = \sigma_+(0)$, $\sigma_- = \sigma_-(0)$, and $\sigma_z = \sigma_z(0)$ denote operators in the Schrödinger picture. Our derivation of the master equation for spontaneous emission proceeded from (1.34) with $\tilde{s}_1(t)$ and $\tilde{s}_2(t)$ given by the expression (2.94) taken in the limit $\Omega \to 0$. The interaction with the laser field has introduce terms modulated at the Rabi frequency. Now, there is no difficulty with substituting the full solutions (2.94) into (1.34) and continuing by performing the time integrals as before. The number of terms to be considered is increased nine fold, however, and we do not want to churn through all of this algebra if it is not really necessary. A quick review of our calculation for the damped harmonic oscillator will show that the oscillatory terms in $\tilde{s}_1(t)$ and $\tilde{s}_2(t)$ only specify the frequencies at which the system interacts with the reservoir; they determine the frequencies at which we evaluate the reservoir coupling constant and density of states. The final result following from (2.94) will then be an equation that contains three terms, each proportional to one of the three damping constants $\gamma(\omega_A), \gamma(\omega_A+\Omega)$, and $\gamma(\omega_A-\Omega)$, where $\gamma(\omega_A)$ is given by (2.21), and $\gamma(\omega_A + \Omega)$ and $\gamma(\omega_A - \Omega)$ are similarly defined with the reservoir coupling constant and density of states evaluated at shifted frequencies. At optical frequencies and reasonable laser intensities $\omega_A \sim 10^{15}$, and $\Omega < 10^{10}$ (this corresponds to 100 times the saturation intensity for sodium). Then, from (2.33),

$$\gamma(\omega_A \pm \Omega) = \gamma(\omega_A)(1 \pm \Omega/\omega_A)^3 \approx \gamma(\omega_A)(1 \pm 3\Omega/\omega_A). \qquad (2.95)$$

Thus, $\gamma(\omega_A \pm \Omega)$ differs from $\gamma = \gamma(\omega_A)$ by less than 0.01%. We therefore neglect Ω compared with ω_A. This is best done in (2.94) rather than at the end of a lot of tedious algebra. Setting Ω to zero in (2.94) is equivalent to deriving the master equation in an interaction picture with H_S replaced by the free Hamiltonian $\tfrac{1}{2}\hbar\omega_A\sigma_z$. Then the damping terms in the master equation for resonance fluorescence are the same as those derived for spontaneous emission. Neglecting thermal effects ($\bar{n} = 0$), the *master equation for resonance fluorescence* is then

$$\dot{\rho} = -i\tfrac{1}{2}\omega_A[\sigma_z, \rho] + i(\Omega/2)\left[e^{-i\omega_A t}\sigma_+ + e^{i\omega_A t}\sigma_-, \rho\right]$$
$$+ \frac{\gamma}{2}(2\sigma_-\rho\sigma_+ - \sigma_+\sigma_-\rho - \rho\sigma_+\sigma_-). \qquad (2.96)$$

In fact, a similar approximation was made, without mention, in our derivation of the scattered field, where we assume σ_- oscillates at the frequency ω_A [Eq. (2.74b)]. Further discussion of these issues, with specific consideration of their relevance in the Scully-Lamb theory of the laser, is given by Carmichael and Walls [2.36, 2.37].

Note 2.3 Recent work by Lewenstein et al. [2.38, 2.39] describes a situation in which the near equality of the damping constants $\gamma(\omega_A)$, $\gamma(\omega_A + \Omega)$, and $\gamma(\omega_A - \Omega)$ does not hold. This happens for an atom inside an optical cavity when the interaction between the atom and the vacuum modes it sees through the cavity mirrors significantly perturbs the free-space interaction between the atom and the vacuum field. Under these conditions the vacuum modes which are filtered by the cavity have a Lorentzian density of states that can vary considerably at the frequencies ω_A, $\omega_A + \Omega$, and $\omega_A - \Omega$. The consequent changes in the three damping constants alter the widths of the peaks in the fluorescence spectrum. Lewenstein et al. formulate their treatment of this effect in terms of non-Markovian equations for the damped atom. This is not necessary, however, if the Lorentzian feature in the density of states is narrower than (or similar in width to) the Rabi frequency, but is still much broader than the linewidths $\gamma(\omega_A)$, $\gamma(\omega_A + \Omega)$, and $\gamma(\omega_A - \Omega)$ (computed with the altered density of states). The method of Carmichael and Walls [2.36, 2.37] is appropriate for these conditions and leads to a Markovian master equation; but one in which the variation of the density of reservoir modes at the three different atomic frequencies is taken into account.

2.3.3 Optical Bloch Equations and Dressed States

Using the quantum regression formula, our derivation of the correlation functions appearing in (2.84) and (2.85) will follow directly from the equations of motion for the operator expectation values $\langle\sigma_-\rangle$, $\langle\sigma_+\rangle$, and $\langle\sigma_z\rangle$. From the master equation (2.96), the equations for expectation values are:

$$\langle\dot{\sigma}_-\rangle = -i\omega_A\langle\sigma_-\rangle - i(\Omega/2)e^{-i\omega_A t}\langle\sigma_z\rangle - \frac{\gamma}{2}\langle\sigma_-\rangle, \tag{2.97a}$$

$$\langle\dot{\sigma}_+\rangle = i\omega_A\langle\sigma_+\rangle + i(\Omega/2)e^{i\omega_A t}\langle\sigma_z\rangle - \frac{\gamma}{2}\langle\sigma_+\rangle, \tag{2.97b}$$

$$\langle\dot{\sigma}_z\rangle = i\Omega e^{-i\omega_A t}\langle\sigma_+\rangle - i\Omega e^{i\omega_A t}\langle\sigma_-\rangle - \gamma(\langle\sigma_z\rangle + 1). \tag{2.97c}$$

These are the *optical Bloch equations* with radiative damping, so called for their relationship to the equations of a spin-$\frac{1}{2}$ particle in a magnetic field [2.40]. They combine the terms describing the atom-field interaction given by (2.88) with the spontaneous decay terms in (2.37).

Note 2.4 When the phase destroying term $(\gamma_p/2)(\sigma_z\rho\sigma_z - \rho)$ in (2.66) is included in the master equation, (2.97a) and (2.97b) have γ replaced by $\gamma + 2\gamma_p$. The energy and phase decay times $1/\gamma$ and $2/(\gamma + 2\gamma_p)$, respectively, are often denoted by T_1 and T_2 in correspondence with the traditional terminology for magnetic systems.

If we neglect the effects of spontaneous decay, which is valid for short times, the optical Bloch equations are equivalent to the classical equations for a magnetic moment m in a rotating magnetic field B. With $\langle \sigma_x \rangle$ and $\langle \sigma_y \rangle$ defined as in (2.9), we can write

$$\dot{m} = B \times m, \tag{2.98}$$

where

$$m \equiv \langle \sigma_x \rangle \hat{x} + \langle \sigma_y \rangle \hat{y} + \langle \sigma_z \rangle \hat{z}, \tag{2.99}$$

and

$$B \equiv -(\Omega \cos \omega_A t)\hat{x} - (\Omega \sin \omega_A t)\hat{y} + \omega_A \hat{z}; \tag{2.100}$$

$\hat{x}$, $\hat{y}$, and $\hat{z}$ are orthogonal unit vectors. A strong intuition for the dynamics in resonance fluorescence can be drawn from this analogy. From (2.98) it follows that

$$\frac{d}{dt}(m \cdot m) = (B \times m) \cdot m + m \cdot (B \times m)$$
$$= 0, \tag{2.101}$$

since m and $B \times m$ are perpendicular vectors. Thus, m is a vector of constant length. In particular, for pure states, with

$$\rho = |\psi\rangle\langle\psi| = \big(c_1|1\rangle + c_2|2\rangle\big)\big(\langle 1|c_1^* + \langle 2|c_2^*\big), \tag{2.102}$$

we have

$$\langle \sigma_- \rangle = \rho_{21} = c_1^* c_2, \tag{2.103a}$$
$$\langle \sigma_+ \rangle = \rho_{12} = c_1 c_2^*, \tag{2.103b}$$
$$\langle \sigma_z \rangle = \rho_{22} - \rho_{11} = |c_2|^2 - |c_1|^2, \tag{2.103c}$$

and

$$m \cdot m = \langle \sigma_x \rangle^2 + \langle \sigma_y \rangle^2 + \langle \sigma_z \rangle^2$$
$$= 4\langle \sigma_- \rangle\langle \sigma_+ \rangle + \langle \sigma_z \rangle^2$$
$$= \big(|c_1|^2 + |c_2|^2\big)^2. \tag{2.104}$$

Thus, for a pure state $m \cdot m = 1$, and (2.101) expresses the requirement that probability be conserved. Here the state of the two-level atom can be represented by a point on the surface of the unit sphere (the *Bloch sphere*) as illustrated in Fig. 2.3(b). Dynamics on the Bloch sphere give a simple interpretation for the solutions (2.93) and (2.94). We define a rotating frame of reference which follows the rotating magnetic field, writing

$$\tilde{m} \equiv R_z(\omega_A t)m = \begin{pmatrix} \cos \omega_A t & \sin \omega_A t & 0 \\ -\sin \omega_A t & \cos \omega_A t & 0 \\ 0 & 0 & 1 \end{pmatrix} m, \tag{2.105}$$

where R_z generates rotations about the z-axis. The motion of $\tilde{m}$ is then determined by a magnetic field frozen in the $\hat{x}$ direction:

$$\dot{\boldsymbol{m}} = (\tilde{\boldsymbol{B}} - \omega_A \hat{z}) \times \tilde{\boldsymbol{m}}, \tag{2.106}$$

where $\tilde{\boldsymbol{B}} \equiv \boldsymbol{R}_z(\omega_A t)\boldsymbol{B}$ and

$$\tilde{\boldsymbol{B}} - \omega_A \hat{z} = -\Omega \hat{x}. \tag{2.107}$$

The modulation at the Rabi frequency shown by (2.93) and (2.94) simply corresponds to the precession of $\tilde{\boldsymbol{m}}$ about the static magnetic field $\tilde{\boldsymbol{B}} - \omega_A \hat{z}$ [Fig. 2.3(c)].

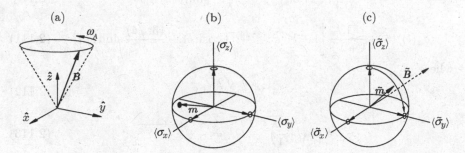

Fig. 2.3 Representation of atomic dynamics on the Bloch sphere: (a) the rotating magnetic field (2.100), (b) the atomic state represented as a point on the Bloch sphere, (c) precession of the atomic state in the rotating frame (2.105).

This simple view of the dynamics no longer provides the complete picture when the dissipative terms are reintroduced. Then (2.97a)–(2.97c) give

$$\frac{d}{dt}(\boldsymbol{m} \cdot \boldsymbol{m}) = 2\big(\langle\sigma_x\rangle\langle\dot{\sigma}_x\rangle + \langle\sigma_y\rangle\langle\dot{\sigma}_y\rangle + \langle\sigma_z\rangle\langle\dot{\sigma}_z\rangle\big)$$

$$= -\gamma\Big[\langle\sigma_x\rangle^2 + \langle\sigma_y\rangle^2 + 2\langle\sigma_z\rangle\big(\langle\sigma_z\rangle + 1\big)\Big]$$

$$= -\gamma(\boldsymbol{m} \cdot \boldsymbol{m} - 1) - \gamma\big(\langle\sigma_z\rangle + 1\big)^2. \tag{2.108}$$

Now the length of $\boldsymbol{m}$ is not conserved. This is not inconsistent with (2.104). Probability is still conserved, but the atomic state has become a mixed state, rather than a pure state; therefore (2.104) no longer gives a valid interpretation for $\boldsymbol{m} \cdot \boldsymbol{m}$. Dynamics cannot be formulated on the Bloch sphere. In fact, evolution proceeds to a steady state, with

$$\boldsymbol{m} \cdot \boldsymbol{m} = 1 - \big(\langle\sigma_z\rangle + 1\big)^2 = 1 - 4\rho_{22}^2, \tag{2.109}$$

which has the state $\boldsymbol{m}$ *within* the unit sphere. Since $\boldsymbol{m} \cdot \boldsymbol{m}$ must be greater than zero, it follows that $\rho_{22} \leq \frac{1}{2}$ in the steady state. Thus, interaction with the laser field can at best give equal probability for finding the atom in its upper and lower states – it cannot produce population inversion. Of course, a higher probability of excitation is possible during transients, which for an intense laser (large enough Ω) closely resemble the precession on the Bloch sphere which we have just described.

Exercise 2.4 Solve the optical Bloch equations (2.97). Show that, for an atom initially in its lower state,

$$e^{\pm i\omega_A t}\langle \sigma_{\mp}(t)\rangle = \pm i\frac{1}{\sqrt{2}}\frac{Y}{1+Y^2}\left[1 - e^{-(3\gamma/4)t}\left(\cosh \delta t + \frac{(3\gamma/4)}{\delta}\sinh \delta t\right)\right]$$

$$\pm i\sqrt{2}Y e^{-(3\gamma/4)t}\frac{(\gamma/4)}{\delta}\sinh \delta t, \tag{2.110}$$

$$\langle \sigma_z(t)\rangle = -\frac{1}{1+Y^2}\left[1 + Y^2 e^{-(3\gamma/4)t}\left(\cosh \delta t + \frac{(3\gamma/4)}{\delta}\sinh \delta t\right)\right], \tag{2.111}$$

where

$$Y \equiv \frac{\sqrt{2}\Omega}{\gamma}, \tag{2.112}$$

$$\delta \equiv \sqrt{\left(\frac{\gamma}{4}\right)^2 - \Omega^2} = \frac{\gamma}{4}\sqrt{1 - 8Y^2}. \tag{2.113}$$

In the limit $\gamma \ll \Omega$, $\gamma t \ll 1$, show that these solutions reproduce the dynamics on the Bloch sphere discussed above.

A complementary view of the atomic dynamics is given by the *dressed-states formalism* whose application to the problem of resonance fluorescence has been championed by Cohen-Tannoudji and Reynaud [2.41]. In this formalism we focus on the eigenstates of H_S, from which a full picture of the dynamics without damping can be constructed in the Schrödinger picture. It is usual to develop the dressed-states formalism around the fully quantized Hamiltonian

$$H_S \equiv \tfrac{1}{2}\hbar\omega_A\sigma_z + \hbar\omega_A a^\dagger a + \hbar(\kappa a\sigma_+ + \kappa^* a^\dagger \sigma_-), \tag{2.114}$$

rather than the time-dependent (semiclassical) Hamiltonian (2.68a). Here $a^\dagger$ and a are creation and annihilation operators for the laser mode, and the free Hamiltonian $\hbar\omega_A a^\dagger a$ generates the time dependence – $a(t) = a(0)e^{-i\omega_A t}$; to make the connection with (2.68a) we must take $\hbar\kappa\langle a\rangle = -dE$.

Without the atom-field interaction the eigenvalues of H_S define the infinite ladder of degenerate energy levels illustrated in Fig. 2.4(a). States $|n\rangle|2\rangle$ and $|n+1\rangle|1\rangle$ correspond to an n-photon Fock state plus an excited atom, and an $(n+1)$-photon Fock state plus an unexcited atom, respectively; both have the energy $(n+\tfrac{1}{2})\hbar\omega_A$. This degeneracy is lifted by the interaction. The size of the resulting level splitting may be found, together with the new energy eigenstates, by diagonalizing the coupled equations

$$H_S\begin{pmatrix} |n\rangle|2\rangle \\ |n+1\rangle|1\rangle \end{pmatrix} = \begin{pmatrix} (n+\tfrac{1}{2})\hbar\omega_A & \sqrt{n+1}\,\hbar\kappa^* \\ \sqrt{n+1}\,\hbar\kappa & (n+\tfrac{1}{2})\hbar\omega_A \end{pmatrix}\begin{pmatrix} |n\rangle|2\rangle \\ |n+1\rangle|1\rangle \end{pmatrix}. \tag{2.115}$$

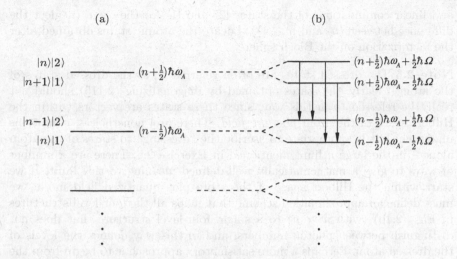

Fig. 2.4 (a) Degenerate ladder of energy levels for the uncoupled atom-field system. (b) Level splitting due to the atom-field interaction. Reading from left to right, the illustrated transitions have frequencies ω_A, $\omega_A - \Omega$, $\omega_A + \Omega$, and ω_A.

The new energy eigenvalues are

$$E_{n,\pm} = (n + \tfrac{1}{2})\hbar\omega_A \pm \sqrt{n+1}\,\hbar|\kappa|. \tag{2.116}$$

If the laser field is in a coherent state with mean photon number $\bar{n} \gg 1$, we may write

$$dE = \hbar|\kappa||\langle a\rangle| = \hbar|\kappa|\sqrt{\bar{n}},$$

and for all the populated eigenstates

$$E_{n,\pm} \approx (n + \tfrac{1}{2})\hbar\omega_A \pm \hbar\left(\frac{d_{12}}{\hbar}E\right) = (n + \tfrac{1}{2})\hbar\omega_A \pm \tfrac{1}{2}\hbar\Omega. \tag{2.117}$$

Transitions between the eigenstates of the interacting atom-field system identify the three frequencies ω_A, $\omega_A + \Omega$, and $\omega_A - \Omega$ encountered in (2.94) [Fig. 2.4(b)]. The three damping constants that arose in our treatment of the fluorescent decay process (Sect. 2.3.2) may now be associated with fluorescent transitions between the states of the coupled atom-field system – the so-called *dressed states*. If we suppress the $n\hbar\omega_A$ which distinguishes states of the Fock hierarchy, the remaining four-level structure gives the *dressed energies* $-\tfrac{1}{2}\hbar(\omega_A \mp \Omega)$ and $+\tfrac{1}{2}\hbar(\omega_A \pm \Omega)$ for the atom.

Exercise 2.5 Construct the eigenvectors corresponding to the eigenvalues (2.116) and hence find explicit expressions for the dressed states as linear combinations of the states $|n\rangle|2\rangle$ and $|n + 1\rangle|1\rangle$. For large n the dressed states approximately factorize as the product of a Fock state for the field and linear combinations of the states $|2\rangle$ and $|1\rangle$ for the atom (neglect the difference between $|n\rangle$ and $|n+1\rangle$). Locate the atomic states obtained after the factorization on the Bloch sphere.

Note 2.5 The dressed states are often referred to as the dressed states of the atom. Clearly, the states obtained by diagonalizing (2.115) should not really be referred to in this way, since these states are vectors within the Hilbert space of the atom *plus the field*. There are, nonetheless, conditions under which it is appropriate to ascribe the "dressing" to states of the atom alone – in the large n limit mentioned in Exercise 2.5. There are a number of ways to give a mathematically well-defined meaning to this limit. If we start within the Hilbert space of the atom plus quantized field mode, we must define an approximation scheme that maps all the four-level structures in Fig. (2.4b) (with $n \approx \bar{n}$) to a single four-level structure that does not distinguish between photon numbers, and in this way defines the levels of the dressed atom. Perhaps a more satisfactory approach is to begin from the semiclassical Hamiltonian (2.68a). This Hamiltonian is time dependent and does not, therefore, define a normal eigenvalue problem. But it is periodic in time. For such a Hamiltonian quasiperiodic solutions to the Schrödinger equation and their associated quasienergies play the role of energy eigenstates and eigenvalues [2.42, 2.43]. It is easy to find these quasiperiodic states and quasienergies for the Hamiltonian (2.68a): first transform to the interaction picture, diagonalize the resulting time-independent Hamiltonian, and then transform back to the Schrödinger picture. The frequencies of the quasiperiodic solutions found in this way, $-\frac{1}{2}(\omega_A \mp \Omega)$ and $+\frac{1}{2}(\omega_A \pm \Omega)$, are those given by the dressed energies of the atom.

2.3.4 The Fluorescence Spectrum

We might expect the spectrum of the fluorescent scattering to show features associated with the three transition frequencies between dressed states, ω_A, $\omega_A + \Omega$, and $\omega_A - \Omega$. Although this seems an obvious conclusion to draw from Fig. 2.4(b), there is really little basis for accepting it a priori. For weak excitation by monochromatic light, the fluorescence spectrum is shown by perturbation theory to also be monochromatic [2.21] – it does not have the linewidth of spontaneous emission. This teaches us that the scattering process is not simply a sequence of absorption and emission events; there is some coherence involved; a view of the quantum dynamics based solely on discrete transitions between atomic energy levels is not to be trusted. Moreover, consider the mean scattered field given by (2.83) and (2.110). For strong excitation this does contain components at the shifted frequencies $\omega_A \pm \Omega$. These decay, however, as transients and in the long-time limit

$$\lim_{t\to\infty} \langle \hat{\boldsymbol{E}}_s^{(+)}(\boldsymbol{r}, t)\rangle = -\frac{\omega_A^2}{4\pi\epsilon_0 c^2 r}(\boldsymbol{d}_{12} \times \hat{r}) \times \hat{r}\left(i\frac{1}{\sqrt{2}}\frac{Y}{1+Y^2}\right)e^{-i\omega_A(t-r/c)}.$$

$$(2.118)$$

Equation (2.118) suggests monochromatic fluorescence, in agreement with the established weak-field result. The dynamical picture is one of coherent reradiation from an induced dipole oscillator, the excitation strength entering only to saturate the oscillator amplitude.

Surely, however, this essentially classical picture is also incomplete. The quantum-mechanical dipole operator lives in a probabilistic world, and therefore we should allow our oscillator amplitude the opportunity to acquire a stochastic component. Then, in general, the fluorescence spectrum should not be calculated from the mean scattered field, but from the Fourier transform of the autocorrelation function (2.71). Using (2.84), for the long-time limit, this gives

$$S(\omega) = f(\boldsymbol{r}) \frac{1}{2\pi} \int_{-\infty}^{\infty} d\tau \, e^{i\omega\tau} \langle \sigma_+(0)\sigma_-(\tau) \rangle_{\text{ss}}, \tag{2.119}$$

where $\langle \sigma_+(0)\sigma_-(\tau) \rangle_{\text{ss}} \equiv \lim_{t\to\infty} \langle \sigma_+(t)\sigma_-(t+\tau) \rangle$. Thus, in a rotating frame, the atomic scatterer decays to the steady state

$$\langle \tilde{\sigma}_\mp \rangle_{\text{ss}} = e^{\pm i\omega_A t} \langle \sigma_\mp \rangle_{\text{ss}} = \pm i \frac{1}{\sqrt{2}} \frac{Y}{1+Y^2}, \tag{2.120a}$$

$$\langle \sigma_z \rangle_{\text{ss}} = -\frac{1}{1+Y^2}. \tag{2.120b}$$

However, fluctuations about this steady state can occur, described by the operators

$$\Delta\tilde{\sigma}_\mp \equiv \tilde{\sigma}_\mp - \langle \tilde{\sigma}_\mp \rangle_{\text{ss}}, \tag{2.121a}$$

$$\Delta\sigma_z \equiv \sigma_z - \langle \sigma_z \rangle_{\text{ss}}. \tag{2.121b}$$

These fluctuations are intrinsic to the quantum mechanics. Now the fluorescence spectrum decomposes into a coherent component, corresponding to (2.118), and an incoherent component arising from quantum fluctuations:

$$S(\omega) = S_{\text{coh}}(\omega) + S_{\text{inc}}(\omega), \tag{2.122}$$

with

$$S_{\text{coh}}(\omega) = f(\boldsymbol{r}) \frac{1}{2\pi} \int_{-\infty}^{\infty} d\tau \, e^{i(\omega-\omega_A)\tau} \langle \tilde{\sigma}_+ \rangle_{\text{ss}} \langle \tilde{\sigma}_- \rangle_{\text{ss}}$$

$$= f(\boldsymbol{r}) \frac{1}{2} \frac{Y^2}{(1+Y^2)^2} \delta(\omega - \omega_A), \tag{2.123}$$

and

$$S_{\text{inc}}(\omega) = f(\boldsymbol{r}) \frac{1}{2\pi} \int_{-\infty}^{\infty} d\tau \, e^{i(\omega-\omega_A)\tau} \langle \Delta\tilde{\sigma}_+(0)\Delta\tilde{\sigma}_-(\tau) \rangle_{\text{ss}}. \tag{2.124}$$

Let I_{coh} and I_{inc} denote the coherent and incoherent intensities obtained by integrating (2.123) and (2.124) over all frequencies:

$$I_{\text{coh}} = f(\boldsymbol{r})\langle\tilde{\sigma}_+\rangle_{\text{ss}}\langle\tilde{\sigma}_-\rangle_{\text{ss}}$$

$$= f(\boldsymbol{r})\frac{1}{2}\frac{Y^2}{(1+Y^2)^2}, \tag{2.125}$$

and

$$I_{\text{inc}} = f(\boldsymbol{r})\langle\Delta\tilde{\sigma}_+\Delta\tilde{\sigma}_-\rangle_{\text{ss}}$$

$$= f(\boldsymbol{r})\big(\langle\tilde{\sigma}_+\tilde{\sigma}_-\rangle_{\text{ss}} - \langle\tilde{\sigma}_+\rangle_{\text{ss}}\langle\tilde{\sigma}_-\rangle_{\text{ss}}\big)$$

$$= f(\boldsymbol{r})\big[\tfrac{1}{2}\big(1+\langle\sigma_z\rangle_{\text{ss}}\big) - \langle\tilde{\sigma}_+\rangle_{\text{ss}}\langle\tilde{\sigma}_-\rangle_{\text{ss}}\big]$$

$$= f(\boldsymbol{r})\frac{1}{2}\frac{Y^4}{(1+Y^2)^2}. \tag{2.126}$$

We can now make a judgment about the qualitative form of the spectrum. At weak laser intensities, the ratio $I_{\text{inc}}/I_{\text{coh}} = Y^2 = 2\Omega^2/\gamma^2$ is very small, and coherent scattering dominates, in agreement with the results from perturbation theory. However, $I_{\text{inc}}/I_{\text{coh}}$ increases with the laser intensity, and the incoherent spectral component will dominate at high laser intensities. Since the relaxation, or regression, of fluctuations around the steady state must surely follow a modulated decay similar to that shown by (2.110) and (2.111), we expect this incoherent spectrum to show sidebands at $\omega_A \pm \Omega$. The general dynamical picture must then be constructed as something of a mixture, showing both elements of coherent reradiation and discrete quantum transitions.

Note 2.6 The face the quantum dynamics shows to us depends on the questions we ask, as is generally the case in quantum mechanics. Illustrating this, we might note that the radiated intensity admits an interpretation in terms of discrete quantum transitions even at weak excitation, where $I_{\text{coh}}(\boldsymbol{r})$ dominates. If $I(\boldsymbol{r}) = I_{\text{coh}}(\boldsymbol{r}) + I_{\text{inc}}(\boldsymbol{r}) = f(\boldsymbol{r})\langle\tilde{\sigma}_+\tilde{\sigma}_-\rangle_{\text{ss}}$ is the total intensity at the position $\boldsymbol{r}$, we can integrate over a sphere of radius r (centered on the atom) to obtain the radiated power:

$$P = 2\epsilon_0 c \int_0^{2\pi} d\phi \int_0^\pi d\theta \sin\theta\, r^2 I(\boldsymbol{r})$$

$$= 2\epsilon_0 c\left(\frac{\omega_A^2 d_{12}}{4\pi\epsilon_0 c^2}\right)^2 \left(\int_0^{2\pi} d\phi \int_0^\pi d\theta \sin^3\theta\right)\langle\tilde{\sigma}_+\tilde{\sigma}_-\rangle_{\text{ss}}$$

$$= \left(\frac{1}{4\pi\epsilon_0}\frac{4\omega_A^3 d_{12}^2}{3\hbar c^3}\right)\hbar\omega_A\,\langle 2|\rho_{\text{ss}}|2\rangle$$

$$= \gamma\,\hbar\omega_A\,\langle 2|\rho_{\text{ss}}|2\rangle. \tag{2.127}$$

The radiated power is just the product of the atomic decay rate, the photon energy carried away per emission, and the probability that the atom is in its excited state. We have an interpretation in terms of discrete spontaneous emission events, despite the fact that the weak-field spectrum is not consistent with these dynamics.

The approach we have outlined for calculating the fluorescence spectrum is essentially the same as that followed by Mollow [2.23] in his original work. It certainly leads to a simple calculation compared to some of those that rederived Mollow's result (see Cresser et al. [2.34] for a review). We need only solve for $\langle \Delta \tilde{\sigma}_+(0) \Delta \tilde{\sigma}_-(\tau) \rangle_{\text{ss}}$ using the optical Bloch equations and the quantum regression formula. From (2.97), (2.120), and (2.121),

$$\frac{d}{dt}\langle \Delta \tilde{\sigma}_- \rangle = -i(\Omega/2)\langle \Delta \sigma_z \rangle - \frac{\gamma}{2}\langle \Delta \tilde{\sigma}_- \rangle, \tag{2.128a}$$

$$\frac{d}{dt}\langle \Delta \tilde{\sigma}_+ \rangle = i(\Omega/2)\langle \Delta \sigma_z \rangle - \frac{\gamma}{2}\langle \Delta \tilde{\sigma}_+ \rangle, \tag{2.128b}$$

$$\frac{d}{dt}\langle \Delta \sigma_z \rangle = i\Omega \langle \Delta \tilde{\sigma}_+ \rangle - i\Omega \langle \Delta \tilde{\sigma}_- \rangle - \gamma \langle \Delta \sigma_z \rangle, \tag{2.128c}$$

and the quantum regression formula gives

$$\frac{d}{d\tau}\langle \Delta \tilde{\sigma}_+(0) \Delta s(\tau) \rangle_{\text{ss}} = M \langle \Delta \tilde{\sigma}_+(0) \Delta s(\tau) \rangle_{\text{ss}}, \tag{2.129}$$

where

$$\Delta s \equiv \begin{pmatrix} \Delta \tilde{\sigma}_- \\ \Delta \tilde{\sigma}_+ \\ \Delta \sigma_z \end{pmatrix}, \tag{2.130}$$

and

$$M \equiv -\frac{\gamma}{2}\begin{pmatrix} 1 & 0 & iY/\sqrt{2} \\ 0 & 1 & -iY/\sqrt{2} \\ i\sqrt{2}Y & -i\sqrt{2}Y & 2 \end{pmatrix}. \tag{2.131}$$

The desired correlation function is the first component of the vector $\langle \Delta \tilde{\sigma}_+(0) \Delta s(\tau) \rangle_{\text{ss}}$. The initial conditions are given by

$$\langle \Delta \tilde{\sigma}_+ \Delta s \rangle_{\text{ss}} = \begin{pmatrix} \langle \tilde{\sigma}_+ \tilde{\sigma}_- \rangle_{\text{ss}} - \langle \tilde{\sigma}_+ \rangle_{\text{ss}}\langle \tilde{\sigma}_- \rangle_{\text{ss}} \\ \langle \tilde{\sigma}_+ \tilde{\sigma}_+ \rangle_{\text{ss}} - \langle \tilde{\sigma}_+ \rangle_{\text{ss}}^2 \\ \langle \tilde{\sigma}_+ \sigma_z \rangle_{\text{ss}} - \langle \tilde{\sigma}_+ \rangle_{\text{ss}}\langle \sigma_z \rangle_{\text{ss}} \end{pmatrix}$$

$$= \begin{pmatrix} \frac{1}{2}(1 + \langle \sigma_z \rangle_{\text{ss}}) - \langle \tilde{\sigma}_+ \rangle_{\text{ss}}\langle \tilde{\sigma}_- \rangle_{\text{ss}} \\ -\langle \tilde{\sigma}_+ \rangle_{\text{ss}}^2 \\ -\langle \tilde{\sigma}_+ \rangle_{\text{ss}}(1 + \langle \sigma_z \rangle_{\text{ss}}) \end{pmatrix},$$

where we have used (2.25), (2.45), and

$$\sigma_+ \sigma_z = |2\rangle\langle 1|(|2\rangle\langle 2| - |1\rangle\langle 1|) = -|2\rangle\langle 1| = -\sigma_+, \tag{2.132a}$$

$$\sigma_- \sigma_z = |1\rangle\langle 2|(|2\rangle\langle 2| - |1\rangle\langle 1|) = |1\rangle\langle 2| = \sigma_-. \tag{2.132b}$$

From the steady-state averages (2.120) we obtain

$$\langle\Delta\tilde{\sigma}_+\Delta s\rangle_{ss} = \frac{1}{2}\frac{Y^2}{(1+Y^2)^2}\begin{pmatrix} Y^2 \\ 1 \\ i\sqrt{2}Y \end{pmatrix}. \tag{2.133}$$

Equation (2.129) can be solve by finding a matrix S to diagonalize M. Multiplying (2.129) on the left by S,

$$\frac{d}{d\tau}S\langle\Delta\tilde{\sigma}_+(0)\Delta s(\tau)\rangle_{ss} = (SMS^{-1})S\langle\Delta\tilde{\sigma}_+(0)\Delta s(\tau)\rangle_{ss}, \tag{2.134}$$

and, formally,

$$\langle\Delta\tilde{\sigma}_+(0)\Delta s(\tau)\rangle_{ss} = S^{-1}\exp(\boldsymbol{\lambda}\tau)S\langle\Delta\tilde{\sigma}_+\Delta s\rangle_{ss}, \tag{2.135}$$

where

$$\boldsymbol{\lambda} \equiv SMS^{-1} = \mathrm{diag}\left(-\frac{\gamma}{2}, -\frac{3\gamma}{4}+\delta, -\frac{3\gamma}{4}-\delta\right) \tag{2.136}$$

is formed from the eigenvalues of M, and the rows (columns) of S (S^{-1}) are the left (right) eigenvectors of M [2.44]; δ is defined in (2.113). After some algebra we obtain the *first-order correlation function for resonance fluorescence*

$$\langle\Delta\tilde{\sigma}_+(0)\Delta\tilde{\sigma}_-(\tau)\rangle_{ss}$$
$$= \frac{1}{4}\frac{Y^2}{1+Y^2}e^{-(\gamma/2)\tau}$$
$$- \frac{1}{8}\frac{Y^2}{(1+Y^2)^2}\left[1-Y^2+(1-5Y^2)\frac{(\gamma/4)}{\delta}\right]e^{-[(3\gamma/4)-\delta]\tau}$$
$$- \frac{1}{8}\frac{Y^2}{(1+Y^2)^2}\left[1-Y^2-(1-5Y^2)\frac{(\gamma/4)}{\delta}\right]e^{-[(3\gamma/4)+\delta]\tau}. \tag{2.137}$$

Explicit expressions for the incoherent spectrum can be calculated from (2.124) and (2.137) as an exercise. In general, the spectrum is given by a sum of three Lorentzian components. It is easy to see that in the strong-field limit, $Y^2 >> 1$ ($\Omega^2 >> \gamma^2$), where incoherent scattering dominates, this calculation gives the well-known Mollow, or Stark, triplet. Figure 2.5 illustrates the dependence of the incoherent component of the fluorescence spectrum on the laser intensity.

2.3.5 Second-Order Coherence

We have identified "coherent" scattering with a monochromatic spectrum. More precisely, a monochromatic spectrum only implies first-order coherence – i.e. when $\langle\Delta\tilde{\sigma}_+(0)\Delta\tilde{\sigma}_-(\tau)\rangle_{ss}$ vanishes the first-order correlation function factorizes:

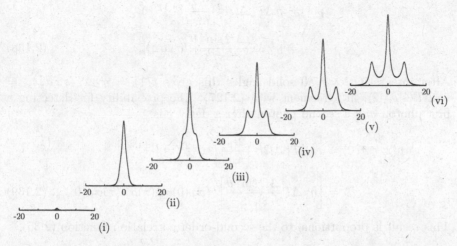

Fig. 2.5 The incoherent fluorescence spectrum. Spectra are plotted as a function of $2(\omega - \omega_A)/\gamma$ for (i) $Y = 0.3$, (ii) $Y = 1.5$, (iii) $Y = 2.7$, (iv) $Y = 3.9$, (v) $Y = 5.1$, and $Y = 6.3$.

$$G_{\mathrm{ss}}^{(1)}(\tau) = f(\boldsymbol{r})\langle \sigma_+(0)\sigma_-(\tau)\rangle_{\mathrm{ss}} = f(\boldsymbol{r})\langle \sigma_+\rangle_{\mathrm{ss}}\langle \sigma_-\rangle_{\mathrm{ss}} e^{-i\omega_A \tau},$$

where $G_{\mathrm{ss}}^{(1)}(\tau) \equiv \lim_{t\to\infty} G^{(1)}(t, t+\tau)$. This guarantees nothing about higher-order correlation functions. Do they factorize in a similar fashion? Is the scattered light in the weak-field limit – where the spectrum is monochromatic – coherent to all orders, as would be the radiation from a classical dipole? It is not difficult to see that it is not. We need look no further than to the second-order correlation function; the scattered light does not have second-order coherence. The lack of second-order coherence is associated with the phenomenon of photon antibunching. It tells us that the fluorescence from a two-level atom is nonclassical, even in the weak-field limit where a model based on classical dipole radiation gives the correct spectrum.

The second-order correlation function is proportional to the probability for the detection of two photons separated by a delay time τ. It is measured in delayed photon coincidence experiments [2.45, 2.46].

Note 2.7 Actual photodetection probabilities depend on such things as the photon counting time and the collection and quantum efficiencies of the detector. In (2.127) we saw that the photon emission rate into a 4π solid angle is $\gamma\langle\sigma_+\sigma_-\rangle_{\mathrm{ss}}$ (the radiated power is $\hbar\omega_A\gamma\langle\sigma_+\sigma_-\rangle_{\mathrm{ss}}$). Consider a detector located at position $\boldsymbol{r}$ which accepts photons over the small solid angle $\Delta\Omega$, and has a detection efficiency η. The single-photon detection probability during a short counting interval $\Delta T \ll \gamma^{-1}$ is the product of the energy density $2\epsilon_0\langle\hat{E}^{(-)}\hat{E}^{(+)}\rangle_{\mathrm{ss}}$, a factor $c/\hbar\omega_A$ which convert this into a photon flux density, the detector area $\Delta\Omega r^2$, photon counting time ΔT, and quantum efficiency η:

$$p(1) = \eta \Delta T (\Delta \Omega r^2) \frac{2\epsilon_0 c}{\hbar \omega_A} G^{(1)}_{ss}(0)$$

$$= \eta \gamma \Delta T \frac{\Delta \Omega \sin^2 \theta}{8\pi/3} \langle \sigma_+ \sigma_- \rangle_{ss}. \tag{2.138}$$

After integration over all solid angles this gives $p(1) = \eta \gamma \Delta T \langle \sigma_+ \sigma_- \rangle_{ss} = \eta \gamma \Delta T \langle 2|\rho_{ss}|2 \rangle$, in agreement with (2.127). The probability for detecting a first photon *and* a second photon after a delay τ is

$$p(2, \tau; 1, 0) = \left[\eta \Delta T (\Delta \Omega r^2) \frac{2\epsilon_0 c}{\hbar \omega_A} \right]^2 G^{(2)}_{ss}(\tau)$$

$$= \left[\eta \gamma \Delta T \frac{\Delta \Omega \sin^2 \theta}{8\pi/3} \right]^2 \langle \sigma_+(0) \sigma_+(\tau) \sigma_-(\tau) \sigma_-(0) \rangle_{ss}. \tag{2.139}$$

This result is proportional to the second-order correlation function (2.85).

In the long-time limit, second-order coherence requires the second-order correlation function to factorize in the form

$$G^{(2)}_{ss}(\tau) = f(\boldsymbol{r})^2 \langle \sigma_+(0) \sigma_+(\tau) \sigma_-(\tau) \sigma_-(0) \rangle_{ss} = \left[f(\boldsymbol{r}) \langle \sigma_+ \rangle_{ss} \langle \sigma_- \rangle_{ss} \right]^2;$$

this factorization must hold in addition to the requirement for first-order coherence stated above. It clearly never holds for $\tau = 0$, since $\langle \sigma_+ \rangle^2_{ss}$ and $\langle \sigma_- \rangle^2_{ss}$ are not zero [from (2.120a)] but $\sigma^2_+ = \sigma^2_- = 0$. The latter simply states that a two-level atom cannot be sequentially raised or lowered twice; two photons cannot be absorbed or emitted simultaneously; the detection of one photon sets the atom in its lower state, after which a second photon cannot be detected until the atom has been reexcited. We might predict then that the probability for detecting two photons is just the probability for detecting the first photon,

$$p(1) \propto f(\boldsymbol{r}) \langle \sigma_+ \sigma_- \rangle_{ss} = f(\boldsymbol{r}) \langle 2|\rho_{ss}|2 \rangle,$$

multiplied by the probability for detecting a second photon at the time $t = \tau$, given that the atom was in its lower state at $t = 0$:

$$p(2, \tau|1, 0) \propto f(\boldsymbol{r}) \langle (\sigma_+ \sigma_-)(\tau) \rangle_{\rho(0)=|1\rangle\langle 1|} = f(\boldsymbol{r}) \langle 2|\rho(\tau)|2 \rangle_{\rho(0)=|1\rangle\langle 1|}.$$

We are suggesting that the second-order correlation function may be factorized as a product of photon detection probabilities, with

$$G^{(2)}_{ss}(\tau) = f(\boldsymbol{r})^2 \langle 2|\rho_{ss}|2 \rangle \langle 2|\rho(\tau)|2 \rangle_{\rho(0)=|1\rangle\langle 1|}. \tag{2.140}$$

This is clearly zero for $\tau = 0$, and gives independent detection events for large τ, as $\rho(\tau) \to \rho_{ss}$. We will use the quantum regression formula to prove this result. (As with the calculation of the fluorescence spectrum, other approaches can be used to obtain the result; Kimble and Mandel, for example, derive (2.140) working entirely in the Heisenberg picture [2.47].)

First, let us consider the formal solution to the Bloch equations for time-dependent expectation values. In a rotating frame, (2.97a)–(2.97c) can be written in the vector form

$$\langle \dot{s} \rangle = M \langle s \rangle + b, \qquad (2.141)$$

where

$$s \equiv \begin{pmatrix} \tilde{\sigma}_- \\ \tilde{\sigma}_+ \\ \sigma_z \end{pmatrix}, \qquad (2.142)$$

M is the 3×3 matrix given by (2.131), and

$$b \equiv -\gamma \begin{pmatrix} 0 \\ 0 \\ 1 \end{pmatrix}. \qquad (2.143)$$

Then

$$\frac{d}{dt} (\langle s \rangle + M^{-1} b) = M (\langle s \rangle + M^{-1} b), \qquad (2.144)$$

and

$$\langle s(t) \rangle = -M^{-1} b + \exp(Mt)(\langle s(0) \rangle + M^{-1} b). \qquad (2.145)$$

Now

$$G_{ss}^{(2)}(\tau) = f(\boldsymbol{r})^2 \langle \sigma_+(0)\sigma_+(\tau)\sigma_-(\tau)\sigma_-(0) \rangle_{ss}$$
$$= f(\boldsymbol{r})^2 \tfrac{1}{2} \big[\langle \sigma_+ \sigma_- \rangle_{ss} + \langle \sigma_+(0)\sigma_z(\tau)\sigma_-(0) \rangle_{ss} \big], \qquad (2.146)$$

where we have used (2.25a). We can calculate the correlation function $\langle \sigma_+(0)\sigma_z(\tau)\sigma_-(0) \rangle_{ss}$ using the quantum regression formula. It is the third component of the vector $\langle \sigma_+ s(\tau)\sigma_- \rangle_{ss}$. To find the equation of motion for this vector, the quantum regression formula applied to a complete set of operators tells us to remove the angular brackets from (2.141) (b is a constant vector multiplied by the expectation of the identity operator), multiply on the left by $\sigma_+(0)$ and on the right by $\sigma_-(0)$, and replace the angular brackets; thus

$$\frac{d}{d\tau} \langle \sigma_+(0) s(\tau)\sigma_-(0) \rangle_{ss} = M \langle \sigma_+(0) s(\tau)\sigma_-(0) \rangle_{ss} + \langle \sigma_+ \sigma_- \rangle_{ss} b$$
$$= M \big[\langle \sigma_+(0) s(\tau)\sigma_-(0) \rangle_{ss} + \langle \sigma_+ \sigma_- \rangle_{ss} M^{-1} b \big]. \qquad (2.147)$$

The formal solution to this equation is

$$\langle \sigma_+(0) s(\tau)\sigma_-(0) \rangle_{ss} = -\langle \sigma_+ \sigma_- \rangle_{ss} M^{-1} b$$
$$+ \exp(M\tau) \big[\langle \sigma_+ s \, \sigma_- \rangle_{ss} + \langle \sigma_+ \sigma_- \rangle_{ss} M^{-1} b \big], \qquad (2.148)$$

with initial conditions

$$\langle \sigma_+ \boldsymbol{s}\, \sigma_- \rangle_{ss} = \begin{pmatrix} \langle \sigma_+ \tilde{\sigma}_- \sigma_- \rangle_{ss} \\ \langle \sigma_+ \tilde{\sigma}_+ \sigma_- \rangle_{ss} \\ \langle \sigma_+ \sigma_z \sigma_- \rangle_{ss} \end{pmatrix}$$

$$= \langle \sigma_+ \sigma_- \rangle_{ss} \begin{pmatrix} 0 \\ 0 \\ -1 \end{pmatrix}, \qquad (2.149)$$

where we have used (2.45) and (2.132). Now (2.148), (2.149), and (2.145) give

$$\langle \sigma_+(0) \boldsymbol{s}(\tau) \sigma_-(0) \rangle_{ss}$$

$$= \langle \sigma_+ \sigma_- \rangle_{ss} \left\{ -\boldsymbol{M}^{-1}\boldsymbol{b} + \exp(\boldsymbol{M}\tau) \left[\begin{pmatrix} 0 \\ 0 \\ -1 \end{pmatrix} + \boldsymbol{M}^{-1}\boldsymbol{b} \right] \right\}$$

$$= \langle \sigma_+ \sigma_- \rangle_{ss} \langle \boldsymbol{s}(\tau) \rangle_{\rho(0)=|1\rangle\langle1|}. \qquad (2.150)$$

Here, we have noted that $\begin{pmatrix} 0 \\ 0 \\ -1 \end{pmatrix}$ is simply the initial condition $\langle \boldsymbol{s}(0) \rangle$ for an

atom prepared in its lower state – i.e. with $\rho(0) = |1\rangle\langle1|$. Substituting the third component of (2.150) into (2.146) establishes our result:

$$G_{ss}^{(2)}(\tau) = f(\boldsymbol{r})^2 \langle \sigma_+ \sigma_- \rangle_{ss} \tfrac{1}{2} \big(1 + \langle \sigma_z(\tau) \rangle_{\rho(0)=|1\rangle\langle1|} \big)$$

$$= f(\boldsymbol{r})^2 \langle 2|\rho_{ss}|2\rangle \langle 2|\rho(\tau)|2\rangle_{\rho(0)=|1\rangle\langle1|}. \qquad (2.151)$$

Note that this calculation is independent of the form of $\boldsymbol{M}$. Thus, while (2.131) only gives $\boldsymbol{M}$ for perfect resonance, (2.151) also holds for nonresonant excitation.

Note 2.8 The factorized result we have obtained in (2.151) actually follows very simply, and quite generally, from the quantum regression formula (1.102):

$$G_{ss}^{(2)}(\tau) = f(\boldsymbol{r})^2 \langle \sigma_+(0)\sigma_+(\tau)\sigma_-(\tau)\sigma_-(0) \rangle_{ss}$$

$$= f(\boldsymbol{r})^2 \mathrm{tr}\big\{ e^{\mathcal{L}\tau} \big[\sigma_-(0)\rho_{ss}\sigma_+(0) \big] \sigma_+(0)\sigma_-(0) \big\}$$

$$= f(\boldsymbol{r})^2 \mathrm{tr}\big\{ e^{\mathcal{L}\tau} \big[|1\rangle\langle2|\rho_{ss}|2\rangle\langle1| \big] |2\rangle\langle2| \big\}$$

$$= f(\boldsymbol{r})^2 \langle 2|\rho_{ss}|2\rangle \langle 2|e^{\mathcal{L}\tau}\big(|1\rangle\langle1| \big) |2\rangle;$$

$\langle 2|e^{\mathcal{L}\tau}\big(|1\rangle\langle1| \big) |2\rangle$ is just a formal expression for $\langle 2|\rho(\tau)|2\rangle_{\rho(0)=|1\rangle\langle1|}$.

Equation (2.111) provides the solution for $\langle \sigma_z(t) \rangle_{\rho(0)=|1\rangle\langle1|}$ from which an explicit expression for $G_{ss}^{(2)}(\tau)$ may be written down. We normalize $G_{ss}^{(2)}(\tau)$ by its factorized form for independent photon detection in the large-delay limit, and write the *second-order correlation function for resonance fluorescence* as

$$g_{ss}^{(2)}(\tau) \equiv \left[\lim_{\tau \to \infty} G_{ss}^{(2)}(\tau) \right]^{-1} G_{ss}^{(2)}(\tau)$$

$$= \left(\langle \sigma_+ \sigma_- \rangle_{ss} \right)^{-1} \left(1 + \langle \sigma_z(\tau) \rangle_{\rho(0)=|1\rangle\langle 1|} \right)$$

$$= 1 - e^{-(3\gamma/4)\tau} \left(\cosh \delta\tau + \frac{3\gamma/4}{\delta} \sinh \delta\tau \right). \qquad (2.152)$$

This expression is plotted in Fig. 2.6. For a field possessing second-order coherence $g_{ss}^{(2)}(\tau) = 1$; the two photons are detected independently for all decay times; in this case a detector responds to the incident light by producing a completely random sequence of photopulses. This picture provides a reference against which the "antibunching" of photopulses is defined. The curves of Fig. 2.6 actually show two nonclassical features – features that are inadmissible in a correlation function generated by a classical stationary stochastic process. Let us look at the definitions of photon antibunching that have been given in terms of each.

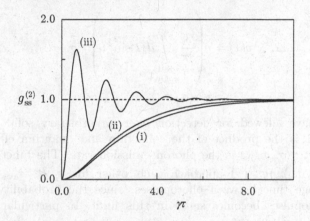

Fig. 2.6 The normalized second-order correlation function (2.152): (i) $8Y^2 = 0.01 \ll 1$ ($\delta \approx \gamma/4$); (ii) $8Y^2 = 1$ ($\delta = 0$); (iii) $8Y^2 = 400 \gg 1$ ($\delta \approx i\Omega$).

2.3.6 Photon Antibunching and Squeezing

All of the curves in Fig. 2.6 satisfy the inequality

$$g_{ss}^{(2)}(0) < 1. \qquad (2.153)$$

This is the definition of photon antibunching given in Refs. [2.28–2.30, 2.32]. The sense of this definition is actually more clearly understood by considering a closely related quantity to $g_{ss}^{(2)}(\tau)$. Imagine a photopulse sequence generated by a fast photodetector – response time much faster than $\min(\gamma^{-1}, \Omega^{-1})$

– monitoring the fluorescence. The quantity we will focus on is the probability density $w_{ss}(\tau)$ for a delay τ between successive photopulses, a quantity we refer to as the *photoelectron waiting-time distribution*. This can be calculated as the probability density that, given a photopulse at time t, there is also a photopulse at time $t + \tau$, *conditioned on the requirement that there are no photopulses in the intervening interval*; thus, the photopulse at time $t + \tau$ is the *next* photopulse in the sequence. For comparison, $\Delta T^{-1}[p(2,\tau;1,0)/p(1)] = \Delta T^{-1}p(1)g_{ss}^{(2)}(\tau)$ [Eqs. (2.138) and (2.139)] gives the probability density for a photopulse at $t + \tau$ *without any restriction on photopulses in the intervening interval*. The distribution $w_{ss}(\tau)$ must satisfy

$$\int_0^\infty d\tau \, w_{ss}(\tau) = 1, \tag{2.154}$$

since the delay between photopulses must take some value between zero and infinity; $\Delta T^{-1}p(1)g_{ss}^{(2)}(\tau)$ does not have to satisfy such a condition.

To clarify the notation we write

$$\Delta T^{-1}p(1) = \eta \left(\frac{3}{8\pi} \int_{\substack{\text{solid} \\ \text{angle}}} d\Omega \, \sin^2\theta \right) \gamma\langle\sigma_+\sigma_-\rangle_{ss}$$
$$= \eta'\gamma\langle\sigma_+\sigma_-\rangle_{ss}, \tag{2.155}$$

where we have allowed for detection over an arbitrary solid angle, and $0 < \eta' \leq 1$ is the product of the collection and quantum efficiencies of the detector; $\gamma\langle\sigma_+\sigma_-\rangle_{ss}$ is the photon emission rate. The functions $w_{ss}(\tau)$ and $(\eta'\gamma\langle\sigma_+\sigma_-\rangle_{ss})g_{ss}^{(2)}(\tau)$ approach each other for $\tau \ll \tau_{av}$, where τ_{av} is the average time between photopulses, since the probability for intervening photopulses becomes small in this limit. In particular, $w_{ss}(0) = (\eta'\gamma\langle\sigma_+\sigma_-\rangle_{ss})g_{ss}^{(2)}(0)$. For longer time intervals, coherent scattering would give the waiting-time distribution

$$w_{ss}(\tau) = \eta'\gamma\langle\sigma_+\sigma_-\rangle_{ss} \exp\left(-\eta'\gamma\langle\sigma_+\sigma_-\rangle_{ss}\tau\right). \tag{2.156}$$

In fact, a calculation of $w_{ss}(\tau)$ for $\eta' \ll 1$ (which holds under the most readily achievable experimental conditions [2.48, 2.49]) produces the result [2.50]

$$w_{ss}(\tau) \approx \eta'\gamma\langle\sigma_+\sigma_-\rangle_{ss} \left[\exp\left(-\eta'\gamma\langle\sigma_+\sigma_-\rangle_{ss}\tau\right) - e^{-(3\gamma/4)\tau} \right.$$
$$\left. \times \left(\cosh\delta\tau + \frac{3\gamma/4}{\delta}\sinh\delta\tau \right) \right] \tag{2.157}$$

for the *photoelectron waiting-time distribution of resonance fluorescence at low detection efficiency*. This expression satisfies (2.154) to lowest order in η'. It should be compared with the expression for $(\eta'\gamma\langle\sigma_+\sigma_-\rangle_{ss})g_{ss}^{(2)}(\tau)$ given by (2.152). The two expressions agree for $\tau \ll (\eta'\gamma\langle\sigma_+\sigma_-\rangle_{ss})^{-1} \approx \tau_{av}$, but $w_{ss}(\tau)$ decays to zero for $\tau > \tau_{av}$ as it becomes more and more unlikely that the *next* photopulse has not yet arrived.

Note 2.9 Equation (2.156) can be derived by considering a random sequence of photopulses, with a probability $\eta'\gamma\langle\sigma_+\sigma_-\rangle_{ss}\Delta t$ for finding a photopulse in any short interval Δt and a probability $1 - \eta'\gamma\langle\sigma_+\sigma_-\rangle_{ss}\Delta t$ for not finding a photopulse in the same interval. The probability for finding no photopulses throughout an interval $\tau = m\Delta t$, and then finding a photopulse in the interval from τ to $\tau + \Delta t$, is just

$$\eta'\gamma\langle\sigma_+\sigma_-\rangle_{ss}\Delta t\left(1 - \eta'\gamma\langle\sigma_+\sigma_-\rangle_{ss}\Delta t\right)^m$$

$$= \eta'\gamma\langle\sigma_+\sigma_-\rangle_{ss}\Delta t \sum_{n=0}^{m} \frac{m!}{(m-n)!n!}\left(-\eta'\gamma\langle\sigma_+\sigma_-\rangle_{ss}\Delta t\right)^n$$

$$= \eta'\gamma\langle\sigma_+\sigma_-\rangle_{ss}\Delta t \sum_{n=0}^{m} m(m-1)\cdots(m-n+1)$$

$$\times \frac{\left(-\eta'\gamma\langle\sigma_+\sigma_-\rangle_{ss}\Delta t\right)^n}{n!}$$

$$= \eta'\gamma\langle\sigma_+\sigma_-\rangle_{ss}\Delta t \sum_{n=0}^{m} \left(1 - \frac{1}{m}\right)\left(1 - \frac{2}{m}\right)\cdots\left(1 - \frac{n-1}{m}\right)$$

$$\times \frac{\left(-\eta'\gamma\langle\sigma_+\sigma_-\rangle_{ss}\Delta t\right)^n}{n!}.$$

On taking the limit $m \to \infty$, $\Delta t \to 0$, with $m\Delta t = \tau$, this gives

$$w_{ss}(\tau)dt = \eta'\gamma\langle\sigma_+\sigma_-\rangle_{ss}dt \exp\left(-\eta'\gamma\langle\sigma_+\sigma_-\rangle_{ss}\tau\right).$$

Now, in what sense does (2.153) imply an "antibunching" of photopulses? Figure 2.7 illustrates the behavior of $w_{ss}(\tau)$ for the light scattered in resonance fluorescence compared with coherent light of the same intensity. There is unit area under both of the curves plotted in the figure [Eq. (2.154)], and both distributions give the same mean time τ_{av} between photopulses. Note, now, that we have the equivalence

$$g_{ss}^{(2)}(0) < 1 \quad \Longleftrightarrow \quad w_{ss}(0) = \left(\eta'\gamma\langle\sigma_+\sigma_-\rangle_{ss}\right)g_{ss}^{(2)}(0) < \eta'\gamma\langle\sigma_+\sigma_-\rangle_{ss}.$$

Thus, (2.153) guarantees that $w_{ss}(0)$ falls below its value for coherent light of the same intensity. Then with increasing τ, $w_{ss}(\tau)$ must first rise above the exponential curve for coherent light, ensuring that both distributions have unit area, and then fall below it once again to ensure that both distributions give the same τ_{av}. We conclude that in comparison with coherent light of the same intensity, on the average, photopulse sequences are redistributed

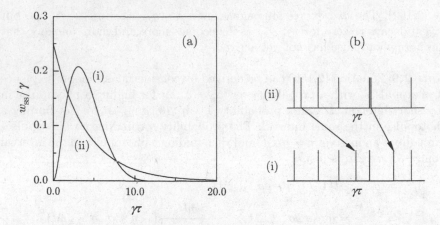

Fig. 2.7 (a) Waiting-time distribution for resonance fluorescence [curve (i)] and coherent scattering of the same intensity [curve (ii)], for $Y^2 = 1$, and $\eta' = 1$. (b) Rearrangement of a typical random photopulse sequence to account for the change in the waiting-time distribution shown in (a).

as illustrated in Fig. 2.7: some photopulses are moved from positions where they separate two very short time intervals, to new positions where they divide some of the very long time intervals into two. The result, as displayed in Fig. 2.7, is that the very shortest and very longest intervals between photopulses become less likely, and the intervals of intermediate length become more likely. A move is made away from photopulse sequences showing bunches and gaps, towards more regimented, evenly spaced, sequences.

Exercise 2.6 For perfect collection and detection efficiencies ($\eta' = 1$) $w_{\rm ss}(\tau)$ can be calculated from [2.50, 2.51]

$$w_{\rm ss}(\tau) = \gamma\langle 2|\bar\rho(\tau)|2\rangle_{\rho(0)=|1\rangle\langle 1|} = \gamma\langle 2|e^{\bar{\mathcal{L}}\tau}\big(|1\rangle\langle 1|\big)|2\rangle,$$

where the action of the superoperator $\bar{\mathcal{L}}$ on an operator $\hat{O}$ is given by

$$\bar{\mathcal{L}}\hat{O} \equiv \mathcal{L}\hat{O} - \gamma\sigma_-\hat{O}\sigma_+,$$

with $\mathcal{L}$ defined by the right-hand-side of (2.96). For these conditions show that the *photoelectron waiting-time distribution of resonance fluorescence at unit detection efficiency* is given by

$$w_{\rm ss}(\tau) = \gamma e^{-(\gamma/2)\tau}\frac{Y^2}{2Y^2 - 1}(1 - \cosh\delta'\tau), \tag{2.158}$$

with

$$\delta' \equiv \frac{\gamma}{2}\sqrt{1 - 2Y^2}. \tag{2.159}$$

Verify that (2.154) is satisfied and that the mean interval between photopulses is $\tau_{\mathrm{av}} = \gamma^{-1}2(1 + Y^2)/Y^2 = \left(\gamma\langle\sigma_+\sigma_-\rangle_{\mathrm{ss}}\right)^{-1} = $ (photon emission rate)$^{-1}$. Plot $w_{\mathrm{ss}}(\tau)$ for $2Y^2 = 1$ and compare it with the exponential $w_{\mathrm{ss}}(\tau) = (\gamma/6)\exp[-(\gamma/6)\tau]$ obtained for coherent light of the same intensity.

The central feature of this definition of photon antibunching is that it is made *in comparison with coherent light of the same intensity*. An alternative definition adopted by Mandel and co-workers [2.48, 2.49] does not make such a comparison. In addition to satisfying (2.153), the curves of Fig 2.6 also have

$$g_{\mathrm{ss}}^{(2)\prime}(0) = 0, \qquad g_{\mathrm{ss}}^{(2)\prime\prime}(0) > 0; \qquad\qquad (2.160)$$

the prime denotes differentiation with respect to τ. Classically, $g_{\mathrm{ss}}^{(2)}(\tau)$ must decrease from its value at $\tau = 0$, or, of course, remain constant if the light is coherent. Stated in terms of $w_{\mathrm{ss}}(\tau)$, no interval between photopulses may be more probable than $\tau = 0$. Mandel and co-workers identify photon antibunching with an initially rising $g_{\mathrm{ss}}^{(2)}(\tau)$. Since the most probable interval between photopulses is then some $\tau \neq 0$, photopulse sequences show a dirth of "tight" bunches in favor of somewhat larger photopulse separations, giving alternative definition to the term "antibunched."

This concept is drawn entirely from a comparison made within the photopulse sequences for the antibunched light – there are more slightly longer photopulse separation times than there are very short separation times. No comparison is made against the reference of coherent light of the same intensity. It is actually possible for photopulse sequences to be bunched in the sense of our previous discussion – with increased probability for short and long photopulse separation times and decreased probability for intermediate separation times – and be antibunched according to this second definition. This possibility is illustrated by Fig. 2.8. The converse also occurs, with (2.153) satisfied and $g_{\mathrm{ss}}^{(2)}(\tau)$ initially decreasing. Such behavior is seen in the forwards fluorescence from a single atom inside a resonant optical cavity [2.53].

The use of two definitions for photon antibunching might be a little confusing; but it is not really a major problem. Both definitions identify non-classical effects. We must remember, however, that strictly these are *distinct* nonclassical effects. Both effects have been demonstrated in experiments on resonance fluorescence [2.33, 2.48, 2.49]. Of course, whenever $g_{\mathrm{ss}}^{(2)}(0) = 0$ [as in (2.152)], the two definitions will be satisfied together. For definiteness we will use "photon antibunching" in the sense of (2.153), which seems to be more in accord with the traditional interpretation of the photon bunching of Hanbury-Brown and Twiss.

Note 2.10 The definition of photon antibunching given by (2.153) is equivalent to the condition for sub-Poissonian photon counting statistics for short counting times. A single-mode field illustrates this point:

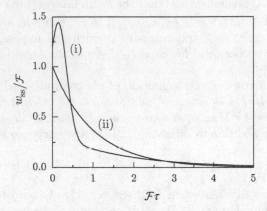

Fig. 2.8 Waiting-time distribution for light that is bunched in the sense of the discussion below (2.153) and antibunched according to the definition (2.160) [curve (i)]. An example of this behavior is shown in Ref. [2.52], Fig. 11(c). $\mathcal{F}$ is the mean photon flux and curve (ii) is the waiting-time distribution for coherent light.

$$g^{(2)}(0) = \langle a^\dagger a \rangle^{-2} \langle a^{\dagger^2} a^2 \rangle$$
$$= \langle a^\dagger a \rangle^{-2} \big[\langle (a^\dagger a)^2 \rangle - \langle a^\dagger a \rangle \big]$$
$$= 1 + \frac{(\langle \hat{n}^2 \rangle - \langle \hat{n} \rangle^2) - \langle \hat{n} \rangle}{\langle \hat{n} \rangle^2},$$

where $\hat{n} \equiv a^\dagger a$ is the photon number operator. Then

$$g^{(2)}(0) - 1 = \frac{Q}{\langle \hat{n} \rangle}, \tag{2.161}$$

where the *Mandel Q parameter*,

$$Q \equiv \frac{(\langle \hat{n}^2 \rangle - \langle \hat{n} \rangle^2) - \langle \hat{n} \rangle}{\langle \hat{n} \rangle}$$
$$= \frac{\langle (\Delta \hat{n})^2 \rangle - \langle \hat{n} \rangle}{\langle \hat{n} \rangle}, \tag{2.162}$$

measures the departure from Poissonian statistics. Clearly, (2.153) is equivalent to the condition for sub-Poissonian statistics, $Q < 1$. On the other hand, when counting times are not short on the scale of the field correlation time, the definition of Q involves integrals over field correlation functions; then (2.153) is no longer equivalent to the condition $Q < 1$.

Before we leave our discussion of photon antibunching, now is a good time to introduce some of the ideas concerning "squeezed" states of the electromagnetic field [2.54]. Walls and Zoller [2.55] pointed out that the light scattered in resonance fluorescence is squeezed in the field quadrature that is in phase with the mean scattered field amplitude. This squeezing is closely related to photon antibunching. We do not want to make a diversion into a detailed discussion of squeezed states here, and anyone who is totally unfamiliar with the subject may find it helpful to refer to the introductory article by Walls [2.56]. We will return to the subject of squeezing in Volume 2 (Chap. 9) and the discussion of background material is postponed until then.

When we write

$$g_{ss}^{(2)}(0) = (\langle \sigma_+\sigma_-\rangle_{ss})^{-2}\langle\sigma_+^2\sigma_-^2\rangle_{ss} \tag{2.163}$$

it is quite obvious that $g_{ss}^{(2)}(0)$ vanishes; we have discussed the simple reason for this above. There is something more to be learned, however, if we look at (2.163) in a slightly different way [2.57]. We may always regard the scattered field as the sum of a coherent component $\langle\hat{E}_s^{(+)}\rangle_{ss}$, which is proportional to $\langle\sigma_-\rangle_{ss}$, and a fluctuating component described by the operator $\Delta\hat{E}_s^{(+)} = \hat{E}_s^{(+)} - \langle\hat{E}_s^{(+)}\rangle_{ss}$, which is proportional to $\Delta\sigma_- = \sigma_- - \langle\sigma_-\rangle_{ss}$. Looked at in this way, (2.163) may be expanded along the same lines as the fluorescence spectrum [Eqs. (2.122)–(2.126)]; after transforming to a rotating frame, we may write

$$g_{ss}^{(2)}(0) - 1 = \left(A^2 + \langle\Delta\tilde{\sigma}_+\Delta\tilde{\sigma}_-\rangle_{ss}\right)^{-2}\left[A^2 4\langle:\Delta\tilde{\sigma}_{\frac{\pi}{2}})^2:\rangle_{ss}\right.$$
$$+ 4A\mathrm{Re}\left(e^{i\frac{\pi}{2}}\langle(\Delta\tilde{\sigma}_+)^2\Delta\tilde{\sigma}_-\rangle_{ss}\right) + \langle(\Delta\tilde{\sigma}_+)^2(\Delta\tilde{\sigma}_-)^2\rangle_{ss}$$
$$\left. - \left(\langle\Delta\tilde{\sigma}_+\Delta\tilde{\sigma}_-\rangle_{ss}\right)^2\right], \tag{2.164}$$

where $\langle : : \rangle$ denotes the normal-ordered average (with $\Delta\tilde{\sigma}_+$ to the left of $\Delta\tilde{\sigma}_-$); using (2.120a), we have defined

$$A \equiv |\langle\tilde{\sigma}_{\mp}\rangle_{ss}| = \frac{1}{\sqrt{2}}\frac{Y}{1+Y^2}, \tag{2.165}$$

and

$$\Delta\tilde{\sigma}_{\frac{\pi}{2}} \equiv \tfrac{1}{2}\left(e^{-i\frac{\pi}{2}}\Delta\tilde{\sigma}_- + e^{i\frac{\pi}{2}}\Delta\tilde{\sigma}_+\right) \tag{2.166}$$

describes fluctuations in the quadrature of the scattered field that is in phase with the mean scattered field amplitude. What is to be gained from this decomposition? To answer this question we must first calculate the steady-state correlations that appear in (2.164):

Exercise 2.7 Show that

$$\langle\Delta\tilde{\sigma}_+\Delta\tilde{\sigma}_-\rangle_{ss} = \frac{1}{2}\frac{Y^4}{(1+Y^2)^2}, \tag{2.167a}$$

$$\langle:(\Delta\tilde{\sigma}_{\frac{\pi}{2}})^2:\rangle_{ss} = \frac{1}{4}\frac{Y^2(Y^2-1)}{(1+Y^2)^2}, \tag{2.167b}$$

$$2\mathrm{Re}\big(e^{i\frac{\pi}{2}}\langle(\varDelta\tilde\sigma_+)^2\varDelta\tilde\sigma_-\rangle_{\mathrm{ss}}\big) = \sqrt{2}\,\frac{Y^5}{(1+Y^2)^3}\,, \tag{2.167c}$$

$$\langle(\varDelta\tilde\sigma_+)^2(\varDelta\tilde\sigma_-)^2\rangle_{\mathrm{ss}} - \big(\langle\varDelta\tilde\sigma_+\varDelta\tilde\sigma_-\rangle_{\mathrm{ss}}\big)^2 = \frac{1}{4}\frac{Y^4(1+4Y^2-Y^4)}{(1+Y^2)^4}\,. \tag{2.167}$$

Now, when (2.165) and (2.167a)–(2.167d) are substituted into (2.164), the answer $g_{\mathrm{ss}}^{(2)}(0) = 0$ must, of course, be recovered for all field strengths Y. The relative importance of the terms within the square bracket changes with Y, however, and it is here that the new insight lies. For weak fields ($Y^2 \ll 1$), the dominant terms in (2.164) are

$$A^2 + \langle\varDelta\tilde\sigma_+\varDelta\tilde\sigma_-\rangle_{\mathrm{ss}} \approx A^2 \approx \tfrac{1}{2}Y^2, \tag{2.168a}$$

$$A^2 4\langle : (\varDelta\tilde\sigma_{\pi/2})^2 : \rangle_{\mathrm{ss}} \approx -\tfrac{1}{2}Y^4, \tag{2.168b}$$

$$\langle(\varDelta\tilde\sigma_+)^2(\varDelta\tilde\sigma_-)^2\rangle_{\mathrm{ss}} - \big(\langle\varDelta\tilde\sigma_+\varDelta\tilde\sigma_-\rangle_{\mathrm{ss}}\big)^2 \approx \tfrac{1}{4}Y^4. \tag{2.168c}$$

For strong fields ($Y^2 \gg 1$) they are

$$A^2 + \langle\varDelta\tilde\sigma_+\varDelta\tilde\sigma_-\rangle_{\mathrm{ss}} \approx \langle\varDelta\tilde\sigma_+\varDelta\tilde\sigma_-\rangle_{\mathrm{ss}} \approx \tfrac{1}{4}, \tag{2.169a}$$

$$\langle(\varDelta\tilde\sigma_+)^2(\varDelta\tilde\sigma_-)^2\rangle_{\mathrm{ss}} - \big(\langle\varDelta\tilde\sigma_+\varDelta\tilde\sigma_-\rangle_{\mathrm{ss}}\big)^2 \approx -\tfrac{1}{4}. \tag{2.169b}$$

(a)

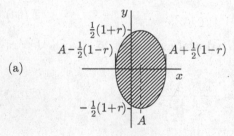

(b)

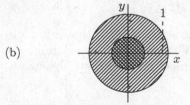

Fig. 2.9 Schematic illustration of the fluctuations in the two quadrature phase amplitudes of (a) a displaced weakly squeezed vacuum state (squeeze parameter $r = A^2$) and (b) a one-photon Fock state. Both states have $g^{(2)}(0) = 0$ (to lowest order in A^2 for the squeezed state). The curves are contours of the Wigner distribution (see Chap. 4).

Observe that the negative term, which is the source of the antibunching – it will produce the -1 on the left-hand side of (2.164) [remember that the $g_{\mathrm{ss}}^{(2)}(0)$ on the left-hand side is zero] – comes from the first term inside the square bracket on the right-hand side of (2.164) for weak fields, and from the third term inside the square bracket for strong fields. These terms, respectively, describe self-homodyning between the incoherent and coherent components of the scattered field, and intensity fluctuations in the incoherent component of the scattered field. Thus, a different physical picture for the fluctuations in the antibunched field is suggested in the weak-field and strong-field limits. A negative value for $\langle : (\Delta\tilde{\sigma}_{\pi/2})^2 : \rangle_{ss}$ is the signature of squeezing; thus, at weak fields photon antibunching arises from the *self-homodyning of squeezed fluorescence*; here photon antibunching is associated with the nonclassical statistics of a *phased* oscillator. Phase information is destroyed in the strong-field limit. For strong excitation the coherent component of the scattered field saturates and the homodyning term in (2.164) becomes unimportant. Photon antibunching in the strong-field limit arises from sub-Poissonian intensity fluctuations in an *unphased* scattered field. For a suggestive illustration we can compare a displaced squeezed vacuum state (weak fields) and a one-photon Fock state (strong fields), as illustrated by Fig. 2.9.

Note 2.11 One scheme for detecting squeezing, described by Mandel [2.58], involves homodyning the scattered light with a strong local oscillator and measuring photon-counting statistics as a function of the local oscillator phase. Squeezing is indicated by a phase dependent variation from super-Poissonian statistics, when the unsqueezed quadrature is selected by the local oscillator phase, to sub-Poissonian statistics, when the squeezed quadrature is selected by the local oscillator phase. Equation (2.164) corresponds to a special case of this procedure where the local oscillator is the coherent fluorescent scattering itself. Under these conditions we do not, of course, have control over the local oscillator amplitude and phase. To convert the expressions we have derived so that they describe a squeezing measurement for the fluorescence in accord with Mandel's scheme we simply replace A by a large local oscillator amplitude B, and replace $\frac{\pi}{2}$ by an adjustable phase ϕ. If the local oscillator intensity is much larger than the fluorescence intensity, the combined field of local oscillator plus fluorescent scattering then gives

$$g_{ss}^{(2)}(0) - 1 \approx \frac{4\langle : (\Delta\tilde{\sigma}_\phi)^2 : \rangle_{ss}}{B^2}. \tag{2.170}$$

Actually, B is not the local oscillator field amplitude, it is only proportional to this amplitude. The proportionality is the same as that between σ_- and $\hat{E}_s^{(+)}$; from (2.138), it is such that the mean number of photons counted during ΔT, for a detection efficiency η and solid angle $\Delta\Omega$, is

$$\langle \hat{n} \rangle = \eta\gamma\Delta T \frac{\Delta\Omega \sin^2\theta}{8\pi/3} B^2. \tag{2.171}$$

Substituting from (2.170) and (2.171) into (2.161), the photon counting distribution is characterized, as either super-Poissonian or sub-Poissonian, by

$$Q_\phi = \eta\gamma\Delta T \frac{\Delta\Omega \sin^2\theta}{8\pi/3} 4\langle :(\Delta\tilde{\sigma}_\phi)^2: \rangle_{ss}. \tag{2.172}$$

When the oscillator phase is $\frac{\pi}{2}$,

$$Q_{\frac{\pi}{2}} = \eta\gamma\Delta T \frac{\Delta\Omega \sin^2\theta}{8\pi/3} \frac{Y^2(Y^2-1)}{(1+Y^2)^2}. \tag{2.173}$$

This gives sub-Poissonian counting statistics for $Y^2 < 1$. An explicit expression for arbitrary ϕ can be calculated as an exercise.

3. Quantum–Classical Correspondence for the Electromagnetic Field I: The Glauber–Sudarshan P Representation

In Chap. 1 we developed a formalism to handle dissipative problems in quantum mechanics. The central result of this formalism was the operator master equation for the reduced density operator ρ of a dissipative system. This equation can be written formally as

$$\dot{\rho} = \mathcal{L}\rho, \tag{3.1}$$

where $\mathcal{L}$ is a generalized Liouvillian, or "superoperator", which acts, not on the states, but on the operators of the system. In a specific application $\mathcal{L}$ is defined by an explicit expression in terms of various commutators involving system operators. While it is generally not possible to solve the operator master equation directly to find $\rho(t)$ in operator form, we have seen that alternative methods of analysis are available to us. We can derive equations of motion for expectation values, and if these form a suitable closed set, solve these equations for time-dependent operator averages. Alternatively, we may choose a representation and take matrix elements of (3.1) to obtain equations of motion for the matrix elements of ρ. We have also seen how equations of motion for one-time operator averages can be used to obtain equations of motion for two-time averages (correlation functions) using the quantum regression formula.

We are now going to meet an entirely new approach to the problem of solving the operator master equation and calculating operator averages and correlation functions. For the present we will only consider the electromagnetic field —i.e. the harmonic oscillator. In Chap. 6 we will generalize the techniques learned here to collections of two-level atoms. This new approach establishes a correspondence between quantum-mechanical operators and ordinary (classical) functions, such that quantities of interest in a quantum-mechanical problem can be calculated using the methods of classical statistical physics. Under this correspondence the operator master equation transforms into a partial differential equation for a quasidistribution function which corresponds to (represents) ρ. For the damped harmonic oscillator this quasidistribution function is a function of the classical phase-space variables q and p, or alternatively, the complex variables $\alpha = (m\omega q + ip)/\sqrt{2\hbar m\omega}$ and $\alpha^* = (m\omega q - ip)/\sqrt{2\hbar m\omega}$ that correspond to the operators a and $a^\dagger$. Operator averages, written in an appropriate order (e.g. normal order), are calculated by integrating functions of these classical variables against the quasidistribution function, in the same manner in which we take classical phase-space averages. This *quantum–classical correspondence* is particularly appealing when the partial differential equation corresponding to the operator master equation is a Fokker–Planck equation. Fokker–Planck equations are familiar from classical statistical physics, and in this context they have been studied extensively [3.1]. When the operator master equation becomes a Fokker–Planck equation, analogies can be drawn between classical fluctuation phenomena and fluctuations generated by the quantum dynamics. This helps us develop an intuition for the effects of quantum fluctuations. Also, mathematical techniques that were developed for analyzing Fokker–Plank equations in their traditional setting can be sequestered to help solve a quantum-mechanical problem.

There are, in fact, many ways in which to set up a quantum–classical correspondence. We will meet a number of these in this book and still more in Volume 2. The original ideas go back to the work of Wigner [3.2]. Wigner, however, was interested in general questions of quantum statistical mechanics, not specifically in quantum-optical applications; wide use of the methods of quantum–classical correspondence for problems in quantum optics only began with the work of Glauber [3.3] and Sudarshan [3.4]. These authors independently developed what is now commonly known as the Glauber–Sudarshan P representation, or simply the P representation, for the electromagnetic field. The representation is based on a correspondence in which *normal-ordered* operator averages are calculated as classical phase-space averages; it has been tailored for the special role played by normal-ordered averages in the theory of photodetection and quantum coherence [3.3, 3.5, 3.6]. The Wigner representation gives the averages of operators written in Weyl, or symmetric, order; other representations exist which use still different ordering conventions.

3.1 The Glauber–Sudarshan P Representation

The Glauber–Sudarshan P representation was introduced primarily for the description of statistical mixtures of coherent states – the closest approach within the quantum theory to the states of the electromagnetic field described by the classical statistical theory of optics. An understanding of this representation can therefore be built on a few simple properties of the coherent states. Formal definition of the P representation can, alternatively, be given without any mention of the coherent states; this is the more useful approach when we want to generalize the methods of quantum–classical correspondence to other representations for the field, and to representations for collections of two-level atoms. We will follow both routes in turn, to define the P representation and then illustrate its use by deriving a Fokker–Planck equation for the damped harmonic oscillator. We first follow the route based on coherent states, where we begin with a review of some of the more important properties of these states. Further discussion of the coherent states can be found in Louisell [3.7] and Sargent, Scully and Lamb [3.8].

3.1.1 Coherent States

The coherent state $|\alpha\rangle$ is the right eigenstate of the annihilation operator a with complex eigenvalue α:

$$a|\alpha\rangle = \alpha|\alpha\rangle, \qquad \langle\alpha|a^\dagger = (a|\alpha\rangle)^\dagger = \alpha^*\langle\alpha|. \tag{3.2}$$

From this definition we may prove the following properties of the coherent states:

Proposition 3.1 *If a harmonic oscillator, with Hamiltonian $H = \hbar\omega a^\dagger a$, has as its initial state the coherent state $|\alpha_0\rangle$, then it remains in a coherent state for all times with the oscillating complex amplitude $\alpha(t) = \alpha_0 e^{-i\omega t}$ – i.e. the time-dependent state of the oscillator is given by*

$$|\Psi(t)\rangle = e^{-(i/\hbar)Ht}|\alpha_0\rangle = e^{-i\omega_0 a^\dagger a t}|\alpha_0\rangle = |e^{-i\omega t}\alpha_0\rangle = |\alpha(t)\rangle. \tag{3.3}$$

Proof. We show that $|\Psi(t)\rangle$ is the right eigenstate of a with eigenvalue $\alpha(t)$:

$$\begin{aligned}
a|\Psi(t)\rangle &= a e^{-i\omega a^\dagger a t}|\alpha_0\rangle \\
&= e^{-i\omega a^\dagger a t}\left(e^{i\omega a^\dagger a t} a e^{-i\omega a^\dagger a t}\right)|\alpha_0\rangle \\
&= \left(e^{-i\omega t}\alpha_0\right)\left(e^{-i\omega a^\dagger a t}|\alpha_0\rangle\right) \\
&= \alpha(t)|\Psi(t)\rangle,
\end{aligned}$$

where we have used (1.40a) and (3.2).   □

Proposition 3.2 *The coherent states are minimum uncertainty states: for a mechanical oscillator with position and momentum operators $\hat{q}$ and $\hat{p}$, respectively,*

$$\Delta q \Delta p = \sqrt{\langle(\hat{q} - \langle\hat{q}\rangle)^2\rangle}\sqrt{\langle(\hat{p} - \langle\hat{p}\rangle)^2\rangle} = \tfrac{1}{2}\hbar, \tag{3.4}$$

where the averages are taken with respect to a coherent state.

Proof. From (1.12a) and (1.12b),

$$\hat{q} = \sqrt{\frac{\hbar}{2m\omega}}(a + a^\dagger), \tag{3.5a}$$

$$\hat{p} = -i\sqrt{\frac{\hbar m\omega}{2}}(a - a^\dagger). \tag{3.5b}$$

Then, for an oscillator in the state $|\alpha\rangle$,

$$
\begin{aligned}
\left\langle \left(\hat{q} - \langle \hat{q} \rangle \right)^2 \right\rangle &= \langle \hat{q}^2 \rangle - \langle \hat{q} \rangle^2 \\
&= \frac{\hbar}{2m\omega} \langle \alpha | (a^2 + aa^\dagger + a^\dagger a + a^{\dagger 2}) | \alpha \rangle - \langle \hat{q} \rangle^2 \\
&= \frac{\hbar}{2m\omega} \left[\langle \alpha | (aa^\dagger - a^\dagger a) | \alpha \rangle + (\alpha + \alpha^*)^2 \right] - \langle \hat{q} \rangle^2 \\
&= \frac{\hbar}{2m\omega} \langle \alpha | [a, a^\dagger] | \alpha \rangle \\
&= \frac{\hbar}{2m\omega},
\end{aligned}
\tag{3.6a}
$$

where we have used (3.2) and the commutation relation (1.10); we assume that the state $|\alpha\rangle$ is normalized. Similarly,

$$
\left\langle \left(\hat{p} - \langle \hat{p} \rangle \right)^2 \right\rangle = \frac{\hbar m \omega}{2}.
\tag{3.6b}
$$

Thus,

$$
\sqrt{\left\langle \left(\hat{q} - \langle \hat{q} \rangle \right)^2 \right\rangle} \sqrt{\left\langle \left(\hat{p} - \langle \hat{p} \rangle \right)^2 \right\rangle} = \tfrac{1}{2} \hbar.
$$

$\square$

Proposition 3.3 *A normalized coherent state can be expanded in terms of the Fock states $|n\rangle$, $n = 0, 1, 2, \ldots$, as*

$$
|\alpha\rangle = e^{-\frac{1}{2}|\alpha|^2} \sum_{n=0}^{\infty} \frac{\alpha^n}{\sqrt{n!}} |n\rangle.
\tag{3.7}
$$

Proof. We write

$$
|\alpha\rangle = \sum_{n=0}^{\infty} c_n |n\rangle
$$

and substitute this expansion into (3.2). Using $a|n\rangle = \sqrt{n}|n-1\rangle$, this gives the relationship

$$
\sum_{n=1}^{\infty} c_n \sqrt{n} |n-1\rangle = \alpha \sum_{n=0}^{\infty} c_n |n\rangle.
$$

Multiplying on the left by $\langle m |$ and using the orthogonality of the Fock states, we have

$$
\sum_{n=1}^{\infty} c_n \sqrt{n}\, \delta_{m,n-1} = \alpha \sum_{n=0}^{\infty} c_n \delta_{m,n},
$$

or

$$
c_{m+1} \sqrt{m+1} = \alpha c_m;
$$

thus,

$$c_n = \frac{\alpha^n}{\sqrt{n!}} c_0.$$

c_0 is determined by the normalization condition $\langle \alpha | \alpha \rangle = 1$:

$$\langle \alpha | \alpha \rangle = |c_0|^2 \sum_{n,m=0}^{\infty} \frac{\alpha^{*n} \alpha^m}{\sqrt{n!m!}} \langle n | m \rangle$$

$$= |c_0|^2 \sum_{n=0}^{\infty} \frac{|\alpha|^{2n}}{n!}$$

$$= |c_0|^2 e^{|\alpha|^2};$$

thus,

$$c_0 = e^{-\frac{1}{2}|\alpha|^2},$$

where the arbitrary phase has been chosen so that c_0 is real.   □

Proposition 3.4 *The coherent states are not orthogonal; the overlap of the states $|\alpha\rangle$ and $|\beta\rangle$ is given by*

$$|\langle \alpha | \beta \rangle|^2 = e^{-|\alpha - \beta|^2}. \tag{3.8}$$

Note that $|\alpha\rangle$ and $|\beta\rangle$ are approximately orthogonal when $|\alpha - \beta|^2$ becomes large.

Proof. Using (3.7)

$$\langle \alpha | \beta \rangle = e^{-\frac{1}{2}|\alpha|^2} e^{-\frac{1}{2}|\beta|^2} \sum_{n,m=0}^{\infty} \frac{\alpha^{*n} \beta^m}{\sqrt{n!m!}} \langle n | m \rangle$$

$$= e^{-\frac{1}{2}|\alpha|^2} e^{-\frac{1}{2}|\beta|^2} \sum_{n=0}^{\infty} \frac{(\alpha^* \beta)^n}{n!}$$

$$= e^{-\frac{1}{2}|\alpha|^2} e^{-\frac{1}{2}|\beta|^2} e^{\alpha^* \beta}.$$

Then

$$|\langle \alpha | \beta \rangle|^2 = e^{-|\alpha|^2} e^{-|\beta|^2} e^{\alpha^* \beta} e^{\alpha \beta^*}$$

$$= e^{-|\alpha - \beta|^2}.$$

□

Proposition 3.5 *The coherent states are complete:*

$$\frac{1}{\pi} \int d^2\alpha \, |\alpha\rangle\langle\alpha| = 1, \tag{3.9}$$

the integration being taken over the entire complex plane.

Proof. From (3.7),

$$\frac{1}{\pi}\int d^2\alpha\,|\alpha\rangle\langle\alpha| = \frac{1}{\pi}\int d^2\alpha\,e^{-|\alpha|^2}\sum_{n,m=0}^{\infty}\frac{\alpha^{*n}\alpha^m}{\sqrt{n!m!}}|n\rangle\langle m|,$$

or, in polar coordinates,

$$\frac{1}{\pi}\int d^2\alpha\,|\alpha\rangle\langle\alpha| = \frac{1}{\pi}\sum_{n,m=0}^{\infty}\frac{|n\rangle\langle m|}{\sqrt{n!m!}}\int_0^{\infty}dr\,e^{-r^2}r^{n+m+1}\int_0^{2\pi}d\phi\,e^{-i(n-m)\phi},$$

where $\alpha = re^{i\phi}$. The integration over ϕ gives zero unless n is equal to m. Thus,

$$\frac{1}{\pi}\int d^2\alpha\,|\alpha\rangle\langle\alpha| = 2\sum_{n=0}^{\infty}\frac{|n\rangle\langle n|}{n!}\int_0^{\infty}dr\,e^{-r^2}r^{2n+1}.$$

After integrating by parts n times,

$$\frac{1}{\pi}\int d^2\alpha\,|\alpha\rangle\langle\alpha| = 2\sum_{n=0}^{\infty}\frac{|n\rangle\langle n|}{n!}\frac{1}{2}n! = \sum_{n=0}^{\infty}|n\rangle\langle n| = 1.$$

The final step follows from the completeness of the Fock states.   □

Proposition 3.6 *The coherent states can be generated from the vacuum state by the action of the creation operator $a^\dagger$:*

$$|\alpha\rangle = e^{-\frac{1}{2}|\alpha|^2}e^{\alpha a^\dagger}|0\rangle. \tag{3.10}$$

Proof. Using $a^\dagger|n\rangle = \sqrt{n+1}|n+1\rangle$, we have

$$e^{-\frac{1}{2}|\alpha|^2}e^{\alpha a^\dagger}|0\rangle = e^{-\frac{1}{2}|\alpha|^2}\sum_{n=0}^{\infty}\frac{\alpha^n}{n!}a^{\dagger n}|0\rangle$$

$$= e^{-\frac{1}{2}|\alpha|^2}\sum_{n=0}^{\infty}\frac{\alpha^n}{n!}\sqrt{n!}\,|n\rangle$$

$$= e^{-\frac{1}{2}|\alpha|^2}\sum_{n=0}^{\infty}\frac{\alpha^n}{\sqrt{n!}}|n\rangle.$$

This is the expression (3.7) for the Fock state expansion of the coherent state $|\alpha\rangle$.   □

3.1.2 Diagonal Representation for the Density Operator Using Coherent States

Using the completeness of the Fock states, a representation for the density operator ρ in terms of these states is obtained by multiplying on the left and right by the unit operator expressed as a sum of outer products:

$$\rho = \left(\sum_{n=0}^{\infty} |n\rangle\langle n| \right) \rho \left(\sum_{m=0}^{\infty} |m\rangle\langle m| \right)$$
$$= \sum_{n,m=0}^{\infty} \rho_{n,m} |n\rangle\langle m|, \tag{3.11}$$

with $\rho_{n,m} \equiv \langle n|\rho|m \rangle$. The Fock states are orthogonal as well as being complete, as is the common situation for a set of basis states. The coherent states are not orthogonal (Proposition 3.4). However, they are complete (Proposition 3.5), and this is all we need to define a representation for ρ analogous to (3.11). From (3.9), we may write

$$\rho = \left(\frac{1}{\pi} \int d^2\alpha \, |\alpha\rangle\langle\alpha| \right) \rho \left(\frac{1}{\pi} \int d^2\beta \, |\beta\rangle\langle\beta| \right)$$
$$= \frac{1}{\pi^2} \int d^2\alpha \int d^2\beta \, |\alpha\rangle\langle\beta| \langle\alpha|\rho|\beta\rangle. \tag{3.12}$$

Glauber has defined what he calls the R *representation*, expanding the density operator in the form [3.3]

$$\rho = \frac{1}{\pi^2} \int d^2\alpha \int d^2\beta \, |\alpha\rangle\langle\beta| \, e^{-\frac{1}{2}|\alpha|^2} e^{-\frac{1}{2}|\beta|^2} R(\alpha^*, \beta), \tag{3.13}$$

where

$$R(\alpha^*, \beta) \equiv e^{\frac{1}{2}|\alpha|^2} e^{\frac{1}{2}|\beta|^2} \langle\alpha|\rho|\beta\rangle$$
$$= e^{\frac{1}{2}|\alpha|^2} e^{\frac{1}{2}|\beta|^2} \left(e^{-\frac{1}{2}|\alpha|^2} \sum_{n=0}^{\infty} \frac{\alpha^{*n}}{\sqrt{n!}} \langle n| \right) \rho \left(e^{-\frac{1}{2}|\beta|^2} \sum_{m=0}^{\infty} \frac{\beta^m}{\sqrt{m!}} |m\rangle \right)$$
$$= \sum_{n,m=0}^{\infty} \frac{\alpha^{*n} \beta^m}{\sqrt{n!m!}} \rho_{n,m}. \tag{3.14}$$

Clearly, this representation follows the familiar methods for specifying an operator in terms of its matrix elements; the exponential factors appearing in (3.13) merely simplify the relationship between the function $R(\alpha^*, \beta)$ and the Fock state matrix elements $\rho_{n,m}$. The P representation is rather different.

The *Glauber–Sudarshan P representation* relies on the fact that the coherent states are not orthogonal. In technical terms they then form an overcomplete basis, and, as a consequence, it is possible to expand ρ as a *diagonal* sum over coherent states:

$$\rho = \int d^2\alpha \, |\alpha\rangle\langle\alpha| P(\alpha). \tag{3.15}$$

This representation for ρ is appealing because the function $P(\alpha)$ plays a role rather analogous to that of a classical probability distribution. First, note that

$$\int d^2\alpha \, P(\alpha) = \int d^2\alpha \, \langle\alpha|\alpha\rangle P(\alpha)$$

$$= \text{tr}\left(\int d^2\alpha \, |\alpha\rangle\langle\alpha| P(\alpha)\right)$$

$$= \text{tr}(\rho)$$

$$= 1, \tag{3.16}$$

where we have inserted $\langle\alpha|\alpha\rangle = 1$ and used the cyclic property of the trace. Thus, $P(\alpha)$ is normalized like a classical probability distribution. Note also that for the expectation values of operators written in normal order (creation operators to the left and annihilation operators to the right), on substituting the expansion (3.15) for ρ,

$$\langle a^{\dagger p} a^q \rangle \equiv \text{tr}\left(\rho a^{\dagger p} a^q \right)$$

$$= \text{tr}\left(\int d^2\alpha \, |\alpha\rangle\langle\alpha| P(\alpha) a^{\dagger p} a^q \right)$$

$$= \int d^2\alpha \, P(\alpha) \langle\alpha| a^{\dagger p} a^q |\alpha\rangle$$

$$= \int d^2\alpha \, P(\alpha) \alpha^{*p} \alpha^q. \tag{3.17}$$

Normal-ordered averages are therefore calculated in the way that averages are calculated in classical statistics, with $P(\alpha)$ playing the role of the probability distribution [(3.16) is a special case of this result with $p = q = 0$]. We will introduce the notation

$$\left(\overline{\alpha^{*p} \alpha^q} \right)_P \equiv \int d^2\alpha \, P(\alpha) \alpha^{*p} \alpha^q, \tag{3.18}$$

and write

$$\langle a^{\dagger p} a^q \rangle = \left(\overline{\alpha^{*p} \alpha^q} \right)_P. \tag{3.19}$$

As mentioned earlier, obtaining normal-ordered averages in this way is particularly useful because measurements in quantum optics have a direct relationship to such normal-ordered quantities, a consequence of the fact that photoelectric detectors work by the absorption of photons.

The analogy between $P(\alpha)$ and a classical probability distribution over coherent states must be made with reservation, however. In the Fock-state representation $\rho_{n,n} = \langle n|\rho|n\rangle$ is an actual probability; it is the probability that the oscillator will be found in the state $|n\rangle$ – the probability that the field mode will be found to contain n photons. But because of the orthogonality of the Fock states, only a limited class of states can be represented by the diagonal matrix elements $\rho_{n,n}$ alone. There exist states whose complete representation requires that at least some nonzero numbers $\rho_{n,m} = \langle n|\rho|m\rangle$, $n \neq m$, be specified in addition to the probabilities $\rho_{n,n}$. The coherent states are not orthogonal, and it is therefore possible to make a diagonal expansion for ρ that is not restricted in the same way; the expansion (3.15) does not automatically require that the off-diagonal coherent state matrix elements vanish. With the help of (3.8), from (3.15) we obtain

$$\langle\alpha|\rho|\beta\rangle = \int d^2\lambda\,\langle\alpha|\lambda\rangle\langle\lambda|\beta\rangle P(\lambda)$$

$$= \int d^2\lambda\, e^{-\frac{1}{2}|\lambda-\alpha|^2} e^{-\frac{1}{2}|\lambda-\beta|^2} P(\lambda). \tag{3.20}$$

There is no need for this to vanish when $\alpha \neq \beta$. There is a price to pay for this versatility, however. We must now accept that $P(\alpha)$ is not strictly a probability. When $\alpha = \beta$, (3.20) gives

$$\langle\alpha|\rho|\alpha\rangle = \int d^2\lambda\, e^{-|\lambda-\alpha|^2} P(\lambda). \tag{3.21}$$

Since $e^{-|\lambda-\alpha|^2}$ is not a δ-function, $\langle\alpha|\rho|\alpha\rangle \neq P(\alpha)$. Only when $P(\lambda)$ is sufficiently broad compared to the Gaussian filter inside the integral in (3.21) does it approximate a probability. Also, although the probability $\langle\alpha|\rho|\alpha\rangle$ must be positive, (3.21) does not require $P(\alpha)$ to be so. Thus, unlike a classical probability, $P(\alpha)$ can take negative values over a limited range [although (3.16) must still be satisfied]. $P(\alpha)$ is not, therefore, a probability distribution, and for this reason it is often referred to as a *quasi*distribution function. We will simply use the word "distribution". In fact, this is quite correct usage if "distribution" is interpreted in the sense of generalized functions. We will see shortly that $P(\alpha)$ is, most generally, a generalized function.

3.1.3 Examples: Coherent States, Thermal States, and Fock States

It is clear from (3.15) that the coherent state $|\alpha_0\rangle$ – density operator $\rho = |\alpha_0\rangle\langle\alpha_0|$ – is represented by the P distribution

$$P(\alpha) = \delta^{(2)}(\alpha - \alpha_0) \equiv \delta(x - x_0)\delta(y - y_0), \tag{3.22}$$

where $\alpha = x + iy$ and $\alpha_0 = x_0 + iy_0$. Can we find a diagonal representation for any density operator? To answer this question we must try to invert (3.15). This is made possible using the relationship

$$\text{tr}\left(\rho e^{iz^*a^\dagger}e^{iza}\right) = \text{tr}\left\{\left[\int d^2\alpha\,|\alpha\rangle\langle\alpha|P(\alpha)\right]e^{iz^*a^\dagger}e^{iza}\right\}$$

$$= \int d^2\alpha\,P(\alpha)\langle\alpha|e^{iz^*a^\dagger}e^{iza}|\alpha\rangle$$

$$= \int d^2\alpha\,P(\alpha)e^{iz^*\alpha^*}e^{iza}. \tag{3.23}$$

Equation (3.23) is just a two-dimensional Fourier transform. The inverse transform gives

$$P(\alpha) = \frac{1}{\pi^2}\int d^2z\,\text{tr}\left(\rho e^{iz^*a^\dagger}e^{iza}\right)e^{-iz^*\alpha^*}e^{-iza}. \tag{3.24}$$

Thus, if the Fourier transform of the function defined by the trace in (3.24) exists for a given density operator ρ, we have our P distribution representing that density operator. A general expression for $P(\alpha)$ in terms of the Fock-state representation of ρ follows by substituting (3.11) into (3.24) and using the cyclic property of the trace:

$$P(\alpha) = \frac{1}{\pi^2}\int d^2z\left(\sum_{n,m=0}^{\infty}\rho_{n,m}\langle m|e^{iz^*a^\dagger}e^{iza}|n\rangle\right)e^{-iz^*\alpha^*}e^{-iza}$$

$$= \frac{1}{\pi^2}\int d^2z\left(\sum_{n,m=0}^{\infty}\sum_{n',m'=0}^{\infty}\rho_{n,m}\langle m|\frac{(iz^*a^\dagger)^{m'}}{m'!}\frac{(iza)^{n'}}{n'!}|n\rangle\right)$$

$$\times e^{-iz^*\alpha^*}e^{-iza}$$

$$= \frac{1}{\pi^2}\int d^2z\left(\sum_{n=0}^{\infty}\sum_{n'=0}^{n}\sum_{m=0}^{\infty}\sum_{m'=0}^{m}\rho_{n,m}\frac{(iz^*)^{m'}}{m'!}\sqrt{\frac{m!}{(m-m')!}}\right.$$

$$\left.\times\frac{(iz)^{n'}}{n'!}\sqrt{\frac{n!}{(n-n')!}}\,\delta_{n-n',m-m'}\right)e^{-iz^*\alpha^*}e^{-iza}.$$

Noting that

$$\sum_{n=0}^{\infty}\sum_{n'=0}^{n}\sum_{m=0}^{\infty}\sum_{m'=0}^{m}\cdots \equiv \sum_{n'=0}^{\infty}\sum_{n-n'=0}^{\infty}\sum_{m'=0}^{\infty}\sum_{m-m'=0}^{\infty}\cdots,$$

and changing the summation indices, with $n' \to n$, $m' \to m$, and $n - n' = m - m' \to k$, we find

$$P(\alpha) = \frac{1}{\pi^2}\int d^2z\left(\sum_{n=0}^{\infty}\sum_{m=0}^{\infty}\sum_{k=0}^{\infty}\rho_{n+k,m+k}\frac{\sqrt{(n+k)!}\sqrt{(m+k)!}}{k!}\right.$$

$$\left.\times\frac{(iz^*)^m}{m!}\frac{(iz)^n}{n!}\right)e^{-iz^*\alpha^*}e^{-iza}. \tag{3.25}$$

Exercise 3.1 Substitute $\rho = |\alpha_0\rangle\langle\alpha_0|$ into (3.24) and the Fock-state representation for this density operator into (3.25); show that both of these equations reproduce the P distribution (3.22) for the coherent state. For the thermal state

$$\rho = (1 - e^{-\hbar\omega/k_B T})e^{-\hbar\omega a^\dagger a/k_B T}, \qquad (3.26)$$

show that (3.25) gives

$$P(\alpha) = \frac{1}{\pi^2}\int d^2z\, e^{-|z|^2\langle\hat{n}\rangle}e^{-iz^*\alpha^*}e^{-iz\alpha}$$

$$= \frac{1}{\pi\langle\hat{n}\rangle}\exp\left(-\frac{|\alpha|^2}{\langle\hat{n}\rangle}\right), \qquad (3.27)$$

where

$$\langle\hat{n}\rangle \equiv \langle a^\dagger a\rangle = \frac{e^{-\hbar\omega/k_B T}}{1 - e^{-\hbar\omega/k_B T}}. \qquad (3.28)$$

Now, consider the P distribution representing a Fock state. We will take $\rho = |l\rangle\langle l|$ where l can be any non-negative integer. From (3.25),

$$P(\alpha) = \frac{1}{\pi^2}\int d^2z\left(\sum_{n=0}^{\infty}\sum_{m=0}^{\infty}\sum_{k=0}^{\infty}\delta_{n+k,l}\delta_{m+k,l}\frac{l!}{k!}\frac{(iz^*)^m}{m!}\frac{(iz)^n}{n!}\right)$$

$$\times\, e^{-iz^*\alpha^*}e^{-iz\alpha}$$

$$= \frac{1}{\pi^2}\int d^2z\left(\sum_{k=0}^{l}\frac{(-1)^k|z|^{2k}}{k!}\frac{l!}{k!(l-k)!}\right)e^{-iz^*\alpha^*}e^{-iz\alpha}, \qquad (3.29)$$

where we have changed the summation index, with $l - k \to k$. Since the summation in (3.29) does not extend to infinity, the expression inside the bracket is a polynomial, and it clearly diverges for $|z| \to \infty$. Thus, this Fourier transform does not exist in the ordinary sense; it would appear that we cannot represent a Fock state using only a diagonal expansion in coherent states. If, however, we write

$$\delta^{(2)}(\alpha) \equiv \frac{1}{\pi^2}\int d^2z\, e^{-iz^*\alpha^*}e^{-iz\alpha} \qquad (3.30)$$

and use the ordinary rules of differentiation inside the integral in (3.29), we may evaluate the integral in terms of derivatives of the δ-function. This gives the P distribution

$$P(\alpha) = \sum_{k=0}^{l}\frac{l!}{k!(l-k)!}\frac{1}{k!}\frac{\partial^{2k}}{\partial\alpha^k\partial\alpha^{*k}}\delta^{(2)}(\alpha). \qquad (3.31)$$

Note 3.1 We will have many occasions to take derivatives with respect to complex conjugate variables. It is convenient to do this by reading the complex variable and its conjugate as two independent variables. This is allowed because

$$
\frac{\partial}{\partial \alpha} \alpha^* = \left(\frac{\partial}{\partial \alpha^*} \alpha \right)^* = \frac{1}{2} \left(\frac{\partial}{\partial x} - i \frac{\partial}{\partial y} \right)(x - iy) = \frac{1}{2} \left(\frac{\partial}{\partial x} x - \frac{\partial}{\partial y} y \right) = 0,
$$

$$(3.32a)$$

and, of course,

$$
\frac{\partial}{\partial \alpha} \alpha = \left(\frac{\partial}{\partial \alpha^*} \alpha^* \right)^* = \frac{1}{2} \left(\frac{\partial}{\partial x} - i \frac{\partial}{\partial y} \right)(x + iy) = \frac{1}{2} \left(\frac{\partial}{\partial x} x + \frac{\partial}{\partial y} y \right) = 1.
$$

$$(3.32b)$$

The mathematical theory that gives precise meaning to (3.31) is the theory of *generalized functions* [3.9–3.11] or *distributions* (in the technical sense of "Schwartz distributions" and "tempered distributions" [3.12, 3.13]). Within this theory the Fourier transform can be formally generalized to cover nonintegrable functions such as polynomials. Such Fourier transforms are not functions in the usual sense; (3.31) does not tell us how to associate a number, $P(\alpha)$, with each value of the variable α. There is certainly no way, then, to interpret $P(\alpha)$ as a probability distribution. It is, however, a "distribution" in the sense defined by the theory of generalized functions. There is no need for us to get deeply involved with the formal theory of generalized functions. Those interested can study this in the books by Lighthill [3.11] and Bremermann [3.13]. Nevertheless, in order to appreciate the sense in which (3.31) provides a diagonal representation for the Fock states we should spend just a little time refreshing our memories about some of the basic properties of generalized functions.

Generalized functions "live" inside integrals. There, they are integrated against some ordinary function from a space of *test functions*. The value of the integral for a given test function is defined as the limit of a sequence of integrals obtained by replacing the generalized function by a sequence of ordinary well-behaved functions. The generalized function is then, in this sense, the limit of a sequence of ordinary functions. Of course, the sequence of functions defining a given generalized function is not unique. For example, for a suitable class of test functions, the δ-function acts inside an integral as the limit of a sequence of Gaussians:

$$
\delta(x) \equiv \lim_{n \to \infty} \sqrt{\frac{n}{\pi}} e^{-nx^2}, \tag{3.33}
$$

where the strict sense of this statement is

$$\int_{-\infty}^{\infty} dx\, \delta(x)\phi(x) \equiv \lim_{n\to\infty} \int_{-\infty}^{\infty} dx\, \sqrt{\frac{n}{\pi}}e^{-nx^2}\phi(x) = \phi(0). \qquad (3.34)$$

Here, the test function $\phi(x)$ must be continuous and grow more slowly at infinity than $Ce^{a|x|}$, with C and a constants. A sequence of functions that decrease faster than Gaussians at infinity would allow us to define the δ-function on a larger space of test functions; most generally, for all continuous functions. Thus, in formal language, generalized functions operate as *functionals*; they associate a number (the limiting value of a sequence of ordinary integrals) with each function from a space of test functions.

The derivative of a generalized function is also a generalized function, defined via the rules of partial integration. Taking $\phi(x) = \psi'(x)$ in (3.34), we can write

$$\int_{-\infty}^{\infty} dx\, \delta(x)\psi'(x)$$

$$= \lim_{n\to\infty} \int_{-\infty}^{\infty} dx\, \sqrt{\frac{n}{\pi}}e^{-nx^2}\psi'(x)$$

$$= \lim_{n\to\infty} \left[\sqrt{\frac{n}{\pi}}e^{-nx^2}\psi(x) \Big|_{-\infty}^{\infty} - \int_{-\infty}^{\infty} dx\left(-2nx\sqrt{\frac{n}{\pi}}e^{-nx^2}\right)\psi(x) \right]$$

$$= -\lim_{n\to\infty} \int_{-\infty}^{\infty} dx\left(-2nx\sqrt{\frac{n}{\pi}}e^{-nx^2}\right)\psi(x). \qquad (3.35)$$

Then, if $\delta'(x)$ is the generalized function defined by the sequence of functions obtained as the derivative of the sequence defining $\delta(x)$ – the functions inside the bracket in (3.35) – the formula for partial integration is preserved:

$$\int_{-\infty}^{\infty} dx\, \delta'(x)\psi(x) = -\int_{-\infty}^{\infty} dx\, \delta(x)\psi'(x) = -\psi'(0). \qquad (3.36)$$

More generally, for the nth derivative of the δ-function, $\delta^{(n)}(x)$, we have

$$\int_{-\infty}^{\infty} dx\, \delta^{(n)}(x)\psi(x) = (-1)^n \int_{-\infty}^{\infty} dx\, \delta(x)\psi^{(n)}(x) = (-1)^n\psi^{(n)}(0), \qquad (3.37)$$

where $\psi^{(n)}(x)$ is the nth derivative of $\psi(x)$. [Do not confuse the notation for the nth derivative of the δ-function with the notation $\delta^{(2)}(\alpha)$ for the two-dimensional δ-function.]

Let us now use (3.37) to see explicitly how (3.31) provides a diagonal representation for the Fock states. We will consider the one-photon state, the simplest example; the general case can be done as an exercise. For $l = 1$, from (3.31),

$$P(\alpha) = \delta^{(2)}(\alpha) + \frac{\partial^2}{\partial\alpha\partial\alpha^*}\delta^{(2)}(\alpha).$$

Substituting into the diagonal expansion (3.15), and using (3.37) (twice for the two-dimensional δ-function),

$$\rho = \int d^2\alpha \, |\alpha\rangle\langle\alpha| P(\alpha)$$

$$= \int d^2\alpha \, |\alpha\rangle\langle\alpha| \left[\delta^{(2)}(\alpha) + \frac{\partial^2}{\partial\alpha\partial\alpha^*} \delta^{(2)}(\alpha) \right]$$

$$= |0\rangle\langle0| + \int d^2\alpha \left(\frac{\partial^2}{\partial\alpha\partial\alpha^*} |\alpha\rangle\langle\alpha| \right) \delta^{(2)}(\alpha)$$

$$= |0\rangle\langle0| + \frac{\partial^2}{\partial\alpha\partial\alpha^*} |\alpha\rangle\langle\alpha| \Big|_{\alpha=0} . \tag{3.38}$$

From this we must recover $\rho = |1\rangle\langle1|$. Using (3.10), we note that

$$\frac{\partial}{\partial\alpha} |\alpha\rangle\langle\alpha| = \frac{\partial}{\partial\alpha} \left(e^{-|\alpha|^2} e^{\alpha a^\dagger} |0\rangle\langle0| e^{\alpha^* a} \right)$$

$$= (a^\dagger - \alpha^*) |\alpha\rangle\langle\alpha|, \tag{3.39a}$$

$$\frac{\partial}{\partial\alpha^*} |\alpha\rangle\langle\alpha| = \frac{\partial}{\partial\alpha^*} \left(e^{-|\alpha|^2} e^{\alpha a^\dagger} |0\rangle\langle0| e^{\alpha^* a} \right)$$

$$= |\alpha\rangle\langle\alpha| (a - \alpha). \tag{3.39b}$$

Then (3.38) readily gives the required result:

$$\rho = |0\rangle\langle0| + \frac{\partial}{\partial\alpha} \Big[|\alpha\rangle\langle\alpha| (a - \alpha) \Big] \Big|_{\alpha=0}$$

$$= |0\rangle\langle0| + \Big[(a^\dagger - \alpha^*) |\alpha\rangle\langle\alpha| (a - \alpha) - |\alpha\rangle\langle\alpha| \Big] \Big|_{\alpha=0}$$

$$= |0\rangle\langle0| + \left(a^\dagger |0\rangle\langle0| a - |0\rangle\langle0| \right)$$

$$= |1\rangle\langle1|.$$

Exercise 3.2 Equation (3.31) is not always the most convenient form to use in calculations. Show that $P(\alpha)$ for the Fock state $|l\rangle$ takes the alternate forms

$$P(\alpha) = \frac{1}{l!} e^{|\alpha|^2} \frac{\partial^{2l}}{\partial\alpha^l \partial\alpha^{*l}} \delta^{(2)}(\alpha), \tag{3.40}$$

and in polar coordinates, with $\alpha = re^{i\theta}$,

$$P(\alpha) = \frac{1}{2\pi r} \frac{l!}{(2l)!} e^{r^2} \frac{\partial^{2l}}{\partial r^{2l}} \delta(r). \tag{3.41}$$

Show that both of these expressions give $\rho = |l\rangle\langle l|$ when substituted into the diagonal expansion for ρ [Eq. (3.15)].

Applications of the P representation in quantum optics have largely been restricted to situations in which $P(\alpha)$ exists as an ordinary function, as it does, for example, for a thermal state [Eq. (3.27)]. With the use of generalized functions it is actually possible to give any density operator a diagonal representation [3.14, 3.15]. As we stated earlier, however, our main objective when introducing the quantum–classical correspondence is to cast the quantum-mechanical theory into a form closely analogous to a classical statistical theory. $P(\alpha)$ is never strictly a probability for observing the coherent state $|\alpha\rangle$, but it can take the *form* of a probability distribution, and when it does, this can be used to aid our intuition – as an example, the phase-independent distribution given by (3.27) essentially corresponds to the classical picture of a field mode subject to thermal fluctuations. Our intuition finds little assistance from a representation in terms of a generalized function. The value of preserving the analogy with a classical statistical system will be further underlined as we now use the P representation to describe the dynamics of the damped harmonic oscillator.

3.1.4 Fokker–Planck Equation
for the Damped Harmonic Oscillator

In Sect. 1.4.1 we derived the master equation for the damped harmonic oscillator:

$$\dot{\rho} = -i\omega_0[a^\dagger a, \rho] + \frac{\gamma}{2}(2a\rho a^\dagger - a^\dagger a\rho - \rho a^\dagger a)$$
$$+ \gamma\bar{n}(a\rho a^\dagger + a^\dagger\rho a - a^\dagger a\rho - \rho aa^\dagger). \tag{3.42}$$

Our goal in this section is to substitute the diagonal representation (3.15) for ρ, and convert the operator master equation into an equation of motion for P. Obviously, we must assume the existence of a time-dependent P distribution, $P(\alpha, t)$, to represent ρ at each instant t.

After substituting for ρ, (3.42) becomes

$$\int d^2\alpha\, |\alpha\rangle\langle\alpha| \frac{\partial}{\partial t} P(\alpha, t)$$
$$= \int d^2\alpha\, P(\alpha, t)\big[-i\omega_0(a^\dagger a|\alpha\rangle\langle\alpha| - |\alpha\rangle\langle\alpha|a^\dagger a)$$
$$+ \frac{\gamma}{2}(2a|\alpha\rangle\langle\alpha|a^\dagger - a^\dagger a|\alpha\rangle\langle\alpha| - |\alpha\rangle\langle\alpha|a^\dagger a)$$
$$+ \gamma\bar{n}(a|\alpha\rangle\langle\alpha|a^\dagger + a^\dagger|\alpha\rangle\langle\alpha|a - a^\dagger a|\alpha\rangle\langle\alpha| - |\alpha\rangle\langle\alpha|aa^\dagger)\big]. \tag{3.43}$$

The central step in our derivation is to replace the action of the operators a and $a^\dagger$ on $|\alpha\rangle\langle\alpha|$ (both to the right and to the left) by multiplication by the complex variables α and α^*, and the action of partial derivatives with respect to these variables. This can be accomplished using (3.2) and (3.39):

$$a|\alpha\rangle\langle\alpha|a^\dagger = \alpha|\alpha\rangle\langle\alpha|\alpha^* = |\alpha|^2|\alpha\rangle\langle\alpha|, \tag{3.44a}$$

$$a^\dagger a|\alpha\rangle\langle\alpha| = a^\dagger\alpha|\alpha\rangle\langle\alpha| = \alpha a^\dagger|\alpha\rangle\langle\alpha| = \alpha\left(\frac{\partial}{\partial\alpha} + \alpha^*\right)|\alpha\rangle\langle\alpha|, \tag{3.44b}$$

$$|\alpha\rangle\langle\alpha|a^\dagger a = |\alpha\rangle\langle\alpha|\alpha^* a = \alpha^*|\alpha\rangle\langle\alpha|a = \alpha^*\left(\frac{\partial}{\partial\alpha^*} + \alpha\right)|\alpha\rangle\langle\alpha|, \tag{3.44c}$$

$$|\alpha\rangle\langle\alpha|aa^\dagger = \left(\frac{\partial}{\partial\alpha^*} + \alpha\right)|\alpha\rangle\langle\alpha|a^\dagger = \left(\frac{\partial}{\partial\alpha^*} + \alpha\right)\alpha^*|\alpha\rangle\langle\alpha|, \tag{3.44d}$$

$$a^\dagger|\alpha\rangle\langle\alpha|a = \left(\frac{\partial}{\partial\alpha} + \alpha^*\right)|\alpha\rangle\langle\alpha|a = \left(\frac{\partial}{\partial\alpha} + \alpha^*\right)\left(\frac{\partial}{\partial\alpha^*} + \alpha\right)|\alpha\rangle\langle\alpha|. \tag{3.44e}$$

Using these results in (3.43), after some cancelation, we find

$$\int d^2\alpha\,|\alpha\rangle\langle\alpha|\frac{\partial}{\partial t}P(\alpha,t) = \int d^2\alpha\,P(\alpha,t)\left[-\left(\frac{\gamma}{2} + i\omega_0\right)\alpha\frac{\partial}{\partial\alpha}\right.$$
$$\left. -\left(\frac{\gamma}{2} - i\omega_0\right)\alpha^*\frac{\partial}{\partial\alpha^*} + \gamma\bar{n}\frac{\partial^2}{\partial\alpha\partial\alpha^*}\right]|\alpha\rangle\langle\alpha|. \tag{3.45}$$

It is a short step to an equation of motion for P. The partial derivatives which now act to the right on $|\alpha\rangle\langle\alpha|$ can be transferred to the distribution $P(\alpha,t)$ by integrating by parts. We will assume that $P(\alpha,t)$ vanishes sufficiently rapidly at infinity to allow us to drop the boundary terms. Then (3.45) becomes

$$\int d^2\alpha\,|\alpha\rangle\langle\alpha|\frac{\partial}{\partial t}P(\alpha,t) = \int d^2\alpha\,|\alpha\rangle\langle\alpha|\left[\left(\frac{\gamma}{2} + i\omega_0\right)\frac{\partial}{\partial\alpha}\alpha\right.$$
$$\left. +\left(\frac{\gamma}{2} - i\omega_0\right)\frac{\partial}{\partial\alpha^*}\alpha^* + \gamma\bar{n}\frac{\partial^2}{\partial\alpha\partial\alpha^*}\right]P(\alpha,t). \tag{3.46}$$

Note 3.2 When integrating by parts α and α^* may be read as independent variables, as in differentiation (Note 3.1). Explicitly, for given functions $f(\alpha)$ and $g(\alpha)$ (whose product vanishes at infinity),

$$\int d^2\alpha\,f(\alpha)\frac{\partial}{\partial\alpha}g(\alpha)$$
$$= \int_{-\infty}^{\infty} dx \int_{-\infty}^{\infty} dy\,f(x,y)\frac{1}{2}\left(\frac{\partial}{\partial x} - i\frac{\partial}{\partial y}\right)g(x,y)$$
$$= \frac{1}{2}\int_{-\infty}^{\infty} dy\left[f(x,y)g(x,y)\Big|_{x=-\infty}^{\infty} - \int_{-\infty}^{\infty} dx\,g(x,y)\frac{\partial}{\partial x}f(x,y)\right]$$
$$- i\frac{1}{2}\int_{-\infty}^{\infty} dx\left[f(x,y)g(x,y)\Big|_{y=-\infty}^{\infty} - \int_{-\infty}^{\infty} dy\,g(x,y)\frac{\partial}{\partial y}f(x,y)\right]$$

$$= -\int_{-\infty}^{\infty} dx \int_{-\infty}^{\infty} dy\, g(x,y) \frac{1}{2}\left(\frac{\partial}{\partial x} - i\frac{\partial}{\partial y}\right) f(x,y)$$

$$= -\int d^2\alpha\, g(\alpha) \frac{\partial}{\partial \alpha} f(\alpha).$$

Similarly,

$$\int d^2\alpha\, f(\alpha) \frac{\partial}{\partial \alpha^*} g(\alpha) = -\int d^2\alpha\, g(\alpha) \frac{\partial}{\partial \alpha^*} f(\alpha).$$

A *sufficient condition* for (3.46) to be satisfied is that the P distribution obeys the equation of motion

$$\frac{\partial P}{\partial t} = \left[\left(\frac{\gamma}{2} + i\omega_0\right) \frac{\partial}{\partial \alpha}\alpha + \left(\frac{\gamma}{2} - i\omega_0\right) \frac{\partial}{\partial \alpha^*}\alpha^* + \gamma\bar{n}\frac{\partial^2}{\partial\alpha\partial\alpha^*}\right] P. \qquad (3.47)$$

We have replaced the operator equation (3.42) by a partial differential equation for P. This is the *Fokker–Planck equation for the damped harmonic oscillator in the P representation.*

Exercise 3.3 The question arises as to whether (3.47) is a *necessary condition* for (3.46) to be satisfied. Multiply both sides of (3.46) on the left by $e^{iz^*a^\dagger}e^{iza}$ and take the trace to show that the necessary condition is that the Fourier transforms of both sides of (3.47) are equal.

3.1.5 Solution of the Fokker–Planck Equation

We will discuss the properties of Fokker–Planck equations in detail in Chap. 5. For the present let us simply illustrate how (3.47) describes the damped harmonic oscillator. We will solve this equation for an initial coherent state $|\alpha_0\rangle$. Thus, we seek the Green function $P(\alpha, \alpha^*, t | \alpha_0, \alpha_0^*, 0)$, with initial condition

$$P(\alpha, \alpha^*, 0 | \alpha_0, \alpha_0^*, 0) = \delta^{(2)}(\alpha - \alpha_0) \equiv \delta(x - x_0)\delta(y - y_0). \qquad (3.48)$$

From now on we display P with two complex conjugate arguments consistent with the interpretation of derivatives and integrals explained below (3.31) and (3.46).

It is convenient to transform to a frame rotating at the frequency ω_0, with

$$\alpha = e^{-i\omega_0 t}\tilde{\alpha}, \qquad \alpha^* = e^{i\omega_0 t}\tilde{\alpha}^*, \qquad (3.49)$$

and

$$P(\alpha, \alpha^*, t) = \tilde{P}(\tilde{\alpha}, \tilde{\alpha}^*, t). \qquad (3.50)$$

We have

$$\frac{\partial \tilde{P}}{\partial t} = \frac{\partial P}{\partial t} + \frac{\partial P}{\partial \alpha}\frac{\partial \alpha}{\partial t} + \frac{\partial P}{\partial \alpha^*}\frac{\partial \alpha^*}{\partial t}$$

$$= \frac{\partial P}{\partial t} - i\omega_0 \left(\alpha \frac{\partial P}{\partial \alpha} - \alpha^* \frac{\partial P}{\partial \alpha^*} \right)$$

$$= \frac{\partial P}{\partial t} - i\omega_0 \left(\frac{\partial}{\partial \alpha}\alpha - \frac{\partial}{\partial \alpha^*}\alpha^* \right) P. \tag{3.51}$$

After substituting for $\partial P/\partial t$ from (3.47),

$$\frac{\partial \tilde{P}}{\partial t} = \left[\frac{\gamma}{2}\left(\frac{\partial}{\partial \tilde{\alpha}}\tilde{\alpha} + \frac{\partial}{\partial \tilde{\alpha}^*}\tilde{\alpha}^* \right) + \gamma \bar{n}\frac{\partial^2}{\partial \tilde{\alpha}\partial \tilde{\alpha}^*} \right] \tilde{P}, \tag{3.52}$$

or, in terms of the real and imaginary parts of $\tilde{\alpha}$,

$$\frac{\partial \tilde{P}}{\partial t} = \left[\frac{\gamma}{2}\left(\frac{\partial}{\partial \tilde{x}}\tilde{x} + \frac{\partial}{\partial \tilde{y}}\tilde{y} \right) + \frac{\gamma \bar{n}}{4}\left(\frac{\partial^2}{\partial \tilde{x}^2} + \frac{\partial^2}{\partial \tilde{y}^2} \right) \right] \tilde{P}, \tag{3.53}$$

where $\tilde{\alpha} = \tilde{x} + i\tilde{y}$. Solutions can now be sought using separation of variables. We write

$$\tilde{P}(\tilde{x}, \tilde{y}, t) = X(\tilde{x}, t)Y(\tilde{y}, t), \tag{3.54}$$

where the functions X and Y satisfy the independent equations

$$\frac{\partial X}{\partial t} = \left(\frac{\gamma}{2}\frac{\partial}{\partial \tilde{x}}\tilde{x} + \frac{\gamma \bar{n}}{4}\frac{\partial^2}{\partial \tilde{x}^2} \right) X, \tag{3.55a}$$

$$\frac{\partial Y}{\partial t} = \left(\frac{\gamma}{2}\frac{\partial}{\partial \tilde{y}}\tilde{y} + \frac{\gamma \bar{n}}{4}\frac{\partial^2}{\partial \tilde{y}^2} \right) Y. \tag{3.55b}$$

These are to be solved for $X(\tilde{x}, t|\tilde{x}_0, 0)$ and $Y(\tilde{y}, t|\tilde{y}_0, 0)$, subject to the initial conditions

$$X(\tilde{x}, 0|\tilde{x}_0, 0) = \delta(\tilde{x} - \tilde{x}_0), \tag{3.56a}$$

$$Y(\tilde{y}, 0|\tilde{y}_0, 0) = \delta(\tilde{y} - \tilde{y}_0). \tag{3.56b}$$

Consider (3.55a). Its solution is found by taking the Fourier transform on both sides of the equation. We find

$$\frac{\partial U}{\partial t} = -\left(\frac{\gamma}{2}\tilde{u}\frac{\partial}{\partial \tilde{u}} + \frac{\gamma \bar{n}}{4}\tilde{u}^2 \right) U, \tag{3.57}$$

where

$$U(\tilde{u}, t|\tilde{x}_0, 0) = \int_{-\infty}^{\infty} d\tilde{x}\, X(\tilde{x}, t|\tilde{x}_0, 0)e^{i\tilde{x}\tilde{u}}, \tag{3.58}$$

and, from (3.56a), the initial condition for U is

$$U(\tilde{u}, 0|\tilde{x}_0, 0) = e^{i\tilde{x}_0\tilde{u}}. \tag{3.59}$$

We then solve (3.57) by the method of characteristics [3.16]. The subsidiary equations are

$$\frac{dt}{1} = \frac{d\tilde{u}}{(\gamma/2)\tilde{u}} = \frac{dU}{-(\gamma\bar{n}/4)\tilde{u}^2 U}, \tag{3.60}$$

with solutions

$$\tilde{u}e^{-(\gamma/2)t} = \text{constant}, \tag{3.61a}$$

$$Ue^{(\bar{n}/4)\tilde{u}^2} = \text{constant}. \tag{3.61b}$$

Thus, U must have the general form

$$U(\tilde{u}, t|\tilde{x}_0, 0) = \phi\big(\tilde{u}e^{-(\gamma/2)t}\big)e^{-(\bar{n}/4)\tilde{u}^2}, \tag{3.62}$$

where ϕ is an arbitrary function. Choosing ϕ to match the initial condition (3.59),

$$U(\tilde{u}, t|\tilde{x}_0, 0) = \exp\big[i\tilde{x}_0\tilde{u}e^{-(\gamma/2)t}\big]\exp\big[-(\bar{n}/4)\tilde{u}^2(1 - e^{-\gamma t})\big]. \tag{3.63}$$

Taking the inverse Fourier transform, we have

$$
\begin{aligned}
X(&\tilde{x}, t|\tilde{x}_0, 0) \\
&= \frac{1}{2\pi}\int_{-\infty}^{\infty} d\tilde{u}\, U(\tilde{u}, t|\tilde{x}_0, 0)e^{-i\tilde{x}\tilde{u}} \\
&= \frac{1}{2\pi}\int_{-\infty}^{\infty} d\tilde{u}\, \exp\big[-i\tilde{u}\big(\tilde{x} - \tilde{x}_0 e^{-(\gamma/2)t}\big)\big] \\
&\quad \times \exp\big[-(\bar{n}/4)\tilde{u}^2(1 - e^{-\gamma t})\big] \\
&= \frac{1}{2\pi}\int_{-\infty}^{\infty} d\tilde{u}\, \cos\big[\tilde{u}\big(\tilde{x} - \tilde{x}_0 e^{-(\gamma/2)t}t\big)\big]\exp\big[-(\bar{n}/4)\tilde{u}^2(1 - e^{-\gamma t})\big] \\
&= \frac{1}{\sqrt{\pi\bar{n}(1 - e^{-\gamma t})}}\exp\left[-\frac{\big(\tilde{x} - \tilde{x}_0 e^{-(\gamma/2)t}\big)^2}{\bar{n}(1 - e^{-\gamma t})}\right].
\end{aligned}
\tag{3.64}
$$

Equation (3.55b) can be solved in a similar fashion, whence,

$$
\tilde{P}(\tilde{x}, \tilde{y}, t|\tilde{x}_0, \tilde{y}_0, 0) \\
= \frac{1}{\pi\bar{n}(1 - e^{-\gamma t})}\exp\left[-\frac{\big(\tilde{x} - \tilde{x}_0 e^{-(\gamma/2)t}\big)^2 + \big(\tilde{y} - \tilde{y}_0 e^{-(\gamma/2)t}\big)^2}{\bar{n}(1 - e^{-\gamma t})}\right],
\tag{3.65}
$$

or, equivalently,

$$
\tilde{P}(\tilde{\alpha}, \tilde{\alpha}^*, t|\tilde{\alpha}_0, \tilde{\alpha}_0^*, 0) = \frac{1}{\pi\bar{n}(1 - e^{-\gamma t})}\exp\left[-\frac{|\tilde{\alpha} - \tilde{\alpha}_0 e^{-(\gamma/2)t}|^2}{\bar{n}(1 - e^{-\gamma t})}\right]. \tag{3.66}
$$

Then the P *distribution for a damped coherent state* is given by

$$P(\alpha, \alpha^*, t | \alpha_0, \alpha_0^*, 0) = \frac{1}{\pi \bar{n}(1 - e^{-\gamma t})} \exp \left[-\frac{|\alpha - \alpha_0 e^{-(\gamma/2)t} e^{-i\omega_0 t}|^2}{\bar{n}(1 - e^{-\gamma t})} \right].$$

(3.67)

$P(\alpha, \alpha^*, t | \alpha_0, \alpha_0^*, 0)$ is a two-dimensional Gaussian distribution. Thus, for this example the P distribution has all the properties of a probability distribution. The mean of the Gaussian gives the oscillating and decaying oscillator amplitude calculated previously directly from the master equation [Eq. (1.78)]:

$$\langle a(t) \rangle = \overline{(\alpha(t))}_P = \alpha_0 e^{-(\gamma/2)t} e^{-i\omega_0 t}.$$

(3.68)

The phase-independent variance describes the thermal fluctuations added to the coherent amplitude by the oscillator's interaction with the reservoir:

$$\langle (a^\dagger a)(t) \rangle - \langle a^\dagger(t) \rangle \langle a(t) \rangle = \overline{((\alpha^* \alpha)(t))}_P - \overline{(\alpha^*(t))}_P \overline{(\alpha(t))}_P$$
$$= \left[\overline{(x^2(t))}_P + \overline{(y^2(t))}_P \right] - \left[\overline{(x(t))}_P^2 + \overline{(y(t))}_P^2 \right]$$
$$= \bar{n}(1 - e^{-\gamma t}).$$

(3.69)

For an initial coherent state, $\langle a^\dagger(t) \rangle \langle a(t) \rangle = |\alpha_0|^2 e^{-\gamma t} = \langle (a^\dagger a)(0) \rangle e^{-\gamma t}$, and therefore (3.69) also agrees with our previous calculation [Eq. (1.80)]. In the long-time limit the coherent amplitude decays to zero and the variance of the fluctuations in each quadrature of the complex amplitude grows to $\bar{n}/2$. A comparison of (3.67) with (3.27) shows that the oscillator reaches a thermal state with mean photon number $\bar{n}$ equal to the mean photon number for a reservoir oscillator of frequency ω_0. Figure 3.1 illustrates these dynamics with $P(\alpha, \alpha^*, t | \alpha_0, \alpha_0^*, 0)$ represented by a single circular contour of radius $\sqrt{(\bar{n}/2)(1 - e^{-\gamma t})}$. For a Gaussian, the mean and variance determine all higher-order moments. Hence, (3.68) and (3.69) determine all of the *normal-ordered* operator averages for the damped oscillator [Eq. (3.19)]. Using the P representation we have put the statistical properties of the *quantum-mechanical* oscillator into a correspondence with a *classical* statistical description in terms of the phase-space variables x and y. (For a mechanical oscillator the coordinate and momentum variables are $q = x\sqrt{2\hbar/m\omega}$ and $p = y\sqrt{2\hbar m\omega}$, respectively.)

3.2 The Characteristic Function
for Normal-Ordered Averages

We now look at an alternative way of defining the P representation and deriving an equation of motion for the P distribution. This second approach leaves the relationship to coherent states somewhat hidden, but introduces a method that can readily be generalized – to define representations based on

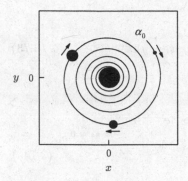

Fig. 3.1 Time evolution of $P(\alpha, \alpha^*, t | \alpha_0, \alpha_0^*, 0)$ [Eq. (3.67)]. The center of the Gaussian distribution follows the spiral curve while the width of the distribution increases with time, as illustrated by the filled circular contours $c_t(\theta) = \alpha_0 e^{-(\gamma/2)t} e^{-i\omega_0 t} + e^{i\theta}\sqrt{(\bar{n}/2)(1 - e^{-\gamma t})}$.

different operator orderings, and to define representations for collections of two-level atoms.

We have recently met two relationships that might suggest the new approach to us. In (3.23) and (3.24), and in Exercise 3.3, we saw that the Fourier transform of $P(\alpha, \alpha^*)$ played an important role. Why not begin from the function appearing on the left-hand side of (3.23) and *define* $P(\alpha, \alpha^*)$ to be its Fourier transform. Indeed, this approach is suggested on the following, more general grounds.

3.2.1 Operator Averages and the Characteristic Function

The function

$$\chi_N(z, z^*) \equiv \mathrm{tr}\big(\rho e^{iz^* a^\dagger} e^{iza}\big) \qquad (3.70)$$

appearing on the left-hand side of (3.23) is a *characteristic function* in the usual sense of statistical physics [3.17]; it determines all *normal-ordered operator averages* via the prescription

$$\langle a^{\dagger p} a^q \rangle \equiv \mathrm{tr}\big(\rho a^{\dagger p} a^q\big)$$
$$= \frac{\partial^{p+q}}{\partial(iz^*)^p \partial(iz)^q} \chi_N(z, z^*)\bigg|_{z=z^*=0}. \qquad (3.71)$$

The definition of a distribution for calculating normal-ordered averages follows quite naturally from this result. If we define $P(\alpha, \alpha^*)$ to be the two-dimensional Fourier transform of $\chi_N(z, z^*)$:

$$P(\alpha, \alpha^*) \equiv \frac{1}{\pi^2} \int d^2z \, \chi_N(z, z^*) e^{-iz^*\alpha^*} e^{-iz\alpha}$$
$$\equiv \frac{1}{\pi^2} \int_{-\infty}^{\infty} d\mu \int_{-\infty}^{\infty} d\nu \, \chi_N(\mu + i\nu, \mu - i\nu) e^{-2i(\mu x - \nu y)}, \quad (3.72)$$

with the inverse relationship

$$\chi_N(z, z^*) = \int d^2\alpha\, P(\alpha, \alpha^*) e^{iz^*\alpha^*} e^{iz\alpha}$$

$$= \int_{-\infty}^{\infty} dx \int_{-\infty}^{\infty} dy\, P(x + iy, x - iy) e^{2i(\mu x - \nu y)}, \qquad (3.73)$$

then, from (3.71) and (3.73),

$$\langle a^{\dagger p} a^q \rangle = \frac{\partial^{p+q}}{\partial(iz^*)^p \partial(iz)^q} \int d^2\alpha\, P(\alpha, \alpha^*) e^{iz^*\alpha^*} e^{iz\alpha} \bigg|_{z=z^*=0}$$

$$= \left(\overline{\alpha^{*p} \alpha^q} \right)_P, \qquad (3.74\text{a})$$

with

$$\left(\overline{\alpha^{*p} \alpha^q} \right)_P \equiv \int d^2\alpha\, P(\alpha, \alpha^*) \alpha^{*p} \alpha^q. \qquad (3.74\text{b})$$

Equation (3.73) is the same as (3.23), and (3.74) reproduces (3.19); the $P(\alpha, \alpha^*)$ defined in this way is the distribution introduced in (3.15) to give a diagonal expansion in terms of coherent states. Let us see how the Fokker–Planck equation for the damped harmonic oscillator can be derived by starting from this new definition of $P(\alpha, \alpha^*)$.

3.2.2 Derivation of the Fokker–Planck Equation Using the Characteristic Function

We will derive an equation of motion for the characteristic function and then use the relationship between $\chi_N(z, z^*, t)$ and $P(\alpha, \alpha^*, t)$ to convert this into an equation of motion for $P(\alpha, \alpha^*, t)$.

From the definition of χ_N,

$$\frac{\partial \chi_N}{\partial t} = \frac{\partial}{\partial t} \mathrm{tr}\left(\rho\, e^{iz^* a^\dagger} e^{iza} \right) = \mathrm{tr}\left(\dot\rho\, e^{iz^* a^\dagger} e^{iza} \right). \qquad (3.75)$$

Then, the master equation (3.42) gives

$$\frac{\partial \chi_N}{\partial t} = \mathrm{tr}\left\{ \left[-i\omega_0(a^\dagger a\rho - \rho a^\dagger a) + \frac{\gamma}{2}(2a\rho a^\dagger - a^\dagger a\rho - \rho a^\dagger a) \right. \right.$$
$$\left. \left. + \gamma\bar{n}(a\rho a^\dagger + a^\dagger \rho a - a^\dagger a\rho - \rho a a^\dagger) \right] e^{iz^* a^\dagger} e^{iza} \right\}. \qquad (3.76)$$

Our aim is to express each of the nine terms on the right-hand side of (3.76) in terms of χ_N and its derivatives with respect to (iz^*) and (iz). For two of the nine terms this can be achieved directly; we may write

$$\mathrm{tr}\left(a\rho a^\dagger e^{iz^* a^\dagger} e^{iza} \right) = \mathrm{tr}\left(\rho a^\dagger e^{iz^* a^\dagger} e^{iza} a \right)$$

$$= \frac{\partial^2}{\partial(iz^*)\partial(iz)} \chi_N, \qquad (3.77)$$

where we have simply used the cyclic property of the trace. The remaining seven terms require a little more algebraic manipulation; but the goal is always the same – to rearrange the terms inside the trace so that $a^\dagger$ is to the left of $e^{iz^*a^\dagger}$ and a is to the right of e^{iza}. Then, $a^\dagger$ and a can be brought down from the exponentials by differentiation with respect to (iz^*) and (iz), respectively. Generally, the rearrangement may require us to pass $a^\dagger$ through the exponential e^{iza}, or a through the exponential $e^{iz^*a^\dagger}$. For this purpose we use

$$e^{iza}a^\dagger e^{-iza} = a^\dagger + iz, \qquad (3.78a)$$

$$e^{-iz^*a^\dagger} a\, e^{iz^*a^\dagger} = a + iz^*. \qquad (3.78b)$$

Equation (3.78a) follows by writing $a^\dagger(iz) = e^{iza}a^\dagger e^{-iza}$, with $a^\dagger(0) = a^\dagger$; then differentiate with respect to (iz):

$$\frac{d}{d(iz)}a^\dagger(iz) = e^{iza}(aa^\dagger - a^\dagger a)e^{-iza} = 1.$$

Thus,

$$a^\dagger(iz) = a^\dagger(0) + iz = a^\dagger + iz.$$

Equation (3.78b) is obtained as the Hermitian conjugate of (3.78a) and the replacement $z^* \to -z^*$.

Now, using (3.78) and the cyclic property of the trace, the remaining terms in (3.76) are:

$$
\begin{aligned}
\operatorname{tr}\!\left(a^\dagger a\rho e^{iz^*a^\dagger} e^{iza}\right) &= \operatorname{tr}\!\left(\rho e^{iz^*a^\dagger} e^{iza}a^\dagger a\right) \\
&= \operatorname{tr}\!\left[\rho e^{iz^*a^\dagger}\!\left(e^{iza}a^\dagger e^{-iza}\right)e^{iza}a\right] \\
&= \operatorname{tr}\!\left[\rho(a^\dagger + iz)e^{iz^*a^\dagger} e^{iza}a\right] \\
&= \left(\frac{\partial}{\partial(iz^*)} + iz\right)\operatorname{tr}\!\left(\rho\, e^{iz^*a^\dagger} e^{iza}a\right) \\
&= \left(\frac{\partial}{\partial(iz^*)} + iz\right)\frac{\partial}{\partial(iz)}\chi_N, \qquad (3.79)
\end{aligned}
$$

$$
\begin{aligned}
\operatorname{tr}\!\left(\rho a^\dagger a\, e^{iz^*a^\dagger} e^{iza}\right) &= \operatorname{tr}\!\left[\rho a^\dagger e^{iz^*a^\dagger}\!\left(e^{-iz^*a^\dagger} a\, e^{iz^*a^\dagger}\right)e^{iza}\right] \\
&= \operatorname{tr}\!\left[\rho a^\dagger e^{iz^*a^\dagger} e^{iza}(a + iz^*)\right] \\
&= \left(\frac{\partial}{\partial(iz)} + iz^*\right)\operatorname{tr}\!\left(\rho a^\dagger e^{iz^*a^\dagger} e^{iza}\right) \\
&= \left(\frac{\partial}{\partial(iz)} + iz^*\right)\frac{\partial}{\partial(iz^*)}\chi_N, \qquad (3.80)
\end{aligned}
$$

$$\text{tr}(\rho a a^\dagger e^{iz^* a^\dagger} e^{iza}) = \text{tr}[\rho(a^\dagger a + 1)e^{iz^* a^\dagger} e^{iza}]$$

$$= \left[\left(\frac{\partial}{\partial(iz)} + iz^*\right)\frac{\partial}{\partial(iz^*)} + 1\right]\chi_N, \qquad (3.81)$$

which follows from (3.80); the last term is left as an exercise:

Exercise 3.4 Show that

$$\text{tr}(a^\dagger \rho a e^{iz^* a^\dagger} e^{iza})$$

$$= \left(1 - |z|^2 + iz\frac{\partial}{\partial(iz)} + iz^*\frac{\partial}{\partial(iz^*)} + \frac{\partial^2}{\partial(iz)\partial(iz^*)}\right)\chi_N. \quad (3.82)$$

After substituting (3.77) and (3.79)–(3.82) into (3.76) the equation of motion for $\chi_N(z, z^*, t)$ is given by

$$\frac{\partial \chi_N}{\partial t} = \left[-\left(\frac{\gamma}{2} + i\omega_0\right)z\frac{\partial}{\partial z} - \left(\frac{\gamma}{2} - i\omega_0\right)z^*\frac{\partial}{\partial z^*} - \gamma\bar{n}zz^*\right]\chi_N. \quad (3.83)$$

To pass to an equation of motion for $P(\alpha, \alpha^*, t)$ we use the Fourier transform relation (3.73) and exchange the differential operator in the variables z and z^* for one in the variables α and α^*:

$$\int d^2\alpha \,\frac{\partial P(\alpha, \alpha^*, t)}{\partial t}\, e^{iz^* \alpha^*} e^{iz\alpha}$$

$$= \int d^2\alpha \, P(\alpha, \alpha^*, t)\left[-\left(\frac{\gamma}{2} + i\omega_0\right)z\frac{\partial}{\partial z} - \left(\frac{\gamma}{2} - i\omega_0\right)z^*\frac{\partial}{\partial z^*}\right.$$

$$\left. - \gamma\bar{n}zz^*\right]e^{iz^* \alpha^*} e^{iz\alpha}$$

$$= \int d^2\alpha \, P(\alpha, \alpha^*, t)\left[-\left(\frac{\gamma}{2} + i\omega_0\right)(i\alpha)\frac{\partial}{\partial(i\alpha)} - \left(\frac{\gamma}{2} - i\omega_0\right)(i\alpha^*)\frac{\partial}{\partial(i\alpha^*)}\right.$$

$$\left. - \gamma\bar{n}\frac{\partial^2}{\partial(i\alpha)\partial(i\alpha^*)}\right]e^{iz^* \alpha^*} e^{iz\alpha}. \qquad (3.84)$$

The action of the derivatives on the right-hand side of (3.84) can be moved from the product of exponentials, $e^{iz^* \alpha^*} e^{iza}$, to $P(\alpha, \alpha^*, t)$ by integrating by parts; we took the same step in passing from (3.45) to (3.46). Once again we assume that $P(\alpha, \alpha^*, t)$ vanishes sufficiently fast at infinity to justify dropping the boundary terms. Then, (3.84) becomes

$$\int d^2\alpha \, e^{iz^* \alpha^*} e^{iza}\frac{\partial P}{\partial t} = \int d^2\alpha \, e^{iz^* \alpha^*} e^{iza}\left[\left(\frac{\gamma}{2} + i\omega_0\right)\frac{\partial}{\partial \alpha}\alpha\right.$$

$$\left. + \left(\frac{\gamma}{2} - i\omega_0\right)\frac{\partial}{\partial \alpha^*}\alpha^* + \gamma\bar{n}\frac{\partial^2}{\partial\alpha\partial\alpha^*}\right]P. \quad (3.85)$$

This is the Fourier transform of the Fokker–Planck equation derived in Sect. 3.1.4. It is precisely the equation derived from (3.46) in Exercise 3.3. Thus, after inverting the Fourier transform we arrive once again at the Fokker–Planck equation (3.47).

4. Quantum–Classical Correspondence for the Electromagnetic Field II: P, Q, and Wigner Representations

The definition of the P representation as the Fourier transform of the normal-ordered characteristic function can be generalized by simply taking different characteristic functions – characteristic functions that give operator averages in other than normal order. Here we will look at two new representations: the Q representation, which is defined in terms of the characteristic function that gives operator averages in antinormal order, and the Wigner representation, defined in terms of the characteristic function that gives operator averages in symmetric, or Weyl, order. This is not a comprehensive list. Cahill and Glauber [4.1], and Agarwal and Wolf [4.2] have introduced formalisms in which whole classes of different representations are defined. In particular, Agarwal and Wolf take the possibilities to their ultimate extreme and develop a very general and elegant formalism which they call the phase-space calculus. These general formalisms are not of much interest, however, when it comes to applications. The P, Q, and Wigner representations are the only examples that have traditionally seen any use in quantum optics. They are special cases within the classes defined by Cahill and Glauber, and Agarwal and Wolf. In Volume 2 we will meet one recent addition to the list which has been used quite extensively, particularly in the treatment of squeezing and related nonclassical effects. This is the positive P representation introduced by Drummond and Gardiner [4.3]. As the name suggests, the positive P representation is closely related to the Glauber–Sudarshan P representation. We postpone its discussion, however, until we have acquired the background needed to appreciate its special purpose and application. Certain properties of the positive P representation are still only partly understood; this representation therefore belongs with the modern research topics that are taken up in Volume 2.

For additional reading on the Q and Wigner representations reference may be made to Louisell [4.4] and Haken [4.5]. Also, Hillery *et al.* provide a comprehensive review with numerous references [4.6].

4.1 The Q and Wigner Representations

4.1.1 Antinormal-Ordered Averages and the Q Representation

If we wish to calculate antinormal-ordered averages, the rather obvious generalization from (3.70) is to define the characteristic function

$$\chi_A(z, z^*) \equiv \mathrm{tr}\big(\rho e^{iza} e^{iz^* a^\dagger}\big). \tag{4.1}$$

Then in place of (3.71), *antinormal-ordered operator averages* are given by

$$\langle a^q a^{\dagger p} \rangle \equiv \mathrm{tr}\big(\rho a^q a^{\dagger p}\big)$$
$$= \frac{\partial^{p+q}}{\partial(iz^*)^p \partial(iz)^q} \chi_A(z, z^*)\bigg|_{z=z^*=0}. \tag{4.2}$$

If we define the distribution $Q(\alpha, \alpha^*)$ as the Fourier transform of $\chi_A(z, z^*)$:

$$Q(\alpha, \alpha^*) \equiv \frac{1}{\pi^2} \int d^2 z\, \chi_A(z, z^*) e^{-iz^* \alpha^*} e^{-iz\alpha}$$
$$\equiv \frac{1}{\pi^2} \int_{-\infty}^{\infty} d\mu \int_{-\infty}^{\infty} d\nu\, \chi_A(\mu + i\nu, \mu - i\nu) e^{-2i(\mu x - \nu y)}, \tag{4.3}$$

with the inverse relationship

$$\chi_A(z, z^*) = \int d^2\alpha\, Q(\alpha, \alpha^*) e^{iz^* \alpha^*} e^{iz\alpha}$$
$$= \int_{-\infty}^{\infty} dx \int_{-\infty}^{\infty} dy\, Q(x + iy, x - iy) e^{2i(\mu x - \nu y)}, \tag{4.4}$$

corresponding to (3.74), we now have

$$\langle a^q a^{\dagger p} \rangle = \frac{\partial^{p+q}}{\partial(iz^*)^p \partial(iz)^q} \int d^2\alpha\, Q(\alpha, \alpha^*) e^{iz^* \alpha^*} e^{iz\alpha}\bigg|_{z=z^*=0}$$
$$= \big(\overline{\alpha^{*p} \alpha^q}\big)_Q, \tag{4.5a}$$

with

$$\big(\overline{\alpha^{*p} \alpha^q}\big)_Q \equiv \int d^2\alpha\, Q(\alpha, \alpha^*) \alpha^{*p} \alpha^q. \tag{4.5b}$$

The Q distribution, so defined, has a very simple relationship to the coherent states. Consider (4.3) with $\chi_A(z, z^*)$ substituted explicitly from (4.1) and the unit operator judiciously introduced in the form (3.9). We find

$$Q(\alpha, \alpha^*) = \frac{1}{\pi^2} \int d^2 z\, \mathrm{tr}\bigg[\rho e^{iza} \bigg(\frac{1}{\pi} \int d^2\lambda\, |\lambda\rangle\langle\lambda|\bigg) e^{iz^* a^\dagger}\bigg] e^{-iz^* \alpha^*} e^{-iz\alpha}$$
$$= \frac{1}{\pi^3} \int d^2 z \int d^2\lambda\, \langle\lambda| e^{iz^* a^\dagger} \rho e^{iza} |\lambda\rangle e^{-iz^* \alpha^*} e^{-iz\alpha}$$
$$= \frac{1}{\pi} \int d^2\lambda\, \langle\lambda|\rho|\lambda\rangle \bigg[\frac{1}{\pi^2} \int d^2 z\, e^{iz^*(\lambda^* - \alpha^*)} e^{iz(\lambda - \alpha)}\bigg]$$

$$= \frac{1}{\pi} \int d^2\lambda \, \langle\lambda|\rho|\lambda\rangle \delta^{(2)}(\lambda - \alpha)$$

$$= \frac{1}{\pi} \langle\alpha|\rho|\alpha\rangle. \tag{4.6}$$

Thus, $\pi Q(\alpha, \alpha^*)$ is the diagonal matrix element of the density operator taken with respect to the coherent state $|\alpha\rangle$. It is therefore strictly a probability – the probability for observing the coherent state $|\alpha\rangle$. This immediately gives us the relationship between Q and P.

From (3.21) and (4.6),

$$Q(\alpha, \alpha^*) = \frac{1}{\pi} \int d^2\lambda \, e^{-|\lambda - \alpha|^2} P(\lambda, \lambda^*). \tag{4.7}$$

Note 4.1 It can be shown that the diagonal matrix elements $\langle\alpha|\rho|\alpha\rangle$ specify the density operator completely. Then the convolution (4.7) forms the basis of formal proofs that every density operator may be given a diagonal representation if P is allowed to be a generalized function. See [4.7] and [4.8] for the details.

Another useful result is the relationship between the characteristic functions $\chi_A(z, z^*)$ and $\chi_N(z, z^*)$. We will make use of this shortly to derive the Fokker–Planck equation for the damped harmonic oscillator in the Q representation. The relationship follows from a special case of the Baker-Hausdorff theorem [4.9]: If $\hat{O}_1$ and $\hat{O}_2$ are two noncommuting operators that both commute with their commutator, then

$$e^{\hat{O}_1 + \hat{O}_2} = e^{\hat{O}_1} e^{\hat{O}_2} e^{-\frac{1}{2}[\hat{O}_1, \hat{O}_2]} = e^{\hat{O}_2} e^{\hat{O}_1} e^{\frac{1}{2}[\hat{O}_1, \hat{O}_2]}. \tag{4.8}$$

Since the commutator of a and $a^\dagger$ is a constant, this result can clearly be applied to the exponentials in the definitions of $\chi_N(z, z^*)$ and $\chi_A(z, z^*)$. It follows from (3.70) and (4.1) that

$$\begin{aligned}
\chi_A(z, z^*) &\equiv \text{tr}\left(\rho e^{iza} e^{iz^* a^\dagger}\right) \\
&= \text{tr}\left(\rho e^{iza + iz^* a^\dagger}\right) e^{-\frac{1}{2}|z|^2} \\
&= \text{tr}\left(\rho e^{iz^* a^\dagger} e^{iza}\right) e^{-|z|^2} \\
&= e^{-|z|^2} \chi_N(z, z^*). \tag{4.9}
\end{aligned}$$

Exercise 4.1 Use (4.9) to derive (4.7) directly from the definitions of the Q and P distributions [Eqs. (4.3) and (3.72)]. Also, use both (3.40) and (3.41) to show that (4.7) gives the correct Q distribution for the Fock state $|l\rangle$ – namely;

$$Q(\alpha, \alpha^*) = \frac{1}{\pi} |\langle \alpha | l \rangle|^2$$

$$= \frac{1}{\pi} e^{-|\alpha|^2} \frac{|\alpha|^{2l}}{l!}. \tag{4.10}$$

An alternative relationship between the Q and P distributions follows from (4.9). Using (4.3) and (4.9),

$$Q(\alpha, \alpha^*) = \frac{1}{\pi^2} \int d^2 z \, \chi_A(z, z^*) e^{-iz^* \alpha^*} c^{-iz\alpha}$$

$$\doteq \frac{1}{\pi^2} \int d^2 z \, e^{-|z|^2} \chi_N(z, z^*) e^{-iz^* \alpha^*} e^{-iz\alpha}.$$

Then, writing $\chi_N(z, z^*)$ as the Fourier transform of $P(\lambda, \lambda^*)$, we have

$$Q(\alpha, \alpha^*)$$

$$= \frac{1}{\pi^2} \int d^2 z \, e^{-|z|^2} \int d^2 \lambda \, P(\lambda, \lambda^*) e^{iz^* \lambda^*} e^{iz\lambda} e^{-iz^* \alpha^*} e^{-iz\alpha}$$

$$= \frac{1}{\pi^2} \int d^2 z \int d^2 \lambda \, P(\lambda, \lambda^*) \left[\exp\left(\frac{\partial^2}{\partial \lambda \partial \lambda^*}\right) e^{iz^* \lambda^*} e^{iz\lambda} \right] e^{-iz^* \alpha^*} e^{-iz\alpha}$$

$$= \frac{1}{\pi^2} \int d^2 z \int d^2 \lambda \left[\exp\left(\frac{\partial^2}{\partial \lambda \partial \lambda^*}\right) P(\lambda, \lambda^*) \right] e^{iz^* (\lambda^* - \alpha^*)} e^{iz(\lambda - \alpha)},$$

where the last line follows after integrating by parts. The integral with respect to z gives a δ-function and we find

$$Q(\alpha, \alpha^*) = \exp\left(\frac{\partial^2}{\partial \alpha \partial \alpha^*}\right) P(\alpha, \alpha^*). \tag{4.11}$$

Note 4.2 If (4.11) is to hold for the coherent state $|\alpha_0\rangle$, (4.7) and (3.22) require that we prove the rather unlikely looking result

$$\exp\left(\frac{\partial^2}{\partial \alpha \partial \alpha^*}\right) \delta^{(2)}(\alpha - \alpha_0) = \frac{1}{\pi} e^{-|\alpha - \alpha_0|^2}.$$

In spite of its unlikely appearance, this result follows from the limit defining the δ-function [Eq. (3.33)] and

$$\exp\left(\frac{\partial^2}{\partial \alpha \partial \alpha^*}\right) \frac{n}{\pi} e^{-n|\alpha|^2} = \frac{1}{\pi} \frac{n}{1+n} e^{-n|\alpha|^2/(1+n)}. \tag{4.12}$$

Equation (4.12) can be proved using the identity (4.46):

$$\exp\left(\frac{\partial^2}{\partial\alpha\partial\alpha^*}\right)e^{-n|\alpha|^2} = e^{-n|\alpha|^2}\sum_{k=0}^{\infty}\frac{(n|\alpha|)^{2k}}{k!}\frac{1}{(1+n)^{k+1}}$$

$$= \frac{1}{1+n}e^{-n|\alpha|^2}e^{n^2|\alpha|^2/(1+n)}$$

$$= \frac{1}{1+n}e^{-n|\alpha|^2/(1+n)}.$$

4.1.2 The Damped Harmonic Oscillator in the Q Representation

A Fokker–Planck equation for the damped harmonic oscillator can be derived in the Q representation by following the same steps as in Sect. 3.2.2. A convenient shortcut is available, however; we can use the relationship (4.9) between $\chi_N(z,z^*)$ and $\chi_A(z,z^*)$ and the equation of motion (3.83) for χ_N to quickly arrive at the equation of motion for χ_A:

$$\frac{\partial\chi_A}{\partial t} = e^{-|z|^2}\frac{\partial\chi_N}{\partial t}$$

$$= e^{-|z|^2}\left[-\left(\frac{\gamma}{2}+i\omega_0\right)z\frac{\partial}{\partial z} - \left(\frac{\gamma}{2}-i\omega_0\right)z^*\frac{\partial}{\partial z^*} - \gamma\bar{n}zz^*\right]\chi_N$$

$$= \left[-\left(\frac{\gamma}{2}+i\omega_0\right)z\left(\frac{\partial}{\partial z}+z^*\right) - \left(\frac{\gamma}{2}-i\omega_0\right)z^*\left(\frac{\partial}{\partial z^*}+z\right)\right.$$

$$\left.- \gamma\bar{n}zz^*\right]e^{-|z|^2}\chi_N$$

$$= \left[-\left(\frac{\gamma}{2}+i\omega_0\right)z\frac{\partial}{\partial z} - \left(\frac{\gamma}{2}-i\omega_0\right)z^*\frac{\partial}{\partial z^*} - \gamma(\bar{n}+1)zz^*\right]\chi_A.$$

$$(4.13)$$

This is the same as the equation of motion for χ_N, except for the replacement $\bar{n}\to\bar{n}+1$. We can therefore write down the corresponding equation of motion for Q directly from (3.47):

$$\frac{\partial Q}{\partial t} = \left[\left(\frac{\gamma}{2}+i\omega_0\right)\frac{\partial}{\partial\alpha}\alpha + \left(\frac{\gamma}{2}-i\omega_0\right)\frac{\partial}{\partial\alpha^*}\alpha^* + \gamma(\bar{n}+1)\frac{\partial^2}{\partial\alpha\partial\alpha^*}\right]Q. \quad (4.14)$$

This is the *Fokker–Planck equation for the damped harmonic oscillator in the Q representation*.

We exploit the relationship between the Fokker–Planck equations in the P and Q representations further to solve (4.14). The Green function $Q(\alpha,\alpha^*,t|\alpha_0,\alpha_0^*,0)$, which has initial condition

$$Q(\alpha,\alpha^*,0|\alpha_0,\alpha_0^*,0) = \delta^{(2)}(\alpha-\alpha_0) \equiv \delta(x-x_0)\delta(y-y_0), \quad (4.15)$$

follows directly from (3.67) in the form

$$Q(\alpha, \alpha^*, t | \alpha_0, \alpha_0^*, 0)$$

$$= \frac{1}{\pi(\bar{n}+1)(1-e^{-\gamma t})} \exp\left[-\frac{|\alpha - \alpha_0 e^{-(\gamma/2)t} e^{-i\omega_0 t}|^2}{(\bar{n}+1)(1-e^{-\gamma t})}\right]. \tag{4.16}$$

It is important to realize that while the Green function in the P representation describes an oscillator that is initially in a coherent state – $P(\alpha, \alpha^*, t | \alpha_0, \alpha_0^*, 0) = P(\alpha, \alpha^*, t)_{\rho(0)=|\alpha_0\rangle\langle\alpha_0|}$ – the Green function in the Q representation does not describe an oscillator initially in a coherent state; a δ-function in the Q representation does not correspond to a coherent state. Indeed, (4.6) tells us that the Q distribution for an initial state $\rho(0) = |\alpha_0\rangle\langle\alpha_0|$ is

$$Q(\alpha, \alpha^*, 0)_{\rho(0)=|\alpha_0\rangle\langle\alpha_0|} = \frac{1}{\pi}\langle\alpha|(|\alpha_0\rangle\langle\alpha_0|)|\alpha\rangle$$

$$= \frac{1}{\pi}|\langle\alpha|\alpha_0\rangle|^2$$

$$= \frac{1}{\pi}e^{-|\alpha-\alpha_0|^2}, \tag{4.17}$$

where we have used (3.8). The time evolution of the Q distribution for this initial state is then calculated using

$$Q(\alpha, \alpha^*, t)_{\rho(0)=|\alpha_0\rangle\langle\alpha_0|}$$

$$= \int d^2\lambda \, Q(\alpha, \alpha^*, t | \lambda, \lambda^*, 0) Q(\lambda, \lambda^*, 0)_{\rho(0)=|\alpha_0\rangle\langle\alpha_0|}. \tag{4.18}$$

Substituting (4.16) and (4.17) into (4.18), and making the change of variable $\lambda e^{-(\gamma/2)t} e^{-i\omega_0 t} \to \lambda$, we have

$$Q(\alpha, \alpha^*, t)_{\rho(0)=|\alpha_0\rangle\langle\alpha_0|}$$

$$= \int d^2\lambda \left\{\frac{1}{\pi(\bar{n}+1)(1-e^{-\gamma t})} \exp\left[-\frac{|\alpha - \lambda e^{-(\gamma/2)t} e^{-i\omega_0 t}|^2}{(\bar{n}+1)(1-e^{-\gamma t})}\right]\right\}$$

$$\times \left\{\frac{1}{\pi}\exp\left[-|\lambda - \alpha_0|^2\right]\right\}$$

$$= \int d^2\lambda \left\{\frac{1}{\pi(\bar{n}+1)(1-e^{-\gamma t})} \exp\left[-\frac{|\alpha - \lambda|^2}{(\bar{n}+1)(1-e^{-\gamma t})}\right]\right\}$$

$$\times \left\{\frac{e^{\gamma t}}{\pi}\exp\left[-|\lambda - \alpha_0 e^{-(\gamma/2)t} e^{-i\omega_0 t}|^2 e^{\gamma t}\right]\right\}. \tag{4.19}$$

This integral is a two-dimensional convolution; therefore, the Fourier transform of the left-hand side is given by the product of the Fourier transforms of the bracketed terms in the integrand; of course, the Fourier transform of the left-hand side is the characteristic function $\chi_A(z, z^*, t)_{\rho(0)=|\alpha_0\rangle\langle\alpha_0|}$. Thus,

$$\chi_A(z, z^*, t)_{\rho(0)=|\alpha_0\rangle\langle\alpha_0|} = \exp\left[-|z|^2(\bar{n}+1)(1-e^{-\gamma t})\right]$$
$$\times \left\{\exp\left[-|z|^2 e^{-\gamma t}\right]e^{2iz^*\alpha_0^*(t)}e^{2iz\alpha_0(t)}\right\}, \quad (4.20)$$

with $\alpha_0(t) \equiv \alpha_0 e^{-(\gamma/2)t}e^{-i\omega_0 t}$. The inverse transform gives the Q *distribution for a damped coherent state*:

$$Q(\alpha, \alpha^*, t)_{\rho(0)=|\alpha_0\rangle\langle\alpha_0|}$$
$$= \frac{1}{\pi[1+\bar{n}(1-e^{-\gamma t})]}\exp\left[-\frac{|\alpha - \alpha_0 e^{-(\gamma/2)t}e^{-i\omega_0 t}|^2}{1+\bar{n}(1-e^{-\gamma t})}\right]. \quad (4.21)$$

Compared with the solution for the P distribution [Eq. (3.67)], the solution (4.21) for the Q distribution shows one simple difference – the phase-independent variance [variance of $x \equiv \mathrm{Re}(\alpha)$ or $y \equiv \mathrm{Im}(\alpha)$] is now $(\bar{n}/2)(1-e^{-\gamma t})+1/2$ rather than $(\bar{n}/2)(1-e^{-\gamma t})$. Thus, the time evolution of the Q distribution can be represented as in Fig. 3.1, but with a circular contour of somewhat larger radius; in particular, the Q distribution has a width at $t = 0$ given by the initial condition (4.17), whereas the P distribution begins as a δ-function; when $\bar{n} = 0$, this initial width is preserved for all times. We find then that the Q distribution has a width even in the absence of thermal fluctuations. We have again set up a correspondence with a classical statistical process; but now there is noise where before there was none. What can this mean? The answer to this question illustrates an important point about the fluctuations at the "classical" end of the quantum–classical correspondence. Although thermal fluctuations from the reservoir are not too quantum mechanical – they should be present in a classical theory of damping also – in general, the fluctuations observed in the distributions derived via the quantum–classical correspondence have a quantum-mechanical origin. They are manifestations of the probabilistic character of quantum mechanics, and arise through the noncommutation of the quantum-mechanical operators. Therefore, the fluctuations that appear in the classical stochastic processes that correspond to a quantum-mechanical system via different operator orderings are different. In our present example, the difference in the variances of the P distribution and the Q distribution arises to preserve the boson commutation relation. From (3.74) and (3.67), we calculate

$$\langle(a^\dagger a)(t)\rangle - \langle a^\dagger(t)\rangle\langle a(t)\rangle = \overline{(\overline{(\alpha^*\alpha)(t)})}_P - \overline{(\alpha^*(t))}_P\overline{(\alpha(t))}_P$$
$$= \bar{n}(1-e^{-\gamma t}), \quad (4.22a)$$

while from (4.5) and (4.21) we calculate

$$\langle(aa^\dagger)(t)\rangle - \langle a^\dagger(t)\rangle\langle a(t)\rangle = \overline{(\overline{(\alpha^*\alpha)(t)})}_Q - \overline{(\alpha^*(t))}_Q\overline{(\alpha(t))}_Q$$
$$= \bar{n}(1-e^{-\gamma t}) + 1. \quad (4.22b)$$

The extra fluctuations in the Q representation, which give the "+1" in (4.22b), are just what are needed to preserve the expectation of the commutator – $\langle[a, a^\dagger](t)\rangle = 1$.

4.1.3 Antinormal-Ordered Averages Using the P Representation

We should not be misled into thinking that the P and Q distributions are inadequate on their own for calculating operator averages in arbitrary order. Of course, an average in antinormal order can first be normal ordered so that moments of the P distribution can be used to calculate the average of the resulting normal-ordered object. Antinormal-ordered averages can also be evaluated, however, directly from the P distribution, without first reordering the operators. Consider (4.2) with $\chi_A(z, z^*)$ written in terms of $\chi_N(z, z^*)$ using (4.9). An arbitrary antinormal-average can be calculated from the relationship

$$
\begin{aligned}
\langle a^q a^{\dagger p} \rangle &= \frac{\partial^{p+q}}{\partial(iz^*)^p \partial(iz)^q} e^{-|z|^2} \chi_N(z, z^*) \Big|_{z=z^*=0} \\
&= \frac{\partial^p}{\partial(iz^*)^p} e^{-|z|^2} \left(iz^* + \frac{\partial}{\partial(iz)} \right)^q \chi_N(z, z^*) \Big|_{z=z^*=0} \\
&= e^{-|z|^2} \left(iz + \frac{\partial}{\partial(iz^*)} \right)^p \left(iz^* + \frac{\partial}{\partial(iz)} \right)^q \chi_N(z, z^*) \Big|_{z=z^*=0} \\
&= \frac{\partial^p}{\partial(iz^*)^p} \left(iz^* + \frac{\partial}{\partial(iz)} \right)^q \chi_N(z, z^*) \Big|_{z=z^*=0}.
\end{aligned}
$$

Substituting for $\chi_N(z, z^*)$ from (3.73), we have

$$
\begin{aligned}
\langle a^q a^{\dagger p} \rangle &= \int d^2\alpha \, P(\alpha, \alpha^*) \frac{\partial^p}{\partial(iz^*)^p} \left(iz^* + \frac{\partial}{\partial(iz)} \right)^q e^{iz^*\alpha^*} e^{iz\alpha} \Big|_{z=z^*=0} \\
&= \int d^2\alpha \, P(\alpha, \alpha^*) \left(\frac{\partial}{\partial\alpha^*} + \alpha \right)^q \alpha^{*p} e^{iz^*\alpha^*} e^{iz\alpha} \Big|_{z=z^*=0}.
\end{aligned}
$$

We now integrate by parts, setting $P(z, z^*)$ and its derivatives to zero at infinity, to arrive at the result

$$
\langle a^q a^{\dagger p} \rangle = \int d^2\alpha \, \alpha^{*p} \left(\alpha - \frac{\partial}{\partial\alpha^*} \right)^q P(\alpha, \alpha^*). \tag{4.23a}
$$

Exercise 4.2 Prove also that

$$
\langle a^q a^{\dagger p} \rangle = \int d^2\alpha \, \alpha^q \left(\alpha^* - \frac{\partial}{\partial\alpha} \right)^p P(\alpha, \alpha^*), \tag{4.23b}
$$

and

$$
\langle a^{\dagger p} a^q \rangle = \int d^2\alpha \, \alpha^{*p} \left(\alpha + \frac{\partial}{\partial\alpha^*} \right)^q Q(\alpha, \alpha^*), \tag{4.24a}
$$

$$
\langle a^{\dagger p} a^q \rangle = \int d^2\alpha \, \alpha^q \left(\alpha^* + \frac{\partial}{\partial\alpha} \right)^p Q(\alpha, \alpha^*). \tag{4.24b}
$$

As an illustration, let us calculate $\langle (aa^\dagger)(t) \rangle$ for the damped harmonic oscillator using (4.23a) and the Green function solution for the P distribution [Eq. (3.67)]. We set $\alpha_0(t) equiv e^{-(\gamma/2)t} e^{-i\omega_0 t}$ and then

$$\langle (aa^\dagger)(t) \rangle$$

$$= \int d^2\alpha \, \alpha^* \left(\alpha - \frac{\partial}{\partial \alpha^*} \right) \left\{ \frac{1}{\pi \bar{n}(1 - e^{-\gamma t})} \exp\left[-\frac{|\alpha - \alpha_0(t)|^2}{\bar{n}(1 - e^{-\gamma t})} \right] \right\}$$

$$= \int d^2\alpha \, \alpha^* \left[\alpha + \frac{\alpha - \alpha_0(t)}{\bar{n}(1 - e^{-\gamma t})} \right] \left\{ \frac{1}{\pi \bar{n}(1 - e^{-\gamma t})} \exp\left[-\frac{|\alpha - \alpha_0(t)|^2}{\bar{n}(1 - e^{-\gamma t})} \right] \right\}$$

$$= \int d^2\alpha \left\{ \alpha^* [\alpha - \alpha_0(t)] \left[1 + \frac{1}{\bar{n}(1 - e^{-\gamma t})} \right] + \alpha^* \alpha_0(t) \right\}$$

$$\times \left\{ \frac{1}{\pi \bar{n}(1 - e^{-\gamma t})} \exp\left[-\frac{|\alpha - \alpha_0(t)|^2}{\bar{n}(1 - e^{-\gamma t})} \right] \right\}.$$

If A is a constant,

$$\int d^2\alpha \, \alpha \frac{A}{\pi} \exp\left[-A|\alpha - \alpha_0(t)|^2 \right] = \alpha_0(t),$$

$$\int d^2\alpha \, |\alpha - \alpha_0(t)|^2 \frac{A}{\pi} \exp\left[-A|\alpha - \alpha_0(t)|^2 \right] = \frac{1}{A}.$$

We can therefore replace α^* by $\alpha^* - \alpha_0^*(t)$ in the first term in the integrand (this adds zero to the integral) and perform the resulting integrals to obtain

$$\langle (aa^\dagger)(t) \rangle = \int d^2\alpha \left[|\alpha - \alpha_0(t)|^2 \frac{\bar{n}(1 - e^{-\gamma t}) + 1}{\bar{n}(1 - e^{-\gamma t})} + \alpha^* \alpha_0(t) \right]$$

$$\times \left\{ \frac{1}{\pi \bar{n}(1 - e^{-\gamma t})} \exp\left[-\frac{|\alpha - \alpha_0(t)|^2}{\bar{n}(1 - e^{-\gamma t})} \right] \right\}$$

$$= \bar{n}(1 - e^{-\gamma t}) + 1 + |\alpha_0(t)|^2$$

$$= \langle (a^\dagger a)(t) \rangle + 1,$$

where the last line follows from (3.68) and (3.69). We have arrived at the result that would be obtained by first writing $aa^\dagger$ in normal order and then using moments of the P distribution to evaluate the normal-ordered operator average.

4.1.4 The Wigner Representation

The Wigner representation is introduced by defining a third characteristic function:

$$\chi_S(z, z^*) \equiv \mathrm{tr}\!\left(\rho e^{iz^* a^\dagger + iza}\right). \tag{4.25}$$

The Wigner distribution $W(\alpha, \alpha^*)$ is the Fourier transform of $\chi_S(z, z^*)$:

$$
\begin{aligned}
W(\alpha, \alpha^*) &\equiv \frac{1}{\pi^2} \int d^2 z \, \chi_S(z, z^*) e^{-iz^* \alpha^*} e^{-iz\alpha} \\
&\equiv \frac{1}{\pi^2} \int_{-\infty}^{\infty} d\mu \int_{-\infty}^{\infty} d\nu \, \chi_S(\mu + i\nu, \mu - i\nu) e^{-2i(\mu x - \nu y)},
\end{aligned} \tag{4.26}
$$

with the inverse relationship

$$
\begin{aligned}
\chi_S(z, z^*) &= \int d^2 \alpha \, W(\alpha, \alpha^*) e^{iz^* \alpha^*} e^{iz\alpha} \\
&= \int_{-\infty}^{\infty} dx \int_{-\infty}^{\infty} dy \, W(x + iy, x - iy) e^{2i(\mu x - \nu y)}.
\end{aligned} \tag{4.27}
$$

The relationship between the Wigner distribution and operator averages is a little more complicated than the relationships that connect the P and Q distributions with operator averages. In terms of position and momentum variables (proportional to x and y respectively) the moments of $W(\alpha, \alpha^*)$ give the averages of operators written in Weyl order [4.10]. Details can be found in the review by Hillery *et al.* [4.6]. The relevant quantities for quantum optics are operator averages corresponding to moments of the complex variables α and α^*. These can be found as follows. The exponential in (4.25) has the expansion

$$
\begin{aligned}
e^{iz^* a^\dagger + iza} &= \sum_{m=0}^{\infty} \frac{1}{m!} (iz^* a^\dagger + iza)^m \\
&= \sum_{m=0}^{\infty} \frac{1}{m!} \sum_{n=0}^{m} \frac{m!}{n!(m-n)!} (iz^*)^n (iz)^{m-n} \left(a^{\dagger n} a^{m-n}\right)_S \\
&= \sum_{n=0}^{\infty} \sum_{m=n}^{\infty} \frac{(iz^*)^n (iz)^{m-n}}{n!(m-n)!} \left(a^{\dagger n} a^{m-n}\right)_S \\
&= \sum_{n=0}^{\infty} \sum_{m=0}^{\infty} \frac{(iz^*)^n (iz)^m}{n!m!} \left(a^{\dagger n} a^m\right)_S,
\end{aligned} \tag{4.28}
$$

where $\left(a^{\dagger n} a^m\right)_S$ denotes the operator product written in symmetric order – the average of $(n+m)!/(n!m!)$ possible orderings of n creation operators and m annihilation operators:

$$(a^\dagger a)_S \equiv \tfrac{1}{2}(a^\dagger a + a a^\dagger), \tag{4.29a}$$

$$(a^{\dagger 2} a)_S \equiv \tfrac{1}{3}(a^{\dagger 2} a + a^\dagger a a^\dagger + a a^{\dagger 2}), \tag{4.29b}$$

$$(a^\dagger a^2)_S \equiv \tfrac{1}{3}(a^\dagger a^2 + a a^\dagger a + a^2 a^\dagger), \tag{4.29c}$$

$$(a^{\dagger 2} a^2)_S \equiv \tfrac{1}{6}(a^{\dagger 2} a^2 + a^\dagger a a^\dagger a + a^\dagger a^2 a^\dagger + a a^{\dagger 2} a + a a^\dagger a a^\dagger + a^2 a^{\dagger 2}), \tag{4.29d}$$

$$\vdots$$

Then, from (4.28) and the definition of $\chi_S(z, z^*)$ [Eq. (4.25)], *symmetric-ordered operator averages* are given by

$$
\begin{aligned}
\langle (a^{\dagger p} a^q)_S \rangle &\equiv \operatorname{tr}\big[\rho(a^{\dagger p} a^q)_S\big] \\
&= \frac{\partial^{p+q}}{\partial(iz^*)^p \partial(iz)^q} \chi_S(z^*, z)\Big|_{z=z^*=0} ;
\end{aligned}
\tag{4.30}
$$

substituting for $\chi_S(z, z^*)$ in terms of $W(\alpha, \alpha^*)$ [Eq. (4.27)] gives

$$
\begin{aligned}
\langle (a^{\dagger p} a^q)_S \rangle &= \frac{\partial^{p+q}}{\partial(iz^*)^p \partial(iz)^q} \int d^2\alpha\, W(\alpha, \alpha^*) e^{iz^*\alpha^*} e^{iz\alpha}\Big|_{z=z^*=0} \\
&= \big(\overline{\alpha^{*p} \alpha^q}\big)_W,
\end{aligned}
\tag{4.31a}
$$

with

$$\big(\overline{\alpha^{*p} \alpha^q}\big)_W \equiv \int d^2\alpha\, W(\alpha, \alpha^*)\alpha^{*p}\alpha^q. \tag{4.31b}$$

Note 4.3 We have defined the Wigner distribution $W(\alpha, \alpha^*)$ to be normalized such that $\int d^2\alpha\, W(\alpha, \alpha^*) = 1$. The Wigner distribution is often defined with a different normalization, such that $\int d^2\alpha\, W(\alpha, \alpha^*) = \pi$. This is the case in [4.4] and [4.6]. With the alternative definition $W(\alpha, \alpha^*)$ is the classical function associated with the density operator ρ by writing it as a power series in symmetric-ordered operators $(a^{\dagger p} a^q)_S$ and replacing each term in this series by $\alpha^{*p}\alpha^q$ (see Sect. 4.3.1).

The quantum–classical correspondence defined in terms of symmetric-ordered operators (also antinormal-ordered operators) is not really the most convenient for applications in quantum optics because it is normal-ordered averages that relate directly to quantities measured with detectors that absorb photons. However, often only low-order moments are of interest and the symmetric ordering is then easily untangled using (4.29a)–(4.29d). More generally, a symmetric-ordered operator can be written in normal order in the following way. With the help of the Baker-Hausdorff theorem [Eq. (4.8)] we write

$$\left(a^{\dagger p} a^q\right)_S = \left.\frac{\partial^{p+q}}{\partial(iz^*)^p \partial(iz)^q} e^{iz^* a^\dagger + iza}\right|_{z=z^*=0}$$

$$= \left.\frac{\partial^{p+q}}{\partial(iz^*)^p (iz)^q} e^{-\frac{1}{2}|z|^2} e^{iz^* a^\dagger} e^{iza}\right|_{z=z^*=0} .$$

It can then be proved by induction that

$$\frac{\partial^{p+q}}{\partial(iz^*)^p \partial(iz)^q} e^{-\frac{1}{2}|z|^2} e^{iz^* a^\dagger} e^{iza}$$

$$= \sum_{k=0}^{\min(p,q)} \frac{1}{2^k} \frac{1}{k!} \frac{p!}{(p-k)!} \frac{q!}{(q-k)!}\left(a^\dagger + \frac{1}{2}iz\right)^{p-k}$$

$$\times e^{-\frac{1}{2}|z|^2} e^{iz^* a^\dagger} e^{iza}\left(a + \frac{1}{2}iz^*\right)^{q-k} , \qquad (4.32)$$

and hence, that

$$\left(a^{\dagger p} a^q\right)_S = \sum_{k=0}^{\min(p,q)} \frac{1}{2^k} \frac{p!}{(p-k)!} \frac{q!}{(q-k)!} a^{\dagger p-k} a^{q-k}. \qquad (4.33)$$

The Baker-Hausdorff theorem also yields the relationship between the characteristic functions $\chi_S(z, z^*)$ and $\chi_N(z, z^*)$, and $\chi_S(z, z^*)$ and $\chi_A(z, z^*)$:

$$\chi_S(z, z^*) \equiv \text{tr}\left(\rho e^{iz^* a^\dagger + iza}\right) = \text{tr}\left(\rho e^{iz^* a^\dagger} e^{iza}\right) e^{-\frac{1}{2}|z|^2} = e^{-\frac{1}{2}|z|^2} \chi_N(z, z^*),$$
$$(4.34a)$$

$$\chi_S(z, z^*) \equiv \text{tr}\left(\rho e^{iz^* a^\dagger + iza}\right) = \text{tr}\left(\rho e^{iza} e^{iz^* a^\dagger}\right) e^{\frac{1}{2}|z|^2} = e^{\frac{1}{2}|z|^2} \chi_A(z, z^*).$$
$$(4.34b)$$

From these results relationships between the distributions $W(\alpha, \alpha^*)$ and $P(\alpha, \alpha^*)$, and $W(\alpha, \alpha^*)$ and $Q(\alpha, \alpha^*)$, analogous to those given in (4.7) and (4.11), can be obtained. The derivations are left as an exercise:

Exercise 4.3 Show that

$$W(\alpha, \alpha^*) = \frac{2}{\pi} \int d^2\lambda \; e^{-2|\lambda - \alpha|^2} P(\lambda, \lambda^*), \qquad (4.35a)$$

$$Q(\alpha, \alpha^*) = \frac{2}{\pi} \int d^2\lambda \; e^{-2|\lambda - \alpha|^2} W(\lambda, \lambda^*), \qquad (4.35b)$$

and that

$$W(\alpha, \alpha^*) = \exp\left(\frac{1}{2} \frac{\partial^2}{\partial\alpha\partial\alpha^*}\right) P(\alpha, \alpha^*), \qquad (4.36a)$$

$$Q(\alpha, \alpha^*) = \exp\left(\frac{1}{2} \frac{\partial^2}{\partial\alpha\partial\alpha^*}\right) W(\alpha, \alpha^*). \qquad (4.36b)$$

From the relationships (4.7) and (4.35), (4.9) and (4.34), and (4.11) and (4.6), the Wigner distribution appears to fall in some sense in between the P and Q distributions. This observation is illustrated explicitly by the example of the damped harmonic oscillator. There is no need for a new calculation to treat this example in the Wigner representation. From a comparison of (4.9) and (4.34a), we immediately conclude that the method of Sect. 4.1.2 will bring us to the following *Fokker–Planck equation for the damped harmonic oscillator in the Wigner representation*:

$$\frac{\partial W}{\partial t} = \left[\left(\frac{\gamma}{2} + i\omega_0\right)\frac{\partial}{\partial \alpha}\alpha + \left(\frac{\gamma}{2} - i\omega_0\right)\frac{\partial}{\partial \alpha^*}\alpha^* + \gamma(\bar{n} + \tfrac{1}{2})\frac{\partial^2}{\partial\alpha\partial\alpha^*}\right]W.$$

(4.37)

Thus, where $\bar{n}$ appears in the Fokker–Planck equation in the P representation [Eq. (3.47)], and $\bar{n} + 1$ appears in the Fokker–Planck equation in the Q representation [Eq. (4.14)], now $\bar{n} + \tfrac{1}{2}$ appears in the Fokker–Planck equation in the Wigner representation. The factor of $\tfrac{1}{2}$ carries over into the solution for a damped coherent state. By referring to (3.67) and (4.16) we see that the Green function $W(\alpha, \alpha^*, t|\alpha_0, \alpha_0^*, 0)$, which has initial condition

$$W(\alpha, \alpha^*, 0|\alpha_0, \alpha_0^*, 0) = \delta^{(2)}(\alpha - \alpha_0) \equiv \delta(x - x_0)\delta(y - y_0),$$
(4.38)

is given by

$$W(\alpha, \alpha^*, t|\alpha_0, \alpha_0^*, 0)$$
$$= \frac{1}{\pi(\bar{n} + \tfrac{1}{2})(1 - e^{-\gamma t})}\exp\left[-\frac{|\alpha - \alpha_0 e^{-(\gamma/2)t}e^{-i\omega_0 t}|^2}{(\bar{n} + \tfrac{1}{2})(1 - e^{-\gamma t})}\right].$$
(4.39)

Then, using (4.35a) and the P distribution for a coherent state [Eq. (3.22)], an initial coherent state $(\rho(0) = |\alpha_0\rangle\langle\alpha_0|)$ is represented by the distribution

$$W(\alpha, \alpha^*, 0)_{\rho(0)=|\alpha_0\rangle\langle\alpha_0|} = \frac{2}{\pi}e^{-2|\alpha-\alpha_0|^2}.$$
(4.40)

By following the steps used to derive (4.21) we find that the *Wigner distribution for a damped coherent state* is given by

$$W(\alpha, \alpha^*, t)_{\rho(0)=|\alpha_0\rangle\langle\alpha_0|}$$
$$= \frac{1}{\pi\left[\tfrac{1}{2} + \bar{n}(1 - e^{-\gamma t})\right]}\exp\left[-\frac{|\alpha - \alpha_0 e^{-(\gamma/2)t}e^{-i\omega_0 t}|^2}{\tfrac{1}{2} + \bar{n}(1 - e^{-\gamma t})}\right].$$
(4.41)

We have now constructed a third correspondence with a classical statistical process. Here the phase-independent variance lies in between those given by the solutions (3.67) and (4.21); the picture of Fig. 3.1 still applies, but now with a circular contour of radius $\sqrt{1/2 + \bar{n}(1 - e^{-\gamma t})}$ representing the distribution. As we observed for the Q distribution, the quantum fluctuations added over and above those coming from the reservoir are required by the commutation relations and the ordering convention underlying the representation. From (4.29a), (4.31), and (4.41), we have

$$\frac{1}{2}\left[\langle(a^\dagger a)(t)\rangle + \langle(aa^\dagger)(t)\rangle\right] - \langle a^\dagger(t)\rangle\langle a(t)\rangle$$

$$= \langle(a^\dagger a)_S(t)\rangle - \langle(a^\dagger)_S(t)\rangle\langle(a)_S(t)\rangle$$

$$= \left(\overline{(\alpha^* \alpha)(t)}\right)_W - \left(\overline{\alpha^*(t)}\right)_W\left(\overline{\alpha(t)}\right)_W$$

$$= \bar{n}(1 - e^{-\gamma t}) + \tfrac{1}{2}. \tag{4.42}$$

This is the average of the expressions in (4.22a) and (4.22b). The factor "$+\frac{1}{2}$" is the contribution obtained from the boson commutation relation by normal ordering the operator $(a^\dagger a)_S = \frac{1}{2}(a^\dagger a + aa^\dagger)$.

4.2 Fun with Fock States

We have followed the treatment of the damped harmonic oscillator prepared in a coherent state throughout our discussions of the P, Q, and Wigner representations. For this example, each of the three distributions has all the properties of a probability distribution, and we can therefore associate the quantum-mechanical problem with each of three classical statistical descriptions. We should remember, however, that the distributions obtained from the quantum–classical correspondence are not guaranteed to have all the properties of a probability distribution. We have already seen in Sect. 3.1.3 that the P distribution for a Fock state is a generalized function, involving derivatives of the δ-function. We now explore the representation of Fock states a little further.

4.2.1 Wigner Distribution for a Fock State

Let us derive the Wigner distribution for the Fock state $|l\rangle$ using (4.35a) and the form of the P distribution given in (3.40). We have

$$W(\alpha, \alpha^*) = \frac{2}{\pi}\int d^2\lambda\, e^{-2|\lambda - \alpha|^2}\frac{1}{l!}e^{|\lambda|^2}\frac{\partial^{2l}}{\partial\lambda^l\partial\lambda^{*l}}\delta^{(2)}(\lambda)$$

$$= \frac{2}{\pi}\frac{1}{l!}\frac{\partial^{2l}}{\partial\lambda^l\partial\lambda^{*l}}e^{-2|\lambda - \alpha|^2}e^{|\lambda|^2}\bigg|_{\lambda=\lambda^*=0}$$

$$= \frac{2}{\pi}\frac{1}{l!}e^{-2|\alpha|^2}\frac{\partial^{2l}}{\partial\lambda^l\partial\lambda^{*l}}e^{-2|\lambda|^2}e^{2\lambda\alpha^*}e^{2\lambda^*\alpha}\bigg|_{\lambda=\lambda^*=0}. \tag{4.43}$$

To evaluate the right-hand side of (4.43) we consider the more general expression (for any complex constants A, B, and C)

$$\frac{\partial^{2l}}{\partial \lambda^l \partial \lambda^{*l}} e^{-A|\lambda|^2} e^{B\lambda} e^{C\lambda^*} = \frac{\partial^l}{\partial \lambda^l} e^{B\lambda} \frac{\partial^l}{\partial \lambda^{*l}} e^{C\lambda^*} e^{-A|\lambda|^2}$$

$$= e^{B\lambda} e^{C\lambda^*} \left(B + \frac{\partial}{\partial \lambda} \right)^l \left(C + \frac{\partial}{\partial \lambda^*} \right)^l e^{-A|\lambda|^2}$$

$$= e^{B\lambda} e^{C\lambda^*} \left(B + \frac{\partial}{\partial \lambda} \right)^l (C - A\lambda)^l e^{-A|\lambda|^2} . \quad (4.44)$$

For $n \le l$, it can be proved by induction that

$$\left(B + \frac{\partial}{\partial \lambda} \right)^l (C - A\lambda)^l$$

$$= \left(B + \frac{\partial}{\partial \lambda} \right)^{l-n} \sum_{k=0}^{n} \frac{n!}{k!(n-k)!} \frac{l!}{(l-k)!} A^k (C - A\lambda)^{l-k} \left(B + \frac{\partial}{\partial \lambda} \right)^{n-k} .$$

$$(4.45)$$

Using this result, with $n = l$, we obtain

$$\frac{\partial^{2l}}{\partial \lambda^l \partial \lambda^{*l}} e^{-A|\lambda|^2} e^{B\lambda} e^{C\lambda^*}$$

$$= e^{B\lambda} e^{C\lambda^*} \sum_{k=0}^{l} (-1)^k \frac{l!}{k!(l-k)!} \frac{l!}{(l-k)!} A^k (C - A\lambda)^{l-k}$$

$$\times \left(B + \frac{\partial}{\partial \lambda} \right)^{l-k} e^{-A|\lambda|^2}$$

$$= e^{-A|\lambda|^2} e^{B\lambda} e^{C\lambda^*} \sum_{k=0}^{l} (-1)^{l-k} \frac{l!}{k!(l-k)!} \frac{l!}{k!} A^{l-k} (B - A\lambda^*)^k (C - A\lambda)^k ,$$

$$(4.46)$$

where in the last line we have changed the summation index, with $l - k \to k$. The right-hand side of (4.43) may now be evaluated using (4.46): setting $A = 1$ and $B^* = C = 2\alpha$, the Wigner distribution for the Fock state $|l\rangle$ is given by

$$W(\alpha, \alpha^*) = \frac{2}{\pi} \frac{1}{l!} e^{-2|\alpha|^2} \sum_{k=0}^{l} (-1)^{l-k} \frac{l!}{k!(l-k)!} \frac{l!}{k!} |2\alpha|^{2k} . \quad (4.47)$$

The distribution (4.47) is an ordinary, well-behaved, function. Nevertheless, it can clearly violate one of the conditions required of a probability distribution – it need not be positive. The one-photon Fock state illustrates this point; for $l = 1$,

$$W(\alpha, \alpha^*) = \frac{2}{\pi} e^{-2|\alpha|^2}(4|\alpha|^2 - 1), \tag{4.48}$$

which is negative for $|\alpha| < \frac{1}{2}$.

Note 4.4 It can be shown that $\chi_S(z, z^*)$ is square integrable and, hence, that its Fourier transform $W(\alpha, \alpha^*)$ is *always* a well-behaved function; there is no need for generalized functions in the Wigner representation. To prove this result we use (4.34) and (4.1) to write

$$\frac{1}{\pi}\int d^2z\,|\chi_S(z, z^*)|^2 = \frac{1}{\pi}\int d^2z\,\chi_N(z, z^*)\chi_A(z, z^*)^*$$

$$= \frac{1}{\pi}\mathrm{tr}\left[\int d^2z\,\chi_N(z, z^*)e^{-iz^*a^\dagger}\rho e^{-iza}\right].$$

Then, introducing the identity in the form (3.9) and using the cyclic property of the trace, and the relationship between $\chi_N(z, z^*)$ and $P(\alpha, \alpha^*)$ [Eq. (3.72)], we find

$$\frac{1}{\pi}\int d^2z\,|\chi_S(z, z^*)|^2 = \frac{1}{\pi^2}\mathrm{tr}\left[\int d^2\alpha\int d^2z\,\chi_N(z, z^*)\langle\alpha|e^{-iz^*a^\dagger}\rho e^{-iza}|\alpha\rangle\right]$$

$$= \frac{1}{\pi^2}\mathrm{tr}\left[\int d^2\alpha\,\langle\alpha|\rho|\alpha\rangle\int d^2z\,\chi_N(z, z^*)e^{-iz^*\alpha^*}e^{-iz\alpha}\right]$$

$$= \mathrm{tr}\left[\rho\int d^2\alpha\,|\alpha\rangle\langle\alpha|P(\alpha, \alpha^*)\right]$$

$$= \mathrm{tr}\,(\rho^2).$$

The last line follows from (3.15). The square integrability of $\chi_S(z, z^*)$ follows because $\mathrm{tr}\,(\rho^2) \leq 1$.

As a simple check on our result for the Fock state Wigner distribution, let us evaluate $(\overline{\alpha^*\alpha})_W$ and show that it gives the symmetric-ordered average

$$\tfrac{1}{2}\langle a^\dagger a + aa^\dagger\rangle = \tfrac{1}{2}\left(2\langle a^\dagger a\rangle + 1\right) = \tfrac{1}{2}(2l + 1). \tag{4.49}$$

From (4.47) we obtain

$$(\overline{\alpha^*\alpha})_W \equiv \int d^2\alpha\,W(\alpha, \alpha^*)\alpha^*\alpha$$

$$= \frac{2}{\pi}\frac{1}{l!}\sum_{k=0}^{l}(-1)^{l-k}\frac{l!}{k!(l-k)!}\frac{l!}{k!}\int d^2\alpha\,e^{-2|\alpha|^2}|2\alpha|^{2k}|\alpha|^2$$

$$= \frac{2}{\pi}\sum_{k=0}^{l}(-1)^{l-k}\frac{l!}{(k!)^2(l-k)!}2^{2k}\int_0^\infty dr\int_0^{2\pi} d\phi\,e^{-2r^2}r^{2(k+1)+1}$$

$$= \frac{2}{\pi}\sum_{k=0}^{l}(-1)^{l-k}\frac{l!}{(k!)^2(l-k)!}2^{2k}2\pi\frac{(k+1)!}{2^{k+3}}.$$

The integral over r has been executed by performing $k + 1$ integrations by parts. The summation on the right-hand side may now be split into two pieces by writing

$$\sum_{k=0}^{l} \cdots \frac{(k+1)!}{(k!)^2} = \sum_{k=1}^{l} \cdots \frac{1}{(k-1)!} + \sum_{k=0}^{l} \cdots \frac{1}{k!}.$$

Then, changing the first summation index, with $k - 1 \to k$, we arrive at the result

$$(\overline{\alpha^* \alpha})_W = \frac{1}{2} \left[2l \sum_{k=0}^{l-1} (-1)^{(l-1)-k} \frac{(l-1)!}{k![(l-1)-k]!} 2^k + \sum_{k=0}^{l} (-1)^{l-k} \frac{l!}{k!(l-k)!} 2^k \right]$$

$$= \tfrac{1}{2} \left[2l(2-1)^{l-1} + (2-1)^l \right]$$

$$= \tfrac{1}{2}(2l + 1).$$

Thus, we recover the symmetric-ordered operator average (4.49) for a Fock state.

4.2.2 Damped Fock State in the P Representation

Nothing in the derivation of the Fokker–Planck equation for the damped harmonic oscillator precludes its use in situations where the distribution is a generalized function, or takes negative values. We certainly lose the correspondence with a classical statistical description under such circumstances, but the mathematics works just fine. The Green function for the appropriate Fokker–Planck equation provides all we need to find the time evolution from an arbitrary initial state; we simply integrate the Green function against the representation for the initial state. This will work even if the initial state is represented by a distribution that is more singular than a δ-function. For an interesting illustration we will calculate the P distribution for a damped harmonic oscillator prepared in the Fock state $|l\rangle$. Recall that a Fock state is represented by a distribution involving derivatives of a two-dimensional δ-function.

The Green function solution to the Fokker–Planck equation in the P representation is given by (3.67). Using this result and the distribution for an initial Fock state [Eq. (3.40)], we have

$$P(\alpha, \alpha^*, t)_{\rho(0)=|l\rangle\langle l|}$$

$$= \int d^2\lambda\, P(\alpha, \alpha^*, t|\lambda, \lambda^*, 0) P(\lambda, \lambda^*, 0)_{\rho(0)=|l\rangle\langle l|}$$

$$= \int d^2\lambda\, \frac{1}{\pi\bar{n}(1 - e^{-\gamma t})} \exp\left[-\frac{|\alpha - \lambda e^{-\frac{\gamma}{2}t}e^{-i\omega_0 t}|^2}{\bar{n}(1 - e^{-\gamma t})}\right]$$

$$\times \frac{1}{l!}e^{|\lambda|^2} \frac{\partial^{2l}}{\partial\lambda^l \partial\lambda^{*l}}\delta(\lambda)$$

$$= \frac{1}{l!} \frac{\partial^{2l}}{\partial\lambda^l \partial\lambda^{*l}} \left\{ \frac{1}{\pi\bar{n}(1 - e^{-\gamma t})} \right.$$

$$\left. \times \exp\left[-\frac{|\alpha - \lambda e^{-\frac{\gamma}{2}t}e^{-i\omega_0 t}|^2}{\bar{n}(1 - e^{-\gamma t})}\right] e^{|\lambda|^2} \right\}\bigg|_{\lambda=\lambda^*=0},$$

where the integration is performed using (3.37). Expanding the function inside the curly bracket,

$$P(\alpha, \alpha^*, t)_{\rho(0)=|l\rangle\langle l|}$$

$$= \frac{1}{\pi\bar{n}(1 - e^{-\gamma t})} \exp\left[-\frac{|\alpha|^2}{\bar{n}(1 - e^{-\gamma t})}\right]$$

$$\times \frac{1}{l!} \frac{\partial^{2l}}{\partial\lambda^l \partial\lambda^{*l}} \left\{ \exp\left[-|\lambda|^2\frac{e^{-\gamma t} - \bar{n}(1 - e^{-\gamma t})}{\bar{n}(1 - e^{-\gamma t})}\right] \right.$$

$$\left. \times \exp\left[\lambda\frac{\alpha^* e^{-(\gamma/2)t}e^{-i\omega_0 t}}{\bar{n}(1 - e^{-\gamma t})}\right] \exp\left[\lambda^*\frac{\alpha e^{-(\gamma/2)t}e^{i\omega_0 t}}{\bar{n}(1 - e^{-\gamma t})}\right] \right\}\bigg|_{\lambda=\lambda^*=0}.$$

The derivatives can be evaluated using (4.46), with

$$A = \frac{e^{-\gamma t} - \bar{n}(1 - e^{-\gamma t})}{\bar{n}(1 - e^{-\gamma t})} \quad \text{and} \quad B^* = C = \frac{\alpha e^{-(\gamma/2)t}e^{i\omega_0 t}}{\bar{n}(1 - e^{-\gamma t})} :$$

the *P distribution for a damped Fock state* is then

$$P(\alpha, \alpha^*, t)_{\rho(0)=|l\rangle\langle l|}$$

$$= \frac{1}{\pi\bar{n}(1 - e^{-\gamma t})} \exp\left[-\frac{|\alpha|^2}{\bar{n}(1 - e^{-\gamma t})}\right] \frac{1}{l!}\left[\frac{e^{-\gamma t} - \bar{n}(1 - e^{-\gamma t})}{\bar{n}(1 - e^{-\gamma t})}\right]^l$$

$$\times \sum_{k=0}^{l}(-1)^{l-k}\frac{l!}{k!(l-k)!}\frac{l!}{k!}\left\{\frac{|\alpha|^2 e^{-\gamma t}}{\bar{n}(1 - e^{-\gamma t})[e^{-\gamma t} - \bar{n}(1 - e^{-\gamma t})]}\right\}^k.$$

$$(4.50)$$

In the long-time limit this expression clearly approaches the Gaussian describing a thermal state with mean photon number $\bar{n}$. This asymptotic solution is, of course, independent of the oscillator's initial state. To follow the evolution of $P(\alpha, \alpha^*, t)_{\rho(0)=|l\rangle\langle l|}$ for short times, it is helpful to rewrite (4.50) in an alternative form. We define

$$A = \frac{e^{-\gamma t} - \bar{n}(1 - e^{-\gamma t})}{\bar{n}(1 - e^{-\gamma t})} \quad \text{and} \quad \lambda = \frac{\alpha e^{-(\gamma/2)t}}{e^{-\gamma t} - \bar{n}(1 - e^{-\gamma t})},$$

and then (4.50) reads

$$P(\alpha, \alpha^*, t)_{\rho(0)=|l\rangle\langle l|}$$

$$= \frac{1}{\pi\bar{n}(1 - e^{-\gamma t})} \exp\left[-\frac{|\alpha|^2}{\bar{n}(1 - e^{-\gamma t})}\right] \frac{1}{l!} \sum_{k=0}^{l} (-1)^{l-k} \frac{l!}{k!(l-k)!}$$

$$\times \frac{l!}{k!} A^{l-k} (-A\lambda^*)^k (-A\lambda)^k.$$

Equation (4.46) may now be used a second time, with $B = C = 0$, to obtain

$$P(\alpha, \alpha^*, t)_{\rho(0)=|l\rangle\langle l|}$$

$$= \frac{1}{\pi\bar{n}(1 - e^{-\gamma t})} \exp\left[-\frac{|\alpha|^2}{\bar{n}(1 - e^{-\gamma t})}\right] \frac{1}{l!} e^{A|\lambda|^2} \frac{\partial^{2l}}{\partial\lambda^l \partial\lambda^{*l}} e^{-A|\lambda|^2}.$$

After resubstituting the explicit expressions for A and λ, we have an *alternative form for the P distribution for a damped Fock state*:

$$P(\alpha, \alpha^*, t)_{\rho(0)=|l\rangle\langle l|}$$

$$= \frac{1}{l!} \exp\left[\frac{|\alpha|^2}{e^{-\gamma t} - \bar{n}(1 - e^{-\gamma t})}\right] \left[\frac{e^{-\gamma t} - \bar{n}(1 - e^{-\gamma t})}{e^{-(\gamma/2)t}}\right]^{2l}$$

$$\times \frac{\partial^{2l}}{\partial\alpha^l \partial\alpha^{*l}} \left\{\frac{1}{\pi\bar{n}(1 - e^{-\gamma t})} \exp\left[-\frac{|\alpha|^2 e^{-\gamma t}}{\bar{n}(1 - e^{-\gamma t})[e^{-\gamma t} - \bar{n}(1 - e^{-\gamma t})]}\right]\right\}.$$

$$(4.51)$$

From this expression

$$P(\alpha, \alpha^*, 0)_{\rho(0)=|l\rangle\langle l|} = \frac{1}{l!} e^{|\alpha|^2} \frac{\partial^{2l}}{\partial\alpha^l \partial\alpha^{*l}} \left\{\lim_{t\to 0+} \left(\frac{1}{\pi\bar{n}\gamma t} e^{-|\alpha|^2/\bar{n}\gamma t}\right)\right\}. \quad (4.52)$$

Equation (4.52) shows explicitly the time-reversed approach $(t \to 0+)$ of $P(\alpha, \alpha^*, t)$ to its initial form in terms of derivatives of a two-dimensional δ-function.

Note that if $\bar{n} \neq 0$, $P(\alpha, \alpha^*, t)$ is actually a well-behaved function for all times $t > 0$. Thermal fluctuations destroy the singular character of the initial Fock state as soon as the interaction with the reservoir is turned on: for short times the singular distribution representing the initial Fock state is replaced by a derivative (of order $2l$) of a very narrow Gaussian whose variance is growing linearly with time. Nonetheless, $P(\alpha, \alpha^*, t)$ remains unacceptable as a classical probability distribution for a finite time after $t = 0$. During the early part of its evolution it takes on negative values – for example, for $l = 1$, (4.50) has the form

$$P(\alpha, \alpha^*, t)_{\rho(0)=|l\rangle\langle l|} = \frac{1}{\pi\bar{n}(1-e^{-\gamma t})} \exp\left[-\frac{|\alpha|^2}{\bar{n}(1-e^{-\gamma t})}\right]$$

$$\times\left\{1 - \frac{e^{-\gamma t}}{\bar{n}(1-e^{-\gamma t})} + \frac{|\alpha|^2 e^{-\gamma t}}{[\bar{n}(1-e^{-\gamma t})]^2}\right\}, \quad (4.53)$$

This distribution takes negative values inside the circle $|\alpha|^2 = \bar{n}(1-e^{-\gamma t})[1-\bar{n}(e^{\gamma t}-1)]$ during the time interval $0 < \gamma t < \ln(\bar{n}+1) - \ln\bar{n}$.

Exercise 4.4 Show that (4.50) gives

$$\langle(a^\dagger a)(t)\rangle = \left(\overline{\alpha^*\alpha(t)}\right)_P = le^{-\gamma t} + \bar{n}(1-e^{-\gamma l}),$$

in agreement with (1.70).

4.2.3 Damped Fock State in the Q and Wigner Representations

We have seen that the Q distribution is proportional to the diagonal matrix elements of ρ in the coherent state basis, and therefore it cannot become negative [Eq. (4.6)]. Indeed, the Green function (4.16) and the distribution (4.10) representing an initial Fock state in the Q representation are everywhere positive; it is clear then that $Q(\alpha, \alpha^*, t)_{\rho(0)=|l\rangle\langle l|}$ for a damped Fock state will be nonnegative at all times. To calculate this distribution explicitly we use (4.16) and (4.10) to write

$$Q(\alpha, \alpha^*, t)_{\rho(0)=|l\rangle\langle l|}$$

$$= \int d^2\lambda\, Q(\alpha, \alpha^*, t|\lambda, \lambda^*, 0) Q(\lambda, \lambda^*, 0)_{\rho(0)=|l\rangle\langle l|}$$

$$= \int d^2\lambda\, \frac{1}{\pi(\bar{n}+1)(1-e^{-\gamma t})} \exp\left[-\frac{|\alpha - \lambda e^{-\frac{\gamma}{2}t}e^{-i\omega_0 t}|^2}{(\bar{n}+1)(1-e^{-\gamma t})}\right] \frac{1}{\pi} e^{-|\lambda|^2} \frac{|\lambda|^{2l}}{l!}$$

$$= \frac{1}{\pi(\bar{n}+1)(1-e^{-\gamma t})} \exp\left[-\frac{|\alpha|^2}{(\bar{n}+1)(1-e^{-\gamma t})}\right]$$

$$\times \frac{1}{\pi}\frac{1}{l!}\int d^2\lambda\, |\lambda|^{2l} \exp\left[-|\lambda|^2 \frac{e^{-\gamma t} + (\bar{n}+1)(1-e^{-\gamma t})}{(\bar{n}+1)(1-e^{-\gamma t})}\right]$$

$$\times \exp\left[\frac{\alpha\lambda^* e^{-(\gamma/2)t}e^{-i\omega_0 t} + \alpha^*\lambda e^{-(\gamma/2)t}e^{i\omega_0 t}}{(\bar{n}+1)(1-e^{-\gamma t})}\right]$$

$$= \frac{1}{\pi(\bar{n}+1)(1-e^{-\gamma t})} \exp\left[-\frac{|\alpha|^2}{(\bar{n}+1)(1-e^{-\gamma t})}\right]$$

$$\times \frac{1}{\pi}\frac{1}{l!}\int_0^\infty dr\, r^{2l+1} \exp\left[-r^2\frac{e^{-\gamma t}+(\bar{n}+1)(1-e^{-\gamma t})}{(\bar{n}+1)(1-e^{-\gamma t})}\right]$$

$$\times \int_0^{2\pi} d\phi \exp\left[\frac{2|\alpha|e^{-(\gamma/2)t}}{(\bar{n}+1)(1-e^{-\gamma t})}r\cos\phi\right],$$

where $r \equiv |\lambda|$, and $\phi \equiv \arg(\lambda) - \arg(\alpha) + \omega_0 t$. The angular integral gives a Bessel function. With this Bessel function expressed in its series representation we find

$$Q(\alpha,\alpha^*,t)_{\rho(0)=|l\rangle\langle l|}$$

$$= \frac{1}{\pi(\bar{n}+1)(1-e^{-\gamma t})} \exp\left[-\frac{|\alpha|^2}{(\bar{n}+1)(1-e^{-\gamma t})}\right]$$

$$\times \frac{1}{\pi}\frac{1}{l!}\int_0^\infty dr\, r^{2l+1} \exp\left[-r^2\frac{e^{-\gamma t}+(\bar{n})+1)(1-e^{-\gamma t})}{(\bar{n}+1)(1-e^{-\gamma t})}\right]$$

$$\times 2\pi\sum_{k=0}^\infty \frac{1}{(k!)^2}\left[\frac{r|\alpha|e^{-(\gamma/2)t}}{(\bar{n}+1)(1-e^{-\gamma t})}\right]^{2k}$$

$$= \frac{1}{\pi(\bar{n}+1)(1-e^{-\gamma t})} \exp\left[-\frac{|\alpha|^2}{(\bar{n}+1)(1-e^{-\gamma t})}\right]$$

$$\times \frac{1}{l!}\sum_{k=0}^\infty \frac{1}{(k!)^2}\left[\frac{|\alpha|e^{-(\gamma/2)t}}{(\bar{n}+1)(1-e^{-\gamma t})}\right]^{2k}$$

$$\times 2\int_0^\infty dr\, r^{2(k+l)+1} \exp\left[-r^2\frac{1+\bar{n}(1-e^{-\gamma t})}{(\bar{n}+1)(1-e^{-\gamma t})}\right].$$

The remaining integral is performed by repeated integration by parts and gives

$$Q(\alpha,\alpha^*,t)_{\rho(0)=|l\rangle\langle l|}$$

$$= \frac{1}{\pi(\bar{n}+1)(1-e^{-\gamma t})} \exp\left[-\frac{|\alpha|^2}{(\bar{n}+1)(1-e^{-\gamma t})}\right]$$

$$\times \frac{1}{l!}\sum_{k=0}^\infty \frac{1}{(k!)^2}\left[\frac{|\alpha|e^{-(\gamma/2)t}}{(\bar{n}+1)(1-e^{-\gamma t})}\right]^{2k}(k+l)!\left[\frac{(\bar{n}+1)(1-e^{-\gamma t})}{1+\bar{n}(1-e^{-\gamma t})}\right]^{k+l+1}.$$

The Q distribution for a damped Fock state is then

$$Q(\alpha, \alpha^*, t)_{\rho(0)=|l\rangle\langle l|}$$

$$= \frac{1}{\pi[1 + \bar{n}(1 - e^{-\gamma t})]} \exp\left[-\frac{|\alpha|^2}{(\bar{n} + 1)(1 - e^{-\gamma t})}\right] \frac{1}{l!} \left[\frac{(\bar{n} + 1)(1 - e^{-\gamma t})}{1 + \bar{n}(1 - e^{-\gamma t})}\right]^l$$

$$\times \sum_{k=0}^{\infty} \frac{(k + l)!}{(k!)^2} \left\{\frac{|\alpha|^2 e^{-\gamma t}}{(\bar{n} + 1)(1 - e^{-\gamma t})[1 + \bar{n}(1 - e^{-\gamma t})]}\right\}^k. \tag{4.54}$$

Again, this expression clearly shows the evolution to a Gaussian distribution describing a thermal state in the long-time limit – now with the increased variance $(\bar{n} \rightarrow \bar{n} + 1)$ discussed below (4.21). Our result does not have the most convenient form, however, since the summation includes an infinite number of divergent terms in the limit $t \rightarrow 0$. Of course, $Q(\alpha, \alpha^*, 0)$ does not diverge; this is prevented by the exponential multiplying the sum. It would be nice to have a form that cancels the divergent sum explicitly to reproduce the Q distribution for the initial Fock state in an obvious way. This can be accomplished using the following result:

$$\sum_{k=0}^{\infty} \frac{(k + l)!}{(k!)^2} x^k = \frac{d^l}{dx^l} \left(\sum_{k=0}^{\infty} \frac{1}{k!} x^{k+l}\right)$$

$$= \frac{d^l}{dx^l} \left(x^l e^x\right)$$

$$= \sum_{k=0}^{l} \frac{l!}{k!(l - k)!} \frac{l!}{(l - k)!} x^{l-k} \frac{d^{l-k}}{dx^{l-k}} e^x$$

$$= e^x \sum_{k=0}^{l} \frac{l!}{k!(l - k)!} \frac{l!}{k!} x^k. \tag{4.55}$$

The third line follows from (4.45), with $A = -1$, $B = C = 0$, and $n = 1$; also, in the last line we have changed the summation index, with $l - k \rightarrow k$. Using (4.55), equation (4.54) may be recast to give an *alternative form for the Q distribution for a damped Fock state*:

$$Q(\alpha, \alpha^*, t)_{\rho(0)=|l\rangle\langle l|}$$

$$= \frac{1}{\pi[1 + \bar{n}(1 - e^{-\gamma t})]} \exp\left[-\frac{|\alpha|^2}{1 + \bar{n}(1 - e^{-\gamma t})}\right] \frac{1}{l!} \left[\frac{(\bar{n} + 1)(1 - e^{-\gamma t})}{1 + \bar{n}(1 - e^{-\gamma t})}\right]^l$$

$$\times \sum_{k=0}^{l} \frac{l!}{k!(l - k)!} \frac{l!}{k!} \left\{\frac{|\alpha|^2 e^{-(\gamma/2)t}}{(\bar{n} + 1)(1 - e^{-\gamma t})[1 + \bar{n}(1 - e^{-\gamma t})]}\right\}^k. \tag{4.56}$$

Equation (4.56) produces the correct initial distribution in an obvious way (only the $k = l$ term in the sum survives), and it also produces the Gaussian form in the long-time limit. It is clearly everywhere positive; for example, for $l = 1$,

$$Q(\alpha, \alpha^*, t)_{\rho(0)=|1\rangle\langle 1|} = \frac{1}{\pi[1 + \bar{n}(1 - e^{-\gamma t})]} \exp\left[-\frac{|\alpha|^2}{1 + \bar{n}(1 - e^{-\gamma t})}\right]$$

$$\times \left\{1 + \frac{|\alpha|^2 e^{-\gamma t}}{[1 + \bar{n}(1 - e^{-\gamma t})]^2}\right\}, \tag{4.57}$$

which is to be compared with the result (4.53) for the corresponding P distribution.

Exercise 4.5 The Wigner distribution can be derived in a similar manner. Show that the *Wigner distribution for a damped Fock state* is given by

$$W(\alpha, \alpha^*, t)_{\rho(0)=|l\rangle\langle l|}$$

$$= \frac{2}{\pi[1 + 2\bar{n}(1 - e^{-\gamma t})]} \exp\left[-\frac{2|\alpha|^2}{1 + 2\bar{n}(1 - e^{-\gamma t})}\right]$$

$$\times \frac{1}{l!} \sum_{k=0}^{l} (-1)^{l-k} \frac{l!}{k!(l-k)!} \frac{l!}{k!} 2^{2k} \left[\frac{(\bar{n} + \frac{1}{2})(1 - e^{-\gamma t})}{1 + 2\bar{n}(1 - e^{-\gamma t})}\right]^k$$

$$\times \sum_{r=0}^{k} \frac{k!}{r!(k-r)!} \frac{k!}{r!} \left\{\frac{|\alpha|^2 e^{-\gamma t}}{(\bar{n} + \frac{1}{2})(1 - e^{-\gamma t})[1 + 2\bar{n}(1 - e^{-\gamma t})]}\right\}^r. \tag{4.58}$$

Like $P(\alpha, \alpha^*, t)_{\rho(0)=|l\rangle\langle l|}$, this distribution can be negative. Analyze its behavior for $l = 1$.

4.3 Two-Time Averages

In Sect. 1.5 we obtained expressions for calculating two-time averages from an operator master equation. We have now seen that the operator master equation can be converted into a partial differential equation – in the case of the damped harmonic oscillator, a Fokker–Planck equation – by setting up a correspondence between ρ and a phase-space distribution function. How can the formal operator expressions given in Sect. 1.5 be cast into phase-space language to allow us to calculate two-time averages at the "classical" end of the quantum–classical correspondence? This is the question we now address. Answering the question in a general way requires that we first develop a little more formalism. The notation of this formalism is itself a bit burdensome, and certainly some of the calculations we eventually perform with it are rather arcane. It is perhaps helpful, then, to look ahead to (4.100a) and (4.100b). These state the result used most widely in applications; namely, that normal-ordered, time-ordered two-time averages, such as those needed to calculation an optical spectrum or intensity correlation function, are given by phase-space integrals in the P representation analogous to those met in classical statistics. The effort expended with the formalism allows us to generalize from this result in two directions: to determine which two-time averages are given by similar phase-space integrals in the Q and Wigner representations, and to see how derivatives of the phase-space distribution must be taken, as in Sec. 4.1.3, if inappropriately ordered operator averages are considered.

4.3.1 Quantum–Classical Correspondence for General Operators

Consider the relationship defined by (3.70) and (3.72) between the operator ρ and the distribution $P(\alpha, \alpha^*)$. There is actually no reason to restrict this relationship to density operators; we can generalize it to set up a correspondence between any system operator $\hat{O}$ and a function $F_{\hat{O}}^{(a)}(\alpha, \alpha^*)$ (we use "function" remembering that this may be a generalized function). As a generalization of the characteristic function $\chi_N(z, z^*)$ we define

$$\tilde{F}_{\hat{O}}^{(a)}(z, z^*) \equiv \pi \mathrm{tr}(\hat{O}e^{iz^*a^\dagger}e^{iza}); \tag{4.59}$$

the generalization of the P distribution is then

$$F_{\hat{O}}^{(a)}(\alpha, \alpha^*) \equiv \frac{1}{\pi^2} \int d^2z \, \tilde{F}_{\hat{O}}^{(a)}(z, z^*)e^{-iz^*\alpha^*}e^{-iz\alpha}, \tag{4.60}$$

with the inverse relationship

$$\tilde{F}_{\hat{O}}^{(a)}(z, z^*) = \int d^2\alpha \, F_{\hat{O}}^{(a)}(\alpha, \alpha^*)e^{iz^*\alpha^*}e^{iza}. \tag{4.61}$$

Taken together (4.59) and (4.60) set up a correspondence between the operator $\hat{O}$ and the phase-space function $F_{\hat{O}}^{(a)}(\alpha, \alpha^*)$. In place of the relationship that gives normal-ordered moments in the P representation [Eqs. (3.71) and (3.74)] we now have the more general result

$$\begin{aligned}
\mathrm{tr}(\hat{O}a^{\dagger p}a^q) &= \frac{1}{\pi}\frac{\partial^{p+q}}{\partial(iz^*)^p\partial(iz)^q}\tilde{F}_{\hat{O}}^{(a)}(z, z^*)\bigg|_{z=z^*=0} \\
&= \frac{1}{\pi}\frac{\partial^{p+q}}{\partial(iz^*)^p\partial(iz)^q}\int d^2\alpha \, F_{\hat{O}}^{(a)}(\alpha, \alpha^*)e^{iz^*\alpha^*}e^{iza}\bigg|_{z=z^*=0} \\
&= \frac{1}{\pi}\int d^2\alpha \, F_{\hat{O}}^{(a)}(\alpha, \alpha^*)\alpha^{*p}\alpha^q. \tag{4.62}
\end{aligned}$$

Within this scheme the P distribution is defined with

$$\chi_N(z, z^*) \equiv \frac{1}{\pi}\tilde{F}_\rho^{(a)}(z, z^*), \tag{4.63a}$$

$$P(\alpha, \alpha^*) \equiv \frac{1}{\pi}F_\rho^{(a)}(\alpha, \alpha^*). \tag{4.63b}$$

We have slipped in some changes here that need an explanation: a factor of π has been added in (4.59) and the subscript N on χ_N has been replaced by the superscript (a) on $\tilde{F}_\rho^{(a)}$. This has been done with the following in mind.

Consider an operator $\hat{A}$ expanded as a power series of terms written in antinormal order:

$$\hat{A} = A(a, a^\dagger) \equiv \sum_{p,q} C_{p,q}^{(a)} a^q a^{\dagger p}, \tag{4.64}$$

where the $C_{p,q}^{(a)}$ are constants. Then, from (4.59),

$$\tilde{F}_{\hat{A}}^{(a)}(z, z^*) = \pi \sum_{p,q} C_{p,q}^{(a)} \mathrm{tr}(a^{\dagger p} e^{iz^* a^\dagger} e^{iza} a^q)$$

$$= \pi \sum_{p,q} C_{p,q}^{(a)} \frac{\partial^{p+q}}{\partial(iz^*)^p \partial(iz)^q} \mathrm{tr}(e^{iz^* a^\dagger} e^{iza}).$$

Introducing the expansion (3.9) for the unit operator,

$$\tilde{F}_{\hat{A}}^{(a)}(z, z^*) = \pi \sum_{p,q} C_{p,q}^{(a)} \frac{\partial^{p+q}}{\partial(iz^*)^p \partial(iz)^q} \mathrm{tr}\left(\frac{1}{\pi} \int d^2\lambda \, |\lambda\rangle\langle\lambda| e^{iz^* a^\dagger} e^{iza}\right)$$

$$= \sum_{p,q} C_{p,q}^{(a)} \frac{\partial^{p+q}}{\partial(iz^*)^p \partial(iz)^q} \int d^2\lambda \, e^{iz^* \lambda^*} e^{iz\lambda}$$

$$= \pi^2 \sum_{p,q} C_{p,q}^{(a)} \frac{\partial^{p+q}}{\partial(iz^*)^p \partial(iz)^q} \delta(z).$$

We substitute this result into (4.60) and integrate by parts to obtain

$$F_{\hat{A}}^{(a)}(\alpha, \alpha^*) = \sum_{p,q} C_{p,q}^{(a)} \int d^2z \left[\frac{\partial^{p+q}}{\partial(iz^*)^p \partial(iz)^q} \delta(z)\right] e^{-iz^* \alpha^*} e^{-iz\alpha}$$

$$= \sum_{p,q} C_{p,q}^{(a)} \int d^2z \, \delta(z) \frac{\partial^{p+q}}{\partial(iz^*)^p \partial(iz)^q} e^{-iz^* \alpha^*} e^{-iz\alpha}.$$

Thus,

$$F_{\hat{A}}^{(a)}(\alpha, \alpha^*) = \sum_{p,q} C_{p,q}^{(a)} \alpha^{*p} \alpha^q = A(\alpha, \alpha^*). \tag{4.65}$$

Equations (4.64) and (4.65) state that, for operators written as an anti-normal-ordered series, $F_{\hat{O}}^{(a)}(\alpha, \alpha^*)$ is obtained by replacing the operators a and $a^\dagger$ in that series by the complex numbers α and α^*, respectively. $F_{\hat{O}}^{(a)}(\alpha, \alpha^*)$ is called the *antinormal-ordered associated function* for the operator $\hat{O}$. The superscript (a) denotes the *antinormal-ordered* associated function. The factor of π in (4.59) leads to the direct association of functions and operators expressed by (4.64) and (4.65), rather than with a $1/\pi$ multiplying the right-hand side of (4.65). We must be careful now not to become confused between our "normals" and "antinormals". In (4.63b) we see that $P(\alpha, \alpha^*)$, which is used to calculate *normal-ordered* averages, is, apart from a factor of π, the *antinormal-ordered* associated function for ρ. This relationship will become clearer as we follow the idea of associated functions a little further.

Analogous definitions of *normal-ordered and symmetrically ordered associated functions* for an operator can be given. We define the normal-ordered associated function $F_{\hat{O}}^{(n)}(\alpha, \alpha^*)$ in terms of its Fourier transform $\tilde{F}_{\hat{O}}^{(n)}(z, z^*)$ introduced as a generalization of (4.1): We define

$$\tilde{F}_{\hat{O}}^{(n)}(z, z^*) \equiv \pi \text{tr}\left(\hat{O} e^{iza} e^{iz^* a^\dagger}\right), \tag{4.66}$$

and

$$F_{\hat{O}}^{(n)}(\alpha, \alpha^*) \equiv \frac{1}{\pi^2} \int d^2 z\, \tilde{F}_{\hat{O}}^{(n)}(z, z^*) e^{-iz^* \alpha^*} e^{-iz\alpha}, \tag{4.67}$$

with the inverse relationship

$$\tilde{F}_{\hat{O}}^{(n)}(z, z^*) = \int d^2 \alpha\, F_{\hat{O}}^{(n)}(\alpha, \alpha^*) e^{iz^* \alpha^*} e^{iz\alpha}. \tag{4.68}$$

In place of the relationship that gives antinormal-ordered moments in the Q representation [Eq. (4.5)], we have

$$\text{tr}\left(\hat{O} a^q a^{\dagger p}\right) = \frac{1}{\pi} \int d^2 \alpha\, F_{\hat{O}}^{(n)}(\alpha, \alpha^*) \alpha^{*p} \alpha^q. \tag{4.69}$$

The Q distribution is proportional to the normal-ordered associated function for ρ:

$$\chi_A(z, z^*) \equiv \frac{1}{\pi} \tilde{F}_{\rho}^{(n)}(z, z^*), \tag{4.70a}$$

$$Q(\alpha, \alpha^*) \equiv \frac{1}{\pi} F_{\rho}^{(n)}(\alpha, \alpha^*). \tag{4.70b}$$

Similarly, the symmetric-ordered associated function $F_{\hat{O}}^{(s)}(\alpha, \alpha^*)$ is defined in terms of its Fourier transform $\tilde{F}_{\hat{O}}^{(s)}(z, z^*)$ introduced as a generalization of (4.25): We define

$$\tilde{F}_{\hat{O}}^{(s)}(z, z^*) \equiv \pi \text{tr}\left(\hat{O} e^{iz^* a^\dagger + iza}\right), \tag{4.71}$$

and

$$F_{\hat{O}}^{(s)}(\alpha, \alpha^*) \equiv \frac{1}{\pi^2} \int d^2 z\, \tilde{F}_{\hat{O}}^{(s)}(z, z^*) e^{-iz^* \alpha^*} e^{-iz\alpha}. \tag{4.72}$$

with the inverse relationship

$$\tilde{F}_{\hat{O}}^{(s)}(z, z^*) = \int d^2 \alpha\, F_{\hat{O}}^{(s)}(\alpha, \alpha^*) e^{iz^* \alpha^*} e^{iz\alpha}. \tag{4.73}$$

In place of the relationship that gives symmetric-ordered moments in the Wigner representation [Eq. (4.31)], we have

$$\text{tr}\left[\hat{O}(a^{\dagger p}a^{q})_{S}\right] = \frac{1}{\pi}\int d^{2}\alpha\, F_{\hat{O}}^{(s)}(\alpha,\alpha^{*})\alpha^{*p}\alpha^{q}. \tag{4.74}$$

The Wigner distribution is proportional to the symmetric-ordered associated function for ρ:

$$\chi_{S}(z,z^{*}) \equiv \frac{1}{\pi}\tilde{F}_{\rho}^{(s)}(z,z^{*}), \tag{4.75a}$$

$$W(\alpha,\alpha^{*}) \equiv \frac{1}{\pi}F_{\rho}^{(s)}(\alpha,\alpha^{*}). \tag{4.75b}$$

Relationships between the various associated functions, and between their Fourier transforms, can be obtained as generalizations of earlier results: equations (4.9) and (4.34) generalize to give

$$\tilde{F}_{\hat{O}}^{(n)}(z,z^{*}) = e^{-\frac{1}{2}|z|^{2}}\tilde{F}_{\hat{O}}^{(s)}(z,z^{*}) = e^{-|z|^{2}}\tilde{F}_{\hat{O}}^{(a)}(z,z^{*}); \tag{4.76}$$

Eqs. (4.7) and (4.35) generalize to give

$$F_{\hat{O}}^{(n)}(\alpha,\alpha^{*}) = \frac{1}{\pi}\int d^{2}\lambda\, e^{-|\lambda-\alpha|^{2}}F_{\hat{O}}^{(a)}(\lambda,\lambda^{*}), \tag{4.77a}$$

$$F_{\hat{O}}^{(s)}(\alpha,\alpha^{*}) = \frac{2}{\pi}\int d^{2}\lambda\, e^{-2|\lambda-\alpha|^{2}}F_{\hat{O}}^{(a)}(\lambda,\lambda^{*}), \tag{4.77b}$$

$$F_{\hat{O}}^{(n)}(\alpha,\alpha^{*}) = \frac{2}{\pi}\int d^{2}\lambda\, e^{-2|\lambda-\alpha|^{2}}F_{\hat{O}}^{(s)}(\lambda,\lambda^{*}); \tag{4.77c}$$

finally, Eqs. (4.11) and (4.36) generalize to give

$$F_{\hat{O}}^{(n)}(\alpha,\alpha^{*}) = \exp\left(\frac{1}{2}\frac{\partial^{2}}{\partial\alpha\partial\alpha^{*}}\right)F_{\hat{O}}^{(s)}(\alpha,\alpha^{*}) = \exp\left(\frac{\partial^{2}}{\partial\alpha\partial\alpha^{*}}\right)F_{\hat{O}}^{(a)}(\alpha,\alpha^{*}). \tag{4.78}$$

We can now understand the relationships between the various associated functions for ρ (the P, Q and Wigner distributions) and the ordered operator averages that are calculated from their moments in a more general context. First, we note the extension of the result expressed by (4.64) and (4.65) to normal-ordered and symmetric-ordered series. For an operator $\hat{N}$ written as a normal-ordered series,

$$\hat{N} = N(a,a^{\dagger}) \equiv \sum_{p,q}C_{p,q}^{(n)}a^{\dagger p}a^{q}, \tag{4.79}$$

the normal-ordered associated function is obtained by replacing a by α and $a^{\dagger}$ by α^{*}:

$$F_{\hat{N}}^{(n)}(\alpha,\alpha^{*}) = \sum_{p,q}C_{p,q}^{(n)}\alpha^{*p}\alpha^{q} = N(\alpha,\alpha^{*}). \tag{4.80}$$

For an operator $\hat{S}$ written as a symmetric-ordered series,

$$\hat{S} = S(a, a^\dagger) \equiv \sum_{p,q} C_{p,q}^{(s)} (a^{\dagger p} a^q)_S, \tag{4.81}$$

the symmetric-ordered associated function is obtained by replacing a by α and $a^\dagger$ by α^*:

$$F_{\hat{S}}^{(s)}(\alpha, \alpha^*) = \sum_{p,q} C_{p,q}^{(s)} \alpha^{*p} \alpha^q = S(\alpha, \alpha^*). \tag{4.82}$$

Now, if $\hat{O}_1$ and $\hat{O}_2$ are arbitrary system operators, and $\hat{N}_2 = N_2(a, a^\dagger) = \hat{O}_2$ is the normal-ordered form of $\hat{O}_2$, we can apply (4.62) to each term in the series expansion of $N_2(a, a^\dagger)$ to obtain

$$\begin{aligned}
\operatorname{tr}(\hat{O}_1 \hat{O}_2) &= \operatorname{tr}[\hat{O}_1 N_2(a, a^\dagger)] \\
&= \frac{1}{\pi} \int d^2\alpha \, F_{\hat{O}_1}^{(a)}(\alpha, \alpha^*) N_2(\alpha, \alpha^*) \\
&= \frac{1}{\pi} \int d^2\alpha \, F_{\hat{O}_1}^{(a)}(\alpha, \alpha^*) F_{\hat{O}_2}^{(n)}(\alpha, \alpha^*),
\end{aligned} \tag{4.83}$$

where the last line follows from (4.80). Equations (3.74) and (4.5), giving normal-ordered and antinormal-ordered operator averages as moments of the P and Q distributions, respectively, are special cases of this more general result. With $\hat{O}_1$ taken as ρ, moments of the *antinormal-ordered* associated function for ρ give the averages of operators $\hat{O}_2$ written in *normal-ordered* form. Alternatively, with $\hat{O}_2$ taken as ρ, moments of the *normal-ordered* associated function for ρ give averages of operators $\hat{O}_1$ written in *antinormal-ordered* form. A similar result can be obtained by writing $\hat{O}_2$ as a symmetric-ordered series and using (4.74) and (4.82):

$$\operatorname{tr}(\hat{O}_1 \hat{O}_2) = \frac{1}{\pi} \int d^2\alpha \, F_{\hat{O}_1}^{(s)}(\alpha, \alpha^*) F_{\hat{O}_2}^{(s)}(\alpha, \alpha^*). \tag{4.84}$$

The relationship (4.31) between symmetric-ordered operator averages and the moments of the Wigner distribution is a special case of this result.

Note 4.5 The association given by (4.79) and (4.80) is easily proved following an argument analogous to that used to establish (4.65). A similar proof of the association given by (4.81) and (4.82) is not so straightforward because partial derivatives with respect to (iz) and (iz^*) act in a rather complicated way on $e^{iz^* a^\dagger + iza}$ (see Sect. 4.3.5). A simple proof can be devised, however, by arguing backwards as follows: Set $F_{\hat{O}}^{(s)}(\alpha, \alpha^*) = \alpha^{*p} \alpha^q$. What, then, is the operator $\hat{O}$ having this symmetric-ordered associated function? The answer to this question can be obtained by converting everything into normal order, using (4.78) to write

$$F_{\hat{O}}^{(n)}(\alpha, \alpha^*) = \exp\left(\frac{1}{2}\frac{\partial^2}{\partial\alpha\partial\alpha^*}\right)\alpha^{*p}\alpha^q$$

$$= \sum_{k=0}^{\min(p,q)} \frac{1}{2^k}\frac{1}{k!}\frac{p!}{(p-k)!}\frac{q!}{(q-k)!}\alpha^{*p-k}\alpha^{q-k}.$$

Then, from (4.79) and (4.80),

$$\hat{O} = \sum_{k=0}^{\min(p,q)} \frac{1}{2^k}\frac{1}{k!}\frac{p!}{(p-k)!}\frac{q!}{(q-k)!}a^{\dagger p-k}a^{q-k}.$$

But (4.33) tells us that this is just the symmetric-ordered operator $(a^{\dagger p}a^q)_S$.

4.3.2 Associated Functions and the Master Equation

We saw how to derive an equation of motion for the P distribution to replace the operator master equation in Sect. 3.2.2. Generally, we will refer to such an equation as a *phase-space equation of motion*. We now see what this equation of motion looks like in the language of our generalized formalism of associated functions for arbitrary operators.

Let us start with a rather formal summary of the derivation of the equation of motion for the P distribution. From the operator master equation (3.1) we write

$$\frac{\partial}{\partial t}\mathrm{tr}\left[\rho(t)e^{iz^*a^{\dagger}}e^{iza}\right] = \mathrm{tr}\left[(\mathcal{L}\rho(t))e^{iz^*a^{\dagger}}e^{iza}\right], \tag{4.85}$$

which, after substituting the explicit form of $\mathcal{L}$ for the damped harmonic oscillator, is just (3.76). In the language of associated functions (4.85) states that

$$\frac{\partial}{\partial t}\tilde{F}_{\rho(t)}^{(a)}(z, z^*) = \tilde{F}_{\mathcal{L}\rho(t)}^{(a)}(z, z^*). \tag{4.86}$$

The Fourier transform of this equation gives the equation of motion for the antinormal-ordered associated function for ρ – the P distribution (multiplied by π):

$$\frac{\partial}{\partial t}F_{\rho(t)}^{(a)}(\alpha, \alpha^*) = F_{\mathcal{L}\rho(t)}^{(a)}(\alpha, \alpha^*). \tag{4.87}$$

Formally, this is the Fokker–Planck equation. But the next step is needed to reveal its explicit form as a partial differential equation; this is the step where most of our effort was spent in Sect. 3.2.2. We must express $F_{\mathcal{L}\rho(t)}^{(a)}(\alpha, \alpha^*)$ in terms of $F_{\rho(t)}^{(a)}(\alpha, \alpha^*)$, with the action of $\mathcal{L}$ on the density operator ρ transformed into the action of some differential operator on the associated function for ρ. Leaving out the details, the aim is to write

$$F_{\mathcal{L}\rho(t)}^{(a)}(\alpha, \alpha^*) = L^{(a)}\left(\alpha, \alpha^*, \frac{\partial}{\partial\alpha}, \frac{\partial}{\partial\alpha^*}\right)F_{\rho(t)}^{(a)}(\alpha, \alpha^*), \tag{4.88}$$

where $L^{(a)}(\alpha, \alpha^*, \frac{\partial}{\partial \alpha}, \frac{\partial}{\partial \alpha^*})$ is a differential operator associated with $\mathcal{L}$. For any particular example this must be found from an explicit calculation similar to the one in Sect. 3.2.2; for the damped harmonic oscillator

$$L^{(a)}\left(\alpha, \alpha^*, \frac{\partial}{\partial \alpha}, \frac{\partial}{\partial \alpha^*}\right)$$
$$= \left(\frac{\gamma}{2} + i\omega_0\right)\frac{\partial}{\partial \alpha}\alpha + \left(\frac{\gamma}{2} - i\omega_0\right)\frac{\partial}{\partial \alpha^*}\alpha^* + \gamma\bar{n}\frac{\partial^2}{\partial \alpha \partial \alpha^*}. \quad (4.89)$$

Now (4.87) becomes

$$\frac{\partial}{\partial t}F^{(a)}_{\rho(t)}(\alpha, \alpha^*) = L^{(a)}\left(\alpha, \alpha^*, \frac{\partial}{\partial \alpha}, \frac{\partial}{\partial \alpha^*}\right)F^{(a)}_{\rho(t)}(\alpha, \alpha^*), \quad (4.90)$$

and setting

$$P(\alpha, \alpha^*, t) = \frac{1}{\pi}F^{(a)}_{\rho(t)}(\alpha, \alpha^*), \quad (4.91)$$

the equation of motion for $P(\alpha, \alpha^*, t)$ is

$$\frac{\partial}{\partial t}P(\alpha, \alpha^*, t) = L^{(a)}\left(\alpha, \alpha^*, \frac{\partial}{\partial \alpha}, \frac{\partial}{\partial \alpha^*}\right)P(\alpha, \alpha^*, t). \quad (4.92)$$

More generally, we may write (4.88), not just for density operators, but for any operator $\hat{O}$. Then, by induction,

$$F^{(a)}_{\mathcal{L}^k\hat{O}}(\alpha, \alpha^*) = \left[L^{(a)}\left(\alpha, \alpha^*, \frac{\partial}{\partial \alpha}, \frac{\partial}{\partial \alpha^*}\right)\right]^k F^{(a)}_{\hat{O}}(\alpha, \alpha^*), \quad (4.93)$$

from which it follows that

$$F^{(a)}_{\exp(\mathcal{L}\tau)\hat{O}}(\alpha, \alpha^*) = e^{L^{(a)}(\alpha, \alpha^*, \frac{\partial}{\partial \alpha}, \frac{\partial}{\partial \alpha^*})\tau} F^{(a)}_{\hat{O}}(\alpha, \alpha^*). \quad (4.94)$$

This result, and (4.83) from the last section, will serve as centerpieces in our conversion of the expressions from Sect. 1.5 for two-time averages into phase-space form.

Of course, we define the differential operators $L^{(n)}(\alpha, \alpha^*, \frac{\partial}{\partial \alpha}, \frac{\partial}{\partial \alpha^*})$ and $L^{(s)}(\alpha, \alpha^*, \frac{\partial}{\partial \alpha}, \frac{\partial}{\partial \alpha^*})$ which govern the dynamics of the Q and the Wigner distributions, respectively, in an analogous manner. For the damped harmonic oscillator $L^{(n)}(\alpha, \alpha^*, \frac{\partial}{\partial \alpha}, \frac{\partial}{\partial \alpha^*})$ is given by (4.89) with the replacement $\bar{n} \to \bar{n} + 1$, and $L^{(s)}(\alpha, \alpha^*, \frac{\partial}{\partial \alpha}, \frac{\partial}{\partial \alpha^*})$ is given by the same expression with the replacement $\bar{n} \to \bar{n} + \frac{1}{2}$.

4.3.3 Normal-Ordered Time-Ordered Averages in the P Representation

We first set ourselves the task of finding a phase-space form in the P representation for the average ($\tau \geq 0$)

$$\langle a^{\dagger p}(t)\hat{N}(t+\tau)a^q(t)\rangle = \text{tr}\{(e^{\mathcal{L}\tau}[a^q\rho(t)a^{\dagger p}])\hat{N}\}, \tag{4.95}$$

where the expression on the right-hand side is obtained from (1.102); $\hat{N}$ can be any system operator written as a normal-ordered series [Eq. (4.79)]. Equation (4.95) provides an expression for calculating a general normal-ordered, time-ordered, two-time average – every $a^\dagger$ to the left of every a, every $a^\dagger(t+\tau)$ to the right of every $a^\dagger(t)$, and every $a(t+\tau)$ to the left of every $a(t)$. These are the averages that most interest us for applications in quantum optics.

Using (4.83) and (4.94), we write the average (4.95) as the phase-space integral

$$\langle a^{\dagger p}(t)\hat{N}(t+\tau)a^q(t)\rangle$$
$$= \frac{1}{\pi}\int d^2\alpha\, F^{(a)}_{\exp(\mathcal{L}\tau)[a^q\rho(t)a^{\dagger p}]}(\alpha,\alpha^*)\, F^{(n)}_{\hat{N}}(\alpha,\alpha^*)$$
$$= \frac{1}{\pi}\int d^2\alpha\left[e^{L^{(a)}(\alpha,\alpha^*,\frac{\partial}{\partial\alpha},\frac{\partial}{\partial\alpha^*})\tau}F^{(a)}_{a^q\rho(t)a^{\dagger p}}(\alpha,\alpha^*)\right]F^{(n)}_{\hat{N}}(\alpha,\alpha^*). \tag{4.96}$$

Then, from (4.60) and (4.59),

$$F^{(a)}_{a^q\rho(t)a^{\dagger p}}(\alpha,\alpha^*) = \frac{1}{\pi^2}\int d^2z\, \tilde{F}^{(a)}_{a^q\rho(t)a^{\dagger p}}(z,z^*)e^{-iz^*\alpha^*}e^{-iz\alpha}$$
$$= \frac{1}{\pi^2}\int d^2z\, \pi\text{tr}\left[a^q\rho(t)a^{\dagger p}e^{iz^*a^\dagger}e^{iza}\right]e^{-iz^*\alpha^*}e^{-iz\alpha}$$
$$= \frac{1}{\pi^2}\int d^2z\, \pi\text{tr}\left[\rho(t)a^{\dagger p}e^{iz^*a^\dagger}e^{iza}a^q\right]e^{-iz^*\alpha^*}e^{-iz\alpha}$$
$$= \frac{1}{\pi^2}\int d^2z\left[\frac{\partial^{p+q}}{\partial(iz^*)^p\partial(iz)^q}\tilde{F}^{(a)}_{\rho(t)}(z,z^*)\right]e^{-iz^*\alpha^*}e^{-iz\alpha}.$$

Substituting for $\tilde{F}^{(a)}_{\rho(t)}(z,z^*)$ from (4.61), we have

$$F^{(a)}_{a^q\rho(t)a^{\dagger p}}(\alpha,\alpha^*) = \frac{1}{\pi^2}\int d^2z\left[\frac{\partial^{p+q}}{\partial(iz^*)^p\partial(iz)^q}\int d^2\lambda\, F^{(a)}_{\rho(t)}(\lambda,\lambda^*)e^{iz^*\lambda^*}e^{iz\lambda}\right]$$
$$\times e^{-iz^*\alpha^*}e^{-iz\alpha}$$
$$= \frac{1}{\pi^2}\int d^2\lambda\, F^{(a)}_{\rho(t)}(\lambda,\lambda^*)\lambda^{*p}\lambda^q\int d^2z\, e^{iz^*(\lambda^*-\alpha^*)}e^{iz(\lambda-\alpha)}$$
$$= \frac{1}{\pi^2}\int d^2\lambda\, F^{(a)}_{\rho(t)}(\lambda,\lambda^*)\lambda^{*p}\lambda^q\delta^{(2)}(\lambda-\alpha)$$
$$= F^{(a)}_{\rho(t)}(\alpha,\alpha^*)\alpha^{*p}\alpha^q. \tag{4.97}$$

We now substitute this result into (4.96) to find ($\tau \geq 0$)

$$\langle a^{\dagger p}(t)\hat{N}(t+\tau)a^q(t)\rangle$$
$$= \frac{1}{\pi}\int d^2\alpha \left[e^{L^{(a)}(\alpha,\alpha^*,\frac{\partial}{\partial\alpha},\frac{\partial}{\partial\alpha^*})\tau}F^{(a)}_{\rho(t)}(\alpha,\alpha^*)\alpha^{*p}\alpha^q\right]F^{(n)}_{\hat{N}}(\alpha,\alpha^*).$$
(4.98)

At first sight, this expression may seem to be a rather useless formal result. However, a little more work casts it into a simple form – a form which might already have been anticipated. In simpler notation, (4.98) reads ($\tau \geq 0$)

$$\langle a^{\dagger p}(t)\hat{N}(t+\tau)a^q(t)\rangle$$
$$= \int d^2\alpha \left[e^{L^{(a)}(\alpha,\alpha^*,\frac{\partial}{\partial\alpha},\frac{\partial}{\partial\alpha^*})\tau}P(\alpha,\alpha^*,t)\alpha^{*p}\alpha^q\right]N(\alpha^*,\alpha), \quad (4.99)$$

where we have used (4.91) and (4.80). Now the action of the propagator $\exp\left[L^{(a)}(\alpha,\alpha^*,\frac{\partial}{\partial\alpha},\frac{\partial}{\partial\alpha^*})\tau\right]$ on the δ-function $\delta^{(2)}(\alpha-\alpha_0)$ generates the Green function for the equation of motion (4.92). This suggests that we should write the operand of the propagator in (4.99) as

$$P(\alpha,\alpha^*,t)\alpha^{*p}\alpha^q = \int d^2\alpha_0\, \delta^{(2)}(\alpha-\alpha_0)P(\alpha_0,\alpha_0^*)\alpha_0^{*p}\alpha_0^q,$$

whence ($\tau \geq 0$), in the P representation a normal-ordered, time-ordered, two-time average is calculated as

$$\langle a^{\dagger p}(t)\hat{N}(t+\tau)a^q(t)\rangle$$
$$= \int d^2\alpha \int d^2\alpha_0\, \alpha_0^{*p}\alpha_0^q N(\alpha,\alpha^*)P(\alpha,\alpha^*,\tau|\alpha_0,\alpha_0^*,0)P(\alpha_0,\alpha_0^*,t)$$
$$= \left(\overline{(\alpha^{*p}\alpha^q)(t)N(t+\tau)}\right)_P, \quad (4.100a)$$

where we have introduced the notation

$$\left(\overline{(\alpha^{*p}\alpha^q)(t)N(t+\tau)}\right)_P$$
$$\equiv \int d^2\alpha \int d^2\alpha_0\, \alpha_0^{*p}\alpha_0^q N(\alpha,\alpha^*)P(\alpha,\alpha^*,t+\tau;\alpha_0,\alpha_0^*,t),$$
(4.100b)

and

$$P(\alpha,\alpha^*,t+\tau;\alpha_0,\alpha_0^*,t) = P(\alpha,\alpha^*,\tau|\alpha_0,\alpha_0^*,0)P(\alpha_0,\alpha_0^*,t) \quad (4.101)$$

is the two-time, or joint, distribution. Thus, the correspondence with a classical statistical description has been extended one step further. Equation (4.100b) is formally equivalent to the formula for calculating two-time averages in a classical statistical theory.

4.3.4 More General Two-Time Averages Using the P Representation

We have seen that antinormal-ordered one-time averages can be calculated using the P representation [Sect. 4.1.3]; although, with some inconvenience, since the expressions for these averages involve derivatives of the P distribution. The situation is similar when we consider two-time averages that are not in normal-ordered time-ordered form. To see how (4.100) must be modified to give these averages we will seek a phase-space expression using the P representation for the general average ($\tau \geq 0$)

$$\langle \hat{O}_{r,q,m}(t)\hat{N}(t+\tau)\hat{O}_{s,p,n}^\dagger(t) \rangle = \mathrm{tr}\{(e^{\mathcal{L}\tau}[\hat{O}_{s,p,n}^\dagger \rho(t)\hat{O}_{r,q,m}])\hat{N}\}, \qquad (4.102)$$

where

$$\hat{O}_{k_1,k_2,k_3} \equiv a^{\dagger k_1} a^{k_2} a^{\dagger k_3}, \qquad (4.103)$$

and $\hat{N}$ is again the arbitrary normal-ordered operator defined by the series expansion (4.79). Once we have a solution to this problem, results for various combinations of normal-ordered and antinormal-ordered operators will follow with little extra effort.

We begin as before, using (4.83) and (4.94) to write

$$\langle \hat{O}_{r,q,m}(t)\hat{N}(t+\tau)\hat{O}_{s,p,n}^\dagger(t) \rangle$$

$$= \frac{1}{\pi}\int d^2\alpha \left[e^{L^{(a)}(\alpha,\alpha^*,\frac{\partial}{\partial\alpha},\frac{\partial}{\partial\alpha^*})\tau} F^{(a)}_{\hat{O}_{s,p,n}^\dagger \rho(t)\hat{O}_{r,q,m}}(\alpha,\alpha^*) \right] F^{(n)}_{\hat{N}}(\alpha,\alpha^*)$$

$$= \frac{1}{\pi}\int d^2\alpha \left[e^{L^{(a)}(\alpha,\alpha^*,\frac{\partial}{\partial\alpha},\frac{\partial}{\partial\alpha^*})\tau} \frac{1}{\pi^2}\int d^2z\, \tilde{F}^{(a)}_{\hat{O}_{s,p,n}^\dagger \rho(t)\hat{O}_{r,q,m}}(z,z^*) \right.$$

$$\left. \times e^{-iz^*\alpha^*} e^{-iz\alpha} \right] F^{(n)}_{\hat{N}}(\alpha,\alpha^*); \qquad (4.104)$$

the second line follows from (4.60). Our aim now is to express the function $\tilde{F}^{(a)}_{\hat{O}_{s,p,n}^\dagger \rho(t)\hat{O}_{r,q,m}}(z,z^*)$ in terms of $\tilde{F}^{(a)}_{\rho(t)}(z,z^*)$ and its derivatives. Using (4.59) and (4.103), we have

$$\tilde{F}^{(a)}_{\hat{O}_{s,p,n}^\dagger \rho(t)\hat{O}_{r,q,m}}(z,z^*) = \pi\mathrm{tr}\left[a^n a^{\dagger p} a^s \rho(t) a^{\dagger r} a^q a^{\dagger m} e^{iz^*a^\dagger} e^{iza} \right]$$

$$= \pi\mathrm{tr}\left[\rho(t) a^{\dagger r} a^q a^{\dagger m} e^{iz^*a^\dagger} e^{iza} a^n a^{\dagger p} a^s \right]$$

$$= \frac{\partial^{m+n}}{\partial(iz^*)^m \partial(iz)^n} \pi\mathrm{tr}\left[\rho(t) a^{\dagger r} a^q e^{iz^*a^\dagger} e^{iza} a^{\dagger p} a^s \right],$$

and then, from (3.78),

$$\tilde{F}^{(a)}_{\hat{O}^\dagger_{n,p,s}\rho(t)\hat{O}_{r,q,m}}(z,z^*)$$

$$= \frac{\partial^{m+n}}{\partial(iz^*)^m\partial(iz)^n}\pi\mathrm{tr}\big[\rho(t)a^{\dagger r}e^{iz^*a^\dagger}e^{iza}(a+iz^*)^qa^{\dagger p}a^s\big]$$

$$= \frac{\partial^{m+n}}{\partial(iz^*)^m\partial(iz)^n}\left(\frac{\partial}{\partial(iz)}+iz^*\right)^q\pi\mathrm{tr}\big[\rho(t)a^{\dagger r}e^{iz^*a^\dagger}e^{iza}a^{\dagger p}a^s\big]$$

$$= \frac{\partial^{m+n}}{\partial(iz^*)^m\partial(iz)^n}\left(\frac{\partial}{\partial(iz)}+iz^*\right)^q\pi\mathrm{tr}\big[\rho(t)a^{\dagger r}(a^\dagger+iz)^pe^{iz^*a^\dagger}e^{iza}a^s\big]$$

$$= \frac{\partial^{m+n}}{\partial(iz^*)^m\partial(iz)^n}\left(\frac{\partial}{\partial(iz)}+iz^*\right)^q\left(\frac{\partial}{\partial(iz^*)}+iz\right)^p\pi\mathrm{tr}\big[\rho(t)a^{\dagger r}e^{iz^*a^\dagger}e^{iza}a^s\big]$$

$$= \frac{\partial^{m+n}}{\partial(iz^*)^m\partial(iz)^n}\left(\frac{\partial}{\partial(iz)}+iz^*\right)^q\left(\frac{\partial}{\partial(iz^*)}+iz\right)^p$$

$$\times \frac{\partial^{r+s}}{\partial(iz^*)^r\partial(iz)^s}\tilde{F}^{(a)}_{\rho(t)}(z,z^*).$$

We write this to reflect the order of the operators in (4.103):

$$\tilde{F}^{(a)}_{\hat{O}^\dagger_{n,p,s}\rho(t)\hat{O}_{r,q,m}}(z,z^*) = \frac{\partial^m}{\partial(iz^*)^m}\left(\frac{\partial}{\partial(iz)}+iz^*\right)^q\frac{\partial^r}{\partial(iz^*)^r}$$

$$\times \frac{\partial^n}{\partial(iz)^n}\left(\frac{\partial}{\partial(iz^*)}+iz\right)^p\frac{\partial^s}{\partial(iz)^s}\tilde{F}^{(a)}_{\rho(t)}(z,z^*).$$

$$(4.105)$$

We now substitute the Fourier transform of $F^{(a)}_{\rho(t)}(\alpha,\alpha^*)$ for $\tilde{F}^{(a)}_{\rho(t)}(z,z^*)$ to obtain

$$\tilde{F}^{(a)}_{\hat{O}^\dagger_{n,p,s}\rho(t)\hat{O}_{r,q,m}}(z,z^*)$$

$$= \frac{\partial^m}{\partial(iz^*)^m}\left(\frac{\partial}{\partial(iz)}+iz^*\right)^q\frac{\partial^r}{\partial(iz^*)^r}\frac{\partial^n}{\partial(iz)^n}\left(\frac{\partial}{\partial(iz^*)}+iz\right)^p\frac{\partial^s}{\partial(iz)^s}$$

$$\times \int d^2\lambda\, F^{(a)}_{\rho(t)}(\lambda,\lambda^*)e^{iz^*\lambda^*}e^{iz\lambda}$$

$$= \int d^2\lambda\, F^{(a)}_{\rho(t)}(\lambda,\lambda^*)\lambda^s\left(\lambda^*+\frac{\partial}{\partial\lambda}\right)^p\lambda^n\lambda^{*r}\left(\lambda+\frac{\partial}{\partial\lambda^*}\right)^q\lambda^{*m}e^{iz^*\lambda^*}e^{iz\lambda}$$

$$= \int d^2\lambda\left[\lambda^{*m}\left(\lambda-\frac{\partial}{\partial\lambda^*}\right)^q\lambda^{*r}\lambda^n\left(\lambda^*-\frac{\partial}{\partial\lambda}\right)^p\lambda^s F^{(a)}_{\rho(t)}(\lambda,\lambda^*)\right]e^{iz^*\lambda^*}e^{iz\lambda},$$

$$(4.106)$$

where the last line follows after repeated integration by parts. When we use this result in (4.104) the integral with respect to z gives a δ-function, $\delta^{(2)}(\alpha-\lambda)$, and the integral with respect to λ is then trivially performed; we find ($\tau \geq 0$)

$$\langle \hat{O}_{r,q,m}(t)\hat{N}(t+\tau)\hat{O}_{n,p,s}^\dagger(t)\rangle$$

$$= \frac{1}{\pi}\int d^2\alpha \left[e^{L^{(a)}(\alpha,\alpha^*,\frac{\partial}{\partial\alpha},\frac{\partial}{\partial\alpha^*})\tau} \alpha^{*m}\left(\alpha - \frac{\partial}{\partial\alpha^*}\right)^q \alpha^{*r} \right.$$

$$\left. \times \alpha^n\left(\alpha^* - \frac{\partial}{\partial\alpha}\right)^p \alpha^s F_{\rho(t)}^{(a)}(\alpha,\alpha^*) \right] F_{\hat{N}}^{(n)}(\alpha,\alpha^*). \quad (4.107)$$

If we proceed, as below (4.98), to express this result in terms of $P(\alpha_0,\alpha_0^*,t)$ and $P(\alpha,\alpha^*,\tau|\alpha_0,\alpha_0^*,0)$, (4.107) becomes ($\tau \geq 0$)

$$\langle \hat{O}_{r,q,m}(t)\hat{N}(t+\tau)\hat{O}_{n,p,s}^\dagger(t)\rangle$$

$$= \int d^2\alpha \int d^2\alpha_0\, N(\alpha,\alpha^*)P(\alpha,\alpha^*,\tau|\alpha_0,\alpha_0^*,0)$$

$$\times \alpha_0^{*m}\left(\alpha_0 - \frac{\partial}{\partial\alpha_0^*}\right)^q \alpha_0^{*r}\alpha_0^n\left(\alpha_0^* - \frac{\partial}{\partial\alpha_0}\right)^p \alpha_0^s P(\alpha_0,\alpha_0^*,t).$$

$$(4.108)$$

The replacement of $a^{\dagger p}$ and a^q by differential operators, below (4.104), may also be performed in the reverse order; this gives an alternative to (4.108) in the form ($\tau \geq 0$)

$$\langle \hat{O}_{r,q,m}(t)\hat{N}(t+\tau)\hat{O}_{n,p,s}^\dagger(t)\rangle$$

$$= \int d^2\alpha \int d^2\alpha_0\, N(\alpha,\alpha^*)P(\alpha,\alpha^*,\tau|\alpha_0,\alpha_0^*,0)$$

$$\times \alpha_0^n\left(\alpha_0^* - \frac{\partial}{\partial\alpha_0}\right)^p \alpha_0^s\alpha_0^{*m}\left(\alpha_0 - \frac{\partial}{\partial\alpha_0^*}\right)^q \alpha_0^{*r} P(\alpha_0,\alpha_0^*,t).$$

$$(4.109)$$

With $p = q = 0$, both of these expressions reproduce the result (4.100) for the average $\langle a^{\dagger m+r}(t)\hat{N}(t+\tau)a^{n+s}(t)\rangle$. When $p \neq 0$, or $q \neq 0$, derivatives of $P(\alpha_0,\alpha_0^*,t)$ are involved, as in (4.23). Equation (4.23a) can be recovered from either (4.108) or (4.109); for example, with $q \neq 0$, $\hat{N} = a^{\dagger p}$, $\tau = 0$, and $r = m = n = p = s = 0$. Similarly, (4.23b) can be recovered with $p \neq 0$, $\hat{N} = a^q$, $\tau = 0$, and $r = q = m = n = s = 0$. There are other combinations of parameters that also recover these earlier results.

A number of results for two-time averages of operators expressed as normal-ordered and antinormal-ordered series now follow from (4.108) and (4.109). We introduce the normal-ordered series

$$\hat{N}_1 = N_1(a,a^\dagger) \equiv \sum_{p,q} C_{1p,q}^{(n)} a^{\dagger p}a^q, \quad (4.110a)$$

$$\hat{N}_2 = N_2(a,a^\dagger) \equiv \sum_{p,q} C_{2pq}^{(n)} a^{\dagger p}a^q, \quad (4.110b)$$

and the antinormal-ordered series

$$\hat{A}_1 = A_1(a, a^\dagger) \equiv \sum_{p,q} C_{1pq}^{(a)} a^q a^{\dagger p}, \tag{4.111a}$$

$$\hat{A}_2 = A_2(a, a^\dagger) \equiv \sum_{p,q} C_{2pq}^{(a)} a^q a^{\dagger p}. \tag{4.111b}$$

Then, applying (4.108) term by term, we prove the following ($\tau \geq 0$):

$$\langle \hat{N}_1(t) \hat{N}(t+\tau) \hat{N}_2(t) \rangle$$

$$= \int d^2\alpha \int d^2\alpha_0 \, N(\alpha, \alpha^*) P(\alpha, \alpha^*, \tau | \alpha_0, \alpha_0^*, 0)$$

$$\times \overleftarrow{N_1}\left(\alpha_0 - \frac{\partial}{\partial\alpha_0^*}, \alpha_0^*\right) \overleftarrow{N_2}\left(\alpha_0, \alpha_0^* - \frac{\partial}{\partial\alpha_0}\right) P(\alpha_0, \alpha_0^*, t), \tag{4.112a}$$

$$\langle \hat{N}_1(t) \hat{N}(t+\tau) \hat{A}_2(t) \rangle$$

$$= \int d^2\alpha \int d^2\alpha_0 \, N(\alpha, \alpha^*) P(\alpha, \alpha^*, \tau | \alpha_0, \alpha_0^*, 0)$$

$$\times \overleftarrow{N_1}\left(\alpha_0 - \frac{\partial}{\partial\alpha_0^*}, \alpha_0^*\right) \overrightarrow{A_2}\left(\alpha_0, \alpha_0^* - \frac{\partial}{\partial\alpha_0}\right) P(\alpha_0, \alpha_0^*, t), \tag{4.112b}$$

$$\langle \hat{A}_1(t) \hat{N}(t+\tau) \hat{N}_2(t) \rangle$$

$$= \int d^2\alpha \int d^2\alpha_0 \, N(\alpha, \alpha^*) P(\alpha, \alpha^*, \tau | \alpha_0, \alpha_0^*, 0)$$

$$\times \overrightarrow{A_1}\left(\alpha_0 - \frac{\partial}{\partial\alpha_0^*}, \alpha_0^*\right) \overleftarrow{N_2}\left(\alpha_0, \alpha_0^* - \frac{\partial}{\partial\alpha_0}\right) P(\alpha_0, \alpha_0^*, t), \tag{4.112c}$$

$$\langle \hat{A}_1(t) \hat{N}(t+\tau) \hat{A}_2(t) \rangle$$

$$= \int d^2\alpha \int d^2\alpha_0 \, N(\alpha, \alpha^*) P(\alpha, \alpha^*, \tau | \alpha_0, \alpha_0^*, 0)$$

$$\times \overrightarrow{A_1}\left(\alpha_0 - \frac{\partial}{\partial\alpha_0^*}, \alpha_0^*\right) \overrightarrow{A_2}\left(\alpha_0, \alpha_0^* - \frac{\partial}{\partial\alpha_0}\right) P(\alpha_0, \alpha_0^*, t). \tag{4.112d}$$

The arrows indicate whether the power series are to be written with the differential operators placed to the right or to the left. Equation (4.109) allows the order of the functions N_1, N_2, A_1, and A_2 to be reversed in these expressions.

Note 4.6 We have not exhausted all combinations of normal-ordered and antinormal-ordered operators here. If $\hat{N}$ is replaced by an antinormal-ordered series [Eq. (4.64)], it can be shown that $N(\alpha, \alpha^*)$ may be replaced in (4.112a)–(4.112d) by either $\overrightarrow{A}(\alpha - \frac{\partial}{\partial\alpha^*}, \alpha^*)$ or $\overrightarrow{A}(\alpha, \alpha^* - \frac{\partial}{\partial\alpha})$. The resulting expressions reproduce (4.23a) and (4.23b), respectively, when $\hat{N}_1 = \hat{N}_2 = \hat{A}_1 = \hat{A}_2 = 1$

and $\hat{A} = a^q a^{\dagger p}$. To prove this, use the relationship between $F_{\hat{A}}^{(n)}(\alpha, \alpha^*)$ and $F_{\hat{A}}^{(a)}(\alpha, \alpha^*)$ given by (4.78).

4.3.5 Two-Time Averages
Using the Q and Wigner Representations

Just as the operator averages corresponding to the moments of the single-time distribution vary from one representation to the other, so too do the averages corresponding to the moments of the two-time, or joint, distribution. In the Q representation a calculation parallel to that of Sect. 4.3.3 shows that antinormal-ordered, reverse-time-ordered, two-time averages are given by ($\tau \geq 0$)

$$\langle a^q(t)\hat{A}(t+\tau)a^{\dagger p}(t)\rangle = \left(\overline{(\alpha^{*p}\alpha^q)(t)A(t+\tau)}\right)_Q, \tag{4.113a}$$

with

$$\left(\overline{(\alpha^{*p}\alpha^q)(t)A(t+\tau)}\right)_Q$$
$$\equiv \int d^2\alpha \int d^2\alpha_0\, \alpha_0^{*p}\alpha_0^q A(\alpha,\alpha^*)Q(\alpha,\alpha^*,t+\tau;\alpha_0,\alpha_0^*,t), \tag{4.113b}$$

and

$$Q(\alpha,\alpha^*,t+\tau;\alpha_0,\alpha_0^*,t) = Q(\alpha,\alpha^*,\tau|\alpha_0,\alpha_0^*,0)Q(\alpha_0,\alpha_0^*,t), \tag{4.114}$$

where $\hat{A}$ is any operator written as a series in antinormal order [Eq. (4.64)]. More general averages not of the antinormal-ordered, reverse-time-ordered form involve derivatives of the Q distribution after the fashion of (4.112a)–(4.112d).

Exercise 4.6 Show that ($\tau \geq 0$)

$$\langle\hat{A}_1(t)\hat{A}(t+\tau)\hat{A}_2(t)\rangle$$
$$= \int d^2\alpha \int d^2\alpha_0\, A(\alpha,\alpha^*)Q(\alpha,\alpha^*,\tau|\alpha_0,\alpha_0^*,0)$$
$$\times \overleftarrow{A}_1\left(\alpha_0,\alpha_0^* + \frac{\partial}{\partial\alpha_0}\right)\overleftarrow{A}_2\left(\alpha_0 + \frac{\partial}{\partial\alpha_0^*},\alpha_0^*\right)Q(\alpha_0,\alpha_0^*,t), \tag{4.115a}$$

$$\langle\hat{A}_1(t)\hat{A}(t+\tau)\hat{N}_2(t)\rangle$$
$$= \int d^2\alpha \int d^2\alpha_0\, A(\alpha,\alpha^*)Q(\alpha,\alpha^*,\tau|\alpha_0,\alpha_0^*,0)$$
$$\times \overleftarrow{A}_1\left(\alpha_0,\alpha_0^* + \frac{\partial}{\partial\alpha_0}\right)\overrightarrow{N}_2\left(\alpha_0 + \frac{\partial}{\partial\alpha_0^*},\alpha_0^*\right)Q(\alpha_0,\alpha_0^*,t), \tag{4.115b}$$

$$\langle \hat{N}_1(t)\hat{A}(t+\tau)\hat{A}_2(t)\rangle$$

$$= \int d^2\alpha \int d^2\alpha_0 \, A(\alpha, \alpha^*) Q(\alpha, \alpha^*, \tau | \alpha_0, \alpha_0^*, 0)$$

$$\times \overrightarrow{N}_1\left(\alpha_0, \alpha_0^* + \frac{\partial}{\partial \alpha_0}\right) \overleftarrow{A}_2\left(\alpha_0 + \frac{\partial}{\partial \alpha_0^*}, \alpha_0^*\right) Q(\alpha_0, \alpha_0^*, t),$$

$$(4.115c)$$

$$\langle \hat{N}_1(t)\hat{A}(t+\tau)\hat{N}_2(t)\rangle$$

$$= \int d^2\alpha \int d^2\alpha_0 \, A(\alpha, \alpha^*) Q(\alpha, \alpha^*, \tau | \alpha_0, \alpha_0^*, 0)$$

$$\times \overrightarrow{N}_1\left(\alpha_0, \alpha_0^* + \frac{\partial}{\partial \alpha_0}\right) \overrightarrow{N}_2\left(\alpha_0 + \frac{\partial}{\partial \alpha_0^*}, \alpha_0^*\right) Q(\alpha_0, \alpha_0^*, t).$$

$$(4.115d)$$

As mentioned in Note 4.6, if $\hat{A}$ is replaced by an operator $\hat{N} = N(a, a^\dagger)$ written as a normal-ordered series, $A(\alpha, \alpha^*)$ may be replaced in these expressions by either $\overrightarrow{N}(\alpha + \frac{\partial}{\partial \alpha^*}, \alpha^*)$ or $\overrightarrow{N}(\alpha, \alpha^* + \frac{\partial}{\partial \alpha})$. From the resulting expressions we can recover (4.24a) and (4.24b) by setting $\hat{A}_1 = \hat{A}_2 = \hat{N}_1 = \hat{N}_2 = 1$ and $N(a, a^\dagger) = a^{\dagger p} a^q$.

We might expect the operator averages that correspond to moments of the two-time distribution in the Wigner representation to be some rather tangled mess. The symmetric-ordered operators related to moments of the one-time distribution are themselves a little imposing beyond the first few orders; how must we distribute the "t's" and "$t+\tau$'s" within the terms of the symmetric operator sums [Eqs. (4.29)] to come up with the two-time operator whose average is given by a double integration like (4.100) or (4.113)? The answer to this question is found by studying Sect. 4.3.3 a little more carefully to find out what really makes the calculation there work. Needless to say, the extension of this calculation to two-time averages calculated in the Wigner representation is going to call for a little more algebraic muscle.

First, note that a sum of averages ($\tau \geq 0$)

$$\sum_{i,j} \langle \hat{O}_i(t)\hat{S}(t+\tau)\hat{O}_j(t)\rangle = \sum_{i,j} \text{tr}\{(e^{\mathcal{L}\tau}[\hat{O}_j\rho(t)\hat{O}_i])\hat{S}\} \qquad (4.116)$$

can be written as a phase-space integral analogous to (4.96):

$$\sum_{i,j} \langle \hat{O}_i(t) \hat{S}(t+\tau) \hat{O}_j(t) \rangle$$

$$= \frac{1}{\pi} \int d^2\alpha \left[e^{L^{(s)}(\alpha, \alpha^*, \frac{\partial}{\partial \alpha}, \frac{\partial}{\partial \alpha^*})\tau} \sum_{i,j} F^{(s)}_{\hat{O}_j \rho(t) \hat{O}_i}(\alpha, \alpha^*) \right] F^{(s)}_{\hat{S}}(\alpha, \alpha^*),$$

$$(4.117)$$

where we have used (4.84) and (4.94), and $\hat{S}$ denotes any operator written as a symmetric-ordered series [Eq. (4.81)]. Now, the point on which the calculation of Sect. 4.3.3 turns is found in the fourth line of the equation below (4.96); if we can substitute $\tilde{F}^{(s)}_{\rho(t)}(z, z^*)$ for $\tilde{F}^{(a)}_{\rho(t)}(z, z^*)$ here we will be able to proceed in a parallel calculation to a result analogous to (4.100) – with W replacing P, and S replacing N. But to connect such a calculation with (4.117) we must answer one question: What operators $\hat{O}_i$ and $\hat{O}_j$ must be chosen so that

$$\sum_{i,j} \tilde{F}^{(s)}_{\hat{O}_j \rho(t) \hat{O}_i}(z, z^*) = \frac{\partial^{p+q}}{\partial (iz^*)^p \partial (iz)^q} \tilde{F}^{(s)}_{\rho(t)}(z, z^*) ?$$

With the answer to this question the two-time operator average obtained from moments of the two-time distribution in the Wigner representation will be the average (4.116).

The key to an answer lies with the following observation. Using (4.71) and the Baker-Hausdorff theorem [Eq. (4.8)], we find

$$\frac{\partial}{\partial (iz)} \tilde{F}^{(s)}_{\rho(t)}(z, z^*)$$

$$= \frac{\partial}{\partial (iz)} \pi \text{tr} \left[\rho(t) e^{iz^* a^\dagger + iza} \right]$$

$$= \frac{\partial}{\partial (iz)} \frac{1}{2} \pi \text{tr} \left[\rho(t) \left(e^{\frac{1}{2}|z|^2} e^{iza} e^{iz^* a^\dagger} + e^{-\frac{1}{2}|z|^2} e^{iz^* a^\dagger} e^{iza} \right) \right]$$

$$= \frac{1}{2} \pi \text{tr} \left\{ \rho(t) \left[\left(a - \frac{1}{2} iz^* \right) e^{iz^* a^\dagger + iza} + e^{iz^* a^\dagger + iza} \left(a + \frac{1}{2} iz^* \right) \right] \right\}$$

$$= \frac{1}{2} \left[\tilde{F}^{(s)}_{a\rho(t)}(z, z^*) + \tilde{F}^{(s)}_{\rho(t)a}(z, z^*) \right], \tag{4.118a}$$

and, in a similar fashion,

$$\frac{\partial}{\partial (iz^*)} \tilde{F}^{(s)}_{\rho(t)}(z, z^*) = \frac{1}{2} \left[\tilde{F}^{(s)}_{a^\dagger \rho(t)}(z, z^*) + \tilde{F}^{(s)}_{\rho(t)a^\dagger}(z, z^*) \right]. \tag{4.118b}$$

Also, if we wish to obtain an answer in a form that preserves the relationship to operators written in symmetric order, we must order the differential operators appearing in (4.118) in a corresponding fashion. Thus, we write

$$\frac{\partial^{p+q}}{\partial (iz^*)^p \partial (iz)^q} = \left(\frac{\partial^p}{\partial (iz^*)^p} \frac{\partial^q}{\partial (iz)^q} \right)_S, \tag{4.119}$$

where the right-hand side is the average of the $(p+q)!/(p!q!)$ orderings of the p differential operators $\partial/\partial(iz^*)$ and the q differential operators $\partial/\partial(iz)$. Now the answer to our question is accessible. To reach it, however, still requires a little combinatorics. The final step is left as an exercise:

Exercise 4.7 Use (4.118a), (4.118b), and (4.119) to show that

$$\frac{\partial^{p+q}}{\partial(iz^*)^p\partial(iz)^q}\tilde{F}^{(s)}_{\rho(t)}(z,z^*) = \frac{1}{2^{p+q}}\sum_{k=0}^{p+q}\binom{p+q}{k}\tilde{F}^{(s)}_{(a^{\dagger p}:\rho(t):a^q)_S^{(k)}}(z,z^*),$$

(4.120)

with

$$\left(a^{\dagger p}:\rho(t):a^q\right)_S^{(k)} \equiv \frac{p!q!}{(p+q)!}\sum_{\{\hat{O}_j\}}\hat{O}_{p+q}\hat{O}_{p+q-1}\cdots\hat{O}_{k+1}\rho(t)\hat{O}_k\cdots\hat{O}_1,$$

(4.121)

where the summation in (4.121) is taken over all different permutations $\hat{O}_1\cdots\hat{O}_{p+q}$ of p creation operators and q annihilation operators – i.e. $\rho(t)$ is placed into each term of $(a^{\dagger p}a^q)_S$ k places from the extreme right.

Equation (4.120) now allows us to follow the steps that led to (4.97) to obtain the corresponding result

$$\frac{1}{2^{p+q}}\sum_{k=0}^{p+q}\binom{p+q}{k}F^{(s)}_{(a^{\dagger p}:\rho(t):a^q)_S^{(k)}}(\alpha,\alpha^*) = F^{(s)}_{\rho(t)}(\alpha,\alpha^*)\alpha^{*p}\alpha^q.$$

(4.122)

The series of operators $\hat{O}_i$ and $\hat{O}_j$ appearing in (4.117) must now be chosen to connect with this result. The choice is fairly obvious from the associated function that appears on the left-hand side of (4.122); we have

$$\frac{1}{2^{p+q}}\sum_{k=0}^{p+q}\binom{p+q}{k}\left\langle\left(a^{\dagger p}(t):\hat{S}(t+\tau):a^q(t)\right)_S^{(k)}\right\rangle$$

$$= \frac{1}{2^{p+q}}\sum_{k=0}^{p+q}\binom{p+q}{k}\frac{p!q!}{(p+q)!}$$
$$\times \sum_{\{\hat{O}_j\}}\left\langle\hat{O}_{p+q}(t)\cdots\hat{O}_{k+1}(t)\hat{S}(t+\tau)\hat{O}_k(t)\cdots\hat{O}_1(t)\right\rangle$$

$$= \frac{1}{2^{p+q}}\sum_{k=0}^{p+q}\binom{p+q}{k}\frac{p!q!}{(p+q)!}$$
$$\times \sum_{\{\hat{O}_j\}}\mathrm{tr}\{(e^{\mathcal{L}\tau}[\hat{O}_k\cdots\hat{O}_1\rho(t)\hat{O}_{p+q}\cdots\hat{O}_{k+1}])\hat{S}\},$$

where we have used (1.102). The order of the subscripts in the sum over permutations of the operator product $a^{\dagger p}a^q$ can be changed with no effect, since operator sequences in every order are covered in the sum. Then

$$\frac{1}{2^{p+q}} \sum_{k=0}^{p+q} \binom{p+q}{k} \Big\langle \big(a^{\dagger p}(t) : \hat{S}(t+\tau) : a^q(t)\big)_S^{(k)} \Big\rangle$$

$$= \frac{1}{2^{p+q}} \sum_{k=0}^{p+q} \binom{p+q}{k} \frac{p!q!}{(p+q)!}$$

$$\times \sum_{\{\hat{O}_j\}} \mathrm{tr}\{(e^{\mathcal{L}\tau}[\hat{O}_{p+q} \cdots \hat{O}_{p+q-k+1}\rho(t)\hat{O}_{p+q-k} \cdots \hat{O}_1])\hat{S}\}.$$

In the operator sequences on the right-hand side of this expression $\rho(t)$ is inserted k places from the extreme *left*, in contrast to its position k places from the extreme *right* in the definition (4.121). This difference is removed, however, by a change of summation index, with $p+q-k \to k$; after making this change we arrive at the desired explicit form for (4.117); using (4.84) and (4.94):

$$\frac{1}{2^{p+q}} \sum_{k=0}^{p+q} \binom{p+q}{k} \Big\langle \big(a^{\dagger p}(t) : \hat{S}(t+\tau) : a^q(t)\big)_S^{(k)} \Big\rangle$$

$$= \mathrm{tr}\left\{\left(e^{\mathcal{L}\tau}\left[\frac{1}{2^{p+q}} \sum_{k=0}^{p+q} \binom{p+q}{k} \big(a^{\dagger p} : \rho(t) : a^q\big)_S^{(k)}\right]\right)\hat{S}\right\}$$

$$= \frac{1}{\pi} \int d^2\alpha \left[e^{L^{(s)}(\alpha,\alpha^*,\frac{\partial}{\partial\alpha},\frac{\partial}{\partial\alpha^*})\tau} \frac{1}{2^{p+q}} \sum_{k=0}^{p+q} \binom{p+q}{k}\right.$$

$$\left.\times F^{(s)}_{(a^{\dagger p}:\rho(t):a^q)_S^{(k)}}(\alpha,\alpha^*)\right] F^{(s)}_{\hat{S}}(\alpha,\alpha^*). \qquad (4.123)$$

Equations (4.122) and (4.123) allow the two-time operator average on the left-hand side of (4.123) to be calculated as a phase-space average with respect to the two-time Wigner distribution. Following the steps leading from (4.98) to (4.100) we obtain the corresponding result ($\tau \geq 0$)

$$\frac{1}{2^{p+q}} \sum_{k=0}^{p+q} \binom{p+q}{k} \Big\langle \big(a^{\dagger p}(t) : \hat{S}(t+\tau) : a^q(t)\big)_S^{(k)} \Big\rangle = \big(\overline{(\alpha^{*p}\alpha^q)(t)S(t+\tau)}\big)_W,$$

$$(4.124a)$$

with

$$\big(\overline{(\alpha^{*p}\alpha^q)(t)S(t+\tau)}\big)_W$$

$$\equiv \int d^2\alpha \int d^2\alpha_0 \, \alpha_0^{*p}\alpha_0^q S(\alpha,\alpha^*)W(\alpha,\alpha^*,t+\tau;\alpha_0,\alpha_0^*,t),$$

$$(4.124b)$$

and

$$W(\alpha, \alpha^*, t + \tau; \alpha_0, \alpha_0^*, t) = W(\alpha, \alpha^*, \tau | \alpha_0, \alpha_0^*, 0) W(\alpha_0, \alpha_0^*, t). \qquad (4.125)$$

We have again managed to construct a relationship between ordered operator two-time averages and two-time averages in the corresponding "classical" statistical system. However, the sum of operator averages appearing on the left-hand side of (4.124a) makes this a rather more formidable relationship than the corresponding relationships for the P and Q representations [Eqs. (4.100) and (4.113)].

To convince ourselves of the consistency of our result we should perhaps show that (4.124) is able to reproduce the expression for calculating one-time averages in the Wigner representation [Eq. (4.31)]. This is clear when we specialize to one-time averages by either taking $p = q = 0$, or $\hat{S} = 1$; in both cases we need only observe that

$$\sum_{k=0}^{p+q} \binom{p+q}{k} = (1+1)^{p+q} = 2^{p+q}.$$

It is less obvious, however, that the single-time result is recovered when τ is set to zero. Then (4.124) becomes

$$\frac{1}{2^{p+q}} \sum_{k=0}^{p+q} \binom{p+q}{k} \left\langle (a^{\dagger p}(t) \colon \hat{S}(t) \colon a^q(t))_S^{(k)} \right\rangle$$

$$= \int d^2\alpha \, \alpha^{*p} \alpha^q S(\alpha, \alpha^*) W(\alpha, \alpha^*, t).$$

If this is to correspond to (4.31), the phase-space function

$$\alpha^{*p} \alpha^q S(\alpha, \alpha^*) = \sum_{p',q'} C_{p',q'}^{(s)} \alpha^{*p+p'} \alpha^{q+q'}$$

that appears with the Wigner distribution in the integrand on the right-hand side must be the symmetric-ordered associated function for the operator that appears on the left-hand side – i.e. for the operator

$$\frac{1}{2^{p+q}} \sum_{k=0}^{p+q} \binom{p+q}{k} \left\langle (a^{\dagger p} \colon \hat{S} \colon a^q)_S^{(k)} \right\rangle$$

$$= \sum_{p',q'} C_{p',q'}^{(s)} \left[\frac{1}{2^{p+q}} \sum_{k=0}^{p+q} \binom{p+q}{k} (a^{\dagger p} \colon (a^{\dagger p'} a^{q'})_S \colon a^q)_S^{(k)} \right].$$

We know that $(a^{\dagger p+p'} a^{q+q'})_S$ is the operator with the symmetric-ordered associated function $\alpha^{*p+p'} \alpha^{q+q'}$; thus, we must show that

$$\frac{1}{2^{p+q}} \sum_{k=0}^{p+q} \binom{p+q}{k} \big(a^{\dagger p} \!: (a^{\dagger p'} a^{q'})_S \!: a^q\big)_S^{(k)} = \big(a^{\dagger p+p'} a^{q+q'}\big)_S. \qquad (4.126)$$

The proof is constructed by using the identity (4.28) to write

$$\frac{1}{2^{p+q}} \sum_{k=0}^{p+q} \binom{p+q}{k} \big(a^{\dagger p} \!: (a^{\dagger p'} a^{q'})_S \!: a^q\big)_S^{(k)}$$

$$= \frac{\partial^{p'+q'}}{\partial(iz^*)^{p'} \partial(iz)^{q'}} \frac{1}{2^{p+q}} \sum_{k=0}^{p+q} \binom{p+q}{k} \big(a^{\dagger p} \!: e^{iz^* a^\dagger + iza} \!: a^q\big)_S^{(k)} \bigg|_{z=z^*=0}.$$

Then, using

$$\frac{\partial}{\partial(iz)} e^{iz^* a^\dagger + iza} = \tfrac{1}{2}\big(a e^{iz^* a^\dagger + iza} + e^{iz^* a^\dagger + iza} a\big), \qquad (4.127a)$$

$$\frac{\partial}{\partial(iz^*)} e^{iz^* a^\dagger + iza} = \tfrac{1}{2}\big(a^\dagger e^{iz^* a^\dagger + iza} + e^{iz^* a^\dagger + iza} a^\dagger\big), \qquad (4.127b)$$

a calculation parallel to the one leading from (4.118) to (4.120) gives

$$\frac{1}{2^{p+q}} \sum_{k=0}^{p+q} \binom{p+q}{k} \big(a^{\dagger p} \!: e^{iz^* a^\dagger + iza} \!: a^q\big)_S^{(k)} = \frac{\partial^{p+q}}{\partial(iz^*)^p \partial(iz)^q} e^{iz^* a^\dagger + iza}.$$

$$(4.128)$$

Substituting this result and making a second use of (4.28), we have

$$\frac{1}{2^{p+q}} \sum_{k=0}^{p+q} \binom{p+q}{k} \big(a^{\dagger p} \!: (a^{\dagger p'} a^{q'})_S \!: a^q\big)_S^{(k)}$$

$$= \frac{\partial^{p+p'+q+q'}}{\partial(iz^*)^{p+p'} \partial(iz)^{q+q'}} e^{iz^* a^\dagger + iza} \bigg|_{z=z^*=0}$$

$$= \big(a^{\dagger p+p'} a^{q+q'}\big)_S.$$

It is possible to derive more general expressions for two-time averages in the Wigner representation – expressions that involve partial derivatives, after the fashion of the results (4.112) and (4.115) for the P and Q representations. We have no use, however, for these expressions later in the book and therefore we will not bother with their derivation here. In general we are interested only in the simple relationships (4.100), (4.113), and (4.124), where two-time operator averages are given by moments of the two-time phase-space distributions. It is important to realize, however, that within each of the three representations we have discussed many two-time averages simply cannot be calculated in terms of a simple "classical" integral; the more complicated expressions such as (4.112) and (4.115) are needed when the ordering is inappropriate for the chosen representation. When calculating single-time averages we always have the option of reordering the operators to suit the representation. Thus, $\langle a^\dagger a \rangle$ can be calculated as $\langle a^\dagger a \rangle = \left(\overline{\alpha^* \alpha} \right)_P$ in the P representation, $\langle a a^\dagger \rangle - 1 = \left(\overline{\alpha^* \alpha} \right)_Q - 1$ in the Q representation, or as $\frac{1}{2} \left(\langle a^\dagger a \rangle + \langle a a^\dagger \rangle \right) - \frac{1}{2} = \left(\overline{\alpha^* \alpha} \right)_W - \frac{1}{2}$ in the Wigner representation. On the other hand, while an average like $\langle a^\dagger(t + \tau) a(t) \rangle$, or $\langle a(t + \tau) a(t) \rangle$, can be calculated as a "classical" integral in the P representation [Eq. (4.100)], we generally do not have commutation relations to tell us how to reorder the operators so that the same result can be obtained as simply in either the Q or the Wigner representations. Applications in quantum optics are ultimately concerned with the normal-ordered time-ordered averages that arise in the theory of photodetection [4.11, 4.12]. Our phase-space results for two-time (more generally multi-time) averages clearly distinguishes the P representation as the most suited to the treatment of problems in quantum optics – results for multi-time averages show this even more clearly than do results for one-time averages.

Note 4.7 The assertion that the P representation is the most suited to problems in quantum optics perhaps requires some qualification. The P representation gains its special status from the theory of photoelectric detection, in which normal-ordered time-ordered averages appear. Therefore questions that are related in an immediate way to the ultimate observation of photons through the photoelectric effect lead in a natural way to a phase-space formulation in terms of the P representation. But there are questions of interest which need not be stated in terms of the photoelectric emission that ultimately completes a measurement process. Certainly then, there are situations in which, as a mathematical tool, the Q or the Wigner representation might be preferred over the P representation. An important consideration in this regard is the fact that the P distribution may be a generalized function. If this is so we do not gain much physical insight, and probably little mathematical assistance, by using the P representation. On the other hand, the Q and Wigner distributions are always well-behaved functions (although the Wigner distribution may take on negative values). For this reason the Q or Wigner representation is often the choice for studies of nonclassical states of the electromagnetic field – for example, squeezed states, in one sense, are related most directly to the Wigner representation.

Having said this, it is still important to reiterate the observation above concerning multi-time averages. When we use a phase-space representation to convert an operator master equation into a Fokker–Planck equation, we do not merely set up a representation for some state of the electromagnetic field; we set up a correspondence between quantum and classical processes *that evolve in time*. When the P representation provides the basis for the quantum–classical correspondence a direct connection exists between all the multi-time correlation functions of the classical process and the multi-time correlation functions of the quantized field that are measured by photoelectric detection. We cannot make a similar general statement connecting the classical multi-time correlation functions and measured multi-time statistics of the quantized field when the Q or Wigner representations provide the basis for the quantum–classical correspondence.

Exercise 4.8 Reproduce the result

$$\langle a^\dagger(0)a^\dagger(\tau)a(\tau)a(0)\rangle_{\text{ss}} = \bar{n}^2(1+e^{-\gamma\tau})$$

from Sect. 1.5.3 using the P representation and the Q representation. From the simple relationship between the Fokker–Planck equations for the damped harmonic oscillator, it follows that (4.113) and (4.124) give

$$\langle a(0)a(\tau)a^\dagger(\tau)a^\dagger(0)\rangle_{\text{ss}} = (\bar{n}+1)^2(1+e^{-\gamma\tau})$$

and

$$\frac{1}{4}\sum_{k=0}^{2}\binom{2}{k}\left\langle (a^\dagger(0):(a^\dagger(\tau)a(\tau))_S:a(0))_S^{(k)}\right\rangle_{ss} = (\bar{n}+\tfrac{1}{2})^2(1+e^{-\gamma\tau}).$$

Reproduce these results using the methods of Sect. 1.5.3.

5. Fokker–Planck Equations and Stochastic Differential Equations

We have seen how the quantum–classical correspondence is used to transform a quantum-mechanical operator description of a dissipative system, such as a damped harmonic oscillator, into the language of classical statistical physics. The distribution that represents the density operator need not satisfy all of the conditions required of a probability density; but in many cases it does, and very often it obeys a Fokker–Planck equation which leads us directly to a treatment using the language and methods of classical statistics. We will shortly discuss the extension of these ideas to the representation of atomic states. However, before moving to this subject, now is a good time to say something about the general properties of Fokker–Planck equations and their connection with stochastic differential equations.

The Fokker–Planck equation has a long history, going back to its use by Fokker in 1915 [5.1], and Planck in 1917 [5.2], to describe Brownian motion. In its traditional context it is an equation for a conditional probability density $P(\boldsymbol{x}, t | \boldsymbol{x}_0, 0)$ of the form

$$
\begin{aligned}
&\frac{\partial P(\boldsymbol{x}, t | \boldsymbol{x}_0, 0)}{\partial t} \\
&= \left(-\sum_{i=1}^{n} \frac{\partial}{\partial x_i} A_i(\boldsymbol{x}) + \frac{1}{2} \sum_{i,j=1}^{n} \frac{\partial^2}{\partial x_i \partial x_j} D_{ij}(\boldsymbol{x}) \right) P(\boldsymbol{x}, t | \boldsymbol{x}_0, 0), \quad (5.1)
\end{aligned}
$$

where $\boldsymbol{x}$ is a vector of n random variables, $x_1, \ldots, x_n$, and the $A_i(\boldsymbol{x})$ and $D_{ij}(\boldsymbol{x})$ are general functions of these variables; the matrix $D_{ij}(\boldsymbol{x})$ is symmetric and positive definite by definition. The conditional probability density is the Green function solution to (5.1), which has initial condition

$$
P(\boldsymbol{x}, 0 | \boldsymbol{x}_0, 0) = \delta(\boldsymbol{x} - \boldsymbol{x}_0) \equiv \delta(x_1 - x_{10}) \cdots \delta(x_n - x_{n0}).
$$

Of course, the unconditioned distribution

$$
P(\boldsymbol{x}, t) = \int d\boldsymbol{x}_0 \, P(\boldsymbol{x}, t | \boldsymbol{x}_0, 0) P(\boldsymbol{x}_0, 0)
$$

also satisfies (5.1). The Fokker–Planck equation is an approximate form of the Chapman-Kolmogorov equation,

$$P(\boldsymbol{x}_3,t_3|\boldsymbol{x}_1,t_1) = \int d\boldsymbol{x}_2\, P(\boldsymbol{x}_3,t_3|\boldsymbol{x}_2,t_2)P(\boldsymbol{x}_2,t_2|\boldsymbol{x}_1,t_1), \qquad (5.2)$$

for a Markov process. The essential content of the approximation leading from (5.2) to (5.1) is the assumption that the stochastic evolution of the state $\boldsymbol{x}(t)$ proceeds via infinitely many infinitesimal jumps, in a *diffusion process*; discontinuous jumps (jumps that are not infinitesimal) add derivatives of all orders to (5.1) (the Kramers–Moyal expansion). Discussion of the derivation and application of the Fokker–Planck equation in the theory of classical stochastic processes can be found in many places, including the books by Gardiner [5.3], van Kampen [5.4], and Risken [5.5]. These books will provide useful references for an expanded coverage of the topics we discuss in this chapter.

A Fokker–Planck equation is always linear in the distribution P. The designation "linear" need not, therefore, be reserved to distinguish between equations that are linear and nonlinear in P, which would be the usual mathematical usage. We will use it to refer to a Fokker–Planck equation in which each $A_i(\boldsymbol{x})$ is a linear function

$$A_i(\boldsymbol{x}) = \sum_{j=1}^{n} A_{ij}x_j, \qquad (5.3)$$

and the $D_{ij}(\boldsymbol{x})$ are all constants:

$$D_{ij}(\boldsymbol{x}) = D_{ij}. \qquad (5.4)$$

A *linear Fokker–Planck equation* can be written in the compact vector notation

$$\frac{\partial P}{\partial t} = \left(-\boldsymbol{x}'^T\boldsymbol{A}\boldsymbol{x} + \tfrac{1}{2}\boldsymbol{x}'^T\boldsymbol{D}\boldsymbol{x}'\right)P, \qquad (5.5)$$

where $\boldsymbol{A}$ and $\boldsymbol{D}$ are $n \times n$ matrices with matrix elements A_{ij} and D_{ij}, respectively, $\boldsymbol{x}$ and $\boldsymbol{x}'$ are the column vectors

$$\boldsymbol{x} \equiv \begin{pmatrix} x_1 \\ \vdots \\ x_n \end{pmatrix}, \qquad \boldsymbol{x}' \equiv \begin{pmatrix} \partial/\partial x_1 \\ \vdots \\ \partial/\partial x_n \end{pmatrix}, \qquad (5.6)$$

and T denotes the transpose.

5.1 One-Dimensional Fokker–Planck Equations

To gain some insight into the physics described by the Fokker–Planck equation, without delving into the details of its derivation for classical stochastic processes, let us spend a little time considering the one-dimensional equation

$$\frac{\partial P}{\partial t} = \left(-\frac{\partial}{\partial x}A(x) + \frac{1}{2}\frac{\partial^2}{\partial x^2}D(x)\right)P. \tag{5.7}$$

We have already met the solution to the linear version of this equation in Sect. 3.1.5. We will now attempt to develop some intuition for the temporal evolution described by this solution, identifying the terms in the Fokker–Planck equation that generate the different features in the evolution.

5.1.1 Drift and Diffusion

The mean and variance of the random variable x are defined, respectively, by

$$\langle x(t) \rangle \equiv \int_{-\infty}^{\infty} dx\, xP(x,t), \tag{5.8}$$

and

$$\sigma^2(t) \equiv \left\langle \left[x(t) - \langle x(t) \rangle\right]^2 \right\rangle = \langle x(t)^2 \rangle - \langle x(t) \rangle^2, \tag{5.9}$$

with

$$\langle x(t)^2 \rangle \equiv \int_{-\infty}^{\infty} dx\, x^2 P(x,t). \tag{5.10}$$

Equations of motion for these moments are obtained from (5.7) in the following way: For the mean of x, we have

$$\langle \dot{x} \rangle = \frac{d}{dt}\int_{-\infty}^{\infty} dx\, xP(x,t)$$

$$= \int_{-\infty}^{\infty} dx\, x\frac{\partial P(x,t)}{\partial t}$$

$$= -\int_{-\infty}^{\infty} dx\, x\frac{\partial}{\partial x}A(x)P(x,t) + \frac{1}{2}\int_{-\infty}^{\infty} dx\, x\frac{\partial^2}{\partial x^2}D(x)P(x,t).$$

Integration by parts gives

$$\langle \dot{x} \rangle = -\left. xA(x)P(x,t)\right|_{-\infty}^{\infty} + \int_{-\infty}^{\infty} dx\, A(x)P(x,t)$$

$$+ \frac{1}{2}x\frac{\partial}{\partial x}D(x)P(x,t)\Big|_{-\infty}^{\infty} - \frac{1}{2}\int_{-\infty}^{\infty} dx\, \frac{\partial}{\partial x}D(x)P(x,t),$$

and then if P and its derivatives vanish sufficiently fast at infinity,

$$\langle \dot{x} \rangle = \langle A(x) \rangle. \tag{5.11}$$

The equation of motion for the variance of x is obtained in a similar manner. We first derive the equation of motion for $\langle x^2 \rangle$:

$$\langle \dot{x^2} \rangle = -\int_{-\infty}^{\infty} dx \, x^2 \frac{\partial}{\partial x} A(x) P(x,t) + \frac{1}{2} \int_{-\infty}^{\infty} dx \, x^2 \frac{\partial^2}{\partial x^2} D(x) P(x,t)$$

$$= -x^2 A(x) P(x,t) \Big|_{-\infty}^{\infty} + \int_{-\infty}^{\infty} dx \, 2x A(x) P(x,t)$$

$$+ \frac{1}{2} x^2 \frac{\partial}{\partial x} D(x) P(x,t) \Big|_{-\infty}^{\infty} - \int_{-\infty}^{\infty} dx \, x \frac{\partial}{\partial x} D(x) P(x,t)$$

$$= 2\langle x A(x)\rangle - x D(x) P(x,t) \Big|_{-\infty}^{\infty} + \int_{-\infty}^{\infty} dx \, D(x) P(x,t)$$

$$= 2\langle x A(x)\rangle + \langle D(x)\rangle. \tag{5.12}$$

Using (5.9), (5.11) and (5.12), we obtain

$$\dot{\sigma^2} = 2\langle x A(x)\rangle - 2\langle x\rangle\langle A(x)\rangle + \langle D(x)\rangle. \tag{5.13}$$

If the Fokker–Planck equation is linear – $A(x) = Ax$, $D(x) = D$, where A and D are constants – (5.11) and (5.13) become

$$\langle \dot{x} \rangle = A\langle x\rangle \tag{5.14}$$

and

$$\dot{\sigma^2} = 2A\sigma^2 + D. \tag{5.15}$$

Then the mean and variance evolve independently, with

$$\langle x(t)\rangle = \langle x(0)\rangle e^{At} \tag{5.16}$$

and

$$\sigma^2(t) = \sigma^2(0) e^{2At} - (D/2A)(1 - e^{2At}). \tag{5.17}$$

The motion of the mean is governed by A; it is generated by the first term on the right-hand side of (5.7) alone. This term is called the *drift term* because it imparts a "drift" to the distribution – the peak of the distribution follows the time-dependent mean [Eq. (5.16)]. The role of the second term on the right-hand side of (5.7) is apparent from the solution for the time-dependent variance. With $D > 0$ ($A < 0$), an initially sharp distribution [$\langle \sigma^2(0)\rangle = 0$] broadens with time [Eq. (5.17)]; in (5.15) D acts as a *source of fluctuations*. The second term on the right-hand side of (5.7) is therefore called the *diffusion term* or *fluctuation term*.

The two pieces of the evolution, drift and diffusion, are seen quite clearly in the solution for the conditional distribution $P(x,t|x_0,0)$. This is given by (3.64), or in the present notation,

$$P(x,t|x_0,0) = \frac{1}{\sqrt{2\pi(D/2A)(e^{2At}-1)}} \exp\left[-\frac{1}{2}\frac{(x-x_0 e^{At})^2}{(D/2A)(e^{2At}-1)}\right]. \tag{5.18}$$

For $A \neq 0$, the evolution of $\langle x \rangle$ corresponds to a drift of the Gaussian distribution as a whole, while if $D > 0$, (5.17) describes the broadening ($A < 0$) of this distribution (Fig. 5.1). If $A = 0$ and $D > 0$ there is no drift; the conditional distribution keeps its initial mean for all times and its variance grows linearly in time. This is the behavior known as Brownian motion:

$$P(x,t|x_0,0) = \frac{1}{\sqrt{2\pi Dt}} \exp\left[-\frac{1}{2}\frac{(x - x_0)^2}{Dt} \right]. \tag{5.19}$$

If $D = 0$ and $A \neq 0$ there is no diffusion; the conditional distribution reduces to a "drifting" δ-function,

$$P(x,t|x_0,0) = \delta(x - x_0 e^{At}). \tag{5.20}$$

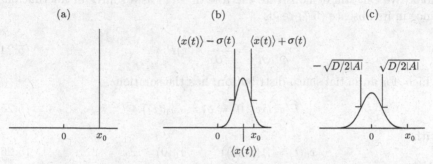

Fig. 5.1 Time evolution of the Gaussian distribution (5.18): (a) initial δ-function, (b) evolution under the combined action of drift and diffusion, (c) steady state.

Of course, if the distribution has an initial width, the drift term does not simply generate a displacement of the initial distribution. Let $D = 0$ and

$$P(x,0) = \frac{1}{\sqrt{2\pi}\sigma(0)} \exp\left[-\frac{1}{2}\frac{(x - x_0)^2}{\sigma^2(0)} \right]. \tag{5.21}$$

Then

$$\begin{aligned}
P(x,t) &= \int_{-\infty}^{\infty} dx'\, P(x,t|x',0) P(x',0) \\
&= \int_{-\infty}^{\infty} dx'\, \delta(x - x' e^{At}) \frac{1}{\sqrt{2\pi}\sigma(0)} \exp\left[-\frac{1}{2}\frac{(x' - x_0)^2}{\sigma^2(0)} \right] \\
&= \frac{1}{\sqrt{2\pi}\sigma(t)} \exp\left[-\frac{1}{2}\frac{(x - x_0(t))^2}{\sigma^2(t)} \right],
\end{aligned} \tag{5.22}$$

where

$$x_0(t) \equiv x_0 e^{At}, \qquad \sigma^2(t) \equiv \sigma^2(0) e^{2At}. \tag{5.23}$$

A simple displacement of the initial distribution to follow the displacement $x_0(t) - x_0$ of the mean would give $P(x,t) = P(x - x_0(t) + x_0, 0)$; Eq. (5.22) does not give this result. The drift term shifts the mean; but it also damps $(A < 0)$ or amplifies $(A > 0)$ any statistical uncertainty (fluctuations) that is present in the initial state. When $A < 0$ and $D > 0$, the balance between the production of fluctuations by the diffusion term, and their damping by the drift term, produces the *steady-state variance* $\sigma^2(\infty) = D/2|A|$ found in (5.17) and (5.18).

When the full nonlinear form of the Fokker–Planck equation is retained, the simple picture of drift and diffusion loses much of its content. To begin with, the mean and variance no longer evolve independently; nor do they even, in general, obey a coupled *pair* of equations – we can expect all of the moments, $\langle x^n \rangle$, $n = 1, 2, \ldots$, to be coupled in an infinite hierarchy of equations. We can still demonstrate the role of $D(x)$ as a source of fluctuations, since in its absence (5.7) reads

$$\frac{\partial P}{\partial t} = -\frac{\partial}{\partial x} A(x) P, \qquad (5.24)$$

which, for an initial sharp distribution, has the solution

$$P(x, t | x_0, 0) = \delta(x - x_0(t)), \qquad (5.25)$$

with

$$\dot{x}_0(t) = A(x_0(t)), \qquad x_0(0) = x_0. \qquad (5.26)$$

This is verified by direct substitution:

$$\frac{\partial}{\partial t} \delta(x - x_0(t))$$

$$= A(x_0(t)) \frac{\partial}{\partial (x_0(t))} \delta(x - x_0(t))$$

$$= \frac{\partial}{\partial (x_0(t))} \Big[A(x_0(t)) \delta(x - x_0(t)) \Big] - \delta(x - x_0(t)) \frac{\partial}{\partial (x_0(t))} A(x_0(t))$$

$$= A(x) \frac{\partial}{\partial (x_0(t))} \delta(x - x_0(t)) - \delta(x - x_0(t)) \frac{\partial}{\partial x} A(x)$$

$$= -A(x) \frac{\partial}{\partial x} \delta(x - x_0(t)) - \delta(x - x_0(t)) \frac{\partial}{\partial x} A(x)$$

$$= -\frac{\partial}{\partial x} A(x) \delta(x - x_0(t)).$$

Thus, an initially sharp distribution remains a sharp distribution. In (5.26), $A(x)$ governs a deterministic motion that is again described by a "drifting" δ-function. On the other hand, if $D(x)$ is nonzero at any point on the trajectory generated by (5.26), the equation of motion for the variance [Eq. (5.13)] shows that the distribution acquires a nonvanishing width. If $D(x)$ remains appropriately small, linearization about the deterministic trajectory is possible, and the picture of a drifting Gaussian can be substituted for the drifting δ-function. However, if both terms are present in the nonlinear Fokker–Planck equation and the diffusion is not small, the distinction between drift and diffusion is rather ambiguous and artificial. We might write (5.7), alternatively, as

$$\frac{\partial P}{\partial t} = \left[-\frac{\partial}{\partial x}\left(A(x) - \frac{1}{2}D'(x) \right) + \frac{1}{2}\frac{\partial}{\partial x}D(x)\frac{\partial}{\partial x} \right] P, \qquad (5.27)$$

where $D'(x) \equiv dD(x)/dx$. Why not call $-\frac{\partial}{\partial x}\left(A(x) - \frac{1}{2}D'(x) \right)P$ the drift term in (5.27)? On its own it generates the drifting δ-function (5.25), but with a modified deterministic equation to replace (5.26). Then $\frac{1}{2}\frac{\partial}{\partial x}D(x)\frac{\partial}{\partial x}P$ is the term adding fluctuations to this picture – the diffusion term. In the full nonlinear case we do best to think in terms of a single, integrated, *nonlinear diffusion process*, rather than in terms of separate drift and diffusion processes.

5.1.2 Steady-State Solution

We will see shortly that linear Fokker–Planck equations can be solved even when they are multidimensional. Nonlinear Fokker–Planck equations are quite a different story. In general even the steady-state solution is impossible to find analytically. The one-dimensional case is rather special in this respect, since it is possible to construct a closed form expression for its steady state solution. There are situations – for example, when potential conditions are satisfied [5.6] – in which the steady-state solution to a multidimensional nonlinear Fokker–Planck equation can be found analytically; nevertheless, these are the exception rather than the rule.

In one-dimension we are looking for a solution $P_{\mathrm{ss}}(x)$ to the equation

$$\frac{d}{dx}\left(-A(x)P_{\mathrm{ss}}(x) + \frac{1}{2}\frac{d}{dx}D(x)P_{\mathrm{ss}}(x) \right) = 0. \qquad (5.28)$$

This gives the first-order differential equation

$$\frac{d}{dx}\big(D(x)P_{\mathrm{ss}}(x) \big) = 2A(x)P_{\mathrm{ss}}(x) + \text{constant}. \qquad (5.29)$$

If $A(x)P_{\mathrm{ss}}(x)$ and $d\big(D(x)P_{\mathrm{ss}}(x)\big)/dx$ vanish at infinity, the constant is zero and we obtain the equation

$$\frac{1}{D(x)P_{\mathrm{ss}}(x)}\frac{d}{dx}\big(D(\dot{x})P_{\mathrm{ss}}(x) \big) = 2\frac{A(x)}{D(x)},$$

with the solution

$$P_{\mathrm{ss}}(x) = \frac{1}{\mathcal{N}}\frac{1}{D(x)}\exp\left(2\int dx\,\frac{A(x)}{D(x)} \right); \qquad (5.30)$$

$\mathcal{N}$ is a constant set by the normalization condition $\int_{-\infty}^{\infty} dx\, P_{ss}(x) = 1$.

This result can be used to further illustrate how ambiguous a distinction between drift and diffusion is for a nonlinear diffusion process. Consider the steady state obtained from (5.30) with

$$A(x) = -A(x^3 - bx), \tag{5.31a}$$

$$D(x) = D, \tag{5.31b}$$

where A, D, and b are constants. If we introduce

$$V(x) = A\left(\tfrac{1}{4}x^4 - \tfrac{1}{2}bx^2\right), \tag{5.32}$$

such that

$$A(x) = -\frac{d}{dx}V(x), \tag{5.33}$$

we find

$$P_{ss}(x) = \frac{1}{\mathcal{N}} \exp\left(-2\frac{V(x)}{D}\right). \tag{5.34}$$

$V(x)$ is the potential underlying the deterministic evolution generated by $A(x)$ when $D = 0$ – a double-well potential with minima at $x = \pm\sqrt{b}$, for $b > 0$; in terms of V, (5.26) can be written in the form

$$\dot{x}_0 = -\frac{d}{dx_0}V(x_0), \tag{5.35}$$

and the speed of the δ-function (5.25) is determined by the local slope of the potential $V(x_0)$. Now, (5.34) is also the steady-state solution for an entirely different Fokker–Planck equation, with linear drift, and an appropriately chosen nonlinear diffusion. The proof of this is left as an exercise:

Exercise 5.1 Show that (5.34) is also the steady-state solution to the Fokker–Planck equation defined by

$$A(x) = Ax, \tag{5.36a}$$

$$D(x) = e^{2V(x)/D} \int dx\, 2Ax e^{-2V(x)/D}. \tag{5.36b}$$

Of course, the Fokker–Planck equations defined by (5.31) and (5.36) are not equivalent; they have the same steady state, but their time-dependent solutions are different. Nevertheless, we do see that the same double-peaked steady-state distribution can be established both by nonlinear drift and constant diffusion, and linear drift and nonlinear diffusion. In a nonlinear diffusion process the roles played by the terms designated as "drift" and "diffusion" are in some sense interchangeable. This observation underlies the subject of noise-induced "phase" transitions, treated at length in the book by Horsthemke and Lefever [5.7].

5.1.3 Linearization and the System Size Expansion

Little progress would be made with Fokker–Planck equation methods if we relied solely on the good fortune of obtaining equations that can be exactly solved. The harmonic oscillator is rather special in giving a linear, and therefore solvable, Fokker–Planck equation. Later in this book we will treat the laser, and in Volume 2, the degenerate parametric oscillator and optical bistability, by the methods of the quantum–classical correspondence. These examples give multidimensional nonlinear equations; two of them do not give Fokker–Planck equations at all – as we will see shortly, the treatment of two-level atoms using the quantum–classical correspondence produces partial derivatives to all orders in the equation of motion for the phase-space distribution. In such situations progress can only be made using approximations. To prepare ourselves for these difficulties, let us spend a little time discussing the method of *system size expansion* applied to a one-dimensional equation. In appropriate circumstances this method can be used to reduce an equation of motion involving partial derivatives beyond second order to a Fokker–Planck equation – usually to a linear Fokker–Planck equation.

The discussion which follows is based on the systematic treatment of fluctuations in classical stochastic systems worked out by van Kampen [5.8]. We begin with the *generalized Fokker–Planck equation*, or what is known in classical stochastic theory as the *Kramers–Moyal expansion* [5.9, 5.10]:

$$\frac{\partial P}{\partial t} = \sum_{k=1}^{\infty} \frac{(-1)^k}{k!} \left(\frac{\partial}{\partial x} \right)^k \left(a_k(x) P \right). \tag{5.37}$$

This equation is formally equivalent to the master equation for a classical jump process, which is itself equivalent to the Chapman-Kolmogorov equation (5.2). Our derivation of such an equation in quantum optics is not grounded in the Chapman-Kolmogorov equation, but proceeds formally from an operator master equation via the methods described in the previous two chapters. Nevertheless, (5.37) provides a general form (in one dimension) for the equation of motion for the phase-space distribution obtained via the quantum–classical correspondence. Two difficulties with this equation usually have to be addressed: First, the appearance of derivatives beyond second order. Second, even if these higher-order derivatives are dropped, this will generally leave a *nonlinear* Fokker–Planck equation; for a multidimensional problem, such an equation will almost certainly be impossible to solve. Both of these difficulties can often be removed on the basis of a "small-noise" approximation.

The central idea is that the picture of the drifting δ-function provided by (5.25) and (5.26) should come pretty close to the exact description if the fluctuations are sufficiently small; all we should need to add is a small, finite width for the drifting distribution. It seems reasonable that this distribution be approximated by a narrow Gaussian, and we have seen that Gaussian distributions are obtained from linear Fokker–Planck equations. The system size expansion follows a systematic path from (5.37) to such a description, basing its development on an expansion in terms of a small parameter related to the inverse of the system "size". The systematic approach offered by the system size expansion leads in a single step to a *linear* Fokker–Planck equation, simultaneously taking care of both of the difficulties mentioned above. This is the consistent thing to do, rather than simply truncating derivatives beyond second order and accepting the nonlinear Fokker–Planck equation that results. As will become clear below, retaining the nonlinearity after truncation brings corrections to the linearized form of the Fokker–Planck that are of the same order as terms that have already been dropped. It is therefore inconsistent not to linearize as well as truncate. There are special circumstances where the lowest order treatment of fluctuations must be nonlinear; these will be shown to us in a natural way by the system size expansion itself.

We must look for an expansion parameter that can take us to the limit of zero fluctuations. What is the rationale for expecting such a limiting procedure to be possible? How can the limit be taken formally? Our interest is with intrinsic fluctuations arising in the microscopic quantum processes that govern the interaction of light with matter. The quantized, or discrete, nature of this interaction is the fundamental source of the fluctuations: photon numbers change discretely, and material states follow suit as photons are exchanged with the optical field. If the number of quanta in the field and the number of interacting material states are large, we might expect the fluctuations associated with individual transitions to be small on the scale of the average behavior. Let us imagine we can scale the "size" of a given system with some *system size parameter* Ω, to obtain a family of systems, all with the same average behavior, but whose fluctuations decrease relative to the mean as Ω is increased. Let x specify a state in microscopic units (numbers of photons, for example), which therefore scales with system size, and let $\bar{x}$ specify the macroscopic state whose average does not change with Ω. We propose a scaling relationship

$$x = \Omega^p \bar{x}. \tag{5.38}$$

This is a generalization of the relationship postulated for a classical jump process [5.8]. In that relationship $p = 1$. We need the more general form, specifically, to include the case $p = 1/2$, which is appropriate for optical field amplitudes.

Consider the example of an optical field amplitude. Let x be the amplitude of an optical cavity mode in units such that x^2 measures the number of photons in the cavity; thus, x corresponds to the variable α in (3.47), (4.14), or (4.37) – forget for the moment the two-dimensional character of the field. The cavity mode interacts with some intracavity medium. The relevant quantity for describing this interaction at the macroscopic level is not the photon number, but the energy density in the medium. We therefore choose $\bar{x}$ to be scaled so that $\bar{x}^2 \sim 1$ corresponds to energy densities in the range typical of the behavior to be studied (for example, the saturation of a two-level atom, the turn on of a parametric oscillator). The size of the cavity can be scaled up, increasing the photon number x^2 corresponding to any fixed energy density $\bar{x}^2$. If n_0 is the photon number at each cavity size corresponding to the reference energy density $\bar{x}^2 = 1$, we would write (5.38) as

$$x = n_0^{1/2}\bar{x}.$$

In this example Ω is a reference photon number and $p = 1/2$.

For a second example let x correspond to the inversion of a two-level medium. The relevant quantity for describing the macroscopic properties of the medium is the inversion density, giving the number of atoms per unit volume available for absorption or emission. Define $\bar{x}$ as the inversion density divided by the atomic density N/V (for N atoms uniformly distributed in a volume V). Systems of increasing size, with fixed atomic density and inversion density $\bar{x}$, have

$$x = N\bar{x}.$$

In this case Ω is a number of atoms and $p = 1$.

The system size expansion now works as follows. We assume that as Ω increases, some mean motion $\bar{x}_0(t)$ is preserved, while fluctuations about this mean decrease. We assume a scaling of the fluctuations such that

$$\bar{x} = \bar{x}_0(t) + \Omega^{-q}\xi, \tag{5.39a}$$

and introduce the change of variable

$$x = \Omega^p \bar{x}_0(t) + \Omega^{p-q}\xi. \tag{5.39b}$$

The new variable ξ is to be of the same order as $\bar{x}_0(t)$, and q must be determined self-consistently from the description of the fluctuations provided by the generalized Fokker–Planck equation (5.37). Setting

$$\bar{P}(\xi,t) = \Omega^{p-q}P\big(\Omega^p\bar{x}_0(t) + \Omega^{p-q}\xi, t\big), \tag{5.40}$$

the generalized Fokker–Planck equation becomes

$$\frac{\partial \bar{P}}{\partial t} = \Omega^{p-q}\left(\frac{\partial P}{\partial \bar{x}_0(t)}\frac{d\bar{x}_0(t)}{dt} + \frac{\partial P}{\partial t}\right)$$

$$= \Omega^q \frac{\partial \bar{P}}{\partial \xi}\frac{d\bar{x}_0(t)}{dt} + \sum_{k=1}^{\infty}\frac{(-1)^k}{k!}\left(\Omega^{q-p}\frac{\partial}{\partial \xi}\right)^k\big(a_k\bar{P}\big).$$

Assuming $P(x,t)$ is normalized with respect to the variable x, $\bar{P}(\xi,t)$ has been defined so that it is normalized with respect to the variable ξ. We now make a Taylor expansion of the functions $a_k(x)$ about the mean motion $\Omega^p\bar{x}_0(t)$:

$$\frac{\partial \bar{P}}{\partial t} = \Omega^q \frac{\partial \bar{P}}{\partial \xi} \frac{d\bar{x}_0(t)}{dt}$$

$$- \Omega^{q-p} \frac{\partial}{\partial \xi} \left[a_1\big(\Omega^p \bar{x}_0(t)\big) + \Omega^{p-q} \xi a_1'\big(\Omega^p \bar{x}_0(t)\big) + \frac{1}{2} \Omega^{2(p-q)} \right.$$

$$\left. \times \xi^2 a_1''\big(\Omega^p \bar{x}_0(t)\big) + \cdots \right] \bar{P}$$

$$+ \frac{1}{2} \Omega^{2(q-p)} \frac{\partial^2}{\partial \xi^2} \left[a_2\big(\Omega^p \bar{x}_0(t)\big) + \Omega^{p-q} \xi a_2'\big(\Omega^p \bar{x}_0(t)\big) + \frac{1}{2} \Omega^{2(p-q)} \right.$$

$$\left. \times \xi^2 a_2''\big(\Omega^p \bar{x}_0(t)\big) + \cdots \right] \bar{P}$$

$$+$$

$$\vdots \tag{5.41}$$

where $'$ denotes differentiation with respect to x.

To take things further we need to know how the functions $a_k\big(\Omega^p \bar{x}_0(t)\big)$ scale with Ω. In the context of classical jump processes this scaling can be argued from the dependence of the a_k – the jump moments – on the transition probability for a jump of given length from an initial state x. Our derivation of the Fokker–Planck equation, starting from an operator master equation, cannot rely on the same argument; indeed, the scaling adopted for a jump process [5.8] must be generalized to include variables corresponding to field amplitudes, for which, as we have already noted, $p = 1/2$ rather than $p = 1$. To cover both values of p we propose

$$a_k\big(\Omega^p \bar{x}_0(t)\big) = \Omega^{k(p-1)+1} \bar{a}_k\big(\bar{x}_0(t)\big). \tag{5.42}$$

This fits all of the examples we will meet later on. Then the expansion (5.41) becomes

$$\frac{\partial \bar{P}}{\partial t} = \Omega^q \left[\frac{d\bar{x}_0(t)}{dt} - \bar{a}_1\big(\bar{x}_0(t)\big) \right] \frac{\partial \bar{P}}{\partial \xi}$$

$$- \frac{\partial}{\partial \xi} \xi \left[\bar{a}_1'\big(\bar{x}_0(t)\big) + \frac{1}{2} \Omega^{-q} \xi \bar{a}_1''\big(\bar{x}_0(t)\big) + O\big(\Omega^{-2q}\big) \right] \bar{P} +$$

$$+ \frac{1}{2} \Omega^{2q-1} \frac{\partial^2}{\partial \xi^2} \left[\bar{a}_2\big(\bar{x}_0(t)\big) + \Omega^{-q} \xi \bar{a}_2'\big(\bar{x}_0(t)\big) + O\big(\Omega^{-2q}\big) \right] \bar{P}$$

$$+ O\big(\Omega^{3q-2}\big), \tag{5.43}$$

where $'$ now denotes differentiation with respect to $\bar{x}$.

Note 5.1 With $p = 1/2$ and $\Omega = n_0$, (5.42) gives the scaling

$$a_k\big(n_0^{1/2} \bar{x}_0(t)\big) = n_0^{1-k/2} \bar{a}_k\big(\bar{x}_0(t)\big)$$

for variables corresponding to field amplitudes. It is a little difficult to justify
this scaling at this early stage, while we do not have explicit examples to refer
to. The ultimate justification is found by referring to the examples to which
we will apply the system size expansion (Chaps. 7, 8, 10 and 14). A general
indication of why the scaling works out this way can be given, however. First
note, that in the case of a jump process, the derivatives enter the generalized
Fokker–Planck equation from shift operators $\exp[\pm\partial/\partial x]$. It is clear then
that the coefficients $a_k(x)$ all scale with the same power of Ω, since each shift
operator produces derivatives of all orders; we will not find different functions
of x multiplying derivatives of different orders. Thus, when $p = 1$ (5.42) has
$a_k(\Omega\bar{x}_0(t))$ scaling as Ω for all k. An example of a jump process is provided
by the inversion dynamics for a medium of two-level atoms [see Sects. (6.2.4),
(6.3.4), and (6.3.5)]. The scaling we have proposed for variables corresponding
to field amplitudes is different, with the power of Ω that scales $a_k(\Omega^{1/2}\bar{x}_0(t))$
depending on k. The reason for this is found in the way derivatives enter
the generalized Fokker–Planck equation using the methods described in the
previous two chapters. The central point is that derivatives always enter, not
alone, but as powers of $(\partial/\partial\alpha + \alpha^*)$ and $(\partial/\partial\alpha^* + \alpha)$ – as, for example, in
(3.44a)–(3.44e). In a one-dimensional version, consider the term $\chi_k(d/dx +
x)^k$, which contributes a derivative of order k; the coefficient χ_k is some
parameter in the master equation which characterizes the strength of the
interaction that generates the term $\chi_k(d/dx+x)^k$ in the generalized Fokker–
Planck equation. We see that the scaling of the coefficient of the derivative of
order k is determined by χ_k. We determine the scaling of χ_k by noting that
whenever the term $\chi_k d^k/dx^k$ enters the generalized Fokker–Planck equation,
it brings with it the first derivative term $\chi_k(d/dx)x^{k-1}$. Assuming that the
coefficient of the first derivative scales as $\chi_k x^{k-1} \sim n_0^{1/2}$, then χ_k must scale
as $n_0^{1-k/2}$; this is the scaling given by (5.42).

We have now reached the point at which we impose self-consistency on
our expansion; we require that (5.43) produce fluctuations ξ of the order $\bar{x}_0(t)$
in the limit of large Ω, as was assumed in the ansatz (5.39a). To avoid the
divergence of the first term on the right-hand side the factor in the square
bracket must vanish identically:

$$\frac{d\bar{x}_0(t)}{dt} = \bar{a}_1(\bar{x}_0(t)). \tag{5.44}$$

This is the *macroscopic law* governing the mean motion of the system; it corresponds to our earlier equation (5.26) which governed the motion of the drifting δ-function in the absence of noise. The self-consistency requirement also
sets the size of q. Assuming that $\bar{a}_1'(\bar{x}_0(t))$ and $\bar{a}_2(\bar{x}_0(t))$ are both nonzero,
we must clearly choose $q = 1/2$. Then the right-hand side of (5.43) becomes
an expansion in powers of $\Omega^{-1/2}$, and in the limit of large Ω, the dominant
terms give the linear Fokker–Planck equation

$$\frac{\partial \bar{P}}{\partial t} = \left[-\bar{a}_1'(\bar{x}_0(t)) \frac{\partial}{\partial \xi} \xi + \frac{1}{2} \bar{a}_2(\bar{x}_0(t)) \frac{\partial^2}{\partial \xi^2} \right] \bar{P}. \tag{5.45}$$

Given a trajectory $\bar{x}_0(t)$ satisfying (5.44), equation (5.45) can be solved for a Gaussian distribution that drifts along this trajectory, accumulating a width as it goes by integration over a time-dependent diffusion.

Exercise 5.2 Show that (5.45) has the Gaussian solution

$$\bar{P}(\xi, t) = \frac{1}{\sqrt{2\pi} \sigma(t)} \exp\left[-\frac{(\xi - \langle \xi(t) \rangle)^2}{2\sigma^2(t)} \right], \tag{5.46}$$

with mean

$$\langle \xi(t) \rangle = \langle \xi(0) \rangle \exp\left[\int_0^t du\, \bar{a}_1'(\bar{x}_0(u)) \right], \tag{5.47a}$$

and variance

$$\sigma^2(t) = \exp\left[2 \int_0^t du\, \bar{a}_1'(\bar{x}_0(u)) \right]$$
$$\times \left\{ \sigma^2(0) + \int_0^t du\, \exp\left[-2 \int_0^u dv\, \bar{a}_1'(\bar{x}_0(v)) \right] \bar{a}_2(\bar{x}_0(u)) \right\}. \tag{5.47b}$$

Since the original construction puts the mean motion in $\bar{x}_0(t)$, this solution is to be taken with $\langle \xi(0) \rangle = 0$.

5.1.4 Limitations of the Linearized Treatment of Fluctuations

We will make extensive use of the truncation and linearization procedure provided by the system size expansion. Much of the remainder of this chapter is therefore devoted to linear Fokker–Planck equations. Linearization has its limitations, however, and now is a good time to note some of these.

The most obvious limitation is that Ω may not be very large. Systems of just a few interacting photons and atoms can be expected to exhibit relatively large quantum fluctuations; for these systems the system size expansion is not a good approximation. In fact, this may well be the most interesting situation, since we do not expect many manifestations of quantum fluctuations to survive at a measurable level in a macroscopic system. In the smallest systems – problems such as single-atom resonance fluorescence – it may actually be easier to deal directly with the operator master equation, and not attempt to use phase-space distributions and Fokker–Planck equations at all. On the other hand, when many atoms and photons are involved – in a laser for example – the quantum–classical correspondence, used in conjunction with the system size expansion, provides a powerful approach. For systems of intermediate size, the phase-space method might be tried, but the system size expansion cannot be used. This is a no-man's-land in which little work has been done. We will have more to say about this subject in Volume 2.

Even when Ω is large there are limitations to what can be done with the linear equation (5.45). Our main interest will be with the linearized treatment of fluctuations about a steady state; therefore, let us focus on this case, setting $\bar{x}_0(t) = \bar{x}_{ss}$, with $\bar{a}_1(\bar{x}_{ss}) = 0$. From (5.46) and (5.47), the solution to the Fokker–Planck equation is

$$\bar{P}(\xi, t) = \frac{1}{\sqrt{2\pi}\sigma(t)} \exp\left(-\frac{\xi^2}{2\sigma^2(t)}\right), \qquad (5.48a)$$

with

$$\sigma^2(t) = \sigma^2(0) \exp\left[2\bar{a}_1'(\bar{x}_{ss})t\right] - \frac{\bar{a}_2(\bar{x}_{ss})}{2\bar{a}_1'(\bar{x}_{ss})}\left\{1 - \exp\left[2\bar{a}_1'(\bar{x}_{ss})t\right]\right\}. \qquad (5.48b)$$

Clearly, the procedure we have followed breaks down after some finite time if $\bar{x}_{ss}$ is not a *stable steady state* $(\bar{a}_1'(\bar{x}_{ss}) < 0)$. Even under conditions of marginal stability $(\bar{a}_1'(\bar{x}_{ss}) = 0)$ the variance grows linearly in time, as in (5.19), and eventually the fluctuations grow to be of the order $\Omega^{1/2}$, invalidating the system size expansion. Unstable states $(\bar{a}_1'(\bar{x}_{ss}) > 0)$ have exponentially growing fluctuations which quickly invalidate the linearized treatment. There are situations, then, for which even the lowest-order treatment of fluctuations must include nonlinearities.

It is sometimes possible to simply include the next term in the system size expansion to overcome a breakdown in the linear theory. The critical point in the bistable system defined by the potential (5.32) provides a good example. We write

$$\bar{a}_1(\bar{x}) = -\bar{A}(\bar{x}^3 - b\bar{x}).$$

Then, if $\bar{b} = 0$, $\bar{a}_1'(0) = \bar{a}_1''(0) = 0$, and in the linear theory the *critical point* $\bar{x}_{ss} = 0$ is unstable; fluctuations grow without limit in the manner of Brownian motion [Eq. (5.19)]. But in reality the critical point is stable; nonlinearities in the potential provide a restoring force to constrain the fluctuations (assuming $\bar{a}_1'''(0) < 0$). In such cases the system size expansion can be extended to include this restoring force to lowest order. If we return to (5.43), with $\bar{x}_0(t) = \bar{x}_{ss} = 0$,

$$\frac{\partial \bar{P}}{\partial t} = -\Omega^{-2q}\frac{\partial}{\partial \xi}\xi\left[\frac{1}{6}\xi^2\bar{a}_1'''(0) + O(\Omega^{-q})\right]\bar{P}$$

$$+ \frac{1}{2}\Omega^{2q-1}\frac{\partial^2}{\partial \xi^2}\left[\bar{a}_2(0) + O(\Omega^{-q})\right]\bar{P}$$

$$+ O(\Omega^{3q-2}). \qquad (5.49)$$

The term proportional to $\bar{a}_1'''(0)$ will constrain the fluctuations. However, the choice $q = \frac{1}{2}$ no longer gives a self-consistent treatment of these fluctuations.

Instead, we require $-2q = 2q - 1$, or $q = \frac{1}{4}$, which leads to the *nonlinear* Fokker–Planck equation

$$\frac{\partial \bar{P}}{\partial \tau} = \left[-\frac{1}{6}\bar{a}_1'''(0)\frac{\partial}{\partial \xi}\xi^3 + \frac{1}{2}\bar{a}_2(0)\frac{\partial^2}{\partial \xi^2} \right]\bar{P}, \tag{5.50}$$

with

$$\tau = \Omega^{-1/2}t. \tag{5.51}$$

The *critical fluctuations* described by (5.50) are much larger than fluctuations around steady states that are stable under the linear requirement $\bar{a}_1'(\bar{x}_{ss}) < 0$. On the scale of $\bar{x}$ they are of order $\Omega^{-1/4}$ rather than $\Omega^{-1/2}$. This is because of the very flat potential at the critical point, as shown in Fig 5.2(a). The time scale on which things evolve is also much slower. This is observed at two levels: First there is the scaling of time by the system size in (5.51). Then, according to the macroscopic law (5.44), a small displacement $\delta\bar{x}_0$ from the critical point relaxes according to the equation

$$\frac{d\big(\delta\bar{x}_0(\tau)\big)}{d\tau} = \frac{1}{6}\bar{a}_1'''(0)\big(\delta\bar{x}_0(\tau)\big)^3. \tag{5.52}$$

This nonlinear dynamic is contained in the drift term in the Fokker–Planck equation. The solution to (5.52) is

$$\delta\bar{x}_0(\tau) = \frac{\delta\bar{x}_0}{\sqrt{1 - \frac{1}{3}\bar{a}_1'''(0)(\delta\bar{x}_0)^2\tau}}. \tag{5.53}$$

The displacement $\delta\bar{x}_0$ relaxes as $\tau^{-\frac{1}{2}}$, compared with the exponential decay for a linear force law. This slowed response at a critical point is known as *critical slowing down*. The classic example of critical behavior in quantum optics is provided by the laser at threshold (Sect. 8.2).

Exercise 5.3 Show that (5.50) has the steady state solution

$$\bar{P}_{ss}(\xi) = \frac{1}{\mathcal{N}\bar{a}_2(0)}\exp\left(\frac{\bar{a}_1'''(0)\xi^4}{12\bar{a}_2(0)}\right). \tag{5.54}$$

Of course, there are many variations on this theme. If more derivatives of $\bar{a}_1$ vanish, a higher order nonlinearity must be retained to constrain the fluctuations. There are also other circumstances in which it is not possible to separate average behavior and fluctuations in the manner achieved above. An extended version of our bistable potential illustrates two important problems of this kind. We now write

$$\bar{a}_1(\bar{x}) = -\bar{A}(\bar{x}^3 - \bar{b}\bar{x} + \bar{c}), \tag{5.55}$$

which corresponds to the potential

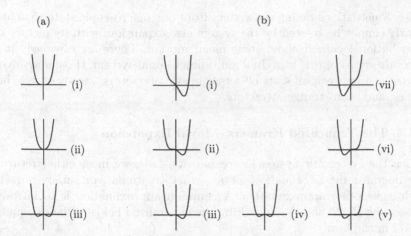

Fig. 5.2 Variation of the potential $\bar{V}(\bar{x}) = \bar{A}(\bar{x}^4/4 - \bar{b}\bar{x}^2/2 + \bar{c}\bar{x})$ with parameters: (a) $\bar{c} = 0$ and (i) $\bar{b} = -1$, (ii) $\bar{b} = 0$, (iii) $\bar{b} = 1$; (b) $\bar{b} = 1$ and (i) $\bar{c} = -1$, (ii) $\bar{c} = -2/3\sqrt{3}$, (iii) $\bar{c} = -1/5$, (iv) $\bar{c} = 0$, (v) $\bar{c} = 1/5$, (vi) $\bar{c} = 2/3\sqrt{3}$, (vii) $\bar{c} = 1$.

$$\bar{V}(\bar{x}) = \bar{A}\left(\tfrac{1}{4}\bar{x}^4 - \tfrac{1}{2}\bar{b}\bar{x}^2 + \bar{c}\bar{x}\right). \tag{5.56}$$

This is the canonical form for the so-called *cusp catastrophe* [5.11, 5.12]. Figure 5.2(b) shows a sequence of potentials that can be accessed with $\bar{c} \neq 0$. For $\bar{c} = \pm 2/3\sqrt{3}$, the equation $\bar{a}_1(\bar{x}) = 0$ that defines the macroscopic steady states has a double root (a root for which $\bar{a}_1(\bar{x}_{\mathrm{ss}}) = \bar{a}_1'(\bar{x}_{\mathrm{ss}}) = 0$). These states are actually unstable to displacements in one direction, as shown by a stability analysis up to second order. Fluctuations are not constrained around the steady state; they lead to a *decay of the unstable state* so that in the long-time limit the distribution will be localized about some other, stable, steady state. This process amplifies the initial fluctuations up to the macroscopic scale, making it impossible to disentangle a mean motion from the fluctuations. A second example of the decay of an unstable state occurs for the steady state at the top of the hump in the double-well potential. Here $\bar{a}_1'(\bar{x}_{ss})$ is positive, and $\bar{x}_{\mathrm{ss}}$ is unstable even in the linear treatment. Initial fluctuations will split the long-time distribution between the two available stable steady states.

Another feature of the bistable system involving macroscopic fluctuations is the process of communication between the sides of the double-well potential. When the depths of the wells are unequal, decay of the unstable state at the top of the potential barrier will first split the distribution between the two sides, producing localized peaks about the two stable macroscopic states. But, in fact, for large Ω, only the absolute minimum of the potential is stable in the presence of fluctuations [the proof of this is given as an exercise (Exercise 5.4)]. Thus, except in the special case where the wells have the same depth, one of the steady states is metastable and decays on a long time scale – often an extremely long time scale – to the other steady state. Fluctuations taking the system from one macroscopic state to another clearly cannot be treated by the system size expansion, with its picture of a distribution localized about some mean motion. There are obviously many generalizations of this idea. In a multidimensional system, these may involve transitions between all sorts of deterministic attractors: steady states, limit cycles, and even strange attractors.

5.1.5 The Truncated Kramers–Moyal Expansion

When the systematic system size expansion fails, we must either return to the operator master equation for an exact treatment, or satisfy ourselves with some other approximation. A common approximation is to introduce the scaling (5.38) and (5.42), writing the generalized Fokker–Planck equation (5.37) in the form

$$\frac{\partial \bar{P}}{\partial t} = \sum_{k=1}^{\infty} \frac{(-1)^k}{k!} \Omega^{1-k} \left(\frac{\partial}{\partial \bar{x}} \right)^k (\bar{a}_k(\bar{x})\bar{P}),$$

and then to truncate this equation at second order to give the nonlinear Fokker–Planck equation

$$\frac{\partial \bar{P}}{\partial t} = \left(-\frac{\partial}{\partial \bar{x}} \bar{a}_1(\bar{x}) + \frac{1}{2} \Omega^{-1} \frac{\partial^2}{\partial \bar{x}^2} \bar{a}_2(\bar{x}) \right) \bar{P}. \tag{5.57}$$

This is not a systematic expansion in inverse powers of Ω because Ω^{-1} controls the sharpness of the peaks in $\bar{P}$, and therefore further Ω-dependence is hidden in the derivatives of the distribution. If the change of variable (5.39a) is now introduced, followed by a Taylor expansion about $\bar{x}_0(t)$, the equations obtained with the system size expansion will be recovered if we retain terms to lowest order as before. Equation (5.57) is often used, however, without taking this extra step. The nonlinear Fokker–Planck equation is taken as a starting point for addressing questions like those we have just raised. It is worth noting that a number of important problems in quantum optics actually produce a nonlinear Fokker–Planck equation in the form of (5.57) without the need for a truncation of higher derivatives. The parametric oscillator is an example of this type. We will discuss this example at some length in Volume 2 (Chaps. 10 and 12).

The steady state solution to (5.57) can be written down for arbitrary functions $\bar{a}_1(\bar{x})$ and $\bar{a}_2(\bar{x})$ following the method of Sect. 5.1.2. Equation (5.30) gives

$$\bar{P}_{\text{ss}}(\bar{x}) = \frac{1}{\mathcal{N}} \frac{\Omega}{\bar{a}_2(\bar{x})} \exp\left(2\Omega \int_0^{\bar{x}} d\bar{x}' \frac{\bar{a}_1(\bar{x}')}{\bar{a}_2(\bar{x}')} \right). \tag{5.58}$$

The extrema of the distribution are given by

$$\frac{d\bar{P}_{\text{ss}}}{d\bar{x}} = \left(-\frac{\bar{a}_2'(\bar{x})}{\bar{a}_2(\bar{x})} + 2\Omega \frac{\bar{a}_1(\bar{x})}{\bar{a}_2(\bar{x})} \right) \bar{P}_{\text{ss}} = 0.$$

These extrema correspond exactly to the steady states given by the macroscopic equation of motion (5.44) if $\bar{a}_2$ is a constant; otherwise, they are shifted from these values; although, when Ω is large, the shift is of the order Ω^{-1}, and therefore much smaller than the fluctuations about the steady state obtained with the system size expansion.

The bistable system defined by (5.55) and (5.56) is widely discussed in the literature using either a nonlinear Fokker–Planck equation like (5.58) or a master equation for a one-step jump process (one-step birth-death master equation) [5.13]. We do not have time to review this subject here. The one result concerning metastable states that we alluded to above is left as an exercise:

Exercise 5.4 It is a little surprising to learn, that in the limit of very small noise, the steady-state distribution for a double well potential with unequal well depths is localized (almost) entirely at the absolute minimum of the potential. Although a second locally stable macroscopic steady state exists at the bottom of the other well, (almost) all of the probability for the system to be found in this state decays over long times – this state is *metastable*. Thus, in passing through the sequence of potentials illustrated in Fig. 5.2(b), a discontinuous transition takes place at $\bar{c} = 0$ between a steady state distribution localized in the well on the left, and one localized in the well on the right. For the potential (5.56), and a constant diffusion $\bar{a}_2(\bar{x}) = \bar{D}$, show that the large Ω limit of the steady state distribution with unequal well depths is given by

$$
\begin{aligned}
\bar{P}_{\text{ss}}(\bar{x}) &= \lim_{\Omega \to \infty} \sqrt{\frac{\Omega \bar{A}(3\bar{x}_{\text{min}}^2 - \bar{b})}{\pi \bar{D}}} \exp\left[-\frac{\Omega \bar{A}(3\bar{x}_{\text{min}}^2 - \bar{b})}{\bar{D}}(\bar{x} - \bar{x}_{\text{min}})^2\right] \\
&= \delta(\bar{x} - \bar{x}_{\text{min}}),
\end{aligned}
\tag{5.59}
$$

where $\bar{x}_{\text{min}}$ is the position of the absolute minimum of the potential. Of course, if $\bar{c}$ is ramped forwards and backwards through the sequence shown in Fig. 5.2(b) on a finite time scale, a dynamic hysteresis will be seen rather than a discontinuous transition.

5.2 Linear Fokker–Planck Equations

We turn now to a detailed look at linear Fokker–Planck equations, equations in the form (5.5). Our first task is to construct the general solution to a multidimensional linear Fokker–Planck equation. From this solution we will derive a number of useful relationships for calculating such things as the covariance matrix and the spectrum of fluctuations. Later we will see how these same results can be obtained, perhaps rather more simply, using stochastic differential equations.

5.2.1 The Green Function

We wish to find the solution $P(\boldsymbol{x}, t|\boldsymbol{x}_0, 0)$ to (5.5) for the initial condition

$$P(\boldsymbol{x}, 0|\boldsymbol{x}_0, 0) = \delta(\boldsymbol{x} - \boldsymbol{x}_0) \equiv \delta(x_1 - x_{10}) \cdots \delta(x_n - x_{n0}). \tag{5.60}$$

It is helpful to first perform a similarity transformation to diagonalize the drift matrix $\boldsymbol{A}$. We write

$$\boldsymbol{y} \equiv \boldsymbol{Sx}, \tag{5.61}$$

where the rows (columns) of $\boldsymbol{S}$ ($\boldsymbol{S}^{-1}$) are the left (right) eigenvectors of $\boldsymbol{A}$ [5.14], such that

$$\hat{\boldsymbol{A}} \equiv \boldsymbol{SAS}^{-1} = \text{diag}(\lambda_1, \ldots, \lambda_n), \tag{5.62}$$

where $\lambda_1, \ldots, \lambda_n$ are the eigenvectors of $\boldsymbol{A}$. We then define

$$\hat{P}(\boldsymbol{y}, t|\boldsymbol{Sx}_0, 0) \equiv P(\boldsymbol{S}^{-1}\boldsymbol{y}, t|\boldsymbol{x}_0, 0); \tag{5.63}$$

in terms of the new variables $\boldsymbol{y}$, (5.5) reads

$$\frac{\partial \hat{P}}{\partial t} = \left(-\boldsymbol{y}'^T \hat{\boldsymbol{A}} \boldsymbol{y} + \tfrac{1}{2} \boldsymbol{y}'^T \hat{\boldsymbol{D}} \boldsymbol{y}'\right) \hat{P}, \tag{5.64}$$

with

$$\boldsymbol{y}' \equiv \begin{pmatrix} \partial/\partial y_1 \\ \vdots \\ \partial/\partial y_n \end{pmatrix} = \left(\boldsymbol{S}^{-1}\right)^T \boldsymbol{x}', \tag{5.65a}$$

$$\hat{\boldsymbol{D}} \equiv \boldsymbol{SDS}^T. \tag{5.65b}$$

The initial condition corresponding to (5.60) is

$$\hat{P}(\boldsymbol{y}, 0|\boldsymbol{Sx}_0, 0) = \det\boldsymbol{S}\, \delta(\boldsymbol{y} - \boldsymbol{Sx}_0). \tag{5.66}$$

Note that this distribution is normalized with respect to the variables $x_1, x_2, \ldots, x_n$, not with respect to the variables $y_1, y_2, \ldots, y_n$.

Equation (5.64) is a linear Fokker–Planck equation with diagonal drift. The Green function solution to equations of this form was derived by Wang and Uhlenbeck [5.15]. A little work generalizing the method of Sect. 3.1.5 leads to their solution. We introduce the Fourier transform

$$\hat{U}(\boldsymbol{u}, t|\boldsymbol{Sx}_0, 0) = \int_{-\infty}^{\infty} dy_1 \cdots dy_n \hat{P}(\boldsymbol{y}, t|\boldsymbol{Sx}_0, 0) \exp\left(i\boldsymbol{y}^T \boldsymbol{u}\right). \tag{5.67}$$

Then the Fourier transform of the Fokker–Planck equation (5.64) is

$$\frac{\partial \hat{U}}{\partial t} = \left(\boldsymbol{u}^T \hat{\boldsymbol{A}} \boldsymbol{u}' - \tfrac{1}{2} \boldsymbol{u}^T \hat{\boldsymbol{D}} \boldsymbol{u}\right) \hat{U}, \tag{5.68}$$

where

$$u \equiv \begin{pmatrix} u_1 \\ \vdots \\ u_n \end{pmatrix} \qquad u' \equiv \begin{pmatrix} \partial/\partial u_1 \\ \vdots \\ \partial/\partial u_n \end{pmatrix}. \tag{5.69}$$

This equation must be solved for the initial condition

$$\hat{U}(u,0|Sx_0,0) = \det S \exp\left[i(Sx_0)^T u\right]. \tag{5.70}$$

We use the method of characteristics [5.16]. Noting the diagonal form of $\hat{A}$, the subsidiary equations are

$$\frac{dt}{1} = \frac{du_1}{-\lambda_1 u_1} = \cdots = \frac{du_n}{-\lambda_n u_n} = \frac{d\hat{U}}{-\frac{1}{2}u^T \hat{D}u\,\hat{U}}, \tag{5.71}$$

and have solutions

$$e^{\hat{A}t}u = c = \text{constant.} \tag{5.72a}$$

Then,

$$\begin{aligned}
\frac{d\hat{U}}{\hat{U}} &= -\tfrac{1}{2}u^T \hat{D}u\,dt \\
&= -\tfrac{1}{2}c^T\left(e^{-\hat{A}^T t}\hat{D}e^{-\hat{A}t}\right)c\,dt \\
&= -\tfrac{1}{2}c^T\left(\hat{D}_{ij}e^{-(\lambda_i+\lambda_j)t}\right)c\,dt,
\end{aligned}$$

where (M_{ij}) denotes the matrix with elements M_{ij}, and we find

$$\begin{aligned}
\ln\hat{U} &= \frac{1}{2}c^T\left(\frac{\hat{D}_{ij}}{\lambda_i+\lambda_j}\left[e^{-(\lambda_i+\lambda_j)t}-1\right]\right)c + \text{constant} \\
&= \frac{1}{2}\tilde{u}\left(\frac{\hat{D}_{ij}}{\lambda_i+\lambda_j}\left[1-e^{(\lambda_i+\lambda_j)t}\right]\right)u + \text{constant.}
\end{aligned}$$

It follows that

$$\hat{U}\exp\left(\tfrac{1}{2}\tilde{u}\hat{Q}(t)u\right) = \text{constant,} \tag{5.72b}$$

where $\hat{Q}(t)$ is the $n \times n$ matrix with elements

$$\hat{Q}_{ij}(t) \equiv -\frac{\hat{D}_{ij}}{\lambda_i+\lambda_j}\left[1-e^{(\lambda_i+\lambda_j)t}\right]. \tag{5.73}$$

Thus, from (5.72a) and (5.72b), the solution for $\hat{U}$ takes the general form

$$\hat{U}(u,t|Sx_0,0) = \phi\left(e^{\hat{A}t}u\right)\exp\left(-\tfrac{1}{2}\tilde{u}\hat{Q}(t)u\right), \tag{5.74}$$

where ϕ is an arbitrary function. Choosing ϕ to match the initial condition (5.70), we find

$$\hat{U}(u,t|Sx_0,0) = \det S \exp\left[i\left(Se^{At}x_0\right)^T u\right]\exp\left(-\tfrac{1}{2}\tilde{u}\hat{Q}(t)u\right). \tag{5.75}$$

In the argument of the first exponential on the right-hand side we have used (5.62) to write

$$(Sx_0)^T(e^{\hat{A}t}u) = (e^{\hat{A}t}Sx_0)^T u = (Se^{At}x_0)^T u.$$

It remains to invert the transformation (5.67). To perform the inversion it is useful to introduce the decomposition

$$\hat{Q}(t) = \hat{R}(t)^T \hat{R}(t). \tag{5.76}$$

Then the Green function solution is

$$\begin{aligned}
\hat{P}&(y, t | Sx_0, 0) \\
&= \frac{\det S}{(2\pi)^n} \int_{-\infty}^{\infty} du_1 \cdots du_n \exp\left[i(Se^{At}x_0)^T u\right] \\
&\quad \times \exp\left(-\tfrac{1}{2}u^T \hat{Q}(t)u\right) \exp\left(-iy^T u\right) \\
&= \frac{\det S}{\sqrt{(2\pi)^n}} \exp\left[-\tfrac{1}{2}(y - Se^{At}x_0)^T \hat{Q}^{-1}(y - Se^{At}x_0)\right] \\
&\quad \times \frac{1}{\sqrt{(2\pi)^n}} \int_{-\infty}^{\infty} du_1 \cdots du_n \exp\left\{-\tfrac{1}{2}\Big[\hat{R}(t)u + i(\hat{R}(t)^T)^{-1} \right. \\
&\quad \times \left. (y - Se^{At}x_0)\Big]^T \Big[\hat{R}(t)u + i(\hat{R}(t)^T)^{-1}(y - Se^{At}x_0)\Big]\right\} \\
&= \frac{\det S}{\sqrt{(2\pi)^n \det \hat{Q}(t)}} \exp\left[-\tfrac{1}{2}(y - Se^{At}x_0)^T \hat{Q}^{-1}(y - Se^{At}x_0)\right].
\end{aligned} \tag{5.77}$$

The integrals leading to the last line are readily performed after transforming to the variables $v \equiv \hat{R}(t)u$. Equations (5.63) and (5.77) now give the general *Green function (conditional distribution) for a linear Fokker–Planck equation*:

$$\begin{aligned}
P&(x, t | x_0, 0) \\
&= \frac{1}{\sqrt{(2\pi)^n \det Q(t)}} \exp\left[-\tfrac{1}{2}(x - e^{At}x_0)^T Q^{-1}(t)(x - e^{At}x_0)\right], \tag{5.78}
\end{aligned}$$

with

$$Q(t) \equiv S^{-1}\hat{Q}(t)(S^{-1})^T. \tag{5.79}$$

This is a multi-dimensional Gaussian, the natural generalization of (5.18). If the eigenvalues $\lambda_1, \ldots, \lambda_n$ all have negative real parts, the distribution (5.78) decays to the general *steady-state distribution for a linear Fokker–Planck equation*:

$$P_{\rm ss}(x) = \frac{1}{\sqrt{(2\pi)^n \det Q_{ss}}} \exp\left(-\tfrac{1}{2}x^T Q_{ss}^{-1}x\right), \tag{5.80}$$

with

$$Q_{\mathrm{ss}} = S^{-1}\hat{Q}_{\mathrm{ss}}(S^{-1})^{T}, \qquad (5.81a)$$

$$(\hat{Q}_{ss})_{ij} = -\frac{\hat{D}_{ij}}{\lambda_i + \lambda_j}. \qquad (5.81b)$$

Note 5.2 The decomposition (5.76) is possible (with $\hat{R}$ a real matrix) because the symmetric matrix $\hat{Q}$ is positive definite, which follows because D is positive definite. For a positive definite matrix M, the quadratic form $z^T M z$ is positive for all nontrivial z; the decomposition $M = N^T N$ expresses the quadratic form as a sum of squares $z^T M z = w^T w$, with $w \equiv Nz$. The requirement that $\hat{Q}$ be positive definite guarantees that the exponential in (5.72b) does not diverge at infinity. This guarantees that the distribution $P(x,t|x_0,0)$ does not diverge at infinity. It may happen that the quantum–classical correspondence leads to a Fokker–Planck equation that does not have positive definite diffusion. In this way, the incompatibility of classical statistics with quantum mechanics is revealed in a particularly direct fashion. There are sometimes technical ways around the problem, but ultimately it arises from the fundamental differences between classical and quantum physics. We will return to this subject as one of our main themes in Volume 2. (In some applications $\hat{Q}$ may be positive *semi*definite. Then it is a singular matrix that generates quadratic forms that are nonnegative, but are not always positive. This corresponds to situations in which there is no diffusion in at least one dimension, and therefore the distribution "drifts" as a δ-function in at least one phase-space direction. To generalize the following mathematics to this case, matrix inverses, which now do not strictly exist, can be interpreted in the sense of a limit in which noise of order ϵ is added in the offending dimensions and then ϵ is taken to zero; in this way Gaussians approach δ-functions.)

5.2.2 Moments of Multi-Dimensional Gaussians

The calculation of averages for a system described by a linear Fokker–Planck equation reduces to the problem of finding the moments of multi-dimensional Gaussian distributions. Consider the normalized distribution

$$P(z) = \frac{1}{\sqrt{(2\pi)^n \det M}} \exp\left[-\tfrac{1}{2}(z - z_0)^T M^{-1}(z - z_0)\right], \qquad (5.82)$$

where z is an n-dimensional vector and M is an $n \times n$ symmetric positive definite matrix. We want to calculate the vector of means

$$\langle z \rangle \equiv \int_{-\infty}^{\infty} dz_1 \cdots dz_n \, z P(z), \qquad (5.83)$$

and the *variances* $(i = j)$ and *correlations* $(i \neq j)$

$$C_{ij} \equiv \left\langle \left(z_i - \langle z_i \rangle\right)\left(z_j - \langle z_j \rangle\right)\right\rangle. \tag{5.84}$$

The $n \times n$ matrix C formed from the C_{ij} is called the *covariance matrix*. It can be expressed in vector notation as

$$
\begin{aligned}
C &\equiv \left\langle \left(z - \langle z \rangle\right)\left(z - \langle z \rangle\right)^T \right\rangle \\
&\equiv \int_{-\infty}^{\infty} dz_1 \cdots dz_n \left(z - \langle z \rangle\right)\left(z - \langle z \rangle\right)^T P(z).
\end{aligned} \tag{5.85}
$$

The integrals needed to evaluate (5.83) and (5.85) are performed using the decomposition introduced in (5.76). For a symmetric, positive definite matrix M, we write

$$M^{-1} = \left(N^{-1}\right)^T N^{-1}. \tag{5.86}$$

Then in terms of the new variables $w \equiv N^{-1}(z - z_0)$, (5.83) becomes

$$
\begin{aligned}
\langle z \rangle &= \det N \int_{-\infty}^{\infty} dw_1 \cdots dw_n \left(z_0 + Nw\right) P(z_0 + Nw) \\
&= z_0 \frac{1}{\sqrt{(2\pi)^n}} \int_{-\infty}^{\infty} dw_1 \cdots dw_n \exp\left(-\tfrac{1}{2} w^T w\right) \\
&\quad + N \frac{1}{\sqrt{(2\pi)^n}} \int_{-\infty}^{\infty} dw_1 \cdots dw_n w \exp\left(-\tfrac{1}{2} w^T w\right) \\
&= z_0.
\end{aligned} \tag{5.87}
$$

Similarly, substituting $\langle z \rangle = z_0$ into (5.85), and using the same change of variables,

$$
\begin{aligned}
C &= \det N \int_{-\infty}^{\infty} dw_1 \cdots dw_n \, N w w^T N^T P(z_0 + Nw) \\
&= N \left[\frac{1}{\sqrt{(2\pi)^n}} \int_{-\infty}^{\infty} dw_1 \cdots dw_n \, w w^T \exp\left(-\tfrac{1}{2} w^T w\right) \right] N^T \\
&= N N^T \\
&= M.
\end{aligned} \tag{5.88}
$$

Moments up to second order are generally all we will need in future applications. In fact, for Gaussian distributions, all higher-order moments can be generated from these; the first two moments completely specify the distribution [5.17].

5.2.3 Formal Solution for Time-Dependent Averages

Using the Green function solution we can construct the time evolution for any initial distribution $P(\boldsymbol{x}, 0)$:

$$P(\boldsymbol{x}, t) = \int_{-\infty}^{\infty} dx_{01} \cdots dx_{0n} \, P(\boldsymbol{x}, t | \boldsymbol{x}_0, 0) P(\boldsymbol{x}_0, 0). \qquad (5.89)$$

Then, since the Green function $P(\boldsymbol{x}, t | \boldsymbol{x}_0, 0)$ [Eq. (5.78)] has the same form as (5.82), we can use (5.89) and (5.87) to obtain the time-dependent means:

$$\begin{aligned}
\langle \boldsymbol{x}(t) \rangle &\equiv \int_{-\infty}^{\infty} dx_1 \cdots dx_n \, \boldsymbol{x} P(\boldsymbol{x}, t) \\
&= \int_{-\infty}^{\infty} dx_1 \cdots dx_n \, \boldsymbol{x} \int_{-\infty}^{\infty} dx_{01} \cdots dx_{0n} \, P(\boldsymbol{x}, t | \boldsymbol{x}_0, 0) P(\boldsymbol{x}_0, 0) \\
&= \int_{-\infty}^{\infty} dx_{01} \cdots dx_{0n} \, P(\boldsymbol{x}_0, 0) \int_{-\infty}^{\infty} dx_1 \cdots dx_n \, \boldsymbol{x} \, P(\boldsymbol{x}, t | \boldsymbol{x}_0, 0) \\
&= \int_{-\infty}^{\infty} dx_{01} \cdots dx_{0n} \, P(\boldsymbol{x}_0, 0) e^{At} \boldsymbol{x}_0 \\
&= e^{At} \langle \boldsymbol{x}(0) \rangle. \qquad (5.90)
\end{aligned}$$

This is the natural generalization of the one-dimensional result (5.16).

The covariance matrix will generally show a dependence on two time arguments. We define the *autocorrelation matrix* $\boldsymbol{C}(t', t)$ to be the $n \times n$ matrix with matrix elements

$$C(t', t)_{ij} \equiv \Big\langle \big[x_i(t') - \langle x_i(t') \rangle \big] \big[x_j(t) - \langle x_j(t) \rangle \big] \Big\rangle, \qquad (5.91)$$

or, in vector notation,

$$\begin{aligned}
\boldsymbol{C}(t', t) &\equiv \Big\langle \big[\boldsymbol{x}(t') - \langle \boldsymbol{x}(t') \rangle \big] \big[\boldsymbol{x}(t) - \langle \boldsymbol{x}(t) \rangle \big]^T \Big\rangle \\
&= \int_{-\infty}^{\infty} dx_1 \cdots dx_n \int_{-\infty}^{\infty} dx_{01} \cdots dx_{0n} \big[\boldsymbol{x} - \langle \boldsymbol{x}(t') \rangle \big] \\
&\quad \times \big[\boldsymbol{x}_0 - \langle \boldsymbol{x}(t) \rangle \big]^T P(\boldsymbol{x}, t' | \boldsymbol{x}_0, t) P(\boldsymbol{x}_0, t); \qquad (5.92)
\end{aligned}$$

here the two-time average is evaluated by integrating against the joint distribution $P(\boldsymbol{x}, t'; \boldsymbol{x}_0, t) = P(\boldsymbol{x}, t' | \boldsymbol{x}_0, t) P(\boldsymbol{x}_0, t)$. We first consider the dependence on the time separation $t' - t$; how do the correlations between variables evaluated at different times behave as the separation of the times increases? For $t' \geq t$, we have

$$C(t',t) = \int_{-\infty}^{\infty} dx_{01} \cdots dx_{0n}\, P(\boldsymbol{x}_0,t)$$

$$\times \left[\int_{-\infty}^{\infty} dx_1 \cdots dx_n \left[\boldsymbol{x} - \langle \boldsymbol{x}(t') \rangle \right] P(\boldsymbol{x},t'|\boldsymbol{x}_0,t) \right] \left[\boldsymbol{x}_0 - \langle \boldsymbol{x}(t) \rangle \right]^T$$

$$= \int_{-\infty}^{\infty} dx_{01} \cdots dx_{0n}\, P(\boldsymbol{x}_0,t) \left[e^{A(t'-t)} \boldsymbol{x}_0 - \langle \boldsymbol{x}(t') \rangle \right]$$

$$\times \left[\boldsymbol{x}_0 - \langle \boldsymbol{x}(t) \rangle \right]^T,$$

where we have used (5.87). Then, using (5.90) and the definition (5.85) of the covariance matrix,

$$C(t',t) = e^{A(t'-t)} \int_{-\infty}^{\infty} dx_{01} \cdots dx_{0n}\, P(\boldsymbol{x}_0,t) \left[\boldsymbol{x}_0 - \langle \boldsymbol{x}(t) \rangle \right] \left[\boldsymbol{x}_0 - \langle \boldsymbol{x}(t) \rangle \right]^T$$

$$= e^{A(t'-t)} \Big\langle \left[\boldsymbol{x}(t) - \langle \boldsymbol{x}(t) \rangle \right] \left[\boldsymbol{x}(t) - \langle \boldsymbol{x}(t) \rangle \right]^T \Big\rangle$$

$$= e^{A(t'-t)} C(t,t), \qquad t' \geq t. \tag{5.93a}$$

By interchanging t and t', (5.93a) can be used to obtain a corresponding result for $t' \leq t$:

$$C(t',t) \equiv \Big\langle \left[\boldsymbol{x}(t) - \langle \boldsymbol{x}(t) \rangle \right] \left[\boldsymbol{x}(t') - \langle \boldsymbol{x}(t') \rangle \right]^T \Big\rangle^T$$

$$= \left[e^{A(t-t')} C(t',t') \right]^T$$

$$= C(t',t') e^{A^T (t-t')}, \qquad t' \leq t. \tag{5.93b}$$

There is a second piece to the time dependence of the autocorrelation matrix contained in its behavior for equal times; we must now look at the time evolution of the covariance matrix $C(t,t)$. We have

$$C(t,t) \equiv \int_{-\infty}^{\infty} dx_1 \cdots dx_n \left[\boldsymbol{x} - \langle \boldsymbol{x}(t) \rangle \right] \left[\boldsymbol{x} - \langle \boldsymbol{x}(t) \rangle \right]^T P(\boldsymbol{x},t)$$

$$= \int_{-\infty}^{\infty} dx_1 \cdots dx_n \left[\boldsymbol{x} - \langle \boldsymbol{x}(t) \rangle \right] \left[\boldsymbol{x} - \langle \boldsymbol{x}(t) \rangle \right]^T$$

$$\times \int_{-\infty}^{\infty} dx_{01} \cdots dx_{0n}\, P(\boldsymbol{x},t|\boldsymbol{x}_0,0) P(\boldsymbol{x}_0,0)$$

$$= \int_{-\infty}^{\infty} dx_{01} \cdots dx_{0n}\, P(\boldsymbol{x}_0,0)$$

$$\times \int_{-\infty}^{\infty} dx_1 \cdots dx_n \left[\boldsymbol{x} - \langle \boldsymbol{x}(t) \rangle \right] \left[\boldsymbol{x} - \langle \boldsymbol{x}(t) \rangle \right]^T P(\boldsymbol{x},t|\boldsymbol{x}_0,0).$$

To carry out the integration over $\boldsymbol{x}$, we write

$$\left[\boldsymbol{x} - \langle\boldsymbol{x}(t)\rangle\right]\left[\boldsymbol{x} - \langle\boldsymbol{x}(t)\rangle\right]^T = \left[\boldsymbol{x} - e^{At}\langle\boldsymbol{x}(0)\rangle\right]\left[\boldsymbol{x} - e^{At}\langle\boldsymbol{x}(0)\rangle\right]^T$$

$$= \left\{(\boldsymbol{x} - e^{At}\boldsymbol{x}_0) + e^{At}\left[\boldsymbol{x}_0 - \langle\boldsymbol{x}(0)\rangle\right]\right\}$$

$$\times \left\{(\boldsymbol{x} - e^{At}\boldsymbol{x}_0) + e^{At}\left[\boldsymbol{x}_0 - \langle\boldsymbol{x}(0)\rangle\right]\right\}^T.$$

The integrals of the four terms arising in this product are then evaluated using

$$\int_{-\infty}^{\infty} dx_1 \cdots dx_n \, (\boldsymbol{x} - e^{At}\boldsymbol{x}_0)(\boldsymbol{x} - e^{At}\boldsymbol{x}_0)^T P(\boldsymbol{x},t|\boldsymbol{x}_0,0) = \boldsymbol{Q}(t),$$

$$\int_{-\infty}^{\infty} dx_1 \cdots dx_n \, (\boldsymbol{x} - e^{At}\boldsymbol{x}_0) P(\boldsymbol{x},t|\boldsymbol{x}_0,0) = 0;$$

these follow from the results of Sect. 5.2.2 and the explicit form for the Green function $P(\boldsymbol{x},t|\boldsymbol{x}_0,0)$ [Eq. (5.78)]. After carrying out the integrals, we find

$$\boldsymbol{C}(t,t) = \int_{-\infty}^{\infty} dx_{01} \cdots dx_{0n} \, P(\boldsymbol{x}_0,0)$$

$$\times \left\{\boldsymbol{Q}(t) + e^{At}\left[\boldsymbol{x}_0 - \langle\boldsymbol{x}(0)\rangle\right]\left[\boldsymbol{x}_0 - \langle\boldsymbol{x}(0)\rangle\right]^T e^{A^T t}\right\}$$

$$= \boldsymbol{Q}(t) + e^{At}\boldsymbol{C}(0,0)e^{A^T t}. \tag{5.94}$$

The matrix $\boldsymbol{C}(0,0)$ appearing on the right-hand side of (5.94) is the co-variance matrix for the initial state; it specifies the initial variances and correlations. If the eigenvalues of $\boldsymbol{A}$ all have negative real parts, the initial state decays to the steady-state distribution (5.80). The contribution to (5.94) coming from the initial covariance matrix decays to zero, and the variances and correlations that survive in the steady state grow in the term $\boldsymbol{Q}(t)$, ultimately taking the form given by (5.81). The picture that unfolds in many dimensions is a simple generalization of the behavior in one dimension following from (5.17).

The solutions given by (5.90), (5.93), and (5.94) are really only useful for formal purposes. For example, rather than calculating $\langle\boldsymbol{x}(t)\rangle$ from (5.90), or $\boldsymbol{C}(t',t)$ from (5.93a), we would normally work directly with the corresponding equations of motion:

$$\frac{d\langle\boldsymbol{x}(t)\rangle}{dt} = \boldsymbol{A}\langle\boldsymbol{x}(t)\rangle,$$

$$\frac{d\boldsymbol{C}(t',t)}{dt'} = \boldsymbol{A}\boldsymbol{C}(t',t), \qquad t' > t.$$

The equation of motion corresponding to (5.94) is not so obvious. When we look back at the definition of $\boldsymbol{Q}(t)$ [Eqs. (5.79), (5.73), and (5.65b)], it is apparent that a little untangling must be done before we can arrive at such an equation.

5.2.4 Equation of Motion for the Covariance Matrix

We first differentiate (5.94) with respect to t to obtain

$$\dot{C} = \dot{Q} + e^{At}\left(AC(0,0) + C(0,0)A^T\right)e^{A^Tt}$$

$$= \dot{Q} - (AQ + QA^T) + AC + CA^T. \tag{5.95}$$

We want to rewrite this equation so that it only involves the matrices C, A, and D. From the definition of Q [Eqs. (5.79), (5.73), and (5.65b)] we write

$$SQS^T = \left(-\frac{\hat{D}_{ij}}{\lambda_i + \lambda_j}\left[1 - e^{(\lambda_i + \lambda_j)t}\right]\right),$$

or, equivalently,

$$\hat{A}SQS^T + SQS^T\hat{A}^T = -\hat{D} + e^{\hat{A}t}\hat{D}e^{\hat{A}^Tt},$$

where the second form follows by recognizing that $\hat{A}$ is the diagonal matrix whose nonzero elements are the eigenvalues $\lambda_1, \ldots, \lambda_n$. Using (5.62) and (5.65b), the second equation gives

$$AQ + QA^T = -D + e^{At}De^{A^Tt}, \tag{5.96}$$

while the first gives

$$\dot{Q} = S^{-1}\left(\hat{D}_{ij}e^{(\lambda_i + \lambda_j)t}\right)(S^{-1})^T$$

$$= S^{-1}e^{\hat{A}t}\hat{D}e^{\hat{A}^Tt}(S^{-1})^T$$

$$= e^{At}De^{A^Tt}. \tag{5.97}$$

From (5.96) and (5.97) we obtain the *equation of motion for* Q:

$$\dot{Q} = AQ + QA^T + D. \tag{5.98}$$

Substituting this result into (5.95) gives the desired *equation of motion for the covariance matrix*:

$$\dot{C} = AC + CA^T + D. \tag{5.99}$$

Notice that Q and C obey the same equation of motion; however, they have different initial conditions, since $Q(0) = 0$, while $C(0,0)$ need not vanish. The difference $C - Q$ obeys an equation similar to (5.98) and (5.99), but without the diffusion matrix D acting as a source. When the eigenvalues of A all have negative real parts, this difference decays to zero in the steady state, as indicated by (5.94).

To solve (5.99) the source term D can be removed by making the transformation

$$\bar{C} \equiv AC + CA^T + D; \tag{5.100}$$

then

$$\dot{\bar{C}} = A\bar{C} + \bar{C}A^T. \tag{5.101}$$

Generally, this gives n^2 linear coupled equations for the matrix elements of $\bar{C}$. If steady-state variances and correlations are all that are required, these may be found by solving the n^2 algebraic equations for the *steady-state covariance matrix* C_∞ obtained directly from (5.99):

$$AC_\infty + C_\infty A^T = -D, \tag{5.102a}$$

where

$$C_\infty \equiv \lim_{t\to\infty} C(t,t). \tag{5.102b}$$

Exercise 5.5 Two harmonic oscillators, both with frequency ω_0, are coupled in the rotating-wave approximation and independently damped by coupling to reservoirs at different temperatures. The Hamiltonian for the coupled oscillator system is

$$H_S = \hbar\omega_0(a^\dagger a + b^\dagger b) + i\hbar g(ab^\dagger - a^\dagger b). \tag{5.103}$$

Show that the Fokker–Planck equation in the P representation is given by

$$
\frac{\partial \tilde{P}}{\partial t} = \left[\frac{\gamma_a}{2}\left(\frac{\partial}{\partial\tilde{\alpha}}\tilde{\alpha} + \frac{\partial}{\partial\tilde{\alpha}^*}\tilde{\alpha}^* \right) + \frac{\gamma_b}{2}\left(\frac{\partial}{\partial\tilde{\beta}}\tilde{\beta} + \frac{\partial}{\partial\tilde{\beta}^*}\tilde{\beta}^* \right) \right.
$$
$$
+ g\left(\frac{\partial}{\partial\tilde{\alpha}}\tilde{\beta} + \frac{\partial}{\partial\tilde{\alpha}^*}\tilde{\beta}^* \right) - g\left(\frac{\partial}{\partial\tilde{\beta}}\tilde{\alpha} + \frac{\partial}{\partial\tilde{\beta}^*}\tilde{\alpha}^* \right)
$$
$$
\left. + \gamma_a\bar{n}_a\frac{\partial^2}{\partial\tilde{\alpha}\partial\tilde{\alpha}^*} + \gamma_b\bar{n}_b\frac{\partial^2}{\partial\tilde{\beta}\partial\tilde{\beta}^*} \right] \tilde{P}, \tag{5.104}
$$

where $\tilde{\ }$ denotes quantities in a frame rotating at the frequency ω_0, γ_a and γ_b are damping constants for the two oscillators, and $\bar{n}_{a,b} = \bar{n}(\omega, T_{a,b})$. Solve (5.102a) to obtain the steady-state expectation values

$$\langle a^\dagger a\rangle_{\rm ss} = \bar{n}_a - (\bar{n}_a - \bar{n}_b)\frac{4g^2/[\gamma_a(\gamma_a + \gamma_b)]}{1 + 4g^2/(\gamma_a\gamma_b)}, \tag{5.105a}$$

$$\langle b^\dagger b\rangle_{\rm ss} = \bar{n}_b + (\bar{n}_a - \bar{n}_b)\frac{4g^2/[\gamma_b(\gamma_a + \gamma_b)]}{1 + 4g^2/(\gamma_a\gamma_b)}, \tag{5.105b}$$

$$\langle ab^\dagger\rangle_{\rm ss} = (\bar{n}_a - \bar{n}_b)\frac{2g/(\gamma_a + \gamma_b)}{1 + 4g^2/(\gamma_a\gamma_b)}. \tag{5.105c}$$

Note that because the reservoirs have different temperatures the coupled oscillator system is maintained away from thermal equilibrium; also that, for $g \to 0$, each oscillator comes to its own independent thermal equilibrium, with $\langle a^\dagger a\rangle_{\rm ss} = \bar{n}_a$ and $\langle b^\dagger b\rangle_{\rm ss} = \bar{n}_b$.

5.2.5 Steady-State Spectrum of Fluctuations

After a system has evolved to a steady state, it is useful to characterize the fluctuations about the steady state in terms of their frequency content. We return here to an earlier comment: a damped electromagnetic field mode is really a *quasi*mode; it possess a linewidth. More generally, if the field mode interacts with some other system, it will not have the simple Lorentzian spectrum obtained from the Fourier transform of (1.116). There will, however, be a broadband component to the spectrum, and a frequency-space decomposition of the steady-state fluctuations determines what it is. When the steady-state fluctuations are described by a linear Fokker–Planck equation, it is straightforward to derive a formal result that accomplishes the decomposition in frequency space.

In the steady state the autocorrelation matrix $C(t', t)$ becomes a function of the time difference $\tau = t' - t$ alone. We define

$$
\boldsymbol{C}_{\mathrm{ss}} \equiv \lim_{t \to \infty} \boldsymbol{C}(t + \tau, t) = \begin{cases} \lim_{t \to \infty} e^{\boldsymbol{A}\tau} \boldsymbol{C}(t, t) & \tau \geq 0 \\ \lim_{t \to \infty} \boldsymbol{C}(t + \tau, t + \tau) e^{-\boldsymbol{A}^T \tau} & \tau \leq 0 \end{cases}
$$

$$
= \begin{cases} e^{\boldsymbol{A}\tau} \boldsymbol{C}_\infty & \tau \geq 0 \\ \boldsymbol{C}_\infty e^{-\boldsymbol{A}^T \tau} & \tau \leq 0 \end{cases}, \tag{5.106}
$$

where we have used (5.93a) and (5.93b) (also $\boldsymbol{C}_{\mathrm{ss}}(0) = \boldsymbol{C}_\infty$). The spectrum of fluctuations is defined by the Fourier transformation of this stationary autocorrelation matrix:

$$
\boldsymbol{T}_{\mathrm{ss}}(\omega) \equiv \frac{1}{2\pi} \int_{-\infty}^{\infty} d\tau \, \boldsymbol{C}_{\mathrm{ss}}(\tau) e^{-i\omega\tau}
$$

$$
= \frac{1}{2\pi} \int_0^{\infty} d\tau \, \exp[(\boldsymbol{A} - i\omega\boldsymbol{I}_n)\tau] \boldsymbol{C}_\infty
$$

$$
+ \frac{1}{2\pi} \int_0^{\infty} d\tau \, \boldsymbol{C}_\infty \exp[(\boldsymbol{A}^T + i\omega\boldsymbol{I}_n)\tau]
$$

$$
= -\frac{1}{2\pi} [(\boldsymbol{A} - i\omega\boldsymbol{I}_n)^{-1} \boldsymbol{C}_\infty + \boldsymbol{C}_\infty (\boldsymbol{A}^T + i\omega\boldsymbol{I}_n)^{-1}]. \tag{5.107}
$$

$\boldsymbol{I}_n$ denotes the $n \times n$ identity matrix, and when evaluating the integrals we have assumed that the eigenvalues of $\boldsymbol{A}$ all have negative real parts (the steady state is stable), so that $\lim_{\tau \to \infty} e^{\boldsymbol{A}\tau} = 0$. We may cast (5.107) into a simpler form, multiplying on the left by $\boldsymbol{A} - i\omega\boldsymbol{I}_n$ and on the right by $\boldsymbol{A}^T + i\omega\boldsymbol{I}_n$, to obtain

$$(\boldsymbol{A} - i\omega \boldsymbol{I}_n)\boldsymbol{T}_{\text{ss}}(\omega)(\boldsymbol{A}^T + i\omega \boldsymbol{I}_n)$$
$$= -\frac{1}{2\pi}\big[\boldsymbol{C}_\infty(\boldsymbol{A}^T + i\omega \boldsymbol{I}_n) + (\boldsymbol{A} - i\omega \boldsymbol{I}_n)\boldsymbol{C}_\infty\big]$$
$$= -\frac{1}{2\pi}(\boldsymbol{C}_\infty \boldsymbol{A}^T + \boldsymbol{AC}_\infty)$$
$$= \frac{1}{2\pi}\boldsymbol{D}.$$

The last step is made using (5.102a). The *steady-state spectrum of fluctuations* is then given by

$$\boldsymbol{T}_{\text{ss}}(\omega) = \frac{1}{2\pi}(\boldsymbol{A} - i\omega \boldsymbol{I}_n)^{-1}\boldsymbol{D}(\boldsymbol{A}^T + i\omega \boldsymbol{I}_n)^{-1}. \qquad (5.108)$$

Equation (5.108) is a very useful result. If the matrices can be multiplied out analytically, each element of $\boldsymbol{T}_{\text{ss}}(\omega)$ is obtained as a ratio of polynomials; and for high-dimensional systems, where analytic manipulation of the matrices is impracticable, the matrix algebra can be implemented directly on a computer. Spectra need not be calculated from (5.108), however. They are often calculated by first deriving explicit expressions for the steady-state autocorrelation functions, as a sum of exponentials. Taking the Fourier transform then gives the spectrum as a sum of Lorentzians. This is the approach we used to calculate the spectrum of the fluorescence from a two-level atom (Sect. 2.3.4); although, in that case the analysis was not based on a Fokker–Planck equation. It is useful to see how (5.108) can be rewritten to explicitly display the Lorentzian structure. To do this we introduce the diagonalized drift matrix $\hat{\boldsymbol{A}}$ [Eq. (5.62)], writing

$$\boldsymbol{T}_{\text{ss}}(\omega) = \frac{1}{2\pi}\big[\boldsymbol{S}^{-1}\hat{\boldsymbol{A}}\boldsymbol{S} - i\omega \boldsymbol{I}_n\big]^{-1}\boldsymbol{D}\big[\boldsymbol{S}^T \hat{\boldsymbol{A}}^T(\boldsymbol{S}^{-1})^T + i\omega \boldsymbol{I}_n\big]^{-1}$$
$$= \frac{1}{2\pi}\boldsymbol{S}^{-1}(\hat{\boldsymbol{A}} - i\omega \boldsymbol{I}_n)^{-1}\hat{\boldsymbol{D}}(\hat{\boldsymbol{A}}^T + i\omega \boldsymbol{I}_n)^{-1}(\boldsymbol{S}^{-1})^T;$$

for the individual matrix element this gives

$$\boldsymbol{T}_{\text{ss}}(\omega)_{ij} = \frac{1}{2\pi}\sum_{kl}(S^{-1})_{ik}(S^{-1})_{lj}\frac{\hat{D}_{kl}}{(\lambda_k - i\omega)(\lambda_l + i\omega)}, \qquad (5.109)$$

where the $(S^{-1})_{ij}$ are the matrix elements of $\boldsymbol{S}^{-1}$.

Exercise 5.6 Use (5.108) to show that the steady-state spectra for the coupled oscillators of Exercise 5.5 are given by

$$\frac{1}{2\pi}\int_{-\infty}^{\infty} d\tau \langle a^\dagger(0)a(\tau)\rangle e^{i\omega\tau}$$
$$= \frac{1}{\pi}\frac{(\gamma_a \bar{n}_a/2)[(\gamma_b/2)^2 + (\omega - \omega_0)^2] + (\gamma_b \bar{n}_b/2)g^2}{(\omega - \omega_0)^4 + [(\gamma_a/2)^2 + (\gamma_b/2)^2 - 2g^2](\omega - \omega_0)^2 + [(\gamma_a\gamma_b/4) + g^2]^2},$$

$$(5.110\text{a})$$

$$\frac{1}{2\pi}\int_{-\infty}^{\infty} d\tau \langle b^{\dagger}(0)b(\tau)\rangle e^{i\omega\tau}$$

$$= \frac{1}{\pi}\frac{(\gamma_b\bar{n}_b/2)[(\gamma_a/2)^2 + (\omega - \omega_0)^2] + (\gamma_a\bar{n}_a/2)g^2}{(\omega - \omega_0)^4 + [(\gamma_a/2)^2 + (\gamma_b/2)^2 - 2g^2](\omega - \omega_0)^2 + [(\gamma_a\gamma_b/4) + g^2]^2}.$$

$$(5.110\text{b})$$

Show that the denominator in these expressions factorizes in the form $[(\omega - \omega_0)^2 + \lambda_+^2][(\omega - \omega_0)^2 + \lambda_-^2]$, where

$$\lambda_{\pm} \equiv -\frac{\gamma_a + \gamma_b}{4} \pm \sqrt{\left(\frac{\gamma_a - \gamma_b}{4}\right)^2 - g^2} \qquad (5.111)$$

are the eigenvalues of the drift matrix $\boldsymbol{A}$ of the Fokker–Planck equation (5.104).

5.3 Stochastic Differential Equations

In classical statistical physics the Fokker–Planck equation provides a dynamical description in terms of an evolving probability distribution which determines the average quantities that would be measured over an ensemble of experiments. An alternative approach to calculating these averages is to find a set of equations whose solutions generate trajectories in phase space, representative of what would be observed in a single experiment. Such trajectories must possess an irregular component modeling processes that are not observed in microscopic detail, but which manifest themselves macroscopically as sources of noise and fluctuations. These *stochastic trajectories* can be generated mathematically by stochastic differential equations – equations of motion that introduce irregularity through fluctuating source terms whose properties are defined in some probabilistic sense. For example, consider the equation

$$\dot{x} = A(x) + B(x)\xi(t), \qquad (5.112)$$

where, at each time t, $\xi(t)$ is chosen from a distribution with zero mean, and some defined variance and correlation properties with respect to its values at earlier times. Assuming there is a sense in which solutions to this equation are defined, it is pretty clear that they are not uniquely determined by an initial choice for x; for a fixed $x(0)$, an infinity of different trajectories must be possible corresponding to different realizations of $\xi(t)$. An ensemble of these trajectories can be averaged at every instant t to obtain the time-dependent averages that might be calculated from a Fokker–Planck equation. Such a mathematical description directly simulates processes as they are observed in the laboratory. If, for a given Fokker–Planck equation, we can find a set of stochastic differential equations that produce the same averages as the Fokker–Planck equation (of course, this must include all multi-time averages), we can speak of an equivalence between descriptions in terms of the Fokker–Planck equation and the stochastic differential equations. Such an equivalence can be set up for every Fokker–Planck equation. Since the stochastic differential equations are sometimes more manageable computationally, we now spend a little time discussing this equivalence. For an in depth study of stochastic differential equations there are many books available [5.3, 5.4, 5.18, 5.19]. The book by Gardiner [5.3] is very useful as a practical guide through the labyrinth of more formal mathematical treatments.

5.3.1 A Comment on Notation

The choice and consistent use of notation in a discussion of stochastic processes can be a bit of a headache. Generally, throughout this book we do not bother to distinguish between random variables and the values they take. Thus, the quantity inside the average $\langle x \rangle$ is a random variable; the conditional distribution $P(x, t|x_0, 0)$ is a function of the possible values taken by a random variable. Time dependence leads us into further notational subtleties. The solutions to Fokker–Planck equations are probability densities showing an explicit dependence on time. Is the time dependence in $\langle x(t) \rangle$ on the averaged quantity, or on the quantity averaged? The former sense seems to be the more natural expression of the explicit time dependence in the probability density. However, now that we are speaking of ensemble averages of trajectories, we are surely averaging time-dependent quantities. While elsewhere we can afford to be a bit sloppy, in this section on stochastic differential equations notational niceties affect the clarity of the presentation. It will perhaps be helpful if we are a little pedantic about notation here.

First, we must distinguish between random variables and the values they take. We use uppercase letters for random variables (or a caret ˆ on Greek characters) and lower case letters for the values they take. Second, when describing a stochastic process we deal either with sequences of random variables in discrete time, or families of random variables in continuous time. Thus, for a process that evolves over the discrete times $t_0, t_1, \ldots$, separated by time step Δt, we define the sequence of random variables $X_0, X_1, \ldots$, and denote the values these random variables take by $x_0, x_1, \ldots$. Note that different random variables describe the statistics at different times. These random variables may be independent or they may be correlated. For a process evolving in continuous time we define the family of random variables X_t, parameterized by the time t, and let x_t denote the values taken by X_t. Conditional probability densities are now written $P(x_n, t_n|x_0, t_0)$ or $P(x_t, t|x_0, 0)$, where

$$P(x_n, t_n|x_0, t_0)dx_n \equiv \mathrm{Prob}\big(x_n \leq X_n < x_n + dx_n | X_0 = x_0\big), \quad (5.113a)$$

$$P(x_t, t|x_0, t_0)dx_t \equiv \mathrm{Prob}\big(x_t \leq X_t < x_t + dx_t | X_0 = x_0\big). \quad (5.113b)$$

The probability density for realizing a particular sequence of outcomes $\{x_0, x_1, \ldots, x_n\}$ in discrete time, given $X_0 = x_0$, is written $P(x_1, t_1; \ldots; x_n, t_n | x_0, t_0)$, where

$$P(x_1, t_1; \ldots; x_n, t_n | x_0, t_0) \prod_{i=1}^{n} dx_i$$

$$\equiv \mathrm{Prob}\big(x_i \leq X_i < x_i + dx_i, \forall\, i = 1, \ldots, n | X_0 = x_0\big). \quad (5.114)$$

5.3.2 The Wiener Process

The Wiener process plays the important role of providing the elementary fluctuating terms that go into the construction of stochastic differential equations. We will discuss the Wiener process in one dimension. Its generalization to many dimensions is a simple extension to a collection of independent one-dimensional processes.

The Wiener process is described by the Fokker–Planck equation with zero drift and unit diffusion. In our new notation,

$$\frac{\partial P(w_t, t | w_0, 0)}{\partial t} = \frac{1}{2} \frac{\partial^2}{\partial w_t^2} P(w_t, t | w_0, 0). \quad (5.115)$$

The Green function solution to this equation provides the familiar description of Brownian motion, or unconstrained diffusion. From (5.19) [or (5.78)], we have

$$P(w_t, t | w_0, 0) = \frac{1}{\sqrt{2\pi t}} \exp\left[-\frac{1}{2} \frac{(w_t - w_0)^2}{t}\right]; \quad (5.116)$$

the initial δ-function distribution evolves with constant mean

$$\langle W_t \rangle = w_0, \quad (5.117a)$$

and a variance increasing linearly in time:

$$\langle (W_t - w_0)^2 \rangle = t. \quad (5.117b)$$

Correlations at unequal times are obtained from (5.93a) and (5.93b) with $A = 0$:

$$\langle (W_{t'} - w_0)(W_t - w_0) \rangle = \begin{cases} \langle (W_t - w_0)^2 \rangle & t' \geq t \\ \langle (W_{t'} - w_0)^2 \rangle & t' \leq t \end{cases}$$

$$= \min(t', t). \quad (5.117c)$$

Now, in what sense can we define irregular trajectories whose ensemble averages reproduce (5.117a)–(5.117c)? Let us begin in discrete time. We define a sequence of random variables $\{W_0, W_1, \ldots, W_n\}$ corresponding to the times $t_0 = 0, t_1, \ldots, t_n$, separated by the time-step Δt. The value the random variable W_i takes is denoted by w_i, $-\infty < w_i < \infty$. The initial condition in

(5.116) requires that $W_0 = w_0$ with unit probability. Then, a discrete trajectory, at resolution Δt, is defined by a sequence $\{w_0, w_1, \ldots, w_n\}$, where each random variable W_i adopts a value w_i chosen from some distribution $P(w_i, t_i)$. We obtain the Wiener process when the w_i, $i = 1, \ldots, n$, are chosen from a series of Gaussian distributions, $P(w_i, t_i | w_{i-1}, t_{i-1})$, each conditioned on the value taken by the random variable one step earlier in time. Specifically, we write

$$P(w_i, t_i | w_{i-1}, t_{i-1}) = \frac{1}{\sqrt{2\pi(t_i - t_{i-1})}} \exp\left[-\frac{1}{2}\frac{(w_i - w_{i-1})^2}{t_i - t_{i-1}}\right]. \qquad (5.118)$$

Then the probability density for a prescribed sequence $\{w_0, w_1, \ldots, w_n\}$ is given by

$$P(w_1, t_1; \cdots; w_n, t_n | w_0, 0) = \frac{1}{\sqrt{(2\pi\Delta t)^n}} \prod_{i=1}^{n} \exp\left[-\frac{1}{2}\frac{(w_i - w_{i-1})^2}{\Delta t}\right]. \qquad (5.119)$$

Note 5.3 The process we construct in this way is Markovian. Fokker–Planck equations describe Markov processes, and the Wiener process is an example of a Markov process. A discrete Markov process is completely defined by the conditional probability connecting the values taken by the random variables at successive times. The probability density for a complete sequence is then constructed as a product of conditional distributions as in (5.119) [5.20].

In order to compare ensemble averages of the trajectories defined in this way with (5.117a)–(5.117c), we first write the distribution (5.119) in standard form. Replacing $w_i - w_{i-1}$ by $(w_i - w_0) - (w_{i-1} - w_0)$, and summing the exponents in the product of exponentials, we rewrite (5.119) as

$$P(w_1, t_1; \cdots; w_n, t_n | w_0, 0)$$
$$= \frac{1}{\sqrt{(2\pi)^n \det M}} \exp\left[-\tfrac{1}{2}(\boldsymbol{w} - w_0\mathbf{1})^T \boldsymbol{M}^{-1}(\boldsymbol{w} - w_0\mathbf{1})\right], \qquad (5.120)$$

where $\boldsymbol{w}$ is the column vector constructed from $w_1, \ldots, w_n$, and $\mathbf{1}$ is the column vector with every entry equal to unity; $\boldsymbol{M}^{-1}$ is the $n \times n$ matrix

$$\boldsymbol{M}^{-1} = \frac{1}{\Delta t}\begin{pmatrix} 2 & -1 & 0 & 0 & \cdots & 0 & 0 & 0 \\ -1 & 2 & -1 & 0 & \cdots & 0 & 0 & 0 \\ 0 & -1 & 2 & -1 & \cdots & 0 & 0 & 0 \\ 0 & 0 & -1 & 2 & \cdots & 0 & 0 & 0 \\ \vdots & \vdots & \vdots & \vdots & \ddots & \vdots & \vdots & \vdots \\ 0 & 0 & 0 & 0 & \cdots & 2 & -1 & 0 \\ 0 & 0 & 0 & 0 & \cdots & -1 & 2 & -1 \\ 0 & 0 & 0 & 0 & \cdots & 0 & -1 & 1 \end{pmatrix}, \qquad (5.121)$$

with $\det M^{-1} = (\det M)^{-1} = 1/(\Delta t)^n$. It can be seen by direct multiplication that (5.121) has the inverse

$$M = \Delta t \begin{pmatrix} 1 & 1 & 1 & 1 & \cdots & 1 \\ 1 & 2 & 2 & 2 & \cdots & 2 \\ 1 & 2 & 3 & 3 & \cdots & 3 \\ 1 & 2 & 3 & 4 & \cdots & 4 \\ \vdots & \vdots & \vdots & \vdots & \ddots & \vdots \\ 1 & 2 & 3 & 4 & \cdots & n \end{pmatrix}. \tag{5.122}$$

Note 5.4 The matrix M^{-1} can be reduced by row operations to the matrix with plus unity everywhere along the diagonal, minus unity in the element of every row to the left of the diagonal, and zeros elsewhere: add row n to row $n-1$, then add row $n-1$ to row $n-2$, and so on. Using the reduced matrix, $\det M^{-1} = 1/(\Delta t)^n$ follows immediately.

Equation (5.120) now has the form of (5.82) and we can use the results of Sect. 5.2.2 to obtain moments. From (5.87) and (5.88) we obtain

$$\langle W_i \rangle = w_0, \tag{5.123a}$$

$$\langle (W_i - w_0)^2 \rangle = i\Delta t = t_i, \tag{5.123b}$$

$$\langle (W_j - w_0)(W_i - w_0) \rangle = \min(j, i)\Delta t = \min(t_j, t_i). \tag{5.123c}$$

These results reproduce the continuous time results (5.117) at the discrete times $0, \Delta t, 2\Delta t, \ldots, n\Delta t$.

Exercise 5.7 The probability that a trajectory wanders a distance $w_i - w_0$ in the first i steps is calculated by integrating (5.119) over all possible intermediate values $w_1, \ldots, w_{i-1}$. Show that

$$P(w_i, t_i | w_0, 0) = \frac{1}{\sqrt{2\pi t_i}} \exp\left[-\frac{1}{2} \frac{(w_i - w_0)^2}{t_i} \right]. \tag{5.124}$$

This is the discrete-time version of (5.116).

To recover the results for the Wiener process in continuous time we must take the limit of infinitely many infinitesimal steps, where the sequence of random variables $W_0, W_1, \ldots$ is replaced by the family W_t. In the limit $\Delta t \to 0$, $i \to \infty$, $j \to \infty$, with $t_i \equiv i\Delta t = t$ and $t_j \equiv j\Delta t = t'$ finite, (5.123a)–(5.123c) reproduce (5.117a)–(5.117c), and (5.124) reproduces (5.116).

Note that with Δt finite, (5.123a)–(5.123c) and (5.124) correspond *exactly* to (5.117a)–(5.117c) and (5.116) at the discrete times $t = 0, \Delta t, 2\Delta t,$ $\ldots, n\Delta t$. The discrete trajectories generated by (5.118) describe the Wiener process at reduced resolution, not reduced accuracy; finite steps do not give an approximation that only approaches the Wiener process in the limit of infinitely many infinitesimal steps. This is in contrast to discrete valued (jump) processes, which do only approach the Wiener process in this limit – for example, the random walk, which replaces the conditional distribution (5.118) with an equal probability for fixed jumps, $+\Delta w$ or $-\Delta w$, at each time step [5.21]. In Volume 2 we will discuss the numerical simulation of stochastic differential equations as an approach to solving nonlinear problems where direct solution of the Fokker–Planck equation is not possible (Chap. 12). In such simulations the Wiener process is implemented in the discrete version described above.

5.3.3 Stochastic Differential Equations

Actually, in the simulation of stochastic differential equations we do not work with the Wiener process itself, but with the differential process, or *Wiener increment*, $dW_t \equiv W_{t+dt} - W_t$. We have seen how irregular trajectories can be generated by W_t. We must now define the sense in which these trajectories are solutions to some differential equation. First, however, we should perhaps consider whether or not we can even define the time derivative of W_t.

Consider the probability that the absolute slope of a trajectory calculated over a short interval Δt is greater than some constant k. Using (5.116), this probability is given by

$$
\begin{aligned}
\operatorname{Prob}&\left(\frac{|W_{t+\Delta t} - W_t|}{\Delta t} > k\right) \\
&= \operatorname{Prob}\left(W_{t+\Delta t} - W_t > k\Delta t\right) + \operatorname{Prob}\left(W_{t+\Delta t} - W_t < -k\Delta t\right) \\
&= \int_{w_t+k\Delta t}^{\infty} dw_{t+\Delta t}\, P(w_{t+\Delta t}, t + \Delta t | w_t, t) \\
&\quad + \int_{-\infty}^{w_t - k\Delta t} dw_{t+\Delta t}\, P(w_{t+\Delta t}, t + \Delta t | w_t, t) \\
&= \frac{2}{\sqrt{2\pi\Delta t}} \int_{w_t+k\Delta t}^{\infty} dw_{t+\Delta t} \exp\left[-\frac{1}{2}\frac{(w_{t+\Delta t} - w_t)^2}{\Delta t}\right] \\
&= \frac{2}{\sqrt{2\pi\Delta t}}\left[\frac{\sqrt{2\pi\Delta t}}{2} - O(k\Delta t)\right].
\end{aligned}
\tag{5.125}
$$

In the limit $\Delta t \to 0$, this tends to unity for any k. Therefore, for Δt short enough, $|w_{t+\Delta t} - w_t|/\Delta t$ is almost certain to be greater than any number we care to choose. For almost all trajectories the time derivative of w_t must be infinite; the Wiener process is therefore not differentiable.

Note 5.5 A series expansion for the integral in (5.125) can be given. The result of the integration is readily appreciated, however, from the following consideration. The required integral is the area under a normalized Gaussian whose width is $\sqrt{\Delta t}$, with a central slice of width $k\Delta t$ omitted (Fig. 5.3). Since, as $\Delta t \to 0$, $k\Delta t$ becomes much smaller than $\sqrt{\Delta t}$, no matter how large k is chosen to be, the result claimed below (5.125) follows; the integral approaches the entire area under the Gaussian.

A similar calculation shows that for any k, no matter how small,

$$\lim_{\Delta t \to 0} \text{Prob}\big(|W_{t+\Delta t} - W_t| > k\big) = 0. \tag{5.126}$$

Trajectories w_t are therefore continuous, but everywhere nondifferentiable. In what sense can they be generated as solutions to a differential equation? Let us again begin by considering discrete time. Consider the sequence of random variables $Z_0, Z_1, \ldots$, with

$$Z_i = Z_0 + \Delta t \sum_{k=0}^{i-1} \Xi_k, \tag{5.127a}$$

or

$$\Delta Z_i = Z_i - Z_{i-1} \equiv \Delta t \Xi_{i-1}. \tag{5.127b}$$

Here Z_0 and $\Xi_0, \Xi_1, \ldots$, are independent random variables, and the Ξ_i are Gaussian distributed with zero mean and variance $1/\Delta t$:

$$P(\xi_i) = \sqrt{\frac{\Delta t}{2\pi}} \exp\big(-\tfrac{1}{2}\Delta t \xi_i^2\big), \tag{5.128}$$

with

$$\langle \Xi_i \rangle = 0, \tag{5.129a}$$

$$\langle \Xi_j \Xi_i \rangle = \begin{cases} 0 & i \neq j \\ 1/\Delta t & i = j. \end{cases} \tag{5.129b}$$

(Ξ_i is a random variable; ξ_i denotes the value taken by the random variable.) From (5.127) and (5.129), we find

$$\langle Z_i \rangle = \langle Z_0 \rangle, \tag{5.130a}$$

$$\langle \big(Z_i - \langle Z_0 \rangle\big)^2 \rangle = (\Delta t)^2 \sum_{k=0}^{i-1} \sum_{k'=0}^{i-1} \langle \Xi_k \Xi_{k'} \rangle$$

$$= i\Delta t$$

$$= t_i, \tag{5.130b}$$

and

$$\langle \big(Z_j - \langle Z_0 \rangle\big)\big(Z_i - \langle Z_0 \rangle\big) \rangle = (\Delta t)^2 \sum_{k=0}^{j-1} \sum_{k'=0}^{i-1} \langle \Xi_{t_k} \Xi_{t_{k'}} \rangle$$

$$= \min(j, i)\Delta t$$

$$= \min(t_j, t_i). \tag{5.130c}$$

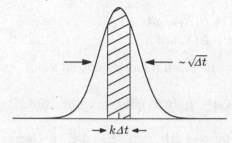

Fig. 5.3 Probability distribution integrated to obtain (5.125). The shaded region which is omitted from the range of integration contributes a term $O(k\Delta t)$.

Equations (5.130) reproduce the results (5.117) for the Wiener process in discrete time. We may now make the identification $Z_i \equiv W_i$ and $\Delta Z_i \equiv \Delta W_i$. Thus, in the limit $\Delta t \to 0$, the Wiener process can be generated as the integral of the *Gaussian white noise* Ξ_t. We write

$$W_t = W_0 + \int_0^t dt' \Xi_{t'}, \tag{5.131a}$$

$$dW_t = \Xi_t dt, \tag{5.131b}$$

where the moments (5.129) become

$$\langle \Xi_t \rangle = 0, \tag{5.132a}$$

$$\langle \Xi_{t'} \Xi_t \rangle = \delta(t' - t). \tag{5.132b}$$

We cannot write down a distribution for Ξ_t as an ordinary function in the continuous time limit because the variance of (5.128) becomes infinite as $\Delta t \to 0$. This infinite variance is carried by the δ-function correlations. The sense of the infinity is well defined, however, by the limiting procedure that led us to (5.131) and (5.132).

Exercise 5.8 Use (5.131a) and (5.132) to arrive directly at (5.117a)–(5.117c).

Equations (5.131a) and (5.131b) define the way in which realizations of the Wiener process are generated as solutions to a differential equation. In the usual notation of differential calculus we would write

$$\dot{W}_t = \Xi_t.$$

However, although this notation is sometimes met in the literature, it is not strictly correct, because, as we have seen, W_t is not differentiable. It is only in the sense of the integral (5.131a) that the equation of motion for the trajectories w_t is defined. We now extend these ideas beyond the Wiener process itself to write an equation of motion for a random variable X_t driven by both deterministic forces and a fluctuating force derived from Ξ_t. This gives the general *stochastic differential equation* in one dimension:

$$dX_t = A(X_t)dt + B(X_t)\Xi_t dt$$
$$= A(X_t)dt + B(X_t)dW_t. \tag{5.133}$$

The interpretation of this equation is to be taken from the integral form

$$X_t = X_0 + \int_0^t dt' A(X_t) + \int_0^t B(X_t)dW_t. \tag{5.134}$$

In numerical simulations the discrete process $\Delta W_i \equiv \Delta t \Xi_i$ ($\Delta t \Xi_i$ is Gaussian distributed with variance Δt) of Wiener increments between t_i and t_{i-1} provides the source of fluctuations.

We have overlooked something here, however. There is a problem of interpretation due to the appearance of the random variable X_t inside the second integral on the right-hand side. The problem is not apparent when we simply write down the standard integral notation in (5.134). But it quickly appears if we think carefully about how the integral is defined as the limit of a sum. We need to know how to interpret the integral in (5.134) when the random variable X_t appears in the integrand. In many applications this dependence is absent, and then the stochastic differential equation is said to involve *additive noise*. When the strength of the fluctuating force depends on the random variable X_t we speak of *multiplicative noise*. Typically, the examples of stochastic differential equations met in quantum optics have multiplicative noise. Often, however, a linearized analysis is performed using the system size expansion discussed in Sect. (5.1.3). In this case the multiplicative noise is approximated by an additive noise, with X_t replaced by $x(t)$ (the solution to the macroscopic law) to determine the noise strength B.

5.3.4 Ito and Stratonovich Integrals

To uncover the ambiguity of the stochastic differential equation (5.134) we consider the integral

$$I \equiv \int_{t_0}^t F_{t'} dW_{t'}, \tag{5.135}$$

where F_t describes some arbitrary stochastic quantity. Let us seek a definition for this integral starting with an approximation for discrete time. We divide the interval $[t_0, t]$ into i subintervals with time-step $\Delta t = (t - t_0)/i$, and $t_k \equiv t_0 + k\Delta t$, $k = 0, \ldots, i$, as illustrated in Fig. 5.4. Within each subinterval, the time

$$\tau_k \equiv t_k + \alpha \Delta t \tag{5.136}$$

lies a fraction α, $0 \le \alpha \le 1$, of a time-step from the lower limit of the subinterval. Now set

$$I_i \equiv \sum_{k=0}^{i-1} F_{\tau_k}\left(W_{t_{k+1}} - W_{t_k}\right). \tag{5.137}$$

One possible choice for F_t is W_t. With this choice we are able to calculate the average value of the integral using (5.117c). We obtain

$$\langle I_i \rangle = \sum_{k=0}^{i-1} \left(\langle W_{\tau_k} W_{t_{k+1}} \rangle - \langle W_{\tau_k} W_{t_k} \rangle \right)$$

$$= \sum_{k=0}^{i-1} \left\{ \left[w_0^2 + \min(\tau_k, t_{k+1}) \right] - \left[w_0^2 + \min(\tau_k, t_k) \right] \right\}$$

$$= \alpha(i\Delta t)$$

$$= \alpha(t - t_0). \tag{5.138}$$

The answer can take any value between 0 and $(t - t_0)$ depending on our choice of α.

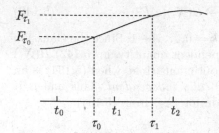

Fig. 5.4 Subdivision of the range of integration in the stochastic integral (5.135).

Different choices for α define different stochastic integrals. Two choices have received wide use. The first sets $\alpha = 0$ and evaluates F_t at the beginning of each interval. This choice gives the definition of the *Ito stochastic integral* [5.22] as the mean square limit of the sequence of approximations

$$I_i^{\text{Ito}} \equiv \sum_{k=0}^{i-1} F_{t_k} \left(W_{t_{k+1}} - W_{t_k} \right). \tag{5.139}$$

The second choice sets $\alpha = \frac{1}{2}$, evaluating F_t at the midpoint of each interval. This gives the definition of the *Stratonovich stochastic integral* [5.23] as the mean square limit of the approximations

$$I_i^{\text{Strat}} \equiv \sum_{k=0}^{i-1} F_{\frac{1}{2}(t_{k+1}+t_k)} \left(W_{t_{k+1}} - W_{t_k} \right). \tag{5.140}$$

For given functions $A(X_t)$ and $B(X_t)$, (5.134) defines two quite different equations depending on the interpretation given to the stochastic integral. The stochastic calculus derived on the basis of the Ito integral is quite different from that based on the Stratonovich integral. The Stratonovich integral leads to conventional rules of calculus; the Ito integral defines a new calculus.

From the point of view of analysis it might seem preferable to always work with Stratonovich integrals and conventional calculus. However, calculations that are quite straightforward using Ito calculus can be quite difficult in the Stratonovich form. The details of this comparison are beyond the scope of this book, however, we get a taste for the differences from the discrete time implementation of (5.134). Let us look briefly at this. A deeper discussion is given in the book by Gardiner [5.24].

The numerical simulations discussed in Volume 2 use the *Cauchy-Euler procedure* to generate approximate realizations of (5.134) interpreted as an Ito equation:

$$X_i = X_0 + \Delta t \sum_{k=0}^{i-1} A(X_k) + \sum_{k=0}^{i-1} B(X_k)(\Delta t \hat{\xi}_k); \qquad (5.141)$$

the solution to (5.141) advances with

$$X_{i+1} = X_i + \Delta t A(X_i) + B(X_i)(\Delta t \Xi_i). \qquad (5.142)$$

From (5.141), X_i depends on X_0 and Ξ_k, $k = 0, \ldots, i-1$. Since the Gaussian random variable Ξ_i is a statistically independent quantity, in (5.142) $B(X_i)$ and Ξ_i are statistically independent. In continuous time, when (5.134) is an *Ito equation*, $B(X_t)$ and dW_t are statistically independent. This makes it rather easy to obtain an evolution equation for $\langle X_t \rangle$ from (5.134):

$$\langle X_t \rangle = \langle X_0 \rangle + \int_0^t dt' \langle A(X_{t'}) \rangle + \int_0^t dt' \langle B(X_{t'}) \rangle \langle dW_{t'} \rangle$$

$$= \langle X_0 \rangle + \int_0^t dt' \langle A(X_{t'}) \rangle, \qquad (5.143a)$$

where we have set $\langle dW_{t'} \rangle = 0$ [Eqs. (5.131b) and (5.132a)]. In differential form,

$$\frac{d}{dt} \langle X_t \rangle = \langle A(X_t) \rangle. \qquad (5.143b)$$

Using the more general statement that $B(X_t)$ and $dW_{t'}$ are statistically independent for $t' \geq t$, we can derive an evolution equation for $\langle X_t^2 \rangle$. From (5.134), we have

$$\langle X_t^2 \rangle = \langle X_0^2 \rangle + 2 \int_0^t t' \langle X_0 A(X_{t'}) \rangle + \int_0^t dt' \int_0^t dt'' \langle A(X_{t'}) A(X_{t''}) \rangle$$

$$+ 2 \int_0^t dt' \int_0^t \langle A(X_{t'}) B(X_{t''}) dW_{t''} \rangle$$

$$+ \int_0^t \int_0^t \langle B(X_{t'}) B(X_{t''}) dW_{t'} dW_{t''} \rangle + 2 \int_0^t \langle X_0 B(X_{t'}) dW_{t'} \rangle.$$

The statistical independence of $B(X_{t'})$ and $dW_{t'}$ allows us to set the last term on the right-hand side to zero. Also, in the second-to-last term $dW_{\max(t',t'')}$

is independent of the other three quantities in the average. The average is therefore zero for $t' \neq t''$. But when the time arguments are equal the product $B(X_{t'})B(X_{t''})$ is independent of $dW_{t'}dW_{t''}$. We can therefore factorize $\langle B(X_{t'})B(X_{t''})dW_{t'}dW_{t''}\rangle$ as $\langle B(X_{t'})B(X_{t''})\rangle\langle dW_{t'}dW_{t''}\rangle$. We now have

$$\langle X_t^2 \rangle = \langle X_0^2 \rangle + 2\int_0^t dt' \langle X_0 A(X_{t'})\rangle + \int_0^t dt' \int_0^t dt'' \langle A(X_{t'})A(X_{t''})\rangle$$

$$+ 2\int_0^t dt' \int_0^{t'} \langle A(X_{t'})B(X_{t''})dW_{t''}\rangle$$

$$+ \int_0^t \int_0^t \langle B(X_{t'})B(X_{t''})\rangle\langle dW_{t'}dW_{t''}\rangle. \tag{5.144a}$$

Differentiating with respect to time, we obtain

$$\frac{d}{dt}\langle X_t^2 \rangle = 2\left\langle \left[X_0 + \int_0^t dt' A(X_{t'}) + \int_0^t B(X_{t'})dW_{t'} \right] A(X_t) \right\rangle$$

$$+ 2\int_0^t dt' \langle B(X_t)B(X_{t'})\rangle\delta(t' - t)$$

$$= 2\langle X_t A(X_t)\rangle + \langle B(X_t)^2 \rangle, \tag{5.144b}$$

where we have used (5.131b) and (5.132b) to write $\langle dW_t dW_{t'}\rangle/dt = dt'\delta(t' - t)$.

Now consider the discrete-time version of (5.134) in the Stratonovich interpretation. In place of (5.142), we have

$$X_{i+1} = X_i + \Delta t A\left(X_{i+\frac{1}{2}}\right) + B\left(X_{i+\frac{1}{2}}\right)(\Delta t \Xi_i), \tag{5.145}$$

where $X_{i+\frac{1}{2}}$ denotes the random variable X_t with $t = \frac{1}{2}(t_{i+1} + t_i)$. We calculate $\langle X_{i+1}\rangle - \langle X_i\rangle$, from which results equivalent to (5.143a) and (5.143b) will follow. Since $B\left(X_{i+\frac{1}{2}}\right)$ is evaluated at a slightly later time than Ξ_i, we should not assume that these two quantities are statistically independent. Instead we must calculate both $B\left(X_{i+\frac{1}{2}}\right)$ and $A\left(X_{i+\frac{1}{2}}\right)$ in some approximate way in terms of Ξ_i, and $A(X_i)$ and $B(X_i)$. The calculation only needs to be correct to lowest order in Δt. We first use linear interpolation to write

$$X_{i+\frac{1}{2}} = X_i + \frac{1}{2}(X_{i+1} - X_i), \tag{5.146}$$

and then expand $A\left(X_{i+\frac{1}{2}}\right)$ and $B\left(X_{i+\frac{1}{2}}\right)$ to first order about X_i. Equation (5.145) becomes

$$X_{i+1} = X_i + \Delta t\left[A(X_i) + \frac{1}{2}(X_{i+1} - X_i)A'(X_i)\right]$$

$$+ \left[B(X_i) + \frac{1}{2}(X_{i+1} - X_i)B'(X_i)\right](\Delta t \Xi_i).$$

Here X_{i+1} appears on both sides of the equation. Solving for X_{i+1} to lowest order in Δt, we have

$$X_{i+1} - X_i = \Delta t A(X_i) + B(X_i)(\Delta t \Xi_i). \tag{5.147a}$$

Substituting this result back into the original equation, we obtain

$$X_{i+1} = X_i + \Delta t A(X_i)\left[1 + \tfrac{1}{2} B'(X_i)(\Delta t \Xi_i) + \tfrac{1}{2} A'(X_i)\right]$$
$$+ B(X_i)(\Delta t \Xi_i)\left[1 + \tfrac{1}{2} B'(X_i)(\Delta t \Xi_i) + \tfrac{1}{2} A'(X_i)\Delta t\right]. \tag{5.147b}$$

It appears that (5.147b) includes four corrections of order Δt^2 when it is compared with (5.147a). But this is not so. After taking the average we see that the term $\tfrac{1}{2} B(X_i)B'(X_i)(\Delta t \Xi_i)^2$ is really of order Δt, since $\langle \Xi_i^2 \rangle = 1/\Delta t$ [Eq. (5.129b)]; we should view $(\Delta t \Xi_i)$ as a term of order $\sqrt{\Delta t}$, so that (5.147b) is an expansion in powers of $\sqrt{\Delta t}$. Then to lowest order we arrive at the result

$$\langle X_{i+1} \rangle - \langle X_i \rangle = \Delta t \left[A(X_i) + \tfrac{1}{2} B(X_i)B'(X_i)\right]. \tag{5.148}$$

This equation of motion for the mean is not the same as the equation of motion obtained from the Ito interpretation of the integral [Eq. (5.143b)]; it includes the additional term $\tfrac{1}{2} B(X_t)B'(X_t)$ on the right-hand side. The additional term reflects the fact that in (5.145), $B(X_{i+\frac{1}{2}})$ and Ξ_i are not statistically independent. In continuous time, when (5.134) is a *Stratonovich equation*, $B(X_t)$ and dW_t *are not statistically independent*.

We must remember then, that for given functions $A(X_t)$ and $B(X_t)$, (5.134) describes different stochastic processes depending on the interpretation of the stochastic integral. This has given rise to extensive debate about the "correct" interpretation [5.25, 5.26]. The debate has content when a stochastic differential equation, formulated in a phenomenological manner, provides the fundamental basis for a stochastic model. The physical argument (for external noise) is then that the Stratonovich interpretation holds, because "real" noise is never exactly white; the unconventional Ito calculus stems from the extreme irregularity of truly white noise. For our purposes, however, this debate is of no importance. We always obtain our stochastic differential equations from a previously derived Fokker–Planck equation. The Fokker–Planck equation unambiguously defines the stochastic process. Stochastic differential equations equivalent to the Fokker–Planck equation can be written in either Ito or Stratonovich form. The equations will look different, but each is to be solved using its own calculus, and a consistent application of the correct calculus to each equation will produce the same result.

5.3.5 Fokker–Planck Equations and Equivalent Stochastic Differential Equations

If we compare the moment equations (5.143b) and (5.144b), with (5.11) and (5.12), a relationship between the Ito stochastic differential equation (5.133) [or (5.134)] and the one-dimensional Fokker–Planck equation (5.7) suggests itself. It seems these may be equivalent if we identify the functions $D(x)$ and $B(x)^2$. This is indeed so. With a little imagination we might even guess the form of the Ito stochastic differential equation equivalent to the multidimensional Fokker–Planck equation (5.1). We state the result, which certainly seems eminently reasonable, without proof; the proof is not difficult, but it requires further excursion into the Ito calculus [5.27]: The *Ito stochastic differential equation equivalent to the multidimensional Fokker–Planck equation* (5.1) is given by

$$dX_t = A(X_t)dt + B(X_t)dW_t, \tag{5.149}$$

where X_t is the column vector formed from the families of random variables $X_{1t}, \ldots, X_{nt}$, $A(x)$ is the column vector of the drifts $A_1(x), \ldots, A_n(x)$, the matrix $B(x)$ is defined by the factorization

$$D(x) = B(x)B(x)^T \tag{5.150}$$

of the positive definite diffusion matrix, and W_t is a column vector of n independent Wiener processes.

For the same Fokker–Planck equation an equivalent Stratonovich stochastic differential equation exists. An educated guess as to its form can be made on the basis of a comparison between (5.148) and (5.143b). The Stratonovich interpretation of the Stochastic integral adds the term $\frac{1}{2}B(X_t)B'(X_t)$ to the evolution equation for $\langle X_t \rangle$. If the same stochastic process is to be described by both Ito and Stratonovich stochastic differential equations, the Stratonovich equation must have a different function $A(X_t)$, so that the addition of this term gives the same evolution equation for $\langle X_t \rangle$. This suggests that

$$A^{\text{Strat}}(X_t) = A(X_t) - \tfrac{1}{2}B(X_t)B'(X_t).$$

This is indeed the correct relationship defining the Stratonovich stochastic differential equation equivalent to the one-dimensional Fokker–Planck equation (5.7); the functions $B(X_t)$ are the same in the equivalent Ito and Stratonovich equations – $B^{\text{Strat}}(X_t) = B(X_t) = \sqrt{D(X_t)}$. More generally, the *Stratonovich stochastic differential equation equivalent to the multidimensional Fokker–Planck equation* (5.1) is

$$dX_t = A^{\text{Strat}}(X_t)dt + B(X_t)dW_t, \tag{5.151a}$$

with

$$A_i^{\text{Strat}}(x) = A_i(x) - \frac{1}{2}\sum_{k,j=1}^{n} B_{kj}(x)\frac{\partial}{\partial x_k}B_{ij}(x). \tag{5.151b}$$

5.3.6 Multi-Dimensional Ornstein–Uhlenbeck Process

To close this section let us see how the results of Sect. 5.2 are obtained in the language of stochastic differential equations. Specifically, we will rederive the results for means, variances, and correlations (Sect. 5.2.3). From these all other moments follow.

The stochastic differential equation corresponding to the linear Fokker–Planck equation (5.5) is

$$d\boldsymbol{X}_t = \boldsymbol{A}\boldsymbol{X}_t dt + \boldsymbol{B} d\boldsymbol{W}_t, \tag{5.152a}$$

with

$$\boldsymbol{D} = \boldsymbol{B}\boldsymbol{B}^T, \tag{5.152b}$$

where $\boldsymbol{A}$ and $\boldsymbol{B}$ are constant matrices. The stochastic process described by (5.152) is known as the *multi-dimensional Ornstein–Uhlenbeck process*. This process involves only additive noise, and therefore the issue concerning the difference between Ito and Stratonovich integrals does not arise. The formal solution to (5.152a) is

$$\boldsymbol{X}_t = \boldsymbol{A}\int_0^t ds\,\boldsymbol{X}_s + \boldsymbol{B}\int_0^t d\boldsymbol{W}_s$$

$$= \boldsymbol{A}\int_0^t ds\,\boldsymbol{X}_s + \boldsymbol{B}(\boldsymbol{W}_t - \boldsymbol{W}_0). \tag{5.153}$$

The equation of motion for the means, and hence its formal solution (5.90), follows trivially on averaging (5.153):

$$\langle\boldsymbol{X}_t\rangle = \boldsymbol{A}\int_0^t ds\,\langle\boldsymbol{X}_s\rangle. \tag{5.154}$$

The autocorrelation matrix is then calculated as

$$\boldsymbol{C}(t',t) \equiv \left\langle (\boldsymbol{X}_{t'} - \langle\boldsymbol{X}_{t'}\rangle)(\boldsymbol{X}_t - \langle\boldsymbol{X}_t\rangle)^T \right\rangle$$

$$= \left\langle \left[\boldsymbol{A}\int_0^{t'} ds'(\boldsymbol{X}_{s'} - \langle\boldsymbol{X}_{s'}\rangle) + \boldsymbol{B}(\boldsymbol{W}_{t'} - \boldsymbol{W}_0) \right] \right.$$

$$\left. \times \left[\boldsymbol{A}\int_0^t ds(\boldsymbol{X}_s - \langle\boldsymbol{X}_s\rangle) + \boldsymbol{B}(\boldsymbol{W}_t - \boldsymbol{W}_0) \right]^T \right\rangle$$

$$= \boldsymbol{A}\int_0^{t'} ds'\int_0^t ds\,\langle(\boldsymbol{X}_{s'} - \langle\boldsymbol{X}_{s'}\rangle)(\boldsymbol{X}_s - \langle\boldsymbol{X}_s\rangle)^T\rangle\boldsymbol{A}^T$$

$$+ \boldsymbol{A}\int_0^{t'} ds'\langle(\boldsymbol{X}_{s'} - \langle\boldsymbol{X}_{s'}\rangle)(\boldsymbol{W}_{\min(s',t)} - \boldsymbol{W}_0)^T\rangle\boldsymbol{B}^T$$

$$+ \boldsymbol{B}\int_0^t ds\,\langle(\boldsymbol{W}_{\min(t',s)} - \boldsymbol{W}_0)(\boldsymbol{X}_s - \langle\boldsymbol{X}_s\rangle)^T\rangle\boldsymbol{A}^T$$

$$+ \boldsymbol{B}\boldsymbol{B}^T\min(t',t). \tag{5.155}$$

We have used the statistical independence of $\boldsymbol{X}_{s'}$ and $\boldsymbol{W}_t - \boldsymbol{W}_{s'}$ for $t > s'$, and of $\boldsymbol{X}_s$ and $\boldsymbol{W}_{t'} - \boldsymbol{W}_s$ for $t' > s$. Also, the last term in (5.155) is obtained using $\langle (\boldsymbol{W}_{t'} - \boldsymbol{W}_0)(\boldsymbol{W}_t - \boldsymbol{W}_0)^T \rangle = \boldsymbol{I}_n \min(t', t)$; this follows from (5.117c) [generalized as in (5.130c) to allow for a distribution over w_0] and the independence of the components of $\boldsymbol{W}_t$. Now, differentiating (5.155) with respect to t', for $t' > t$,

$$
\begin{aligned}
\frac{d}{dt'} \boldsymbol{C}(t', t) &= \boldsymbol{A} \Big\langle (\boldsymbol{X}_{t'} - \langle \boldsymbol{X}_{t'} \rangle) \\
&\quad \times \Big[\int_0^t ds (\boldsymbol{X}_s - \langle \boldsymbol{X}_s \rangle)^T \boldsymbol{A}^T + (\boldsymbol{W}_t - \boldsymbol{W}_0)^T \boldsymbol{B}^T \Big] \Big\rangle \\
&= \boldsymbol{A} \langle (\boldsymbol{X}_{t'} - \langle \boldsymbol{X}_{t'} \rangle)(\boldsymbol{X}_t - \langle \boldsymbol{X}_t \rangle)^T \rangle \\
&= \boldsymbol{A} \boldsymbol{C}(t', t), \qquad t' > t,
\end{aligned}
\tag{5.156a}
$$

where the term in the square bracket is set equal to $(\boldsymbol{X}_t - \langle \boldsymbol{X}_t \rangle)^T$ using (5.153) and (5.154). Similarly, differentiation with respect to t, for $t' < t$, gives

$$
\frac{d}{dt} \boldsymbol{C}(t', t) = \boldsymbol{C}(t', t) \boldsymbol{A}^T, \qquad t' < t.
\tag{5.156b}
$$

These differential equations give the formal solutions (5.93a) and (5.93b). Finally, setting $t' = t$ in (5.155) and differentiating with respect to t, we have

$$
\begin{aligned}
&\frac{d}{dt} \boldsymbol{C}(t, t) \\
&= \boldsymbol{A} \Big\langle (\boldsymbol{X}_t - \langle \boldsymbol{X}_t \rangle) \Big[\int_0^t ds (\boldsymbol{X}_s - \langle \boldsymbol{X}_s \rangle)^T \boldsymbol{A}^T + (\boldsymbol{W}_t - \boldsymbol{W}_0)^T \boldsymbol{B}^T \Big] \Big\rangle \\
&\quad + \Big\langle \Big[\boldsymbol{A} \int_0^t ds' (\boldsymbol{X}_{s'} - \langle \boldsymbol{X}_{s'} \rangle) + \boldsymbol{B}(\boldsymbol{W}_t - \boldsymbol{W}_0) \Big] (\boldsymbol{X}_t - \langle \boldsymbol{X}_t \rangle)^T \Big\rangle \boldsymbol{A}^T \\
&\quad + \boldsymbol{B} \boldsymbol{B}^T \\
&= \boldsymbol{A} \langle (\boldsymbol{X}_t - \langle \boldsymbol{X}_t \rangle)(\boldsymbol{X}_t - \langle \boldsymbol{X}_t \rangle)^T \rangle + \langle (\boldsymbol{X}_t - \langle \boldsymbol{X}_t \rangle)(\boldsymbol{X}_t - \langle \boldsymbol{X}_t \rangle)^T \rangle \boldsymbol{A}^T \\
&\quad + \boldsymbol{B} \boldsymbol{B}^T.
\end{aligned}
$$

From (5.152b) and the definition of $\boldsymbol{C}(t, t)$, this gives

$$
\frac{d}{dt} \boldsymbol{C}(t, t) = \boldsymbol{A} \boldsymbol{C}(t, t) + \boldsymbol{C}(t, t) \boldsymbol{A}^T + \boldsymbol{D},
\tag{5.157}
$$

which reproduces the equation of motion (5.99) for the covariance matrix.

6. Quantum–Classical Correspondence for Two-Level Atoms

After our brief diversion we now return to the theme of Chaps. 3 and 4, namely, the transformation of an operator description for a quantum-optical system into the language of classical statistics. So far we have met methods that accomplish this task for systems described entirely in terms of harmonic oscillator creation and annihilation operators. At least we have seen that a Fokker–Planck equation description is possible for the damped harmonic oscillator, in a variety of versions defined by representations based on different operator orderings. We noted also that there is no guarantee that a system of interacting bosons can be described using a Fokker–Planck equation; although, as attested to by the example of the laser (Chap. 8), there are certainly nontrivial examples that can. The methods used to derive phase-space equations for systems of bosons can be generalized to the treatment of two-level atoms, or more generally, multi-level atomic systems. We now develop the representation for atomic states that is needed for our treatment of the laser.

6.1 Haken's Representation and the Damped Two-Level Atom

A variety of phase-space distributions are available for representing atomic states. We will briefly mention some of this variety later on. In general, however, our attention will focus on the representation introduced by Haken and co-workers as a direct extension of the Glauber–Sudarshan P representation. Like the P representation, this is a representation based on a characteristic function in normal order. We must define here what we mean by normal order. The two-level atom is described by pseudo-spin operators σ_-, σ_+, and σ_z (Sect. 2.1). By normal order we mean an ordered operator product $\sigma_+^p \sigma_z^r \sigma_-^q$, with every σ_+ to the left, every σ_- to the right, and σ_z sandwiched in between. Averages for such ordered operators can be calculated from a normal-ordered characteristic function and corresponding distribution defined in the manner of (3.70)–(3.74). This representation for atomic states was introduced by Haken, Risken, and Weidlich in their theory of the laser [6.1]. It is discussed in this context in Haken's book on laser theory [6.2]. A treatment of atomic damping using this representation, developed for three-level, rather than two-level atoms, can be found in the book by Louisell [6.3]. Both of these authors consider a collection of many atoms which, of course, is what must be done to develop a theory of the laser. We begin by considering a single atom and then extend our results to many atoms. Our first objective is to derive a phase-space equation of motion equivalent to the master equation for the damped two-level atom.

6.1.1 The Characteristic Function and Associated Distribution

We introduce the normal-ordered characteristic function

$$\chi_N(\xi, \xi^*, \eta) \equiv \mathrm{tr}\big(\rho e^{i\xi^* \sigma_+} e^{i\eta \sigma_z} e^{i\xi \sigma_-}\big), \tag{6.1}$$

where ξ is a complex variable, ξ^* is its complex conjugate, and η is real. From this characteristic function we can calculate the normal-ordered operator averages

$$\langle \sigma_+^p \sigma_z^r \sigma_-^q \rangle \equiv \mathrm{tr}\big(\rho \sigma_+^p \sigma_z^r \sigma_-^q\big)$$
$$= \frac{\partial^{p+r+q}}{\partial (i\xi^*)^p \partial (i\eta)^r \partial (i\xi)^q} \chi_N \bigg|_{\xi=\xi^*=\eta=0}. \tag{6.2}$$

The distribution $P(v, v^*, m)$ is defined as the three-dimensional Fourier transform of $\chi_N(\xi, \xi^*, \eta)$:

$$P(v, v^*, m) \equiv \frac{1}{2\pi^3} \int d^2\xi \int d\eta \, \chi_N e^{-i\xi^* v^*} e^{-i\xi v} e^{-i\eta m}$$
$$\equiv \frac{1}{2\pi^3} \int_{-\infty}^{\infty} dw \int_{-\infty}^{\infty} ds \int_{-\infty}^{\infty} d\eta \, \chi_N(w + is, w - is, \eta)$$
$$\times e^{-2i(w\vartheta - s\varphi)} e^{-i\eta m}, \tag{6.3}$$

with the inverse relationship

$$\chi_N(\xi, \xi^*, \eta) = \int d^2 v \int dm \, P(v, v^*, m) e^{i\xi^* v^*} e^{i\xi v} e^{i\eta m}$$
$$= \int_{-\infty}^{\infty} d\vartheta \int_{-\infty}^{\infty} d\varphi \int_{-\infty}^{\infty} dm \, P(\vartheta + i\varphi, \vartheta - i\varphi, m)$$
$$\times e^{2i(w\vartheta - s\varphi)} e^{i\eta m}. \tag{6.4}$$

Then, from (6.2) and (6.4),

$$\langle \sigma_+^p \sigma_z^r \sigma_-^q \rangle = \frac{\partial^{p+r+q}}{\partial (i\xi^*)^p \partial (i\eta)^r \partial (i\xi)^q} \int d^2 v \int dm \, P(v, v^*, m)$$
$$\times e^{i\xi^* v^*} e^{i\xi v} e^{i\eta m} \bigg|_{\xi=\xi^*=\eta=0}$$
$$= \big(\overline{v^{*p} m^r v^q}\big)_P, \tag{6.5a}$$

with

$$(\overline{v^{*p}m^r v^q})_P \equiv \int d^2v \int dm\, P(v, v^*, m) v^{*p} m^r v^q. \tag{6.5b}$$

6.1.2 Some Operator Algebra

The derivation of a phase-space equation of motion for the damped two-level atom is carried out in essentially the same way as the derivation of the Fokker–Planck equation for the damped harmonic oscillator [Sect. 3.2.2]. Because, however, of the different operator algebra obeyed by σ_-, σ_+, and σ_z, the strategy for obtaining the equation of motion for the characteristic function $\chi_N(\xi, \xi^*, \eta)$ is slightly different. Actually, the algebraic form of the equation of motion we are going to derive is not unique. This is because of the relationship that exists between products of the Pauli spin operators and linear combinations of these operators. Use of this relationship will be a necessary part of the calculation when we consider many atoms, and we therefore pattern the single-atom calculation after the approach that is required in the many-atom case, even though a closer parallel with the harmonic oscillator example could be maintained for one atom. Our strategy will be to arrange all terms in the master equation involving products of σ_-, σ_+, and σ_z – for example $\sigma_- \rho \sigma_+$ – so that the operator products can be replaced by a sum of operators taken from the set σ_-, σ_+, σ_z, and 1. This is obviously possible, because a two-state basis $|1\rangle$, $|2\rangle$ has only four outer products, $|2\rangle\langle 2|$, $|1\rangle\langle 1|$, $|2\rangle\langle 1|$, and $|1\rangle\langle 2|$; clearly, any operator can be expanded in terms of these. Specifically, collecting results together from (2.25), (2.45), and (2.132), we have

$$\sigma_+^2 = |2\rangle\langle 1|2\rangle\langle 1| = 0, \tag{6.6a}$$

$$\sigma_-^2 = |1\rangle\langle 2|1\rangle\langle 2| = 0, \tag{6.6b}$$

$$\sigma_z^2 = \big(|2\rangle\langle 2| - |1\rangle\langle 1|\big)\big(|2\rangle\langle 2| - |1\rangle\langle 1|\big) = |2\rangle\langle 2| + |1\rangle\langle 1| = 1, \tag{6.6c}$$

$$\sigma_+\sigma_z = |2\rangle\langle 1|\big(|2\rangle\langle 2| - |1\rangle\langle 1|\big) = -|2\rangle\langle 1| = -\sigma_+, \tag{6.6d}$$

$$\sigma_-\sigma_z = |1\rangle\langle 2|\big(|2\rangle\langle 2| - |1\rangle\langle 1|\big) = |1\rangle\langle 2| = \sigma_-, \tag{6.6e}$$

$$\sigma_+\sigma_- = |2\rangle\langle 1|1\rangle\langle 2| = |2\rangle\langle 2| = \tfrac{1}{2}(1 + \sigma_z), \tag{6.6f}$$

$$\sigma_-\sigma_+ = |1\rangle\langle 2|2\rangle\langle 1| = |1\rangle\langle 2| = \tfrac{1}{2}(1 - \sigma_z). \tag{6.6g}$$

In addition to these relations we will need the following three identities:

Proposition 6.1

$$e^{i\xi\sigma_-}\sigma_z e^{-i\xi\sigma_-} = \sigma_z + 2i\xi\sigma_-. \tag{6.7}$$

Proof. The left-hand side of (6.7) may be viewed as the formal solution to an operator equation of motion with $i\xi$ as the independent variable [$i\xi\sigma_-$ replaces $(-iH/\hbar)t$ in the formal solution of a Heisenberg equation of motion]. Define

$$\sigma_z(i\xi) \equiv e^{i\xi\sigma_-}\sigma_z e^{-i\xi\sigma_-}$$

with $\sigma_z(0) = \sigma_z$. Then, differentiating with respect to $(i\xi)$, we obtain

$$\frac{d\sigma_z(i\xi)}{d(i\xi)} = e^{i\xi\sigma_-}(\sigma_-\sigma_z - \sigma_z\sigma_-)e^{-i\xi\sigma_-}$$
$$= e^{i\xi\sigma_-}2\sigma_- e^{-i\xi\sigma_-}$$
$$= 2\sigma_-,$$

where the commutator is taken from (2.11). Thus,

$$\sigma_z(i\xi) = \sigma_z(0) + 2i\xi\sigma_- = \sigma_z + 2i\xi\sigma_-.$$

$\square$

Proposition 6.2

$$e^{i\eta\sigma_z}\sigma_- e^{-i\eta\sigma_z} = e^{-2i\eta}\sigma_-. \tag{6.8}$$

Proof. Following the same approach, we define

$$\sigma_-(i\eta) \equiv e^{i\eta\sigma_z}\sigma_- e^{-i\eta\sigma_z}$$

with $\sigma_-(0) = \sigma_-$. Differentiating with respect to $(i\eta)$, we have

$$\frac{d\sigma_-(i\eta)}{d(i\eta)} = e^{i\eta\sigma_z}(\sigma_z\sigma_- - \sigma_-\sigma_z)e^{-i\eta\sigma_z}$$
$$= e^{i\eta\sigma_z}(-2\sigma_-)e^{-i\eta\sigma_z}$$
$$= -2\sigma_-(i\eta).$$

Integration of this equation gives

$$\sigma_-(i\eta) = e^{-2i\eta}\sigma_-(0) = e^{-2i\eta}\sigma_-.$$

$\square$

Proposition 6.3

$$e^{i\xi\sigma_-}\sigma_+ e^{-i\xi\sigma_-} = \sigma_+ - i\xi\sigma_z - (i\xi)^2\sigma_-. \tag{6.9}$$

Proof. Define

$$\sigma_+(i\xi) \equiv e^{i\xi\sigma_-}\sigma_+ e^{-i\xi\sigma_-}$$

with $\sigma_+(0) = \sigma_+$. Differentiating with respect to $(i\xi)$, we obtain

$$\frac{d\sigma_+(i\xi)}{d(i\xi)} = e^{i\xi\sigma_-}(\sigma_-\sigma_+ - \sigma_+\sigma_-)e^{-i\xi\sigma_-}$$

$$= -e^{i\xi\sigma_-}\sigma_z e^{-i\xi\sigma_-}$$

$$= -\sigma_z - 2i\xi\sigma_-.$$

The commutator is taken from (2.11) and the last line follows from (6.7). Integrating this equation gives

$$\sigma_+(i\xi) = \sigma_+(0) - i\xi\sigma_z - (i\xi)^2\sigma_-$$

$$= \sigma_+ - i\xi\sigma_z - (i\xi)^2\sigma_-.$$

□

We now have all the pieces we need to derive a phase-space equation of motion for the damped two-level atom.

6.1.3 Phase-Space Equation of Motion for the Damped Two-Level Atom

The master equation for a radiatively damped two-level atom is given by (2.26):

$$\dot{\rho} = -i\tfrac{1}{2}\omega_A[\sigma_z, \rho] + \frac{\gamma}{2}(\bar{n} + 1)(2\sigma_-\rho\sigma_+ - \sigma_+\sigma_-\rho - \rho\sigma_+\sigma_-)$$

$$+ \frac{\gamma}{2}\bar{n}(2\sigma_+\rho\sigma_- - \sigma_-\sigma_+\rho - \rho\sigma_-\sigma_+). \tag{6.10}$$

Our first task is to derive an equation of motion for the characteristic function χ_N. Using (6.1) and (6.10), we have

$$\frac{\partial \chi_N}{\partial t}$$

$$\equiv \frac{\partial}{\partial t}\left[\mathrm{tr}(\rho e^{i\xi^*\sigma_+}e^{i\eta\sigma_z}e^{i\xi\sigma_-})\right]$$

$$= \mathrm{tr}\left(\dot{\rho}e^{i\xi^*\sigma_+}e^{i\eta\sigma_z}e^{i\xi\sigma_-}\right)$$

$$= \mathrm{tr}\left\{\left[-i\tfrac{1}{2}\omega_A(\sigma_z\rho - \rho\sigma_z) + \frac{\gamma}{2}(\bar{n} + 1)(2\sigma_-\rho\sigma_+ - \sigma_+\sigma_-\rho - \rho\sigma_+\sigma_-)\right.\right.$$

$$\left.\left. + \frac{\gamma}{2}\bar{n}(2\sigma_+\rho\sigma_- - \sigma_-\sigma_+\rho - \rho\sigma_-\sigma_+)\right]e^{i\xi^*\sigma_+}e^{i\eta\sigma_z}e^{i\xi\sigma_-}\right\}. \tag{6.11}$$

Now, as in Sect. 3.2.2, our aim is to express each term on the right-hand side of (6.11) in terms of χ_N and its derivatives. As mentioned above, the strategy here will be a little different from the one followed in Sect. 3.2.2. We wish to eliminate all quadratic terms in the operators σ_-, σ_+, and σ_z using

(6.6a)–(6.6g). All but two of these can be removed immediately; using (6.6f) and (6.6g), (6.11) becomes

$$
\frac{\partial \chi_N}{\partial t}
$$

$$
= \text{tr}\Big\{\Big[-i\tfrac{1}{2}\omega_A(\sigma_z\rho - \rho\sigma_z) + \frac{\gamma}{2}(\bar{n}+1)\big(\sigma_-\rho\sigma_+ - \tfrac{1}{2}\sigma_z\rho - \tfrac{1}{2}\rho\sigma_z - \rho\big)
$$

$$
+ \frac{\gamma}{2}\bar{n}\big(2\sigma_+\rho\sigma_- + \tfrac{1}{2}\sigma_z\rho + \tfrac{1}{2}\rho\sigma_z - \rho\big)\Big]e^{i\xi^*\sigma_+}e^{i\eta\sigma_z}e^{i\xi\sigma_-}\Big\}. \tag{6.12}
$$

To reexpress (6.12) in terms of χ_N and its derivatives we proceed as follows.

Consider first the terms involving $\sigma_z\rho$ and $\rho\sigma_z$. These are treated in a straightforward fashion, using (6.7) to pass σ_z through the exponentials $e^{i\xi^*\sigma_+}$ and $e^{i\xi\sigma_-}$ so that it is positioned next to $e^{i\eta\sigma_z}$. It may then be brought down from the exponential by differentiating with respect to $(i\eta)$. We obtain

$$
\text{tr}\big(\sigma_z\rho e^{i\xi^*\sigma_+}e^{i\eta\sigma_z}e^{i\xi\sigma_-}\big) = \text{tr}\big(\rho e^{i\xi^*\sigma_+}e^{i\eta\sigma_z}e^{i\xi\sigma_-}\sigma_z\big)
$$

$$
= \text{tr}\big[\rho e^{i\xi^*\sigma_+}e^{i\eta\sigma_z}\big(e^{i\xi\sigma_-}\sigma_z e^{-i\xi\sigma_-}\big)e^{i\xi\sigma_-}\big]
$$

$$
= \text{tr}\big[\rho e^{i\xi^*\sigma_+}e^{i\eta\sigma_z}(\sigma_z + 2i\xi\sigma_-)e^{i\xi\sigma_-}\big]
$$

$$
= \Big(\frac{\partial}{\partial(i\eta)} + 2i\xi\frac{\partial}{\partial(i\xi)}\Big)\chi_N, \tag{6.13}
$$

and

$$
\text{tr}\big(\rho\sigma_z e^{i\xi^*\sigma_+}e^{i\eta\sigma_z}e^{i\xi\sigma_-}\big) = \text{tr}\big[\rho e^{i\xi^*\sigma_+}\big(e^{-i\xi^*\sigma_+}\sigma_z e^{i\xi^*\sigma_+}\big)e^{i\eta\sigma_z}e^{i\xi\sigma_-}\big]
$$

$$
= \text{tr}\big[\rho e^{i\xi^*\sigma_+}(\sigma_z + 2i\xi^*\sigma_+)e^{i\eta\sigma_z}e^{i\xi\sigma_-}\big]
$$

$$
= \Big(\frac{\partial}{\partial(i\eta)} + 2i\xi^*\frac{\partial}{\partial(i\xi^*)}\Big)\chi_N. \tag{6.14}
$$

The treatment of the term involving $\sigma_-\rho\sigma_+$ is no more complicated; but in accordance with our general strategy, it must begin with some method for replacing the quadratic dependence on atomic operators by a linear dependence. This is not the only way to proceed. We could write

$$
\text{tr}\big(\sigma_-\rho\sigma_+ e^{i\xi^*\sigma_+}e^{i\eta\sigma_z}e^{i\xi\sigma_-}\big) = \text{tr}\big(\rho\sigma_+ e^{i\xi^*\sigma_+}e^{i\eta\sigma_z}e^{i\xi\sigma_-}\sigma_-\big)
$$

$$
= \frac{\partial^2}{\partial(i\xi^*)\partial(i\xi)}\chi_N, \tag{6.15}
$$

which is completely analogous to the treatment of the corresponding term for the damped harmonic oscillator [Eq. (3.77)]. We will see shortly, however, that while (6.15) works for a single atom, it does not work for many atoms. For the single atom case we therefore have a choice: we can use (6.15) or the approach that generalizes to many atoms. Because of choices like this the equation of motion we derive for a single atom is not unique.

The procedure that generalizes to many atoms first uses (6.8) to pass σ_- through the exponential $e^{i\eta\sigma_z}$. This sets σ_- and σ_+ next to one another so that we may replace their product by $\sigma_+\sigma_- = \frac{1}{2}(1+\sigma_z)$. Thus,

$$
\begin{aligned}
\operatorname{tr}\left(\sigma_-\rho\sigma_+ e^{i\xi^*\sigma_+}e^{i\eta\sigma_z}e^{i\xi\sigma_-}\right) &= \operatorname{tr}\left(\rho e^{i\xi^*\sigma_+}\sigma_+ e^{i\eta\sigma_z}\sigma_- e^{i\xi\sigma_-}\right) \\
&= \operatorname{tr}\left[\rho e^{i\xi^*\sigma_+}\sigma_+\left(e^{i\eta\sigma_z}\sigma_- e^{-i\eta\sigma_z}\right)e^{i\eta\sigma_z}e^{i\xi\sigma_-}\right] \\
&= \operatorname{tr}\left[\rho e^{i\xi^*\sigma_+}\sigma_+\left(e^{-2i\eta}\sigma_-\right)e^{i\eta\sigma_z}e^{i\xi\sigma_-}\right] \\
&= e^{-2i\eta}\operatorname{tr}\left[\rho e^{i\xi^*\sigma_+}\tfrac{1}{2}(1+\sigma_z)e^{i\eta\sigma_z}e^{i\xi\sigma_-}\right] \\
&= e^{-2i\eta}\frac{1}{2}\left(1+\frac{\partial}{\partial(i\eta)}\right)\chi_N.
\end{aligned}
\tag{6.16}
$$

The philosophy is the same for the final term – $\sigma_+\rho\sigma_-$ – but the algebra is now a little more complicated. We first use (6.9) and its Hermitian conjugate (taken with $\xi \to -\xi$) to write

$$
\begin{aligned}
&\operatorname{tr}\left(\sigma_+\rho\sigma_- e^{i\xi^*\sigma_+}e^{i\eta\sigma_z}e^{i\xi\sigma_-}\right) \\
&= \operatorname{tr}\left(\rho\sigma_- e^{i\xi^*\sigma_+}e^{i\eta\sigma_z}e^{i\xi\sigma_-}\sigma_+\right) \\
&= \operatorname{tr}\left[\rho e^{i\xi^*\sigma_+}\left(e^{-i\xi^*\sigma_+}\sigma_- e^{i\xi^*\sigma_+}\right)e^{i\eta\sigma_z}\left(e^{i\xi\sigma_-}\sigma_+ e^{-i\xi\sigma_-}\right)e^{i\xi\sigma_-}\right] \\
&= \operatorname{tr}\left[\rho e^{i\xi^*\sigma_+}\left(\sigma_- - i\xi^*\sigma_z - (i\xi^*)^2\sigma_+\right)e^{i\eta\sigma_z}\left(\sigma_+ - i\xi\sigma_z - (i\xi)^2\sigma_-\right)e^{i\xi\sigma_-}\right].
\end{aligned}
$$

We now pass $\left(\sigma_+ - i\xi\sigma_z - (i\xi)^2\sigma_-\right)$ through the exponential $e^{i\eta\sigma_z}$ using (6.8):

$$
\begin{aligned}
&\operatorname{tr}\left(\sigma_+\rho\sigma_- e^{i\xi^*\sigma_+}e^{i\eta\sigma_z}e^{i\xi\sigma_-}\right) \\
&= \operatorname{tr}\Big\{\rho e^{i\xi^*\sigma_+}\left(\sigma_- - i\xi^*\sigma_z - (i\xi^*)^2\sigma_+\right) \\
&\quad \times \Big[\left(e^{i\eta\sigma_z}\sigma_+ e^{-i\eta\sigma_z}\right) - i\xi\sigma_z - (i\xi)^2\left(e^{i\eta\sigma_z}\sigma_- e^{-i\eta\sigma_z}\right)\Big]e^{i\eta\sigma_z}e^{i\xi\sigma_-}\Big\} \\
&= \operatorname{tr}\Big[\rho e^{i\xi^*\sigma_+}\left(\sigma_- - \xi^*\sigma_z - (i\xi^*)^*\sigma_+\right) \\
&\quad \times \left(e^{2i\eta}\sigma_+ - i\xi\sigma_z - (i\xi)^2 e^{-2i\eta}\sigma_-\right)e^{i\eta\sigma_z}e^{i\xi\sigma_-}\Big] \\
&= \operatorname{tr}\Big\{\rho e^{i\xi^*\sigma_+}\Big[e^{2i\eta}\tfrac{1}{2}(1-\sigma_z) + (i\xi)(i\xi^*) + (i\xi)^2(i\xi^*)^2 e^{-2i\eta}\tfrac{1}{2}(1+\sigma_z) \\
&\quad - (i\xi)\left(1+(i\xi)(i\xi^*)e^{-2i\eta}\right)\sigma_- - (i\xi^*)\left(e^{2i\eta}+(i\xi)(i\xi^*)\right)\sigma_+\Big] \\
&\quad \times e^{i\eta\sigma_z}e^{i\xi\sigma_-}\Big\},
\end{aligned}
$$

where in the last step all operator products are reexpressed as sums using (6.6a)–(6.6g). In this form the operators σ_z and σ_+ appear in the appropriate places with respect to the exponentials so that they may be brought down from the exponents by differentiation. On the other hand, the operator σ_- is not in the appropriate place. It must be passed back through the exponential $e^{i\eta\sigma_z}$ using (6.8) (taken with $\eta \to -\eta$) to set it beside $e^{i\xi\sigma_-}$. After taking this final step, we obtain

$$tr(\sigma_+\rho\sigma_-e^{i\xi^*\sigma_+}e^{i\eta\sigma_z}e^{i\xi\sigma_-})$$

$$= [\tfrac{1}{2}(e^{2i\eta} + (i\xi)^2(i\xi^*)^2e^{-2i\eta} + 2(i\xi)(i\xi^*))$$

$$- \tfrac{1}{2}(e^{2i\eta} - (i\xi)^2(i\xi^*)^2e^{-2i\eta})\frac{\partial}{\partial(i\eta)}$$

$$- i\xi(e^{2i\eta} + (i\xi)(i\xi^*))\frac{\partial}{\partial(i\xi)} - i\xi^*(e^{2i\eta} + (i\xi)(i\xi^*))\frac{\partial}{\partial(i\xi^*)}]\chi_N. \quad (6.17)$$

We may now use (6.13), (6.14), (6.16), and (6.17) to substitute for the various terms in (6.12). After a little algebra the equation of motion for χ_N reads

$$\frac{\partial\chi_N}{\partial t} = D\left(\xi, \xi^*, \eta, \frac{\partial}{\partial\xi}, \frac{\partial}{\partial\xi^*}, \frac{\partial}{\partial\eta}\right)\chi_N, \quad (6.18)$$

where $D\left(\xi, \xi^*, \eta, \frac{\partial}{\partial\xi}, \frac{\partial}{\partial\xi^*}, \frac{\partial}{\partial\eta}\right)$ is the differential operator

$$D\left(\xi, \xi^*, \eta, \frac{\partial}{\partial\xi}, \frac{\partial}{\partial\xi^*}, \frac{\partial}{\partial\eta}\right)$$

$$\equiv -i\omega_A\left(\xi\frac{\partial}{\partial\xi} - \xi^*\frac{\partial}{\partial\xi^*}\right)$$

$$+ \frac{\gamma}{2}(\bar{n}+1)\left[(e^{-2i\eta} - 1)\left(1 - i\frac{\partial}{\partial\eta}\right) - \xi\frac{\partial}{\partial\xi} - \xi^*\frac{\partial}{\partial\xi^*}\right]$$

$$+ \frac{\gamma}{2}\bar{n}\left[(e^{2i\eta} - 1)\left(1 + i\frac{\partial}{\partial\eta}\right) + (\xi\xi^*)^2e^{-2i\eta}\left(1 - i\frac{\partial}{\partial\eta}\right)\right.$$

$$\left. -2(e^{2i\eta} - \xi\xi^* - \tfrac{1}{2})\left(\xi\frac{\partial}{\partial\xi} + \xi^*\frac{\partial}{\partial\xi^*}\right) - 2\xi\xi^*\right]. \quad (6.19)$$

The equation of motion for the distribution P now follows from a calculation analogous to that in Sect. 3.2.2. We first substitute χ_N as the Fourier transform of P [Eq. (6.4)] to obtain

$$\int d^2v \int dm \frac{\partial P(v, v^*, m)}{\partial t} e^{i\xi^*v^*}e^{i\xi v}e^{i\eta m}$$

$$= \int d^2v \int dm\, P(v, v^*, m)D\left(\xi, \xi^*, \eta, \frac{\partial}{\partial\xi}, \frac{\partial}{\partial\xi^*}, \frac{\partial}{\partial\eta}\right)e^{i\xi^*v^*}e^{i\xi v}e^{i\eta m}. \quad (6.20)$$

The action of the differential operators on the exponentials in (6.20) allows the replacement

$$\frac{\partial}{\partial\xi} \rightarrow iv, \qquad \frac{\partial}{\partial\xi^*} \rightarrow iv^*, \qquad -i\frac{\partial}{\partial\eta} \rightarrow m.$$

Then the terms in ξ, ξ^*, and $e^{\pm 2i\eta}$ in (6.19), may be passed to the right of the terms in v, v^*, and m, which have replaced the differential operators, and to the left of the exponentials; with this rearrangement we can make the substitution

$$\xi \to -i\frac{\partial}{\partial v}, \qquad \xi^* \to -i\frac{\partial}{\partial v^*}, \qquad e^{\pm 2i\eta} \to e^{\pm 2\frac{\partial}{\partial m}}.$$

Now, in (6.20), $D\left(\xi, \xi^*, \eta, \frac{\partial}{\partial \xi}, \frac{\partial}{\partial \xi^*}, \frac{\partial}{\partial \eta}\right)$ has been replaced by the differential operator

$$L^+\left(v, v^*, m, \frac{\partial}{\partial v}, \frac{\partial}{\partial v^*}, \frac{\partial}{\partial m}\right)$$

$$\equiv -i\omega_A\left(\frac{\partial}{\partial v}v - \frac{\partial}{\partial v^*}v^*\right)$$

$$+ \frac{\gamma}{2}(\bar{n}+1)\left[(1+m)\left(e^{-2\frac{\partial}{\partial m}} - 1\right) - v\frac{\partial}{\partial v} - v^*\frac{\partial}{\partial v^*}\right]$$

$$+ \frac{\gamma}{2}\bar{n}\left[(1-m)\left(e^{2\frac{\partial}{\partial m}} - 1\right) + (1+m)\frac{\partial^4}{\partial v^2 \partial v^{*2}}e^{-2\frac{\partial}{\partial m}}\right.$$

$$\left. -2\left(v\frac{\partial}{\partial v} + v^*\frac{\partial}{\partial v^*}\right)\left(e^{2\frac{\partial}{\partial m}} + \frac{\partial^2}{\partial v \partial v^*} - \frac{1}{2}\right) + 2\frac{\partial^2}{\partial v \partial v^*}\right].$$

$$(6.21)$$

Finally, we integrate each term by parts the required number of times to pass all of the differential operators from the product of exponentials onto P. Each derivative changes sign in the process and (6.20) becomes

$$\int d^2v \int dm\, e^{i\xi^* v^*} e^{i\xi v} e^{i\eta m} \frac{\partial P}{\partial t}$$

$$= \int d^2v \int dm\, e^{i\xi^* v^*} e^{i\xi v} e^{i\eta m} L\left(v, v^*, m, \frac{\partial}{\partial v}, \frac{\partial}{\partial v^*}, \frac{\partial}{\partial m}\right)P,$$

$$(6.22)$$

where L is the adjoint of the differential operator L^+:

$$L\left(v, v^*, m, \frac{\partial}{\partial v}, \frac{\partial}{\partial v^*}, \frac{\partial}{\partial m}\right)$$

$$\equiv i\omega_A\left(\frac{\partial}{\partial v}v - \frac{\partial}{\partial v^*}v^*\right)$$

$$+ \frac{\gamma}{2}(\bar{n}+1)\left[\left(e^{2\frac{\partial}{\partial m}} - 1\right)(1+m) + \frac{\partial}{\partial v}v + \frac{\partial}{\partial v^*}v^*\right]$$

$$+ \frac{\gamma}{2}\bar{n}\left[\left(e^{-2\frac{\partial}{\partial m}} - 1\right)(1-m) + \frac{\partial^4}{\partial v^2 \partial v^{*2}}e^{2\frac{\partial}{\partial m}}(1+m)\right.$$

$$\left. +2\left(e^{-2\frac{\partial}{\partial m}} + \frac{\partial^2}{\partial v \partial v^*} - \frac{1}{2}\right)\left(\frac{\partial}{\partial v}v + \frac{\partial}{\partial v^*}v^*\right) + 2\frac{\partial^2}{\partial v \partial v^*}\right].$$

$$(6.23)$$

Equation (6.22) is the Fourier transform of the desired equation of motion. Inverting the Fourier transform, we arrive at the *phase-space equation of motion for a radiatively damped two-level atom*:

$$\frac{\partial P}{\partial t} = L\left(v, v^*, m, \frac{\partial}{\partial v}, \frac{\partial}{\partial v^*}, m\right) P, \tag{6.24}$$

with $L\left(v, v^*, m, \frac{\partial}{\partial v}, \frac{\partial}{\partial v^*}, \frac{\partial}{\partial m}\right)$ given by (6.23).

The first lesson to be learned from this calculation is fairly clear. Equation (6.24) has the form of (5.37); it has the form of a *generalized* Fokker–Planck equation, with derivatives beyond second order – up to infinite order in the inversion variable m. The quantum–classical correspondence provides a representation for quantum-mechanical states in terms of phase-space variables, but there is no guarantee that the phase-space dynamics will be described by a Fokker–Planck equation. In fact, as we will see shortly, (6.24) is not strictly even a generalized Fokker–Planck equation, since its solution does not have the properties of a classical probability density. It is certainly a much more complicated equation than the Fokker–Planck equation obtained for the damped harmonic oscillator. The difference has arisen, of course, from the different algebras obeyed by a and $a^\dagger$, and σ_-, σ_+, and σ_z. Techniques for analyzing Fokker–Planck equations have nothing to offer with respect to an equation like (6.24), and, indeed, we appear to have made the problem of the damped two-level atom more complicated by using the quantum–classical correspondence.

We saw in Chap. 2 that the damped two-level atom is readily analyzed using the optical Bloch equations and the quantum regression formula. The practical use for the representation defined in Sect. 6.1.1 comes from its application to collections of many atoms, where we will find a way around the higher derivatives and again recover a Fokker–Planck equation. The single-atom calculation is useful, however, first because it dispenses with some of the algebraic manipulations required to treat the many-atom case; but, more importantly, because we already have an exact solution to the single-atom master equation. Equations (6.1) and (6.3) may be used directly to construct the distribution $P(v, v^*, m, t)$ corresponding to the density matrix given by solutions to the optical Bloch equations. This distribution must satisfy (6.24), and its form should teach us something about the significance of the higher-order derivatives in (6.24). In the many-atom calculation we will only be able to remove these derivatives by making an approximation, and an understanding of their significance is important to an understanding of that approximation. Thus, before tackling the extension to many-atoms, let us see what can be learned from the direct construction of $P(v, v^*, m, t)$.

Exercise 6.1 The differential operator (6.23) does not include terms describing nonradiative dephasing processes, such as elastic collisions. When these processes are included the master equation has the additional term $(\gamma_p/2)(\sigma_z \rho \sigma_z - \rho)$ [Eq. (2.66)]. Show that this adds the term

$$L_{\text{dephase}} \equiv \gamma_p \left[\frac{\partial}{\partial v} v + \frac{\partial}{\partial v^*} v^* + \frac{\partial^2}{\partial v \partial v^*} e^{-2\frac{\partial}{\partial m}} (1 + m) \right] \tag{6.25}$$

to the operator appearing on the right-hand side of (6.24).

6.1.4 A Singular Solution to the Phase-Space Equation of Motion

We will construct the distribution $P(v, v^*, m, t)$ for a damped two-level atom explicitly using the solution for $\rho(t)$ obtained in Chap. 2. We will then show that this distribution satisfies the phase-space equation of motion (6.24).

We begin by evaluating $\chi_N(\xi, \xi^*, \eta)$ in terms of the operator averages $\langle \sigma_- \rangle$, $\langle \sigma_+ \rangle$, and $\langle \sigma_z \rangle$. Using (6.6a)–(6.6c), we write

$$e^{i\xi^* \sigma_+} = \sum_{n=0}^{\infty} \frac{(i\xi^*)^n}{n!} \sigma_+^n = 1 + i\xi^* \sigma_+, \tag{6.26a}$$

$$e^{i\xi \sigma_-} = \sum_{n=0}^{\infty} \frac{(i\xi)^n}{n!} \sigma_-^n = 1 + i\xi \sigma_-, \tag{6.26b}$$

$$e^{i\eta \sigma_z} = \sum_{n=0}^{\infty} \frac{(i\eta)^n}{n!}$$

$$= \sum_{k=0}^{\infty} \frac{(-1)^k \eta^{2k}}{2k!} + i\sigma_z \sum_{k=0}^{\infty} \frac{(-1)^k \eta^{2k+1}}{(2k+1)!}$$

$$= \cos \eta + i\sigma_z \sin \eta. \tag{6.26c}$$

Then, from the definition of $\chi_N(\xi, \xi^*, \eta)$ [Eq. (6.1)], we have

$$\chi_N(\xi, \xi^*, \eta) = \text{tr}\left[\rho(1 + i\xi^* \sigma_+)(\cos \eta + i\sigma_z \sin \eta)(1 + i\xi \sigma_-) \right]$$

$$= \tfrac{1}{2}(1 + \langle \sigma_z \rangle)(e^{i\eta} - \xi^* \xi e^{-i\eta}) + \tfrac{1}{2}(1 - \langle \sigma_z \rangle)e^{-i\eta}$$

$$+ \langle \sigma_- \rangle i\xi e^{-i\eta} + \langle \sigma_+ \rangle i\xi^* e^{-i\eta}. \tag{6.27}$$

We should still be able to calculate all operator averages by taking derivatives of $\chi_N(\xi, \xi^*, \eta)$ [Eq. (6.2)]. All we have done is simplify the form of $\chi_N(\xi, \xi^*, \eta)$ by using the relationships (6.6) *a priori*. Thus, (6.27) shows once again that all operator averages for the two-level atom can be expressed in terms of expectation values of σ_-, σ_+, σ_z, and 1 alone.

Exercise 6.2 Verify that (6.2) and (6.27) produce the correct operator averages up to second order; show that

$$\frac{\partial}{\partial (i\xi^*)} \chi_N \bigg|_{\xi=\xi^*=\eta=0} = \langle \sigma_+ \rangle, \qquad \frac{\partial^2}{\partial (i\xi^*)^2} \chi_N \bigg|_{\xi^*=\xi=\eta=0} = 0 = \langle \sigma_+^2 \rangle,$$

$$\frac{\partial}{\partial (i\xi)} \chi_N \bigg|_{\xi=\xi^*=\eta=0} = \langle \sigma_- \rangle, \qquad \frac{\partial^2}{\partial (i\xi)^2} \chi_N \bigg|_{\xi^*=\xi=\eta=0} = 0 = \langle \sigma_-^2 \rangle,$$

$$\frac{\partial}{\partial (i\eta)} \chi_N \bigg|_{\xi=\xi^*=\eta=0} = \langle \sigma_z \rangle, \qquad \frac{\partial^2}{\partial (i\eta)^2} \chi_N \bigg|_{\xi^*=\xi=\eta=0} = 1 = \langle \sigma_z^2 \rangle,$$

$$\frac{\partial^2}{\partial(i\xi^*)\partial(i\eta)}\chi_N\bigg|_{\xi^*=\xi=\eta=0} = -\langle\sigma_+\rangle = \langle\sigma_+\sigma_z\rangle,$$

$$\frac{\partial^2}{\partial(i\eta)\partial(i\xi)}\chi_N\bigg|_{\xi^*=\xi=\eta=0} = -\langle\sigma_-\rangle = \langle\sigma_z\sigma_-\rangle,$$

$$\frac{\partial^2}{\partial(i\xi^*)\partial(i\xi)}\chi_N\bigg|_{\xi^*=\xi=\eta=0} = \tfrac{1}{2}(1+\langle\sigma_z\rangle) = \langle\sigma_+\sigma_-\rangle.$$

Clearly, this agreement extends to operator products of arbitrary powers. For example, from (6.27), $\partial^n\chi_N/\partial(i\xi^*)^n = \partial^n\chi_N/\partial(i\xi)^n = 0$ for $n > 1$; all averages evaluated using (6.2) with $p > 1$ or $q > 1$ are therefore zero, as required by (6.6a) and (6.6b).

We now construct the distribution $P(v,v^*,m,t)$ by taking the Fourier transform of (6.27). The transform with respect to the variable η is straightforward; we have

$$
\begin{aligned}
P(v,v^*,m) &\equiv \frac{1}{2\pi^3}\int d^2\xi \int d\eta\, \chi_N\, e^{-i\xi^*v^*}e^{-i\xi v}e^{-i\eta m}\\
&= \frac{1}{2\pi^3}\int d^2\xi \int d\eta\, \Big\{\tfrac{1}{2}(1+\langle\sigma_z\rangle)(e^{i\eta}-\xi^*\xi e^{-i\eta})\\
&\quad +\tfrac{1}{2}(1-\langle\sigma_z\rangle)e^{-i\eta}\\
&\quad +\langle\sigma_-\rangle i\xi e^{-i\eta} + \langle\sigma_+\rangle i\xi^* e^{-i\eta}\Big\}e^{-i\xi^*v^*}e^{-i\xi v}e^{-i\eta m}\\
&= \frac{1}{\pi^2}\int d^2\xi\, \Big\{\tfrac{1}{2}(1+\langle\sigma_z\rangle)\big(\delta(m-1)-\xi^*\xi\delta(m+1)\big)\\
&\quad +\tfrac{1}{2}(1-\langle\sigma_z\rangle)\delta(m+1)\\
&\quad +\langle\sigma_-\rangle i\xi\,\delta(m+1) + \langle\sigma_+\rangle i\xi^*\delta(m+1)\Big\}e^{-i\xi^*v^*}e^{-i\xi v}.
\end{aligned}
$$

$$\text{(6.28)}$$

The remaining two-dimensional Fourier transform presents a difficulty, since it does not exist in the usual sense. We face a similar situation here to the one we encountered when deriving the Glauber–Sudarshan P representation for a Fock state [see the discussion below (3.28)]: The resolution of the difficulty is to allow $P(v,v^*,m)$ to be a generalized function. Specifically, we can evaluate the Fourier transform in (6.28) if we introduce derivatives of the δ-function, writing

$$\frac{1}{\pi^2}\int d^2\xi\, e^{-i\xi^*v^*}e^{-i\xi v} = \delta^{(2)}(v), \tag{6.29a}$$

$$
\begin{aligned}
\frac{1}{\pi^2}\int d^2\xi\,(i\xi)e^{-i\xi^*v^*}e^{-i\xi v} &= -\frac{\partial}{\partial v}\frac{1}{\pi^2}\int d^2\xi\, e^{-i\xi^*v^*}e^{-i\xi v}\\
&= -\frac{\partial}{\partial v}\delta^{(2)}(v), \tag{6.29b}
\end{aligned}
$$

$$\frac{1}{\pi^2}\int d^2\xi\,(i\xi^*)e^{-i\xi^*v^*}e^{-i\xi v} = -\frac{\partial}{\partial v^*}\frac{1}{\pi^2}\int d^2\xi\, e^{-i\xi^*v^*}e^{-i\xi v}$$

$$= -\frac{\partial}{\partial v^*}\delta^{(2)}(v), \tag{6.29c}$$

$$\frac{1}{\pi^2}\int d^2\xi\,(i\xi^*)(i\xi)e^{-i\xi^*v^*}e^{-i\xi v} = -\frac{\partial^2}{\partial v^*\partial v}\frac{1}{\pi^2}\int d^2\xi\, e^{-i\xi^*v^*}e^{-i\xi v}$$

$$= -\frac{\partial^2}{\partial v^*\partial v}\delta^{(2)}(v). \tag{6.29d}$$

Then, for the general time-dependent density operator

$$\rho(t) = \tfrac{1}{2}\big(1 + \langle\sigma_z(t)\rangle\big)|2\rangle\langle2| + \tfrac{1}{2}\big(1 - \langle\sigma_z(t)\rangle\big)|1\rangle\langle1|$$
$$+ \langle\sigma_-(t)\rangle|2\rangle\langle1| + \langle\sigma_+(t)\rangle|1\rangle\langle2|, \tag{6.30}$$

the corresponding *distribution* $P(v,v^*,m,t)$ *for a (radiatively damped) two-level atom* is given by

$$P(v,v^*,m,t)$$
$$= \tfrac{1}{2}\big(1 + \langle\sigma_z(t)\rangle\big)\delta(m-1)\delta^{(2)}(v) + \tfrac{1}{2}\big(1 - \langle\sigma_z(t)\rangle\big)\delta(m+1)\delta^{(2)}(v)$$

$$- \langle\sigma_-(t)\rangle\delta(m+1)\frac{\partial}{\partial v}\delta^{(2)}(v) - \langle\sigma_+(t)\rangle\delta(m+1)\frac{\partial}{\partial v^*}\delta^{(2)}(v)$$

$$+ \tfrac{1}{2}\big(1 + \langle\sigma_z(t)\rangle\big)\delta(m+1)\frac{\partial^2}{\partial v\partial v^*}\delta^{(2)}(v). \tag{6.31}$$

This is a highly singular distribution. The singular character in the polarization variable v can be traced to the requirement that $\sigma_-^2 = \sigma_+^2 = 0$. It is the vanishing of these operator products that gives $\chi_N(\xi,\xi^*,\eta)$ its truncated, polynomial form in the variables ξ and ξ^*, and this then requires derivatives of the δ-function to appear in the Fourier transform. The singular character enables the distribution to reproduce the manifestly nonclassical moments required in the variables v and v^*: while $\langle\sigma_-\rangle = \big(\bar{v}\big)_P$ and $\langle\sigma_+\rangle = \big(\overline{v^*}\big)_P$ are generally nonzero, all higher-order moments $\langle\sigma_-^q\rangle = \big(\overline{v^q}\big)_P$, $q > 1$, and $\langle\sigma_+^p\rangle = \big(\overline{v^{*p}}\big)_P$, $p > 1$, must vanish; such behavior cannot be reproduced if $P(v,v^*,m)$ is an ordinary probability density.

Exercise 6.3 The phenomenon of photon antibunching in resonance fluorescence (Sects. 2.3.5 and 2.3.6) provides a good example of a situation that calls for nonclassical behavior in moments of the polarization. Show that (6.5) and (6.31) give

$$\langle\sigma_+\sigma_-\rangle = \big(\overline{v^*v}\big)_P = \tfrac{1}{2}\big(1 + \langle\sigma_z\rangle\big), \tag{6.32a}$$
$$\langle\sigma_+^2\sigma_-^2\rangle = \big(\overline{v^{*2}v^2}\big)_P = 0. \tag{6.32b}$$

From these equations the result $g^{(2)}(0) = 0$ follows. Show also that (6.5) and (6.31) reproduce all of the moments given in Exercise 6.2.

The distribution (6.31) is also singular in the inversion variable m; although, the δ-function singularity does not signify a nonclassical character as do derivatives of the δ-function. What we have here is consistent with a classical process that involves only discrete states, with the inversion restricted to the values $m = \rho_{22} - \rho_{11} = +1$, for the atom in its upper state, and $m = \rho_{22} - \rho_{11} = -1$, for the atom in its lower state. Indeed, if we integrate $P(v, v^*, m)$ over the polarization variable v, we are left with the reduced distribution

$$P(m, t) = \tfrac{1}{2}\big(1 + \langle \sigma_z(t) \rangle\big)\delta(m - 1) + \tfrac{1}{2}\big(1 - \langle \sigma_z(t) \rangle\big)\delta(m + 1)$$
$$= \rho_{22}(t)\delta(m - 1) + \rho_{11}(t)\delta(m + 1), \qquad (6.33)$$

where ρ_{22} and ρ_{11} give the probabilities for finding the atom in its upper ($m = +1$) and lower ($m = -1$) states, respectively. In classical statistical physics the dynamics for such a discrete state system would be given by a jump process describing transitions between the two states. A closer look at the phase-space equation of motion (6.24) confirms the relationship to a jump process. The differential operators $e^{\pm 2\frac{\partial}{\partial m}}$ appearing in (6.23) describe transitions between discrete inversion states. These are displacement, or shift operators, which generate steps of ± 2 units in m, just what is required for transitions between atomic states with $m = \pm 1$. To show this we write $e^{\pm 2\frac{\partial}{\partial m}}$ as a power series; then it acts on a function $g(m)$ to give the Taylor series expansion for the shifted function $g(m \pm 2)$:

$$e^{\pm 2\frac{\partial}{\partial m}} g(m) = \sum_{k=0}^{\infty} \frac{(\pm 2)^k}{k!} \frac{\partial^k}{\partial m^k} g(m) = g(m \pm 2). \qquad (6.34)$$

Armed with these observations let us now return to the phase-space equation of motion (6.24) and show that the distribution (6.31) does, indeed, satisfy this equation. We will approach the demonstration in steps which bring out something of the structure of the dynamics. First, we explicitly display the confinement to discrete inversion states, writing

$$P(v, v^*, m, t) = P_{|2\rangle}(v, v^*, t)\delta(m - 1) + P_{|1\rangle}(v, v^*, t)\delta(m + 1), \qquad (6.35)$$

with

$$\int d^2v\, P_{|2\rangle}(v, v^*, t) = \rho_{22}(t), \qquad (6.36a)$$

$$\int d^2v\, P_{|1\rangle}(v, v^*, t) = \rho_{11}(t). \qquad (6.36b)$$

This form is consistent with the explicitly constructed distributions (6.31) and (6.33), and with the observation that the operators $e^{\pm 2\frac{\partial}{\partial m}}$ are shift operators. The action of $L(v, v^*, m, \frac{\partial}{\partial v}, \frac{\partial}{\partial v^*}, \frac{\partial}{\partial m})$ on the variable m in (6.35) follows from the relationships

$$\left(e^{2\frac{\partial}{\partial m}} - 1\right)(1 + m)\delta(m - 1)$$
$$= (3 + m)\delta(m + 1) - (1 + m)\delta(m - 1)$$
$$= 2\delta(m + 1) - 2\delta(m - 1), \tag{6.37a}$$

$$\left(e^{-2\frac{\partial}{\partial m}} - 1\right)(1 - m)\delta(m + 1)$$
$$= (3 - m)\delta(m - 1) - (1 - m)\delta(m + 1)$$
$$= 2\delta(m - 1) - 2\delta(m + 1), \tag{6.37b}$$

$$\left(e^{2\frac{\partial}{\partial m}} - 1\right)(1 + m)\delta(m + 1)$$
$$= (3 + m)\delta(m + 3) - (1 + m)\delta(m + 1) = 0, \tag{6.37c}$$

$$\left(e^{-2\frac{\partial}{\partial m}} - 1\right)(1 - m)\delta(m - 1)$$
$$= (3 - m)\delta(m - 3) - (1 - m)\delta(m - 1) = 0, \tag{6.37d}$$

and

$$e^{-2\frac{\partial}{\partial m}}\delta(m - 1) = \delta(m - 3), \tag{6.37e}$$

$$e^{-2\frac{\partial}{\partial m}}\delta(m + 1) = \delta(m - 1). \tag{6.37f}$$

Each of these relationships preserves the restriction to inversion states $m = \pm 1$, except for (6.37e). This equation permits the shift operator $e^{\pm 2\frac{\partial}{\partial m}}$ that appears in the last line of (6.23), without an accompanying multiplicative factor $(1 - m)$, to couple to the states $m = 3, 5, \ldots$. This coupling must be suppressed by the v-dependence of $P_{|2\rangle}$. Now, substituting (6.35) into (6.24), and using (6.37a)–(6.37f), we equate coefficients of the δ-functions to obtain the equations

$$\frac{\partial P_{|2\rangle}}{\partial t} = \left[\left(\frac{\gamma}{2} + i\omega_A\right)\frac{\partial}{\partial v}v + \left(\frac{\gamma}{2} - i\omega_A\right)\frac{\partial}{\partial v^*}v^* + \gamma\bar{n}\frac{\partial^2}{\partial v \partial v^*}\right.$$
$$\left. + \gamma\bar{n}\frac{\partial^2}{\partial v \partial v^*}\left(\frac{\partial}{\partial v}v + \frac{\partial}{\partial v^*}v^*\right) - \gamma(\bar{n} + 1)\right]P_{|2\rangle}$$
$$+ \left[\gamma\bar{n}\left(\frac{\partial}{\partial v}v + \frac{\partial}{\partial v^*}v^*\right) + \gamma\bar{n}\right]P_{|1\rangle}, \tag{6.38a}$$

$$\frac{\partial P_{|1\rangle}}{\partial t} = \left[\left(\frac{\gamma}{2} + i\omega_A\right)\frac{\partial}{\partial v}v + \left(\frac{\gamma}{2} - i\omega_A\right)\frac{\partial}{\partial v^*}v^* + \gamma\bar{n}\frac{\partial^2}{\partial v \partial v^*}\right.$$
$$\left. + \gamma\bar{n}\frac{\partial^2}{\partial v \partial v^*}\left(\frac{\partial}{\partial v}v + \frac{\partial}{\partial v^*}v^*\right) - \gamma\bar{n}\right]P_{|1\rangle}$$
$$+ \left[\gamma\bar{n}\frac{\partial^4}{\partial v^2 \partial v^{*2}} + \gamma(\bar{n} + 1)\right]P_{|2\rangle}, \tag{6.38b}$$

and

$$\left(\frac{\partial}{\partial v}v + \frac{\partial}{\partial v^*}v^*\right)P_{|2\rangle} = 0,$$ (6.38c)

Equation (6.38c) suppresses coupling to inversion states other than $m = \pm 1$.

The connection between the inversion dynamics and a classical jump process can now be made more explicit. If we integrate (6.38a) and (6.38b) over v, and require $P_{|1\rangle}$, $P_{|2\rangle}$, and their derivatives to vanish sufficiently fast at infinity, we obtain the equations

$$\dot{\rho}_{22} = -\gamma(\bar{n} + 1)\rho_{22} + \gamma\bar{n}\rho_{11},$$ (6.39a)
$$\dot{\rho}_{11} = -\gamma\bar{n}\rho_{11} + \gamma(\bar{n} + 1)\rho_{22}.$$ (6.39b)

These are the equations of a random telegraph process [6.4]. They are often referred to as the *Einstein equations* in recognition of their use by Einstein in his phenomenological theory of spontaneous and stimulated emission [6.5, 6.6]. We derived these equations directly from the operator master equation in Sect. 2.2.3.

Instead of averaging over v we might average over the inversion to obtain an equation for the polarization dynamics. We do this by adding (6.38a) and (6.38b):

$$\frac{d}{dt}\left(P_{|1\rangle} + P_{|2\rangle}\right) = \left[\left(\frac{\gamma}{2} + i\omega_A\right)\frac{\partial}{\partial v}v + \left(\frac{\gamma}{2} - i\omega_A\right)\frac{\partial}{\partial v^*}v^* + \gamma\bar{n}\frac{\partial^2}{\partial v\partial v^*}\right.$$
$$\left.+\gamma\bar{n}\frac{\partial^2}{\partial v\partial v^*}\left(\frac{\partial}{\partial v}v + \frac{\partial}{\partial v^*}v^*\right)\right]\left(P_{|1\rangle} + P_{|2\rangle}\right)$$
$$+ \gamma\bar{n}\left(\frac{\partial}{\partial v}v + \frac{\partial}{\partial v^*}v^*\right)P_{|1\rangle} + \gamma\bar{n}\frac{\partial^4}{\partial v^2\partial v^{*2}}P_{|2\rangle}.$$ (6.40)

The most important thing to observe here is the way in which this equation differs from the Fokker–Planck equation for the damped harmonic oscillator. The first line on the right-hand side describes a damped harmonic oscillator; however, there are two deviations from this simple form. First, we have not obtained a closed equation for the polarization dynamics; the last two terms on the right-hand side of (6.40) couple to the individual inversion states. Second, there are third-order derivatives added to the differential operator in the square bracket. Under certain conditions these complications can be removed – for example, by setting $\bar{n}$ to zero. However, this still does not recover the simple classical picture we reached in our treatment of the damped harmonic oscillator. It is still necessary to consider solutions in the highly singular form given by (6.31). The equation of motion might reduce to a simple form, but it must be solved for an initial state that is represented by a generalized function.

With the phase-space equation of motion written as the coupled equations (6.38a) and (6.38b), it is now just a short step to show that the distribution (6.31) solves these equations. We first note that (6.38c) requires

$$P_{|2\rangle}(v, v^*, t) = \rho_{22}(t)\delta^{(2)}(v), \tag{6.41}$$

consistent with (6.31), where the time-dependent function that multiplies the δ-function has been determined using (6.36a). The fact that (6.41) solves (6.38c) follows trivially from $v\delta^{(2)}(v) = v^*\delta^{(2)}(v) = 0$. Equation (6.38c) also has solutions as ordinary functions – for example, in the form $f(t)/vv^*$ – these, however, are not normalizable, and are not therefore consistent with (6.36a). Together with (6.41) we take $P_{|1\rangle}(v, v^*, t)$ in the form [Eq. (6.31)]

$$P_{|1\rangle}(v, v^*, t) = \rho_{11}(t)\delta^{(2)}(v) - \rho_{21}(t)\frac{\partial}{\partial v}\delta^{(2)}(v) - \rho_{12}(t)\frac{\partial}{\partial v^*}\delta^{(2)}(v)$$

$$+ \rho_{22}(t)\frac{\partial^2}{\partial v\partial v^*}\delta^{(2)}(v). \tag{6.42}$$

Then, substituting $P_{|2\rangle}$ and $P_{|1\rangle}$ into (6.38a) and (6.38b), and using the relationships $v\delta(v) = v^*\delta(v) = 0$ and $v\partial\delta^{(2)}(v)/\partial v = v^*\partial\delta^{(2)}(v)/\partial v^* = -\delta^{(2)}(v)$, we arrive at the equations

$$\left[\dot{\rho}_{22} + \gamma(\bar{n}+1)\rho_{22} - \gamma\bar{n}\right]\delta^{(2)}(v) = 0, \tag{6.43a}$$

and

$$\left[\dot{\rho}_{11} + \gamma\bar{n}\rho_{11} - \gamma(\bar{n}+1)\rho_{22}\right]\delta^{(2)}(v)$$

$$- \left[\dot{\rho}_{21} + \gamma(\bar{n}+\tfrac{1}{2})\rho_{21} + i\omega_A\rho_{21}\right]\frac{\partial}{\partial v}\delta^{(2)}(v)$$

$$- \left[\dot{\rho}_{12} + \gamma(\bar{n}+\tfrac{1}{2})\rho_{12} - i\omega_A\rho_{12}\right]\frac{\partial}{\partial v^*}\delta^{(2)}(v)$$

$$+ \left[\dot{\rho}_{22} + \gamma(\bar{n}+1)\rho_{22} - \gamma\bar{n}\rho_{11}\right]\frac{\partial^2}{\partial v\partial v^*}\delta^{(2)}(v) = 0. \tag{6.43b}$$

By requiring the coefficients of $\delta^{(2)}(v)$, $\partial\delta^{(2)}(v)/\partial v$, $\partial\delta^{(2)}(v)/\partial v^*$, and $\partial^2\delta^{(2)}(v)/\partial v\partial v^*$ to vanish, we reproduce the matrix element equations (2.36) derived directly from the operator master equation. Thus, the distribution (6.31) satisfies the equation of motion (6.24), so long as the expectations $\langle\sigma_-(t)\rangle = \rho_{21}(t)$, $\langle\sigma_+(t)\rangle = \rho_{12}(t)$, and $\langle\sigma_z(t)\rangle = \rho_{22}(t) - \rho_{11}(t)$ obey the optical Bloch equations.

6.2 Normal-Ordered Representation for a Collection of Two-Level Atoms

We have now seen that although the quantum–classical correspondence is easily extended formally to the description of atomic states, it does not lead to a classical statistical picture for the damped two-level atom in any useful sense. Difficulties arise both with the representation of states, and with their dynamical evolution. First, the distribution is always a generalized function, more singular than a δ-function in the polarization variable, and with a δ-function singularity restricting the inversion to discrete states. Second, the phase-space equation of motion is not a Fokker–Planck equation. The inversion dynamics are described by a jump process, which introduces partial derivatives up to infinite order, as in the Kramers–Moyal expansion; the phase-space equation of motion is therefore really a pair of coupled partial differential equations. Even after averaging over inversion states, partial derivatives beyond second order remain in the polarization variable.

Despite the poor prognosis, the representation discussed in Sect. 6.1 is actually very useful. We will use it later in this book to analyze quantum fluctuations in the laser, and again in Volume 2 to treat certain problems in cavity QED. The difficulties we have observed can be removed in the treatment of a macroscopic medium, a collection of $N \gg 1$ two-level atoms.

6.2.1 Collective Atomic Operators

We consider a collection of N two-level atoms with spatial positions r_j, $j = 1, \ldots, N$. A complete microscopic description for this system requires the specification of the state of each atom. For many purposes, however, such detail is not needed, a description of collective properties defined by the sum over all atoms is adequate. If the atoms can be considered as identical, then individual atomic properties can even be deduced from such course-grained information. We will consider a description in terms of the *collective atomic operators*

$$J_\pm \equiv \sum_{j=1}^{N} e^{\pm i\phi_j} \sigma_{j\pm}, \qquad J_z \equiv \sum_{j=1}^{N} \sigma_{jz}, \qquad (6.44)$$

where σ_{j-}, σ_{j+}, and σ_{jz} are pseudo-spin operators for atom j, and ϕ_j is an arbitrary phase. We will see shortly (Sect. 6.3.2), that when the atoms are identical, a closed dynamical description in terms of collective operators alone can be formulated. The representation we develop in the following sections presupposes such conditions.

Note 6.1 When two-level atoms interact with optical fields, generally the field distribution will be spatially dependent. Then different atoms find themselves in different local environments and the atoms are not identical. If the interaction is with a single plane, traveling-wave mode, with wavevector k_0, the only difference between atoms is the phase of the field, $\exp[-i(\omega_0 t - k_0 \cdot r_j)]$, at the site of each atom. This difference can be removed by setting $\phi_j = k_0 \cdot r_j$ in (6.44); atomic states described by the phased operators $\exp(-ik_0 \cdot r_j)\sigma_{j-}$, $\exp(ik_0 \cdot r_j)\sigma_{j+}$, and σ_{jz} are then identical. For this reason quantum-statistical theories for intracavity interactions have a preference for ring cavities over standing-wave cavities [6.7, 6.8].

From the single-atom commutators (2.11), it is straightforward to show that collective atomic operators obey the same commutation relations:

$$[J_+, J_-] = J_z, \qquad [J_\pm, J_z] = \mp 2J_\pm. \tag{6.45}$$

This is expected, as both are formally angular momentum operators. The algebraic properties of collective operators are quite different, however, when we consider the relationships that allowed products of single-atom operators to be written as sums [Eqs. (6.6)]. There does exist a generalization of these relationships. However, now the set of operators needed to express all higher-order operator products is larger than just J_-, J_+, J_z, and 1. Some of the generalized relationships are easily deduced. When $N > 1$ it is clearly possible for two or more quanta to be absorbed or emitted simultaneously; they can be absorbed or emitted by different atoms. We expect, however, no more than a maximum of N simultaneous absorptions or emissions. Thus, we expect to replace (6.6a) and (6.6b) by

$$J_+^{N+1} = J_-^{N+1} = 0. \tag{6.46}$$

Formally, this follows by applying $\sigma_{j+}^2 = \sigma_{j-}^2 = 0$ for each atom. The only nonvanishing terms in J_+^N and J_-^N are products of N operators for *different* atoms:

$$J_+^N = N! \prod_{j=1}^{N} e^{i\phi_j} \sigma_{j+}, \qquad J_-^N = N! \prod_{j=1}^{N} e^{-i\phi_j} \sigma_{j-}. \tag{6.47}$$

These operators raise and lower the entire collection of atoms between the ground state, with all atoms in their lower state, and the fully excited state, with all atoms in their upper state. Every term in J_+^{N+1} and J_-^{N+1} must contain a second or higher power of at least one single atom operator; from this (6.46) follows.

It is also straightforward to deduce the generalization of (6.6c). We first write this equation in the form

$$(\sigma_z + 1)(\sigma_z - 1) = 0, \tag{6.48}$$

or, multiplying by $(\frac{1}{2}\hbar\omega_A)^2$,

$$\left(H_A + \tfrac{1}{2}\hbar\omega_A\right)\left(H_A - \tfrac{1}{2}\hbar\omega_A\right) = 0. \tag{6.49}$$

The left-hand side of (6.49) is the polynomial $(H_A + E_1)(H_A + E_2)$ formed from the energy eigenvalues in such a way that the action of the polynomial on each energy eigenstate, and therefore on an arbitrary state, is zero. The energy eigenvalues for a collection of N identical two-level atoms are given by

$$\left(N_{|2\rangle} - N_{|1\rangle}\right)\tfrac{1}{2}\hbar\omega_A = M\hbar\omega_A, \tag{6.50}$$

where $N_{|2\rangle}$ and $N_{|1\rangle}$ are the numbers of atoms in states $|2\rangle$ and $|1\rangle$, respectively, with

$$N_{|2\rangle} + N_{|1\rangle} = N, \tag{6.51}$$

and $M = -N/2, -N/2 + 1, \ldots, N/2$; $2M$ runs in steps of two units over all possible inversion states. Now,

$$H_A = \tfrac{1}{2}\hbar\omega_A J_z, \tag{6.52}$$

and the generalization of (6.48) is

$$(J_z + N)(J_z + N - 2)\cdots(J_z - N) = 0. \tag{6.53}$$

This is a polynomial in J_z of order $N + 1$, allowing powers of J_z greater than N to be expressed as a sum over the operators $1, J_z, J_z^2, \ldots, J_z^N$.

Our interest is with *normal-ordered products of collective atomic operators*, products of the form $J_+^p J_z^r J_-^q$. Equations (6.46) and (6.53) state that all such products can be expressed in terms of those with $p, q, r \leq N$, i.e. in terms of $(N + 1)^3$ operators. In fact the number of operators needed to construct arbitrary normal-ordered operator products is somewhat smaller. Further relationships between collective operator products exist. We will not attempt to construct a general algorithm giving them all, but can easily see that more exist. Consider the operator $(q \leq N)$

$$J_-^q = \sum_{\{j_1,\cdots,j_q\}} \left[\exp(-i\phi_{j_1})\sigma_{j_1-}\right]\cdots\left[\exp(-i\phi_{j_q})\sigma_{j_q-}\right], \tag{6.54}$$

where the sum covers all sets of nonrepeating atomic labels $j_1, \ldots, j_q$. Clearly, the action of J_-^q on any state $|\psi\rangle$ of the N-atom system gives a state with at least q atoms in their lower states, and at most $N - q$ atoms in their upper states. The state $J_-^q|\psi\rangle$ must be a superposition of energy eigenstates having energies $-N\frac{1}{2}\hbar\omega_A, -(N-2)\frac{1}{2}\hbar\omega_A, \ldots, (N-2q)\frac{1}{2}\hbar\omega_A$. The possible inversion states in $J_-^q|\psi\rangle$ are correspondingly $2M = N_{|2\rangle} - N_{|1\rangle} = -N, -N+2, \ldots, N-2q$. Thus,

$$(J_z + N)(J_z + N - 2)\cdots(J_z - N + 2q)J_-^q = 0; \tag{6.55a}$$

from the conjugate relation,

$$J_+^p(J_z + N)(J_z + N - 2)\cdots(J_z - N + 2p) = 0. \tag{6.55b}$$

Equation (6.55a) allows us to construct $J_z^r J_-^q$, for $r > N-q$, in terms of $J_z^r J_-^q$, $r = 0, \ldots, N - q$; Eq. (6.55b) allows us to construct $J_+^p J_z^r$, for $r > N - p$, in terms of $J_+^p J_z^r$, $r = 0, \ldots, N - p$.

In fact, all operator products $J_+^p J_z^r J_-^q$ with $p + r + q > N$ can be expressed as sums of the operators $J_+^p J_z^r J_-^q$, $p + r + q \leq N$. The total number of independent normal-ordered collective operator products is then

$$\sum_{q=0}^{N}\sum_{r=0}^{N-q}\sum_{p=0}^{N-q-r} 1$$

$$= \sum_{q=0}^{N}\sum_{r=0}^{N-q}(N-q-r+1)$$

$$= \sum_{q=0}^{N}(N-q+1)(N-q+1) - \sum_{q=0}^{N}\tfrac{1}{2}(N-q)(N-q+1)$$

$$= \sum_{k=0}^{N}(k+1)^2 - \tfrac{1}{2}k(k+1)$$

$$= \tfrac{1}{6}(N+1)(N+2)(N+3). \tag{6.56}$$

The proof that operator products with $p+r+q \leq N$ determine all others follows from two observations:

1. Collective operator products can be expressed as sums of products between single-atom operators for *different* atoms, since products of single-atom operators for the same atom can be replaced by sums using (6.6). Products between single-atom operators for different atoms can appear, at most, up to order N.
2. In any collective operator product, single-atom operator products appear in a symmetric fashion with respect to permutations between atoms.

From these observations all products of collective atomic operators can be expanded as sums over the operators

$$\hat{\Sigma}_{n,k,m} \equiv \frac{1}{n!k!m!} \sum_{j_1,\ldots,j_{n+k+m}} \exp\left[i(\phi_{j_1} + \cdots + \phi_{j_n})\right]$$

$$\times \exp\left[-i(\phi_{j_{n+k+1}} + \cdots + \phi_{j_{n+k+m}})\right]$$

$$\times (\sigma_{j_1+} \cdots \sigma_{j_n+})(\sigma_{j_{n+1}z} \cdots \sigma_{j_{n+k}z})(\sigma_{j_{n+k+1}-} \cdots \sigma_{j_{n+k+m}-}),$$
$$\tag{6.57}$$

where $n+k+m \leq N$; the summation is over all permutations of $n+k+m$ nonrepeating atomic labels. To generate all of these operators in a collective operator product expansion we must consider $J_+^p J_z^r J_-^q$, $p+r+q \leq N$. These expansions give $\tfrac{1}{6}(N+1)(N+2)(N+3)$ linear relationships expressing the operators $J_+^p J_z^r J_-^q$, $p+r+q \leq N$, in terms of the operators $\hat{\Sigma}_{n,k,m}$, $n+k+m \leq N$. Inverting these relationships gives the operators $\hat{\Sigma}_{n,k,m}$, $n+k+m \leq N$, in terms of the operators $J_+^p J_z^r J_-^q$, $p+r+q \leq N$. The operators $\hat{\Sigma}_{n,k,m}$ then determine all remaining collective operator products, $p+r+q > N$.

Exercise 6.4 For $N = 2$, $\tfrac{1}{6}(N+1)(N+2)(N+3) = 10$. There are twenty-seven collective operator products with $p, q, r \leq N$. Show that the following seventeen relationships hold:

$$\left.\begin{array}{ll} J_+^2 J_z = -2J_+^2, & J_z J_-^2 = -2J_-^2, \\ J_+^2 J_- = J_+(J_z + 2), & J_+ J_-^2 = (J_z + 2)J_-, \\ J_+ J_z^2 = -2J_+ J_z, & J_z^2 J_- = -2J_z J_-, \\ \multicolumn{2}{c}{J_+ J_z J_- = -2J_+ J_- + \frac{1}{2}(J_z + 2)J_z,} \end{array}\right\} \tag{6.58a}$$

$$\left.\begin{array}{ll} J_+^2 J_z^2 = 4J_+^2, & J_z^2 J_-^2 = 4J_-^2, \\ J_+^2 J_z J_- = -2J_+(J_z + 2), & J_+ J_z J_-^2 = -2(J_z + 2)J_-, \\ \multicolumn{2}{c}{J_+^2 J_-^2 = (J_z + 2)J_z,} \\ \multicolumn{2}{c}{J_+ J_z^2 J_- = 4J_+ J_- - (J_z + 2)J_z,} \end{array}\right\} \tag{6.58b}$$

$$\left.\begin{array}{ll} J_+^2 J_z^2 J_- = 4J_+(J_z + 2), & J_+ J_z^2 J_-^2 = 4(J_z + 2)J_-, \\ \multicolumn{2}{c}{J_+^2 J_z J_-^2 = -2(J_z + 2)J_z,} \end{array}\right\} \tag{6.58c}$$

$$J_+^2 J_z^2 J_-^2 = 4(J_z + 2)J_z. \tag{6.58d}$$

The ten operator products with $p + r + q \leq N$ are the unit operator, plus

$$\left.\begin{array}{l} J_+ = e^{i\phi_1}\sigma_{1+} + e^{i\phi_2}\sigma_{2+}, \\ J_z = \sigma_{1z} + \sigma_{2z}, \\ J_- = e^{-i\phi_1}\sigma_{1-} + e^{-i\phi_2}\sigma_{2-}, \end{array}\right\} \tag{6.59a}$$

$$\left.\begin{array}{l} J_+^2 = 2e^{i(\phi_1+\phi_2)}\sigma_{1+}\sigma_{2+}, \\ J_z^2 = 2(1 + \sigma_{1z}\sigma_{2z}), \\ J_-^2 = 2e^{-i(\phi_1+\phi_2)}\sigma_{1-}\sigma_{2-}, \end{array}\right\} \tag{6.59b}$$

$$\left.\begin{array}{l} J_+ J_z = -J_+ + e^{i\phi_1}\sigma_{1+}\sigma_{2z} + e^{i\phi_2}\sigma_{2+}\sigma_{1z}, \\ J_+ J_- = \frac{1}{2}(J_z + 2) + e^{i(\phi_1-\phi_2)}\sigma_{1+}\sigma_{2-} + e^{i(\phi_2-\phi_1)}\sigma_{2+}\sigma_{1-}, \\ J_z J_- = -J_- + e^{-i\phi_1}\sigma_{1-}\sigma_{2z} + e^{-i\phi_2}\sigma_{2-}\sigma_{1z}. \end{array}\right\} \tag{6.59c}$$

6.2.2 Direct Product States, Dicke States, and Atomic Coherent States

The master equation for a single damped two-level atom was readily analyzed using matrix element equations (Sects. 2.2.3 and 2.3.3). We have seen that a phase-space approach to this problem leads to a rather complicated picture. For many-atom systems the situation tends to be reversed. The large set of basis states needed for a many-atom system yields a large number of matrix element equations. These are generally not solvable analytically. There are special cases where analytical solutions are possible [6.9, 6.10], and for moderate numbers of atoms the numerical solutions of matrix element equations can be useful [6.11, 6.12] – indeed, with rapid advances in supercomputing the potential for numerical solutions is only beginning to be explored. For studying the true large N limit, however, phase-space methods like the one we describe shortly provide the most manageable and insightful approach. We will make little use, then, of matrix element equations when treating the laser, for example. Nevertheless, we should say something about the different basis states that are commonly used to derive such equations.

An obvious basis for a collection of two-level atoms is provided by the states

$$|\mu_1\rangle_1|\mu_2\rangle_2 \cdots |\mu_j\rangle_j \cdots |\mu_N\rangle_N,$$

where for the jth atom μ_j can take the values 1 or 2 to denote the lower state $|1\rangle_j$ or upper state $|2\rangle_j$, respectively. There are 2^N such states. We adopt a more compact notation, defining the *direct product states* by

$$|\boldsymbol{u}; M\rangle \equiv \prod_{k\in u} \left(e^{\frac{1}{2}i\phi_k}|2\rangle_k\right) \prod_{k\notin u} \left(e^{-\frac{1}{2}i\phi_k}|1\rangle_k\right), \qquad (6.60)$$

where $\boldsymbol{u} = \{j_1, \ldots, j_{N/2+M}\}$ denotes a vector of $N/2 + M$ nonrepeating atomic labels, and $N/2 + M$ is the number of atoms in their upper states ($2M = N_{|2\rangle} - N_{|1\rangle}$ is the inversion); the phases are simply introduced for convenience in view of the arbitrary phases included in (6.44). These states are energy eigenstates, or eigenstates of the inversion operator J_z:

$$J_z|\boldsymbol{u}; M\rangle = 2M|\boldsymbol{u}; M\rangle. \qquad (6.61)$$

The labels $\boldsymbol{u}$ distinguish between states that are $[N!/(N/2 + M)!(N/2 - M)!]$-fold degenerate with respect to the inversion eigenvalue (summing this degeneracy over $M = -N/2, \ldots, N/2$ gives back 2^N states).

Direct product states do not behave in a very convenient way under the action of the collective operators J_+ and J_-. We have

$$J_+|\boldsymbol{u}; M\rangle = \begin{cases} \sum_{k\notin u} |\boldsymbol{u}_{+k}; M+1\rangle & -N/2 \leq M < N/2 \\ 0 & M = N/2, \end{cases} \qquad (6.62a)$$

$$J_-|\boldsymbol{u}; M\rangle = \begin{cases} \sum_{k\in u} |\boldsymbol{u}_{-k}; M-1\rangle & -N/2 < M \leq N/2 \\ 0 & M = -N/2, \end{cases} \qquad (6.62b)$$

where $\boldsymbol{u}_{+k} \equiv \{j_1, \ldots, j_{N/2+M+1}\}$ and $\boldsymbol{u}_{-k} \equiv \{j_1, \ldots, j_{N/2+M-1}\}$, respectively, add and delete one atomic label in the vector $\boldsymbol{u}$ labeling atoms in their upper states. Thus, J_+ and J_- generate transitions up and down the ladder of inversion states; however, they generally connect one initial product state to many final states. In his treatment of superradiance, Dicke [6.13] introduced alternative basis states chosen for their simple behavior under the action of J_+ and J_-. These states are often referred to as *Dicke states*; although, as Dicke noted himself, they are formally the eigenstates of the total spin constructed from the sum of N spin-$\frac{1}{2}$ operators, and therefore familiar from the theory of angular momentum.

The Dicke states are simultaneous eigenstates of the commuting operators J_z and

$$\hat{J}^2 = \left(\tfrac{1}{2}J_x\right)^2 + \left(\tfrac{1}{2}J_y\right)^2 + \left(\tfrac{1}{2}J_z\right)^2$$
$$= \tfrac{1}{2}(J_+J_- + J_-J_+) + \tfrac{1}{4}J_z^2. \tag{6.63}$$

We denote these states by $|\lambda, J, M\rangle$. From the theory of angular momentum [6.14, 6.15], we have

$$\hat{J}^2|\lambda, J, M\rangle = J(J+1)|\lambda, J, M\rangle, \tag{6.64a}$$

$$J_z|\lambda, J, M\rangle = 2M|\lambda, J, M\rangle, \tag{6.64b}$$

with allowed values $J = \left(0, \tfrac{1}{2}\right), \left(1, \tfrac{3}{2}\right), \ldots, N/2$ and $M = -J, -J+1, \ldots, J$. The total number of allowed (J, M) values is $(N/2+1)^2$ for N even, and $[(N+1)/2+1](N+1)/2$ for N odd. For $N > 2$ this is less than 2^N. Therefore, states labeled by J and M alone must still be degenerate, and λ distinguishes amongst the degenerate states. The action of J_+ and J_- now connects a single state to a single state:

$$J_+|\lambda, J, M\rangle = \begin{cases} \sqrt{(J-M)(J+M+1)}|\lambda, J, M+1\rangle & -J \leq M < J \\ 0 & M = J, \end{cases} \tag{6.65a}$$

$$J_-|\lambda, J, M\rangle = \begin{cases} \sqrt{(J+M)(J-M+1)}|\lambda, J, M-1\rangle & -J < M \leq J \\ 0 & M = -J. \end{cases} \tag{6.65b}$$

What is the degeneracy for a given J and M? First note that the degeneracy must be the same for all states with the same J. This follows because any state $|\lambda', J, M'\rangle$ produces a set of states with the same $\lambda = \lambda'$, covering all possible M values, under the action of J_+ and J_- – the states $|\lambda', J, M\rangle$, $M = -J, -J+1, \ldots, J$. We may now use an iterative argument to deduce the degeneracy $d_D(J)$ of the Dicke states for a given J, from the degeneracy

$$d_p(M) = \frac{N!}{(N/2+M)!(N/2-M)!} \tag{6.66}$$

of the direct product states for a given M:

1. There is one product state with $M = -N/2$; corresponding to this state there is one Dicke state with $M = -N/2$ and $J = N/2$; thus

$$d_D(N/2) = 1.$$

2. There are $d_p(-N/2+1) = N$ product states with $M = -N/2+1$; there are $d_D(N/2) = 1$ Dicke states with $M = -N/2+1$ and $J = -N/2$; the

only other value of J allowing $M = -N/2 + 1$ is $J = N/2 - 1$, so there are

$$d_D(N/2 - 1) = d_p(-N/2 + 1) - d_D(N/2)$$
$$= N - 1$$

Dicke states with $M = -N/2 + 1$ and $J = N/2 - 1$.

3. There are $d_p(-N/2+2) = \frac{1}{2}N(N-1)$ product states with $M = -N/2+2$; there are $d_D(N/2) = 1$ Dicke states with $M = -N/2+2$ and $J = N/2$, and $d_D(N/2-1) = N-1$ Dicke states with $M = -N/2+2$ and $J = N/2-1$; the only other value of J allowing $M = -N/2 + 2$ is $J = N/2 - 2$, so there are

$$d_D(N/2 - 2) = d_p(-N/2 + 2) - [d_D(N/2 - 1) + d_D(N/2)]$$
$$= d_p(-N/2 + 2) - d_p(-N/2 + 1)$$
$$= \frac{1}{2}N(N - 3)$$

Dicke states with $M = -N/2 + 2$ and $J = N/2 - 2$.

4. Iterating this argument gives

$$d_D(N/2 - k) = d_p(-N/2 + k) - [d_D(N/2 - k + 1)$$
$$+ d_D(N/2 - k + 2) + \cdots + d_D(N/2)]$$
$$= d_p(-N/2 + k) - d_p(-N/2 + k - 1)$$
$$= \frac{N!}{(N - k)!k!} - \frac{N!}{(N - k + 1)!(k - 1)!}$$
$$= \frac{N!(N - 2k + 1)}{(N - k + 1)!k!}.$$

Setting $J = N/2 - k$, we finally obtain

$$d_D(J) = \frac{N!(2J + 1)}{(N/2 + J + 1)!(N/2 - J)!}. \tag{6.67}$$

Of course, summing (6.67) over all states with the same value of M gives

$$\sum_{J=M}^{N/2} d_D(J) = d_p(M); \tag{6.68}$$

with degeneracies accounted for, the direct product and Dicke bases both contain 2^N states.

Dicke states may be expressed as superpositions of direct product states, and an explicit Dicke basis may be constructed in an iterative fashion following closely the steps of the argument giving $d_D(J)$:

1. Construct states with $J = N/2$:
   - set

$$|1, N/2, -N/2\rangle = |\{\ \}; -N/2\rangle = \prod_{j=1}^{N}\left(e^{-\frac{1}{2}i\phi_j}|1\rangle_j\right); \qquad (6.69)$$

   - apply J_+, $N/2 + M$ times to $|1, N/2, -N/2\rangle$, to construct states

$$|1, N/2, M\rangle = \frac{1}{\sqrt{d_p(M)}}\sum_{u}|u; M\rangle, \qquad (6.70)$$

   for $M = -N/2, -N/2 + 1, \ldots, N/2$, where the summation extends over the $d_p(M)$ labels u differentiating between product states with inversion $2M$.

2. Construct states with $J = N/2 - 1$:
   There is only $d_D(N/2) = 1$ state with $M = -N/2 + 1$ included in (6.70); the state $|1, N/2, -N/2 + 1\rangle$
   - construct $d_D(N/2 - 1) = d_p(-N/2 + 1) - d_D(N/2) = N - 1$ mutually orthogonal states, orthogonal to $|1, N/2, -N/2+1\rangle$, as linear combinations of the $d_p(-N/2 + 1) = N$ product states $|u; -N/2 + 1\rangle$, to obtain states $|\lambda, N/2 - 1, -N/2 + 1\rangle$, $\lambda = 1, \ldots, N - 1$;
   - apply J_+, $N/2 + M - 1$ times to each of these states to construct states $|\lambda, N/2 - 1, M\rangle$, $M = -N/2 + 1, -N/2 + 2, \ldots, N/2 - 1$, $\lambda = 1, \ldots, N - 1$.

3. Construct states with $J = N/2 - 2$:
   We now have $d_D(N/2) + d_D(N/2 - 1) = N$ states with $M = -N/2 + 2$; the states $|1, N/2, -N/2+2\rangle$ and $|\lambda, N/2 - 1, -N/2+2\rangle$, $\lambda = 1, \ldots, N - 1$
   - construct $d_D(N/2 - 2) = d_p(-N/2 + 2) - [d_D(N/2) + d_D(N/2 - 1)] = \frac{1}{2}N(N - 3)$ mutually orthogonal states, orthogonal to $|1, N/2, -N/2+2\rangle$ and $|\lambda, N/2 - 1, -N/2 + 2\rangle$, $\lambda = 1, \ldots, N - 1$, as linear combinations of the $d_p(-N/2 + 2) = \frac{1}{2}N(N - 1)$ product states $|u; -N/2 + 2\rangle$ to obtain states $|\lambda, N/2 - 2, -N/2 + 2\rangle$, $\lambda = 1, \ldots, \frac{1}{2}N(N - 3)$;
   - apply J_+, $N/2 + M - 2$ times to each of these states to construct states $|\lambda, N/2 - 2, -N/2 + 2\rangle$, $M = -N/2 + 2, -N/2 + 3, \ldots, N/2 - 2$, $\lambda = 1, \ldots, \frac{1}{2}N(N - 3)$.

4. Construct states with $J < N/2 - 2$:
   - iterate the procedure.

Exercise 6.5 For $N = 2$ define

$$\left.\begin{aligned}
|\downarrow\downarrow\rangle &\equiv |\{\ \}; -1\rangle = e^{-\frac{1}{2}i(\phi_1+\phi_2)}|1\rangle_1|1\rangle_2, \\
|\uparrow\downarrow\rangle &\equiv |\{2\}; 0\rangle = e^{\frac{1}{2}i(\phi_1-\phi_2)}|2\rangle_1|1\rangle_2, \\
|\downarrow\uparrow\rangle &\equiv |\{1\}; 0\rangle = e^{-\frac{1}{2}i(\phi_1-\phi_2)}|1\rangle_1|2\rangle_2, \\
|\uparrow\uparrow\rangle &\equiv |\{1,2\}; 1\rangle = e^{\frac{1}{2}i(\phi_1+\phi_2)}|2\rangle_1|2\rangle_2.
\end{aligned}\right\} \qquad (6.71)$$

Use the above procedure to construct the Dicke basis

$$|1,1,-1\rangle = |\downarrow\downarrow\rangle, \quad |1,1,0\rangle = \tfrac{1}{\sqrt{2}}\big(|\downarrow\uparrow\rangle + |\uparrow\downarrow\rangle\big), \quad |1,1,1\rangle = |\uparrow\uparrow\rangle, \Bigg\}$$
$$|1,0,0\rangle = \tfrac{1}{\sqrt{2}}\big(|\downarrow\uparrow\rangle - |\uparrow\downarrow\rangle\big). \tag{6.72}$$

For $N = 3$ define

$$
\begin{aligned}
|\downarrow\downarrow\downarrow\rangle &\equiv |\{\ \};-3/2\rangle = e^{-\frac{1}{2}i(\phi_1+\phi_2+\phi_3)}|1\rangle_1|1\rangle_2|1\rangle_3,\\
|\downarrow\downarrow\uparrow\rangle &\equiv |\{3\};-1/2\rangle = e^{-\frac{1}{2}i(\phi_1+\phi_2-\phi_3)}|1\rangle_1|1\rangle_2|2\rangle_3,\\
|\downarrow\uparrow\downarrow\rangle &\equiv |\{2\};-1/2\rangle = e^{-\frac{1}{2}i(\phi_1-\phi_2+\phi_3)}|1\rangle_1|2\rangle_2|1\rangle_3,\\
|\uparrow\downarrow\downarrow\rangle &\equiv |\{1\};-1/2\rangle = e^{\frac{1}{2}i(\phi_1-\phi_2-\phi_3)}|2\rangle_1|1\rangle_2|1\rangle_3,\\
|\downarrow\uparrow\uparrow\rangle &\equiv |\{2,3\};1/2\rangle = e^{-\frac{1}{2}i(\phi_1-\phi_2-\phi_3)}|1\rangle_1|2\rangle_2|2\rangle_3,\\
|\uparrow\downarrow\uparrow\rangle &\equiv |\{1,3\};1/2\rangle = e^{\frac{1}{2}i(\phi_1-\phi_2+\phi_3)}|2\rangle_1|1\rangle_2|2\rangle_3,\\
|\uparrow\uparrow\downarrow\rangle &\equiv |\{1,2\};1/2\rangle = e^{\frac{1}{2}i(\phi_1+\phi_2-\phi_3)}|2\rangle_1|2\rangle_2|1\rangle_3,\\
|\uparrow\uparrow\uparrow\rangle &\equiv |\{1,2,3\};3/2\rangle = e^{\frac{1}{2}i(\phi_1+\phi_2+\phi_3)}|2\rangle_1|1\rangle_2|2\rangle_3.
\end{aligned} \tag{6.73}
$$

Show that a valid Dicke basis is given by

$$
\begin{aligned}
|1,3/2,-3/2\rangle &= |\downarrow\downarrow\downarrow\rangle,\\
|1,3/2,-1/2\rangle &= \tfrac{1}{\sqrt{3}}\big(|\downarrow\downarrow\uparrow\rangle + |\downarrow\uparrow\downarrow\rangle + |\uparrow\downarrow\downarrow\rangle\big),\\
|1,3/2,1/2\rangle &= \tfrac{1}{\sqrt{3}}\big(|\uparrow\uparrow\downarrow\rangle + |\uparrow\downarrow\uparrow\rangle + |\downarrow\uparrow\uparrow\rangle\big),\\
|1,3/2,3/2\rangle &= |\uparrow\uparrow\uparrow\rangle,\\
|1,1/2,-1/2\rangle &= \tfrac{1}{\sqrt{2}}\big(|\downarrow\downarrow\uparrow\rangle - |\uparrow\downarrow\downarrow\rangle\big)\\
|2,1/2,-1/2\rangle &= \tfrac{1}{\sqrt{6}}\big(|\downarrow\downarrow\uparrow\rangle - 2|\downarrow\uparrow\downarrow\rangle + |\uparrow\downarrow\downarrow\rangle\big),\\
|1,1/2,1/2\rangle &= \tfrac{1}{\sqrt{2}}\big(|\uparrow\uparrow\downarrow\rangle - |\downarrow\uparrow\uparrow\rangle\big),\\
|2,1/2,1/2\rangle &= \tfrac{1}{\sqrt{6}}\big(|\uparrow\uparrow\downarrow\rangle - 2|\uparrow\downarrow\uparrow\rangle + |\downarrow\uparrow\uparrow\rangle\big).
\end{aligned} \tag{6.74}
$$

Dicke states find their main use in the treatment of problems such as superradiance, superfluorescence, and cooperative resonance fluorescence, where $\hat{J}^2$ is conserved and the dynamics in subspaces with different J (and λ) are not coupled. This is not the case for the applications we discuss later in the book. For the problems we will be interested in the Dicke states do not provide a very convenient basis. For $\hat{J}^2$-conserving situations a further set of states has been introduced. These are variously called *atomic coherent states*, *Bloch states*, and *coherent spin states* [6.14, 6.15]. In the limit $N \to \infty$ they may be connected formally to the coherent states of the harmonic oscillator. Within each subspace with fixed J and λ they are defined by rotating the

Dicke state $|\lambda, J, J\rangle$ through an angle (θ, ψ) in angular momentum space. We denote them by $|\lambda, J; \theta, \psi\rangle$. Their expansion in terms of Dicke states takes the form

$$
|\lambda, J; \theta, \psi\rangle
$$
$$
= \sum_{M=-J}^{J} \binom{2J}{J+M}^{\frac{1}{2}} \cos^{J-M}\left(\tfrac{1}{2}\theta\right) \sin^{J+M}\left(\tfrac{1}{2}\theta\right) e^{-i(J+M)\psi} |\lambda, J, M\rangle.
\tag{6.75}
$$

An alternative parameterization $|\lambda, J; z\rangle$ with $z = \tan(\tfrac{1}{2}\theta)e^{i\psi}$ is also used. A geometrical relationship between the two parameterizations is obtained by representing (θ, ψ) by a point on the unit sphere and z by a point in the complex plane drawn tangent to this sphere. Since we make no use of these states we will not investigate them further. Formal properties are discussed by Radcliffe [6.14] and Arecchi et al. [6.15]. Applications to superradiance, superfluorescence, and cooperative resonance fluorescence can be found in Refs. [6.11, 6.16-6.18]. (Arecchi et al. [6.15] also provide a nice discussion of Dicke states and their relationship to direct product states in group-theoretic language.)

6.2.3 The Characteristic Function and Associated Distribution

The normal-ordered representation for N two-level atoms is formally defined in an analogous fashion to the single-atom representation. Restricting our attention to collective operator averages, corresponding to (6.1) we define the characteristic function

$$
\chi_N(\xi, \xi^*, \eta) \equiv \mathrm{tr}\left(\rho e^{i\xi^* J_+} e^{i\eta J_z} e^{i\xi J_-}\right),
\tag{6.76}
$$

from which normal-ordered averages are calculated:

$$
\langle J_+^p J_z^r J_-^q \rangle = \mathrm{tr}\left(\rho J_+^p J_z^r J_-^q\right)
$$
$$
= \frac{\partial^{p+r+q}}{\partial(i\xi^*)^p \partial(i\eta)^r \partial(i\xi)^q} \chi_N \bigg|_{\xi=\xi^*=\eta=0}.
\tag{6.77}
$$

The distribution $P(v, v^*, m)$ is defined as in (6.3), with the inverse relationship (6.4). Then (6.5) becomes

$$
\langle J_+^p J_z^r J_-^q \rangle = \int d^2 v \int dm\, P(v, v^*, m) v^{*p} m^r v^q
$$
$$
= \overline{(v^{*p} m^r v^q)}_P.
\tag{6.78}
$$

Now, the central question! Can we expect $P(v, v^*, m)$ to be a well-behaved nonsingular function, recognizing that the single-atom distribution takes a highly singular form?

6.2.4 Nonsingular Approximation for the P Distribution

Following the approach of Sect. 6.1.4 we can evaluate the characteristic function formally in terms of matrix elements of the density operator. Either direct product states or Dicke states may be used to evaluate the trace. The important features for our discussion are contained in (6.46) and (6.61) [alternatively (6.64b)]. From these equations it follows that $\chi_N(\xi, \xi^*, \eta)$ takes the form

$$\chi_N(\xi, \xi^*, \eta) = \sum_{p=0}^{N} \sum_{q=0}^{N} \sum_{M=-N/2}^{N/2} C_{p,q,M} \frac{(i\xi^*)^p}{p!} \frac{(i\xi)^q}{q!} e^{2iM\eta}, \qquad (6.79)$$

where the $C_{p,q,M}$ are constants determined by the matrix elements of ρ. If the trace is evaluated using direct product states,

$$C_{p,q,M} = \sum_{\boldsymbol{u}} \langle \boldsymbol{u}; M | J_-^q \rho J_+^p | \boldsymbol{u}; M \rangle. \qquad (6.80)$$

The Fourier transform of (6.79) gives the distribution

$$P(v, v^*, m)$$
$$= \sum_{p=0}^{N} \sum_{q=0}^{N} \sum_{M=-N/2}^{N/2} \frac{(-1)^{p+q} C_{p,q,M}}{p!q!} \delta(m - 2M) \frac{\partial^{p+q}}{\partial v^{*p} \partial v^q} \delta^{(2)}(v), \quad (6.81)$$

where we have evaluated the Fourier transform over the complex variable ξ in the manner leading to (6.29):

$$\frac{1}{\pi^2} \int d^2\xi \, (i\xi^*)^p (i\xi)^q e^{-i\xi^* v^*} e^{-i\xi v}$$
$$= (-1)^{p+q} \frac{\partial^{p+q}}{\partial v^{*p} \partial v^q} \frac{1}{\pi} \int d^2\xi \, e^{-i\xi^* v^*} e^{-i\xi v}$$
$$= (-1)^{p+q} \frac{\partial^{p+q}}{\partial v^{*p} \partial v^q} \delta^{(2)}(v). \qquad (6.82)$$

The distribution (6.81) is still strictly singular, regardless of the state of the system (for any $C_{p,q,M}$). If N is very large, however, there is a sense in which the singular behavior in both the inversion and polarization variables can be approximated by a well-behaved function. Consider first the dependence on m. We again have a distribution confined to discrete inversion states, with M taking the values $2M = -N, -N+2, \ldots, N$; integrating (6.81) over v gives

$$P(m) = \int d^2v \, P(v, v^*, m)$$
$$= \sum_{M=-N/2}^{N/2} p_M \delta(m - 2M), \qquad (6.83)$$

where

$$p_M \equiv C_{0,0,M} = \sum_{u} \langle u; M | \rho | u; M \rangle \qquad (6.84)$$

is the probability for the system to be found in any of the $d_p(M)$ config-urations with inversion $2M$. Thus, strictly, the distribution over inversion states takes the singular form illustrated by Fig. 6.1(a). But if N is very large, and the width of the p_M distribution is large compared with the sep-aration, $\Delta M = 2$, between neighboring states, and small compared to the range $-N \leq 2M \leq N$ of allowable states, we might approximate the sin-gular distribution by a smooth envelope function [Fig. 6.1(b)], fitted with $P(2M) \approx p_M / \Delta M = p_M / 2$ to preserve the normalization $\int dm\, P(m) = 1$.

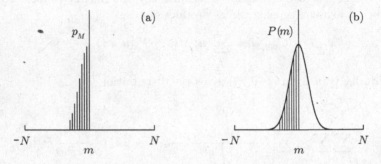

Fig. 6.1 (a) Discrete distribution over inversion states illustrating the singular form of the exact phase-space distribution. (b) Smooth approximation to the singular distribution in the limit of large N.

In practice this approximation is imposed at the level of the phase-space equation of motion. Shortly we will see that, just as for one atom, the inversion dynamics evolve as a jump process over the discrete states $m = -N, -N+2, \ldots, N$, with transitions generated by the shift operators $e^{\pm 2\frac{\partial}{\partial m}}$. Reduction of the phase-space equation of motion to Fokker–Planck form will then be made using the system size expansion [Sect. 5.1.3], which truncates the exponential derivatives at second order (truncates the Kramers–Moyal expansion). A slightly different view of this truncation can be given in the following way: The inversion dynamics obey the birth-death equation (Sect. 6.3.4)

$$\dot{p}_M = f(p_M, p_{M+1}, p_{M-1}), \qquad (6.85)$$

where f is a linear function of the probabilities for occupying each of three neighboring states. We want to replace the right-hand side by a differential operator acting on a *smooth* interpolation function $P(m)$ that fits the discrete distribution at $m = 2M$, $2M + 2$, and $2M - 2$, and also allows $P(m+2)$ and $P(m-2)$ to be calculated in terms of $P(m)$ and its derivatives (Fig. 6.2). The simplest such function is the parabola

$$2P(m) = p_M + \tfrac{1}{4}(p_{M+1} - p_{M-1})(m - 2M)$$
$$+ \tfrac{1}{8}(p_{M+1} - 2p_M + p_{M-1})(m - 2M)^2, \qquad (6.86)$$

where the factor two on the left-hand side is for normalization. The action of $e^{2\pm\frac{\partial}{\partial m}}$ on (6.86) truncates at the second-order derivative. This is the truncation used to reduce the phase-space equation of motion to a Fokker–Planck equation. The truncation is self-consistent with a polynomial interpolation between the three inversion states that appear on the right-hand side of (6.85).

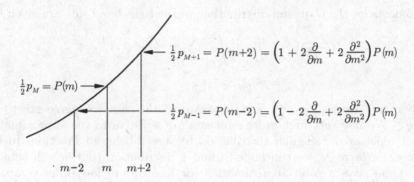

Fig. 6.2 Parabolic interpolation between discrete inversion states with p_{M+1} and p_{M-1} determined by $p_M = 2P(m)$ and the first two derivatives of $P(m)$.

The distribution (6.81) is also singular in the polarization variable v. Indeed, it appears to be highly singular when written in this exact form, which involves high-order derivatives of the δ-function. Derivatives of δ-functions can be misleading, however. The singular form is required by the truncation of the p and q summations in (6.79); $\chi_N(\xi, \xi^*, \eta)$ diverges for $\xi \to \infty$ and its Fourier transform therefore only exists as a generalized function. But as $N \to \infty$ the sum of derivatives of the δ-function can approach a well-behaved function. An example illustrating this possibility was given in Note 4.2. Using the definition of the δ-function as the limit of a sequence of Gaussians [Eq. (3.33)] we have

$$\exp\left(\frac{\partial^2}{\partial v^* \partial v}\right)\delta(v) = \sum_{k=0}^{\infty} \frac{\partial^{2k}}{\partial v^{*k} \partial v^k}\left(\lim_{n\to\infty} \frac{n}{\pi}e^{-n|v|^2}\right)$$

$$= \lim_{n\to\infty} \frac{1}{\pi}\frac{n}{1+n}e^{-n|v|^2/(1+n)}$$

$$= \frac{1}{\pi}e^{-|v|^2}. \qquad (6.87)$$

If an infinite sum of singular functions can give a nonsingular function, we might expect that a sum truncated at some large N will closely approximate a nonsingular function. Thus, we might write

$$\sum_{k=0}^{N} \frac{1}{k!} \frac{\partial^{2k}}{\partial v^{*k} \partial v^k} \delta(v) \approx \frac{1}{\pi} e^{-|v|^2}. \tag{6.88}$$

The approximate equality must hold in the sense of moment calculations. From the distribution on the left-hand side moments can be calculated using (3.37):

$$\left(\overline{v^{*p} v^q}\right)_P = \begin{cases} p! \delta_{p,q} & p, q \leq N \\ 0 & p, q > N. \end{cases} \tag{6.89}$$

Moments for the Gaussian distribution on the right-hand side are given by

$$\left(\overline{v^{*p} v^q}\right)_P = \int_0^\infty dr \int_0^{2\pi} d\phi \, r^{p+q+1} e^{i(q-p)} \frac{1}{\pi} e^{-r^2}$$

$$= p! \delta_{p,q}, \tag{6.90}$$

where we have set $v = re^{i\phi}$. For this example, the moments agree exactly for $p, q \leq N$; only very high-order moments are sensitive to the approximation that replaces the singular distribution by a well-behaved function. In this sense, for large N, we can hope to find a distribution that is well-behaved in v and gives a good approximation for moments of low order compared with N. Again, the selection of a nonsingular approximation to the exact distribution will be made at the level of the phase-space equation of motion by truncating derivatives in v and v^* to obtain a Fokker–Planck equation.

6.2.5 Two-Time Averages

Phase-space expressions for calculating two-time averages for collective atomic operators are derived following methods similar to those used in Sect. 4.3. We first generalize the notion of the phase-space distribution to set up a correspondence between an arbitrary system operator $\hat{O}$ and an associated function $F_{\hat{O}}(v, v^*, m)$, writing

$$\tilde{F}_{\hat{O}}(\xi, \xi^*, \eta) \equiv \text{tr}\big(\hat{O} e^{i\xi^* J_+} e^{i\eta J_z} e^{i\xi J_-}\big), \tag{6.91}$$

and

$$F_{\hat{O}}(v, v^*, m) \equiv \frac{1}{2\pi^3} \int d^2\xi \int d\eta \, \tilde{F}_{\hat{O}}(\xi, \xi^*, \eta) e^{-i\xi^* v^*} e^{-i\xi v} e^{-i\eta m}. \tag{6.92}$$

The inverse of the Fourier transform (6.92) gives

$$\tilde{F}_{\hat{O}}(\xi, \xi^*, \eta) = \int d^2v \int dm \, F_{\hat{O}}(v, v^*, m) e^{i\xi^* v^*} e^{i\xi v} e^{i\eta m}. \tag{6.93}$$

These expressions generalize (6.76), (6.3), and (6.4), respectively; in the new notation, we have

$$\chi_N(\xi, \xi^*, \eta) \equiv \tilde{F}_\rho(\xi, \xi^*, \eta), \tag{6.94a}$$

$$P(v, v^*, m) \equiv F_\rho(v, v^*, m). \tag{6.94b}$$

In place of the results (6.77) and (6.78) for calculating normal-ordered averages, we now have

$$\text{tr}(\hat{O} J_+^p J_z^r J_-^q) = \int d^2v \int dm \, F_{\hat{O}}(v, v^*, m) v^{*p} m^r v^q. \tag{6.95}$$

The phase-space equation of motion for the master equation (3.1) is written formally as

$$\frac{\partial}{\partial t} F_{\rho(t)}(v, v^*, m) = F_{\mathcal{L}\rho(t)}(v, v^*, m), \tag{6.96a}$$

with

$$F_{\mathcal{L}\rho(t)}(v, v^*, m) = L\left(v, v^*, m, \frac{\partial}{\partial v}, \frac{\partial}{\partial v^*}, \frac{\partial}{\partial m}\right) F_{\rho(t)}(v, v^*, m). \tag{6.96b}$$

Then we can show that (see Sect. 4.3.2)

$$F_{\exp(\mathcal{L}\tau)\hat{O}}(v, v^*, m) = e^{L\left(v, v^*, m, \frac{\partial}{\partial v}, \frac{\partial}{\partial v^*}, \frac{\partial}{\partial m}\right)\tau} F_{\hat{O}}(v, v^*, m). \tag{6.97}$$

For a single damped two-level atom $L\left(v, v^*, m, \frac{\partial}{\partial v}, \frac{\partial}{\partial v^*}, \frac{\partial}{\partial m}\right)$ is given by (6.23). Shortly, we will see explicitly how the approximations we have just discussed lead to a Fokker–Planck equation for a (many-atom) damped two-level medium. In this case $L\left(v, v^*, m, \frac{\partial}{\partial v}, \frac{\partial}{\partial v^*}, \frac{\partial}{\partial m}\right)$ will only involve up to second-order derivatives.

Note 6.2 The relationship between phase-space functions and operator power series expansions described in Sect. 4.3.1 is not directly transferable to atomic systems.

We consider averages in the form $\langle \hat{O}_1(t)\hat{O}_2(t + \tau)\rangle$, $\tau \geq 0$, where $\hat{O}_1$ and $\hat{O}_2$ may each be any one of the operators J_-, J_+, and J_z. We seek phase-space expressions for calculating these averages analogous to those of Sects. 4.3.3 and 4.3.4. From (1.97), (6.95), and (6.97), we may write

$$\langle \hat{O}_1(t)\hat{O}_2(t + \tau)\rangle$$

$$= \text{tr}\{(e^{\mathcal{L}\tau}[\rho(t)\hat{O}_1])\hat{O}_2\}$$

$$= \int d^2v \int dm \, F_{\exp(\mathcal{L}\tau)\rho(t)\hat{O}_1}(v, v^*, m)\vartheta_2$$

$$= \int d^2v \int dm \left[e^{L\left(v, v^*, m, \frac{\partial}{\partial v}, \frac{\partial}{\partial v^*}, \frac{\partial}{\partial m}\right)\tau} F_{\rho(t)\hat{O}_1}(v, v^*, m)\right]\vartheta_2, \tag{6.98}$$

where ϑ_2 is the phase-space variable replacing $\hat{O}_2$ through the correspondence $(J_+, J_z, J_-) \leftrightarrow (v^*, m, v)$. We introduce the Green function solution to the phase-space equation of motion by writing

$$F_{\rho(t)\hat{O}_1}(v, v^*, m)$$
$$= \int d^2 v_0 \int dm_0 \, F_{\rho(t)\hat{O}_1}(v_0, v_0^*, m_0) \delta^{(2)}(v - v_0)\delta(m - m_0),$$
(6.99)

so that (6.98) takes the form

$$\langle \hat{O}_1(t)\hat{O}_2(t+\tau)\rangle = \int d^2 v \int dm \int d^2 v_0 \int dm_0 \, \vartheta_2 F_{\rho(t)\hat{O}_1}(v_0, v_0^*, m_0)$$
$$\times P(v, v^*, m, t+\tau | v_0, v_0^*, m_0, t),$$
(6.100)

where $P(v, v^*, m, t + \tau | v_0, v_0^*, m_0, t)$ is the Green function solution to the phase-space equation of motion (6.96). It remains for us to find the explicit form for $F_{\rho(t)\hat{O}_1}(v_0, v_0^*, m_0)$. The objective is to express $F_{\rho(t)\hat{O}_1}(v_0, v_0^*, m_0)$ in terms of a differential operator acting on the distribution $F_{\rho(t)}(v_0, v_0^*, m_0) = P(v_0, v_0^*, m_0, t)$. Whether or not any derivatives appear will depend on the operator $\hat{O}_1$. From (6.91) and (6.92) we have

$$F_{\rho(t)\hat{O}_1}(v_0, v_0^*, m_0) = \frac{1}{2\pi^3}\int d^2\xi \int d\eta \, \mathrm{tr}\big[\rho(t)\hat{O}_1 e^{i\xi^* J_+} e^{i\eta J_z} e^{i\xi J_-}\big]$$
$$\times e^{-i\xi^* v_0^*} e^{-i\xi v_0} e^{-i\eta m_0}.$$
(6.101)

We now consider the three possible choices for $\hat{O}_1$ separately.

First, we take $\hat{O}_1 = J_+$ in (6.101). This is the simplest case. We differentiate $\mathrm{tr}[\cdots]$ with respect to $(i\xi^*)$ to bring $\hat{O}_1 = J_+$ down from the exponential. Then

$$F_{\rho(t)J_+}(v_0, v_0^*, m_0) = \frac{1}{2\pi^3}\int d^2\xi \int d\eta \left[\frac{\partial}{\partial(i\xi^*)}\tilde{F}_{\rho(t)}(\xi, \xi^*, \eta)\right]$$
$$\times e^{-i\xi^* v_0^*} e^{-i\xi v_0} e^{-i\eta m_0}$$
$$= v_0^* F_{\rho(t)}(v_0, v_0^*, m_0)$$
$$= v_0^* P(v_0, v_0^*, m_0, t),$$
(6.102)

where the second line follows from (6.92) after a single integration by parts.

Second, we take $\hat{O}_1 = J_z$ in (6.101). We write

$$F_{\rho(t)J_z}(v_0, V_0^*, m_0)$$
$$= \frac{1}{2\pi^3}\int d^2\xi \int d\eta \, \mathrm{tr}\big[\rho(t)e^{i\xi^* J_+}\big(e^{-i\xi^* J_+}J_z e^{i\xi^* J_+}\big)e^{i\eta J_z} e^{i\xi J_-}\big]$$
$$\times e^{-i\xi^* v_0^*} e^{-i\xi v_0} e^{-i\eta m_0}.$$
(6.103)

Now J_-, J_+, and J_z obey the same commutation relations as the single-atom operators σ_-, σ_+, and σ_z. Therefore the identities (6.7)–(6.9) also hold for collective atomic operators. In particular, corresponding to (6.7), we have

$$e^{-i\xi^* J_+}J_z e^{i\xi^* J_+} = J_z + 2i\xi^* J_+$$
(6.104)

[we have actually taken the Hermitian conjugate of (6.7) with $\xi \to -\xi$]. Using (6.104), (6.103) gives

$$
\begin{aligned}
F_{\rho(t)J_z}&(v_0, v_0^*, m_0) \\
&= \frac{1}{2\pi^3} \int d^2\xi \int d\eta \, \mathrm{tr}\big[\rho(t)e^{i\xi^* J_+}(J_z + 2i\xi^* J_+)e^{i\eta J_z}e^{i\xi J_-}\big] \\
&\quad \times e^{-i\xi^* v_0^*} e^{-i\xi v_0} e^{-i\eta m_0} \\
&= \frac{1}{2\pi^3} \int d^2\xi \int d\eta \left[\left(\frac{\partial}{\partial(i\eta)} + 2i\xi^* \frac{\partial}{\partial(i\xi^*)}\right)\tilde{F}_{\rho(t)}(\xi, \xi^*, \eta)\right] \\
&\quad \times e^{-i\xi^* v_0^*} e^{-i\xi v_0} e^{-i\eta m_0} \\
&= \left(m_0 - 2\frac{\partial}{\partial v_0^*}v_0^*\right) F_{\rho(t)}(v_0, v_0^*, m_0) \\
&= \left(m_0 - 2\frac{\partial}{\partial v_0^*}v_0^*\right) P(v_0, v_0^*, m_0, t).
\end{aligned}
\tag{6.105}
$$

Here we have first integrated by parts, and then written $i\xi^* e^{-i\xi^* v_0^*} = -\partial e^{-i\xi^* v_0^*}/\partial v_0^*$.

Finally, we take $\hat{O}_1 = J_-$ in (6.101). This last case is more complicated algebraically, but follows the same principles as above. The operator J_- must be passed through the exponentials inside the trace in (6.101) so that it can be brought down from the exponent by a partial derivative. We first write

$$
\begin{aligned}
F_{\rho(t)J_-}&(v_0, v_0^*, m_0) \\
&= \frac{1}{2\pi^3} \int d^2\xi \int d\eta \, \mathrm{tr}\big[\rho(t)e^{i\xi^* J_+}\big(e^{-i\xi^* J_+}J_- e^{i\xi^* J_+}\big)e^{i\eta J_z}e^{i\xi J_-}\big] \\
&\quad \times e^{-i\xi^* v_0^*} e^{-i\xi v_0} e^{-i\eta m_0}.
\end{aligned}
\tag{6.106}
$$

Corresponding to the identity (6.9) for single-atom operators, we have

$$
e^{-i\xi^* J_+} J_- e^{i\xi^* J_+} = J_- - i\xi^* J_z - (i\xi^*)^2 J_+,
\tag{6.107}
$$

[taking the Hermitian conjugate of (6.9) with $\xi \to -\xi$]. Also, corresponding to the identity (6.8) (with $\eta \to -\eta$),

$$
e^{-i\eta J_z} J_- e^{i\eta J_z} = e^{2i\eta} J_-.
\tag{6.108}
$$

Now, (6.106) becomes

$$F_{\rho(t)J_-}(v_0, v_0^*, m_0)$$

$$= \frac{1}{2\pi^3} \int d^2\xi \int d\eta \, \mathrm{tr}\big[\rho(t) e^{i\xi^* J_+} \big(J_- - i\xi^* J_z - (i\xi^*)^2 J_+\big)$$

$$\times e^{i\eta J_z} e^{i\xi J_-}\big] e^{-i\xi^* v_0^*} e^{-i\xi v_0} e^{-i\eta m_0}$$

$$= \frac{1}{2\pi^3} \int d^2\xi \int d\eta \, \mathrm{tr}\big\{\rho(t) e^{i\xi^* J_+} \big[e^{i\eta J_z} \big(e^{2i\eta} J_-\big)$$

$$- \big(i\xi^* J_z + (i\xi^*)^2 J_+\big) e^{i\eta J_z}\big] e^{i\xi J_-}\big\} e^{-i\xi^* v_0^*} e^{-i\xi v_0} e^{-i\eta m_0}$$

$$= \frac{1}{2\pi^3} \int d^2\xi \int d\eta \bigg[\bigg(e^{2i\eta} \frac{\partial}{\partial(i\xi)} - i\xi^* \frac{\partial}{\partial(i\eta)} - (i\xi^*)^2 \frac{\partial}{\partial(i\xi^*)}\bigg)$$

$$\times \tilde{F}_{\rho(t)}(\xi, \xi^*, \eta)\bigg] e^{-i\xi^* v_0^*} e^{-i\xi v_0} e^{-i\eta m_0}.$$

Integrating by parts and replacing $e^{2i\eta}$, $i\xi^*$, and $(i\xi^*)^2$ by partial derivatives acting on the product of exponentials, we have

$$F_{\rho(t)J_-}(v_0, v_0^*, m_0) = \bigg(e^{-2\frac{\partial}{\partial m_0}} v_0 + \frac{\partial}{\partial v_0^*} m_0 - \frac{\partial^2}{\partial v_0^{*2}} v_0^*\bigg) F_{\rho(t)}(v_0, v_0^*, m_0)$$

$$= \bigg(e^{-2\frac{\partial}{\partial m_0}} v_0 + \frac{\partial}{\partial v_0^*} m_0 - \frac{\partial^2}{\partial v_0^{*2}} v_0^*\bigg) P(v_0, v_0^*, m_0, t).$$

$$(6.109)$$

Equations (6.100) and (6.102) give three two-time averages that can be calculated by integrating against a two-time, or joint, distribution as in classical statistics ($\tau \geq 0$):

$$\langle J_+(t) J_+(t+\tau)\rangle = \overline{\big(v^*(t) v^*(t+\tau)\big)}_P, \qquad (6.110a)$$

$$\langle J_+(t) J_z(t+\tau)\rangle = \overline{\big(v^*(t) m(t+\tau)\big)}_P, \qquad (6.110b)$$

$$\langle J_+(t) J_-(t+\tau)\rangle = \overline{\big(v^*(t) v(t+\tau)\big)}_P, \qquad (6.110c)$$

where we define (ϑ_1 and ϑ_2 are either v^*, m, or v)

$$\overline{\big(\vartheta_1(t)\vartheta_2(t+\tau)\big)}_P$$

$$\equiv \int d^2v \int dm \int d^2v_0 \int dm_0 \, \vartheta_{10}\vartheta_2 P(v, v^*, m, t+\tau; v_0, v_0^*, m_0, t), \quad (6.111)$$

and

$$P(v, v^*, m, t+\tau; v_0, v_0^*, m_0, t)$$

$$= P(v, v^*, m, t+\tau | v_0, v_0^*, m_0, t) P(v_0, v_0^*, m_0, t)$$

$$(6.112)$$

is the two-time distribution. The complex conjugates of (6.110a)–(6.110c) give three more averages calculated in a similar manner ($\tau \geq 0$):

$$\langle J_-(t+\tau)J_-(t)\rangle = \overline{(v(t+\tau)v(t))}_P, \tag{6.113a}$$

$$\langle J_z(t+\tau)J_-(t)\rangle = \overline{(m(t+\tau)v(t))}_P, \tag{6.113b}$$

$$\langle J_+(t+\tau)J_-(t)\rangle = \overline{(v^*(t+\tau)v(t))}_P. \tag{6.113c}$$

There are six more averages obtained from (6.100) using (6.105) and (6.109). These involve derivatives of the distribution $P(v_0, v_0^*, m_0, t)$. Using (6.105) we find ($\tau \geq 0$)

$$\langle J_z(t)\hat{O}_2(t+\tau)\rangle$$

$$= \int d^2v \int dm \int d^2v_0 \int dm_0\, \vartheta_2 \left[\left(m_0 - 2\frac{\partial}{\partial v_0^*}v_0^*\right)P(v_0, v_0^*, m_0, t)\right]$$

$$\times P(v, v^*, m, t+\tau|v_0, v_0^*, m_0, t), \tag{6.114}$$

and, using (6.109) ($\tau \geq 0$),

$$\langle J_-(t)\hat{O}_2(t+\tau)\rangle$$

$$= \int d^2v \int dm \int d^2v_0 \int dm_0\, \vartheta_2 \left[\left(e^{-2\frac{\partial}{\partial m_0}}v_0 + \frac{\partial}{\partial v_0^*}m_0 - \frac{\partial^2}{\partial v_0^{*2}}v_0^*\right)\right.$$

$$\times P(v_0, v_0^*, m_0, t)\bigg] P(v, v^*, m, t+\tau|v_0, v_0^*, m_0, t), \tag{6.115}$$

where $\hat{O}_2$ and ϑ_2 are, respectively, J_+, J_z, or J_-, and v^*, m, or v. The complex conjugates of these equations give a further six averages ($\tau \geq 0$):

$$\langle \hat{O}_2(t+\tau)J_z(t)\rangle$$

$$= \int d^2v \int dm \int d^2v_0 \int dm_0\, \vartheta_2 \left[\left(m_0 - 2\frac{\partial}{\partial v_0}v_0\right)P(v_0, v_0^*, m_0, t)\right]$$

$$\times P(v, v^*, m, t+\tau|v_0, v_0^*, m_0, t), \tag{6.116}$$

and

$$\langle \hat{O}_2(t+\tau)J_+(t)\rangle$$

$$= \int d^2v \int dm \int d^2v_0 \int dm_0\, \vartheta_2 \left[\left(e^{-2\frac{\partial}{\partial m_0}}v_0^* + \frac{\partial}{\partial v_0}m_0 - \frac{\partial^2}{\partial v_0^2}v_0\right)\right.$$

$$\times P(v_0, v_0^*, m_0, t)\bigg] P(v, v^*, m, t+\tau|v_0, v_0^*, m_0, t), \tag{6.117}$$

It is important to stress again here, as we did in Sect. 4.3 [below (4.128)], that it is not possible to calculate every two-time average in terms of a "classical" integral. For example, the above results show that J_z must be evaluated at the later time $t+\tau$ if an average involving J_z is to be calculated by direct integration against the joint distribution. It is also necessary that J_+ and J_- appear in normal order, to the left and the right of J_z, respectively. Clearly, more general results than those given above can be derived using the same methods. In particular, a general expression for normal-ordered time-ordered averages similar to (4.100a) can be obtained:

Exercise 6.6 Beginning from (1.102), show that ($\tau \geq 0$)

$$\langle J_+^p(t) \hat{N}(t+\tau) J_-^q(t) \rangle = \overline{\left(\overline{(v^{*p}v^q)}(t) N(t+\tau) \right)}_P, \qquad (6.118a)$$

with

$$\left(\overline{(v^{*p}v^q)}(t) N(t+\tau) \right)_P \equiv \int d^2v \int dm \int d^2v_0 \int dm_0 \, v_0^{*p} v_0^q N(v, v^*, m)$$
$$\times P(v, v^*, m, t+\tau; v_0, v_0^*, m_0, t), \qquad (6.118b)$$

where $\hat{N}$ is the normal-ordered power series

$$\hat{N} \equiv \sum_{p,q,r} C_{p,q,r} J_+^p J_z^r J_-^q, \qquad (6.119)$$

and

$$N(v, v^*, m) = \sum_{p,q,r} C_{p,q,r} v^{*p} m^r v^q. \qquad (6.120)$$

6.2.6 Other Representations

Equations (6.76) and (6.3) defining the normal-ordered representation for N two-level atoms can be generalized in the manner described in Sect. 4.1; by starting from different characteristic functions, representations giving antinormal-ordered and symmetric-ordered averages can be defined. In practice, only the *Wigner representation* (symmetric-ordered averages) has been used for applications in quantum optics.

The Wigner representation is rather awkward to use for two-level atoms, in comparison with what we learned about it as a representation for the electromagnetic field. This is because disentangling the operators that appear in the exponent in the characteristic function is more complicated when these operators are J_-, J_+, and J_z, rather than a and $a^\dagger$; the commutator of a and $a^\dagger$ is a constant, whereas the commutators of angular momentum operators are other angular momentum operators. Nevertheless, the Wigner representation has been used successfully; in particular, for the laser and optical bistability [6.19, 6.20]. Actually, if the Fokker–Planck form for the phase-space equation of motion is imposed *a priori*, it is quite straightforward to construct the appropriate drift and diffusion terms in the Wigner representation [6.20]. Of course, the resulting Fokker–Planck equation is an approximate equation of motion. The algebraic difficulty is met when we attempt to derive an exact phase-space equation of motion along the lines of

Sect. 6.1.3 (Sect. 6.3.4 for the N-atom system). Since we will not make explicit use of the Wigner representation for atoms we will not spend any time on the details.

One further representation deserves mention. In Sect. 6.2.2 we noted that a convenient set of basis states for $\hat{J}^2$-conserving systems is provided by the atomic coherent states. A diagonal representation for states within the subspace spanned by Dicke states of fixed J and λ can be defined, in close analogy to the Glauber–Sudarshan representation for the electromagnetic field [Eq. (3.15)]. In this *atomic coherent state representation*,

$$\rho = \int d^2z\, P(z,z^*)|\lambda,J;z\rangle\langle\lambda,J;z|, \qquad (6.121)$$

where normal-ordered operator averages are given by

$$\langle J_+^p J_z^r J_-^q \rangle$$
$$= \int d^2z\, P(z,z^*)(1+|z|^2)^{-N} \frac{\partial^{p+q}}{\partial z^p \partial z^{*q}}\left(N/2 - z^*\frac{\partial}{\partial z^*}\right)^r (1+|z|^2)^N.$$
$$(6.122)$$

Alternatively, this representation can be expressed in terms of the parameterization $|\lambda,J;\theta,\psi\rangle$ for atomic coherent states, where $P(\theta,\psi)$ is then a distribution over the unit sphere. In (6.122) we do not have a simple classical relationship between operator averages and corresponding moments of the distribution $P(z,z^*)$; however, the atomic coherent state representation does led to equations of motion in the Fokker–Planck form for a number of interesting problems. More details about this representation and its use can be found in Refs. [6.14-6.18].

6.3 Fokker–Planck Equation for a Radiatively Damped Two-Level Medium

6.3.1 Master Equation for Independently Damped Two-Level Atoms

We will illustrate the use of the normal-ordered representation for a collection of two-level atoms by considering spontaneous emission in a two-level medium.

In this system identical two-level atoms each couple to the modes of the electromagnetic field in the manner described in Sect. 2.2.1. Within our general formulation for a system S coupled to a reservoir R (Sect. 1.3) we have

$$H_S = \sum_{j=1}^{N} H_S^j, \tag{6.123a}$$

$$H_{SR} = \sum_{j=1}^{N} H_{SR}^j, \tag{6.123b}$$

where H_S^j and H_{SR}^j are defined as in (2.15a) and (2.15c), with σ_-, σ_+, and σ_z replaced by σ_{j-}, σ_{j+}, and σ_{jz}, and the dipole coupling constants $\kappa_{k,\lambda}$ replaced by

$$\kappa_{k,\lambda}^j \equiv -ie^{ik\cdot r_j}\sqrt{\frac{\omega_k}{2\hbar\epsilon_0 V}}\,\hat{e}_{k,\lambda}\cdot d_{21}. \tag{6.124}$$

The reservoir Hamiltonian H_R is given by (2.15b); thus, each atom couples to the same electromagnetic field (the same reservoir), but at different spatial positions; the different spatial positions produce the different phases that appear in the coupling constants (6.124). In the notation of (1.32) and (1.33), we have

$$s_{1j} = \sigma_{j-}, \qquad s_{2j} = \sigma_{j+}, \tag{6.125a}$$

$$\Gamma_{1j} = \Gamma_j^\dagger \equiv \sum_{k,\lambda} \kappa_{k,\lambda}^{j*} r_{k,\lambda}^\dagger, \qquad \Gamma_{2j} = \Gamma_j \equiv \sum_{k,\lambda} \kappa_{k,\lambda}^j r_{k,\lambda}. \tag{6.125b}$$

In the interaction picture

$$\tilde{s}_{1j}(t) = \sigma_{j-}e^{-i\omega_A t}, \tag{6.126a}$$

$$\tilde{s}_{2j}(t) = \sigma_{j+}e^{i\omega_A t}, \tag{6.126b}$$

and

$$\tilde{\Gamma}_{1j} = \tilde{\Gamma}_j^\dagger = \sum_{k,\lambda} \kappa_{k,\lambda}^{j*} r_{k,\lambda}^\dagger e^{i\omega_k t}, \tag{6.127a}$$

$$\tilde{\Gamma}_{2j} = \tilde{\Gamma}_j = \sum_{k,\lambda} \kappa_{k,\lambda}^j r_{k,\lambda} e^{-i\omega_k t}. \tag{6.127b}$$

The master equation in the Born approximation [Eq. (1.34)] for the reduced density operator of the N-atom system reads [compare (1.42)]

$$\dot{\tilde{\rho}} = -\sum_{j,l=1}^{N}\int_0^t dt'\Big\{\big[\sigma_{j-}\sigma_{l-}\tilde{\rho}(t') - \sigma_{l-}\tilde{\rho}(t')\sigma_{j-}\big]e^{-i\omega_A(t+t')}\langle\tilde{\Gamma}_j^\dagger(t)\tilde{\Gamma}_l^\dagger(t')\rangle_R$$

$$+ \big[\sigma_{j+}\sigma_{l+}\tilde{\rho}(t') - \sigma_{l+}\tilde{\rho}(t')\sigma_{j+}\big]e^{i\omega_A(t+t')}\langle\tilde{\Gamma}_j(t)\tilde{\Gamma}_l(t')\rangle_R$$

$$+ \big[\sigma_{j-}\sigma_{l+}\tilde{\rho}(t') - \sigma_{l+}\tilde{\rho}(t')\sigma_{j-}\big]e^{-i\omega_A(t-t')}\langle\tilde{\Gamma}_j^\dagger(t)\tilde{\Gamma}_l(t')\rangle_R$$

$$+ \big[\sigma_{j+}\sigma_{l-}\tilde{\rho}(t') - \sigma_{l-}\tilde{\rho}(t')\sigma_{j+}\big]e^{i\omega_A(t-t')}\langle\tilde{\Gamma}_j(t)\tilde{\Gamma}_l^\dagger(t')\rangle_R\Big\} + \text{h.c.}. \tag{6.128}$$

Equation (6.128) will be greatly simplified if we can set all reservoir correlation functions to zero for $j \neq l$, and retain only the nonvanishing correlations given by (1.45) and (1.46) for $j = l$. Effectively, this amounts to coupling the individual atoms to N *statistically independent* reservoirs. Can this be justified in view of the fact that all atoms interact with the same electromagnetic field? Yes it can. Since the atoms are located at different positions r_j, the question is one of spatial correlations in the electromagnetic field. The factor determining these correlations is the spatially-dependent phase appearing in the coupling constant (6.124). The atoms interact most strongly with a narrow band of frequencies centered about the resonant frequency ω_A. The scale for measuring the spatial variation of the phases in the coupling constants $\kappa^j_{k,\lambda}$ is then set by the wavelength $\lambda_A = 2\pi c/\omega_A$. If the atoms are separated by large distances compared to this wavelength, correlations between the reservoir operators $\tilde{\Gamma}_j$, $\tilde{\Gamma}^\dagger_j$ and $\tilde{\Gamma}_l$, $\tilde{\Gamma}^\dagger_l$, for $j \neq l$, can be shown to vanish; then atoms see statistically independent reservoirs. We can then pass immediately (see Sect. 2.2.1) to the *master equation for N independent radiatively damped two-level atoms*:

$$\dot{\rho} = -i\tfrac{1}{2}\omega_A \sum_{j=1}^{N}[\sigma_{jz}, \rho] + \frac{\gamma}{2}(\bar{n}+1)\sum_{j=1}^{N}(2\sigma_{j-}\rho\sigma_{j+} - \sigma_{j+}\sigma_{j-}\rho - \rho\sigma_{j+}\sigma_{j-})$$
$$+ \frac{\gamma}{2}\bar{n}\sum_{j=1}^{N}(2\sigma_{j+}\rho\sigma_{j-} - \sigma_{j-}\sigma_{j+}\rho - \rho\sigma_{j-}\sigma_{j+}). \tag{6.129}$$

Equation (6.129) is a rather obvious generalization of the master equation (2.26) for a single radiatively damped atom. It has the solution

$$\rho(t) = \prod_{j=1}^{N}\rho_j(t), \tag{6.130}$$

where $\rho_j(t)$ is the density operator for the jth independent atom. Since the atoms are identical, each $\rho_j(t)$ obeys the matrix element equations (2.36). We therefore already know the exact solution to this problem. Nevertheless, (6.129) will form an important constituent in the master equations for the laser and optical bistability. In these systems additional interactions generate correlations between the atoms, and the master equation cannot be solved by such a simple matrix element approach; we will need to use phase-space methods. The solvable problem defined by (6.129) provides us with a good illustration of how these methods work for atomic variables.

Note 6.3 If the assumption of statistical independence between the reservoirs seen by the different atoms is not justified, the problem of spontaneous emission in a collective atomic sample becomes much more complicated. Most generally, the detailed spatial distribution of the atoms is important; we must deal with the complicated spatial interference of radiation from the atoms, plus the communication between atoms by way of this radiation. The modification to (6.129) required under these conditions can be found in the work of Lehmberg [6.21] and Agarwal [6.22]. There is one other set of conditions, however, that lead to an essentially simple description. If all atoms reside in a volume that is small compared to λ_A^3, the phases in (6.124) differ very little for field modes with frequencies $\sim \omega_A$ and we may drop the subscripts j and l from the reservoir operators in (6.128). The summations then replace single-atom operators σ_{j-}, σ_{j+}, and σ_{jz} by collective operators J_-, J_+, and J_z. The resulting master equation reads

$$\dot{\rho} = -i\tfrac{1}{2}\omega_A[J_z, \rho] + \frac{\gamma}{2}(\bar{n}+1)(2J_-\rho J_+ - J_+J_-\rho - \rho J_+J_-)$$
$$+ \frac{\gamma}{2}\bar{n}(2J_+\rho J_- - J_-J_+\rho - \rho J_-J_+). \tag{6.131}$$

This equation describes Dicke superradiance and superfluorescence [6.13, 6.16–6.18]. Actually, we are being a little glib here, since, as the atoms are brought closer together, an interaction energy between the atomic dipoles becomes important which adds level shift terms that depend on the spatial arrangement of the atoms to (6.131) [6.21, 6.22].

6.3.2 Closed Dynamics for Normally-Ordered Averages of Collective Operators

We propose to use a phase-space representation for ρ defined in terms of collective atomic operators (Sect. 6.2.3). It is clear that a master equation like (6.131) can be converted to a phase-space equation of motion using such a representation. However, the right-hand side of (6.129) cannot be expressed solely in terms of collective atomic operators. Using (6.6f) and (6.6g), we are able to write (6.129) in the form

$$\dot{\rho} = -i\tfrac{1}{2}\omega_A[J_z, \rho] + \frac{\gamma}{2}\left(2\sum_{j=1}^{N}\sigma_{j-}\rho\sigma_{j+} - \tfrac{1}{2}J_z\rho - \tfrac{1}{2}\rho J_z - N\rho\right)$$
$$+ \gamma\bar{n}\left(\sum_{j=1}^{N}\sigma_{j-}\rho\sigma_{j+} + \sum_{j=1}^{N}\sigma_{j+}\rho\sigma_{j-} - N\rho\right). \tag{6.132}$$

The terms involving $\sigma_{j-}\rho\sigma_{j+}$ and $\sigma_{j+}\rho\sigma_{j-}$ cannot be rewritten in terms of collective atomic operators. But the phase-space representation we propose to use only generates collective operator averages. It might seem that there is an inconsistency here. It is useful therefore, before deriving the phase-space equation of motion corresponding to (6.132), to see explicitly that, in spite of the presence of single-atom operators in (6.132), a closed set of equations involving only the normal-ordered collective operator averages does exist.

Exercise 6.7 Show that (6.131) conserves the magnitude of the total pseudo-spin [Eq. (6.63)] while (6.132) does not. It follows that matrix element equations derived from (6.131) using a Dicke state basis are only coupled within each subspace defined by a fixed J and λ; there is no coupling between subspaces. Matrix element equations for (6.132) couple subspaces with different J and λ.

We will derive a coupled set of equations for the averages of all normal-ordered operator products. Consider the average $\langle J_+^p J_z^r J_-^q \rangle$. From the master equation (6.132) we obtain

$$
\frac{d}{dt}\langle J_+^p J_z^r J_-^q \rangle = -i\frac{1}{2}\omega_A \left(\langle J_+^p J_z^r J_-^q J_z \rangle - \langle J_z J_+^p J_z^r J_-^q \rangle \right)
$$

$$
+ \frac{\gamma}{2}\left(2\sum_{j=1}^N \langle \sigma_{j+} J_+^p J_z^r J_-^q \sigma_{j-} \rangle - \frac{1}{2}\langle J_+^p J_z^r J_-^q J_z \rangle \right.
$$

$$
\left. - \frac{1}{2}\langle J_z J_+^p J_z^r J_-^q \rangle - N\langle J_+^p J_z^r J_-^q \rangle \right)
$$

$$
+ \gamma\bar{n}\left(\sum_{j=1}^N \langle \sigma_{j+} J_+^p J_z^r J_-^q \sigma_{j-} \rangle + \sum_{j=1}^N \langle \sigma_{j-} J_+^p J_z^r J_-^q \sigma_{j+} \rangle \right.
$$

$$
\left. - N\langle J_+^p J_z^r J_-^q \rangle \right). \tag{6.133}
$$

Our task is to write each term on the right-hand side in terms of normal-ordered averages of collective operators. We first use the commutation relations (6.45) to write

$$
J_-^q J_z^r = J_-^{q-1}(J_z + 2)J_- J_z^{r-1}
$$

$$
= J_-^{q-2}(J_z + 4)J_-^2 J_z^{r-1}
$$

$$
= (J_z + 2q)J_-^q J_z^{r-1}
$$

$$
= (J_z + 2q)^r J_-^q, \tag{6.134a}
$$

and, from the Hermitian conjugate,

$$
J_z^r J_+^p = J_+^p (J_z + 2p)^r. \tag{6.134b}
$$

Using these identities, we obtain

$$
\langle J_+^p J_z^r J_-^q J_z \rangle = \langle J_+^p J_z^r (J_z + 2q)J_-^q \rangle, \tag{6.135a}
$$

$$
\langle J_z J_+^p J_z^r J_-^q \rangle = \langle J_+^p (J_z + 2p)J_z^r J_-^q \rangle, \tag{6.135b}
$$

where the right-hand sides are now in normal order. The terms in (6.133) involving sums over single-atom operators involve a little more effort.

The algebraic manipulations required here are very similar to those used to convert the master equation for a single damped two-level atom into phase-space form (Sect. 6.1.3). The sums remaining in (6.133) are on terms that are quadratic in single-atom operators. The two single-atom operators in each product are, however, separated by collective operators; consequently, we are unable to use (6.6a)–(6.6g) directly to reduce the quadratic dependence to a linear dependence which is summable. We must first use commutation relations to pass the single-atom operators through the collective operators, then use (6.6a)–(6.6g). After this the sums will replace linear combinations of single-atom operators by the corresponding linear combinations of collective operators, and a final reordering into normal order gives the desired result. We will perform the simpler of the two remaining calculations in detail and leave the second as an exercise.

Consider the average $\langle \sigma_{j+} J_+^p J_z^r J_-^q \sigma_{j-} \rangle = \langle J_+^p \sigma_{j+} J_z^r \sigma_{j-} J_-^q \rangle$. We first pass σ_{j-} to the left through J_z^r. Using single-atom commutation relations, we have

$$J_z^r \sigma_{j-} = J_z^{r-1} \sigma_{j-} (J_z - 2)$$
$$= \sigma_{j-} (J_z - 2)^r. \tag{6.136}$$

Then

$$\sum_{j=1}^N \langle \sigma_{j+} J_+^p J_z^r J_-^q \sigma_{j-} \rangle = \sum_{j=1}^N \langle J_+^p \sigma_{j+} \sigma_{j-} (J_z - 2)^r J_-^q \rangle$$

$$= \sum_{j=1}^N \langle J_+^p \tfrac{1}{2} (1 + \sigma_{jz}) (J_z - 2)^r J_-^q \rangle$$

$$= \tfrac{1}{2} \langle J_+^p (N + J_z)(J_z - 2)^r J_-^q \rangle, \tag{6.137}$$

where the second line follows from (6.6f). The last term in (6.133) – $\langle \sigma_{j-} J_+^p J_z^r J_-^q \sigma_{j+} \rangle$ – is evaluated in a similar way, but requires rather more algebra:

Exercise 6.8 Show that

$$\sum_{j=1}^N \langle \sigma_{j-} J_+^p J_z^r J_-^q \sigma_{j+} \rangle$$
$$= \langle J_+^p \left[\tfrac{1}{2}(N - J_z) - p - q \right] (J_z + 2)^r J_-^q \rangle + pq(3 - p - q)\langle J_+^{p-1} J_z^r J_-^{q-1} \rangle$$
$$+ pq(p-1)(q-1)\langle J_+^{p-2}(N + J_z)(J_z - 2)^r J_-^{q-2} \rangle. \tag{6.138}$$

We now use (6.135), (6.137), and (6.138) to write the moment equations (6.133) in the form

$$\frac{d}{dt}\langle J_+^p J_z^r J_-^q \rangle = -\left[i\omega_A(q-p) + N\frac{\gamma}{2}(2\bar{n}+1) \right]\langle J_+^p J_z^r J_-^q \rangle$$

$$+ \frac{\gamma}{2}\left[\langle J_+^p (N+J_z)(J_z-2)^r J_-^q \rangle \right.$$

$$\left. - \langle J_+^p (J_z+p+q) J_z^r J_-^q \rangle \right]$$

$$+ \gamma\bar{n}\left[\langle J_+^p \tfrac{1}{2}(N+J_z)(J_z-2)^r J_-^q \rangle \right.$$

$$+ \langle J_+^p \left(\tfrac{1}{2}(N-J_z) - p - q\right)(J_z+2)^r J_-^q \rangle$$

$$+ pq(3-p-q)\langle J_+^{p-1} J_z^r J_-^{q-1} \rangle$$

$$\left. + pq(p-1)(q-1)\langle J_+^{p-2} \tfrac{1}{2}(N+J_z)(J_z-2)^r J_-^{q-2} \rangle \right].$$

$$(6.139)$$

Equation (6.139) defines a coupled hierarchy of linear equations for normal-ordered collective operator averages. We have been able to obtain a closed set of equations because the atoms are all identical. A variation in resonant frequencies, replacing ω_A by ω_{Aj} inside the sum in (6.129), would not change this situation; we can transform to the interaction picture and define $J_\pm$ as in (6.44) with $\phi_j = \omega_{Aj}t$. However, if, for example, each atom had a different decay rate γ_j, we would not obtain a closed set of equations for the collective operator averages. As stated in Note 6.1, a common situation in which members of an atomic population are not identical arises when the atoms interact with a spatially varying field mode – a Gaussian beam or standing wave for example. Actually, it is not necessary that the distinction between atoms enter the equations of motion explicitly. The atoms may be distinguished by selecting a non-permutationally-symmetric initial condition. For such initial conditions collective operators alone will not be adequate to completely describe the subsequent evolution.

We observed from the permutational symmetry of the collective operators that there are actually only $\frac{1}{6}(N+1)(N+2)(N+3)$ independent normal-ordered collective operator products [Eq. (6.56)]. Thus, if this symmetry is used, (6.139) defines a closed set of $\frac{1}{6}(N+1)(N+2)(N+3)$ equations. Sarkar and Satchell [6.12] have used the permutational symmetry to numerically solve matrix element equations for absorptive optical bistability in the bad-cavity limit. They were able to reduce the 2^{2N} matrix element equations obtained from a naive use of the direct product state basis to a set of $\frac{1}{6}(N+1)(N+2)(N+3)$ equations for independent matrix elements.

Note 6.4 There is a trap for the unwary in the consideration of permutational symmetry and identical atoms. It is easy to be confused by what we have learned from quantum mechanics courses about indistinguishable particles [6.23]. From this background we might expect that we only have to deal with symmetric superpositions. (There are no antisymmetric superpositions – except when $N = 2$ – since in our system the eigenvalue that distin-

ρ	$|\downarrow\downarrow\downarrow\rangle$	$|\downarrow\downarrow\uparrow\rangle$	$|\downarrow\uparrow\downarrow\rangle$	$|\uparrow\downarrow\downarrow\rangle$	$|\uparrow\uparrow\downarrow\rangle$	$|\uparrow\downarrow\uparrow\rangle$	$|\downarrow\uparrow\uparrow\rangle$	$|\uparrow\uparrow\uparrow\rangle$
$|\downarrow\downarrow\downarrow\rangle$	$|a|^2$	ab^*			ac^*			ad^*
$|\downarrow\downarrow\uparrow\rangle$	a^*b	$|b|^2$			bc^*			bd^*
$|\downarrow\uparrow\downarrow\rangle$								
$|\uparrow\downarrow\downarrow\rangle$								
$|\uparrow\uparrow\downarrow\rangle$	a^*c	b^*c			$|c|^2$			cd^*
$|\uparrow\downarrow\uparrow\rangle$								
$|\downarrow\uparrow\uparrow\rangle$								
$|\uparrow\uparrow\uparrow\rangle$	a^*d	b^*d			c^*d			$|d|^2$

Fig. 6.3 Permutational symmetry of the matrix elements of the density operator for a three-atom system. For the pure state (6.140) the matrix elements within each block are equal to the value shown in the upper left-hand corner. For a mixed state the matrix elements within each of the four central squares are equal along the diagonal, and off the diagonal, but the matrix elements along the diagonal are not equal to those off the diagonal.

guishes between single-particle states only takes two values.) For N two-level atoms the symmetric superpositions are the $N+1$ Dicke states $|1, N/2, M\rangle$, $M = -N/2, -N/2+1, \ldots, N/2$. If these where the only states considered the density operator would have $(N+1)^2$ independent matrix elements rather than $\frac{1}{6}(N+1)(N+2)(N+3)$ as claimed. The resolution of this apparent inconsistency lies in the fact that a *dissipative* quantum system evolves into a *mixed* state, rather than a pure state. The permutational symmetry requirements on a pure state $|\psi\rangle$ impose a larger number of relationships between matrix elements of the density operator than are demanded for permutational symmetry of the density operator matrix elements themselves; extra relationships are needed for the density operator to factorize in the form $|\psi\rangle\langle\psi|$. A system of three atoms illustrates this point. The symmetric superpositions appear as the first four states listed in (6.74). The most general pure state constructed from these is

$$|\psi\rangle = a|\downarrow\downarrow\downarrow\rangle + b\frac{1}{\sqrt{3}}\left(|\downarrow\downarrow\uparrow\rangle + |\downarrow\uparrow\downarrow\rangle + |\uparrow\downarrow\downarrow\rangle\right)$$
$$+ c\frac{1}{\sqrt{3}}\left(|\uparrow\uparrow\downarrow\rangle + |\uparrow\downarrow\uparrow\rangle + |\downarrow\uparrow\uparrow\rangle\right) + d|\uparrow\uparrow\uparrow\rangle. \qquad (6.140)$$

Figure 6.3 displays the relationships between matrix elements of the density operator. Matrix elements within each block are equal and given by the value shown in the upper left-hand corner of that block. The number of independent relationships represented by each block is one less than the number of elements in the block. Thus, there are 48 relationships and $2^{2N} - 48 = 64 - 48 = 16 = (N + 1)^2$ independent density matrix elements. However, if the permutational symmetry is only applied to density matrix elements (not to the wavefunction $|\psi\rangle$), the nine elements in each of the four squares at the center of Fig. 6.3 need not all be equal; instead, the three elements along each diagonal are equal and the six off the diagonal are equal; the three need not equal the six. Take the square labeled by $|b|^2$ for example. Down the diagonal we require

$$\langle \downarrow\downarrow\uparrow |\rho| \downarrow\downarrow\uparrow \rangle = \langle \downarrow\uparrow\downarrow |\rho| \downarrow\uparrow\downarrow \rangle = \langle \uparrow\downarrow\downarrow |\rho| \uparrow\downarrow\downarrow \rangle, \qquad (6.141a)$$

and off the diagonal we require

$$\langle \downarrow\downarrow\uparrow |\rho| \downarrow\uparrow\downarrow \rangle = \langle \downarrow\downarrow\uparrow |\rho| \uparrow\downarrow\downarrow \rangle = \langle \downarrow\uparrow\downarrow |\rho| \downarrow\downarrow\uparrow \rangle$$
$$= \langle \downarrow\uparrow\downarrow |\rho| \uparrow\downarrow\downarrow \rangle = \langle \uparrow\downarrow\downarrow |\rho| \downarrow\downarrow\uparrow \rangle = \langle \uparrow\downarrow\downarrow |\rho| \downarrow\uparrow\downarrow \rangle. \qquad (6.141b)$$

All of these relationships follow by interchanging a pair of single atom labels. It is not possible, however, to establish equality between the matrix elements of (6.141a) and those of (6.141b) by such an exchange. Now each of the squares in Fig. 6.3 represents 7, rather than 8, relationships between matrix elements, and the number of independent density matrix elements demanded by permutational symmetry is $64 - 44 = 20 = \frac{1}{6}(N + 1)(N + 2)(N + 3)$.

6.3.3 Operator Averages Without Quantum Fluctuations

We can use the moment equations (6.139) to illustrate the sense in which quantum fluctuations become a small perturbation on deterministic dynamics in the limit of large N. We will first develop a treatment to lowest order, not including quantum fluctuations, and compare it with exact results based on the factorized density operator (6.130). After this we will derive a phase-space equation of motion that includes quantum fluctuations to first order in $1/N$.

Let us define variables scaled by the system size, as in (5.38). We write

$$J_- = N\bar{J}_-, \qquad J_+ = N\bar{J}_+, \qquad J_z = N\bar{J}_z. \qquad (6.142)$$

Equation (6.139) defines a coupled hierarchy for the normal-ordered averages of $\bar{J}_-$, $\bar{J}_+$, and $\bar{J}_z$:

$$\frac{d}{dt}\langle \bar{J}_+^p \bar{J}_z^r \bar{J}_-^q \rangle$$
$$= -\left[i\omega_A(q - p) + (p + q)\frac{\gamma}{2}(\bar{n} + 1) \right] \langle \bar{J}_+^p \bar{J}_z^r \bar{J}_-^q \rangle$$

$$+ \frac{\gamma}{2} \langle \bar{J}_+^p (1 + \bar{J}_z) N [(\bar{J}_z - 2/N)^r - \bar{J}_z^r] \bar{J}_-^q \rangle$$

$$+ \gamma \bar{n} \{ \langle \bar{J}_+^p \tfrac{1}{2} (1 + \bar{J}_z) N [(\bar{J}_z - 2/N)^r - \bar{J}_z^r] \bar{J}_-^q \rangle$$

$$+ \langle \bar{J}_+^p [\tfrac{1}{2}(1 - \bar{J}_z) - (p+q)/N] N [(\bar{J}_z + 2/N)^r - \bar{J}_z^r] \bar{J}_-^q \rangle$$

$$+ N^{-2} pq(3 - p - q) \langle \bar{J}_+^{p-1} \bar{J}_z^r \bar{J}_-^{q-1} \rangle$$

$$+ N^{-3} pq(p-1)(q-1) \langle \bar{J}_+^{p-2} \tfrac{1}{2}(1 + \bar{J}_z)(\bar{J}_z - 2/N)^r \bar{J}_-^{q-2} \rangle \}. \quad (6.143)$$

We have

$$N[(\bar{J}_z \mp 2/N)^r - \bar{J}_z^r] = \mp 2r \bar{J}_z^{r-1} + O(1/N), \quad (6.144)$$

and for $p, q \ll N$, to lowest order we find

$$\frac{d}{dt} \langle \bar{J}_+^p \bar{J}_z^r \bar{J}_-^q \rangle$$

$$= - \left[(p + q + 2r) \frac{\gamma}{2} (2\bar{n} + 1) + i(q - p)\omega_A \right] \langle \bar{J}_+^p \bar{J}_z^r \bar{J}_-^q \rangle - r\gamma \langle \bar{J}_+^p \bar{J}_z^{r-1} \bar{J}_-^q \rangle. \quad (6.145)$$

Equation (6.145) defines a coupled hierarchy for the operator averages $\langle \bar{J}_+^p \bar{J}_z^r \bar{J}_-^q \rangle$, $k = 0, 1, \ldots, r$. The p and q dependence on the right-hand side is easily removed and the resulting equations solved by induction:

Exercise 6.9 Show by induction (or otherwise) that

$$\langle \left(\bar{J}_+^p \bar{J}_z^r \bar{J}_-^q \right)(t) \rangle = e^{-(p+q)(\gamma/2)(2\bar{n}+1)t} e^{i(q-p)\omega_A t} \sum_{k=0}^r (-1)^k \frac{r!}{(r-k)! k!}$$

$$\times e^{-(r-k)\gamma(2\bar{n}+1)t} \left[\frac{1 - e^{-\gamma(2\bar{n}+1)t}}{2\bar{n}+1} \right]^k \langle \left(\bar{J}_+^p \bar{J}_z^{r-k} \bar{J}_-^q \right)(0) \rangle. \quad (6.146)$$

In (6.146) we easily recognize the solution that preserves the initial factorization

$$\langle \left(\bar{J}_+^p \bar{J}_z^r \bar{J}_-^q \right)(0) \rangle = \langle \bar{J}_+(0) \rangle^p \langle \bar{J}_z(0) \rangle^r \langle \bar{J}_-(0) \rangle^q$$

$$= \langle e^{i\phi_j} \sigma_{j+}(0) \rangle^p \langle \sigma_{jz}(0) \rangle^r \langle e^{-i\phi_j} \sigma_-(0) \rangle^q : \quad (6.147)$$

For this initial state, (6.146) gives

$$\langle \left(\bar{J}_+^p \bar{J}_z^r \bar{J}_-^q \right)(t) \rangle = \langle \bar{J}_+(t) \rangle^p \langle \bar{J}_z(t) \rangle^r \langle \bar{J}_-(t) \rangle^q$$

$$= \langle e^{i\phi_j} \sigma_{j+}(t) \rangle^p \langle \sigma_{jz}(t) \rangle^r \langle e^{-i\phi_j} \sigma_-(t) \rangle^q, \quad (6.148)$$

where

$$\langle e^{\pm i\phi_j}\sigma_{j\pm}(t)\rangle = e^{-(\gamma/2)(2\bar{n}+1)t}e^{\mp i\omega_A t}\langle e^{\pm i\phi_j}\sigma_{j\pm}(0)\rangle, \qquad (6.149a)$$

$$\langle\sigma_{jz}(t)\rangle = e^{-\gamma(2\bar{n}+1)t}\langle\sigma_{jz}(0)\rangle$$
$$- (2\bar{n}+1)^{-1}\left[1 - e^{-\gamma(2\bar{n}+1)t}\right], \qquad (6.149b)$$

are the solutions for a single radiatively damped two-level atom [solutions to (2.37)]. The index j on single-atom operators denotes *any* atom; all atoms are identical.

This lowest-order treatment neglects quantum fluctuations, and even though there are no correlations between different atoms in the simple example we are discussing – no *unlike-atom correlations* – quantum fluctuations are present due to correlations between operators for the same atom – *like-atom correlations*. To illustrate that these like-atom correlations exist, and to show the form they take, let us derive exact results for the quadratic operator averages.

In each of the following we use the independence stated by (6.130) to factorize averages involving operator products for different atoms, and the relations (6.6) to reduce products between operators for the same atom to linear functions of single-atom operators. We have

$$\langle\bar{J}_+^2\rangle - \langle\bar{J}_+\rangle^2 = N^{-2}\left\langle\left(\sum_{j=1}^N e^{i\phi_j}\sigma_{j+}\right)\left(\sum_{k=1}^N e^{i\phi_k}\sigma_{k+}\right)\right\rangle - \langle\bar{J}_+\rangle^2$$

$$= N^{-2}\sum_{j=1}^N\sum_{k\neq j}\langle e^{i\phi_j}\sigma_{j+}\rangle\langle e^{i\phi_k}\sigma_{k+}\rangle - \langle\bar{J}_+\rangle^2$$

$$= N^{-2}\left[N(N-1)\langle e^{i\phi_j}\sigma_{j+}\rangle^2\right] - \langle\bar{J}_+\rangle^2$$

$$= -N^{-1}\langle\bar{J}_+\rangle^2, \qquad (6.150a)$$

and, similarly,

$$\langle\bar{J}_-^2\rangle - \langle\bar{J}_-\rangle^2 = -N^{-1}\langle\bar{J}_-\rangle^2, \qquad (6.150b)$$

$$\langle\bar{J}_z^2\rangle - \langle\bar{J}_z\rangle^2 = N^{-1}\left(1 - \langle\bar{J}_z\rangle^2\right), \qquad (6.150c)$$

$$\langle\bar{J}_+\bar{J}_-\rangle - \langle\bar{J}_+\rangle\langle\bar{J}_-\rangle = N^{-1}\left[\tfrac{1}{2}\left(1 + \langle\bar{J}_z\rangle\right) - \langle\bar{J}_+\rangle\langle\bar{J}_-\rangle\right], \qquad (6.150d)$$

$$\langle\bar{J}_+\bar{J}_z\rangle - \langle\bar{J}_+\rangle\langle\bar{J}_z\rangle = -N^{-1}\left(\langle\bar{J}_+\rangle + \langle\bar{J}_+\rangle\langle\bar{J}_z\rangle\right), \qquad (6.150e)$$

$$\langle\bar{J}_z\bar{J}_-\rangle - \langle\bar{J}_z\rangle\langle\bar{J}_-\rangle = -N^{-1}\left(\langle\bar{J}_-\rangle + \langle\bar{J}_z\rangle\langle\bar{J}_-\rangle\right). \qquad (6.150f)$$

We see that corrections due to quantum fluctuations are of order N^{-1}. The same approach can be taken to calculate normal-ordered collective operator moments of all orders; although, in the general case the bookkeeping becomes rather complicated:

$$\langle \bar{J}_+^p \bar{J}_z^r \bar{J}_-^q \rangle$$

$$= N^{-(p+r+q)} \left\langle \left\{ \sum_{\{j_1,\ldots,j_p\}} [\exp(i\phi_{j_1})\sigma_{j_1+}] \cdots [\exp(i\phi_{j_p})\sigma_{j_p+}] \right\} \right.$$

$$\times \left\{ \sum_{\{k_1,\ldots,k_r\}} \sigma_{k_1 z} \cdots \sigma_{k_r z} \right\}$$

$$\left. \times \left\{ \sum_{\{l_1,\ldots,l_q\}} [\exp(-i\phi_{l_1})\sigma_{l_1-}] \cdots [\exp(-i\phi_{l_q})\sigma_{l_q-}] \right\} \right\rangle$$

$$= N^{-(p+r+q)} \left\{ \left[N(N-1) \cdots \right.\right.$$

$$\left. \cdots (N-p-r-q+1)\langle e^{i\phi}\sigma_+\rangle^p \langle \sigma_z \rangle^r \langle e^{-i\phi}\sigma_-\rangle^q \right]$$

$$+ \left(\begin{array}{c} \text{terms with two} \\ \text{atomic labels equal} \end{array} \right) + \left(\begin{array}{c} \text{terms with three} \\ \text{atomic labels equal} \end{array} \right) + \cdots \right\}$$

$$= \left[1 - N^{-1}\tfrac{1}{2}(p+r+q)(p+r+q-1) + O(N^{-2}) + \cdots \right]$$

$$\times \langle \bar{J}_+\rangle^p \langle \bar{J}_z\rangle^r \langle \bar{J}_-\rangle^q$$

$$+ N^{-(p+r+q)} \left[\left(\begin{array}{c} \text{terms with two} \\ \text{atomic labels equal} \end{array} \right) + \left(\begin{array}{c} \text{terms with three} \\ \text{atomic labels equal} \end{array} \right) + \cdots \right].$$

$$(6.151)$$

The first term on the right-hand side comes from the factorization of single-atom averages in terms with all atomic labels different. When we restrict our attention to low-order moments, $\frac{1}{2}(p+r+q)(p+r+q-1) \ll N$, corrections due to quantum fluctuations enter in powers of N^{-1}. First-order corrections will be given by $-N^{-1}\frac{1}{2}(p+r+q)(p+r+q-1)\langle \bar{J}_+\rangle^p \langle \bar{J}_z\rangle^r \langle \bar{J}_-\rangle^q$ plus a contribution from the terms with the atomic labels equal. (There are fewer of these, by a factor of order N, than there are terms with all atomic labels different.) We will now see how the normal-ordered phase-space representation is used to derive a Fokker–Planck equation including the first-order corrections to the factorized dynamics.

Note 6.5 Of course, it is not always the case that the atoms in an atomic population are statistically independent. This is not so, for example, when the atoms interact with a common field mode, as in a laser or a passive bistable system. Here, the interaction of the atoms with the field mode introduces correlations between different atoms. Then unlike-atom correlations do not vanish, and (6.150a)–(6.150f) no longer hold. We can still use the phase-space approach, though, to find the expressions that replace these results. In Volume 2 we will go through the calculation explicitly for the case of absorptive optical bistability (Sect. 15.2.4).

6.3.4 Phase-Space Equation of Motion for Independently Damped Two-Level Atoms

We wish to derive the phase-space equation of motion equivalent to the master equation (6.132) using the representation defined in Sect. 6.2.3. The calculation follows essentially the same steps as the single-atom calculation. Our first task is to derive an equation of motion for the characteristic function. Corresponding to (6.12) we obtain

$$
\frac{\partial \chi_N}{\partial t} = \mathrm{tr}\Bigg\{ \Bigg[-i\tfrac{1}{2}\omega_A (J_z\rho - \rho J_z)
$$

$$
+ \frac{\gamma}{2}(\bar{n}+1)\Bigg(2\sum_{j=1}^{N} \sigma_{j-}\rho\sigma_{j+} - \tfrac{1}{2}J_z\rho - \tfrac{1}{2}\rho J_z - N\rho \Bigg)
$$

$$
+ \frac{\gamma}{2}\bar{n}\Bigg(2\sum_{j=1}^{N} \sigma_{j+}\rho\sigma_{j-} + \tfrac{1}{2}J_z\rho + \tfrac{1}{2}\rho J_z - N\rho \Bigg) \Bigg] e^{i\xi^* J_+} e^{i\eta J_z} e^{i\xi J_-} \Bigg\}.
$$

$$
\tag{6.152}
$$

Since collective atomic operators obey the same commutation relations as single-atom operators, most terms on the right-hand side can be evaluated as in Sect. 6.1.3; from (6.13) and (6.14),

$$
\mathrm{tr}\big(J_z\rho e^{i\xi^* J_+} e^{i\eta J_z} e^{i\xi J_-} \big) = \Bigg(\frac{\partial}{\partial(i\eta)} + 2i\xi\frac{\partial}{\partial(i\xi)} \Bigg)\chi_N, \tag{6.153a}
$$

$$
\mathrm{tr}\big(\rho J_z e^{i\xi^* J_+} e^{i\eta J_z} e^{i\xi J_-} \big) = \Bigg(\frac{\partial}{\partial(i\eta)} + 2i\xi^*\frac{\partial}{\partial(i\xi^*)} \Bigg)\chi_N; \tag{6.153b}
$$

we only need to give special consideration to the two terms that have not been expressed directly in terms of collective atomic operators.

The treatment of these terms follows the principles used in (6.137) and (6.138) to obtain closed equations for the collective operator normal-ordered moments. If single-atom operators can be placed next to each other, their product can be reduced to a summable form using (6.6a)–(6.6g). We saw how this is done in Sect. 6.1.3. In that section the method used to derive the phase-space equation of motion for a single damped two-level atom was not actually unique; and therefore the form of the resulting equation of motion was not unique. We chose the method that preserves a close correspondence between the single-atom phase-space equation of motion and the many-atom equation of motion. In the many-atom calculation we do not have a choice about how to proceed. We must reduce all quadratic dependence on single-atom operators to a linear dependence if we are to perform the sums and obtain a description in terms of collective operator averages alone.

Note 6.6 In the single-atom limit there is no difference between the master equations (6.129) and (6.131). For many atoms these equations are quite different and their corresponding phase-space equations of motion must reflect this difference. It is in the treatment of the terms $\sum_{j=1}^{N} \sigma_{j\mp}\rho\sigma_{j\pm}$ and $J_{\mp}\rho J_{\pm}$ that the difference arises. The later may be treated after the fashion of (6.15); the former requires the method leading to (6.16) and (6.17), the method we now use to reduce the remaining sums in (6.152) to their phase space form.

Following the derivation of (6.16), we use (6.8) and (6.6f) to obtain

$$
\mathrm{tr}\left(\sum_{j=1}^{N} \sigma_{j-}\rho\sigma_{j+} e^{i\xi^* J_+} e^{i\eta J_z} e^{i\xi J_-} \right)
$$

$$
= \mathrm{tr}\left[\sum_{j=1}^{N} \rho e^{i\xi^* J_+} \sigma_{j+} e^{i\eta\sigma_{jz}} \sigma_{j-} \left(\prod_{k\neq j} e^{i\eta\sigma_{kz}} \right) e^{i\xi J_-} \right]
$$

$$
= e^{-2i\eta}\, \mathrm{tr}\left[\sum_{j=1}^{N} \rho e^{i\xi^* J_+} \tfrac{1}{2}(1+\sigma_{jz}) e^{i\eta\sigma_{jz}} \left(\prod_{k\neq j} e^{i\eta\sigma_{kz}} \right) e^{i\xi J_-} \right]
$$

$$
= e^{-2i\eta}\, \mathrm{tr}\left[\rho e^{i\xi^* J_+} \tfrac{1}{2}(N+J_z) e^{i\eta J_z} e^{i\xi J_-} \right]
$$

$$
= e^{-2i\eta} \frac{1}{2}\left(N + \frac{\partial}{\partial(i\eta)} \right) \chi_N. \tag{6.154}
$$

A similar calculation, following the derivation of (6.17), gives

$$
\mathrm{tr}\left(\sum_{j=1}^{N} \sigma_{j+}\rho\sigma_{j-} e^{i\xi^* J_+} e^{i\eta J_z} e^{i\xi J_-} \right)
$$

$$
= \Big[(N/2)\big(e^{2i\eta} + (i\xi)^2(i\xi^*)^2 e^{-2i\eta} + 2(i\xi)(i\xi^*) \big)
$$

$$
- \tfrac{1}{2}\big(e^{2i\eta} - (i\xi)^2(i\xi^*)^2 e^{-2i\eta} \big) \frac{\partial}{\partial(i\eta)}
$$

$$
- i\xi\big(e^{2i\eta} + (i\xi)(i\xi^*) \big) \frac{\partial}{\partial(i\xi)} - i\xi^*\big(e^{2i\eta} + (i\xi)(i\xi^*) \big) \frac{\partial}{\partial(i\xi^*)} \Big] \chi_N.
$$

$$
\tag{6.155}
$$

The derivation of the phase-space equation of motion is now concluded in exactly the same manner as in Sect. 6.1.3. We go immediately to the result. Comparing (6.23), we obtain the *phase-space equation of motion for N independent radiatively damped two-level atoms*:

$$\frac{\partial P}{\partial t} = L\left(v, v^*, m, \frac{\partial}{\partial v}, \frac{\partial}{\partial v^*}, \frac{\partial}{\partial m}\right) P, \qquad (6.156)$$

with

$$
\begin{aligned}
L&\left(v, v^*, m, \frac{\partial}{\partial v}, \frac{\partial}{\partial v^*}, \frac{\partial}{\partial m}\right) \\
&\equiv i\omega_A\left(\frac{\partial}{\partial v}v - \frac{\partial}{\partial v^*}v^*\right) \\
&+ \frac{\gamma}{2}(\bar{n}+1)\left[\left(e^{2\frac{\partial}{\partial m}}-1\right)(N+m) + \frac{\partial}{\partial v}v + \frac{\partial}{\partial v^*}v^*\right] \\
&+ \frac{\gamma}{2}\bar{n}\left[\left(e^{-2\frac{\partial}{\partial m}}-1\right)(N-m) + \frac{\partial^4}{\partial v^2 \partial v^{*2}}e^{2\frac{\partial}{\partial m}}(N+m)\right. \\
&\left. +2\left(e^{-2\frac{\partial}{\partial m}} + \frac{\partial^2}{\partial v \partial v^*} - \frac{1}{2}\right)\left(\frac{\partial}{\partial v}v + \frac{\partial}{\partial v^*}v^*\right) + 2N\frac{\partial^2}{\partial v \partial v^*}\right].
\end{aligned}
$$

$$(6.157)$$

Exercise 6.10 Show that with dephasing processes included the term $\gamma_p\left(\sum_{j=1}^N \sigma_{jz}\rho\sigma_{jz} - N\rho\right)$ must be added to the master equation and the operator

$$L_{\text{dephase}} \equiv \gamma_p\left[\frac{\partial}{\partial v}v + \frac{\partial}{\partial v^*}v^* + \frac{\partial^2}{\partial v \partial v^*}e^{-2\frac{\partial}{\partial m}}(N+m)\right] \qquad (6.158)$$

must then be added to (6.157).

The shift operators $e^{\pm 2\frac{\partial}{\partial m}}$ appearing in (6.157) lead to a description for the inversion dynamics in terms of a jump process, evolving between discrete states, as in the single-atom case (Sect. 6.1.4). In place of (6.35) and (6.36) we may write

$$P(v, v^*, m, t) = \sum_{M=-N/2}^{N/2} P_M(v, v^*, t)\delta(m - 2M), \qquad (6.159)$$

with

$$\int d^2v\, P_M(v, v^*, t) = p_M \equiv \sum_{\boldsymbol{u}}\langle \boldsymbol{u}; M|\rho(t)|\boldsymbol{u}; M\rangle; \qquad (6.160)$$

p_M gives the probability for the system to adopt the inversion state $m = 2M$, $M = -N/2, -N/2+1, \ldots, N/2$, with $N/2 - M$ atoms in their upper states and $N/2 + M$ atoms in their lower states. Substituting the expansion (6.159) into (6.156), and applying the shift operators, we equate coefficients of the δ-functions to obtain the coupled equations

$$\frac{\partial P_M}{\partial t} = \left[\left(\frac{\gamma}{2} + i\omega_A \right) \frac{\partial}{\partial v} v + \left(\frac{\gamma}{2} - i\omega_A \right) \frac{\partial}{\partial v^*} v^* + N\gamma\bar{n} \frac{\partial^2}{\partial v \partial v^*} \right.$$

$$\left. + \gamma\bar{n} \frac{\partial^2}{\partial v \partial v^*} \left(\frac{\partial}{\partial v} v + \frac{\partial}{\partial v^*} v^* \right) - \gamma(N/2 + M + N\bar{n}) \right] P_M$$

$$+ (N/2 + M + 1) \left[\gamma\bar{n} \frac{\partial^4}{\partial v^2 \partial v^{*2}} + \gamma(\bar{n} + 1) \right] P_{M+1}$$

$$+ \left[\gamma\bar{n} \left(\frac{\partial}{\partial v} v + \frac{\partial}{\partial v^*} v^* \right) + \gamma\bar{n}(N/2 - M + 1) \right] P_{M-1}, \quad (6.161\text{a})$$

and

$$\left(\frac{\partial}{\partial v} v + \frac{\partial}{\partial v^*} v^* \right) P_{N/2} = 0. \quad (6.161\text{b})$$

For $N = 1$ these reproduce (6.38). After integrating over the polarization variable v we arrive at rate equations describing the evolution of population between discrete inversion states:

$$\dot{p}_M = -\gamma(N/2 + M + N\bar{n})p_M + \gamma(\bar{n} + 1)(N/2 + M + 1)p_{M+1}$$

$$+ \gamma\bar{n}(N/2 - M + 1)p_{M-1}. \quad (6.162)$$

The set of coupled partial differential equations (6.161) is equivalent to the phase-space equation of motion (6.156). The exact solution to this equation of motion is going to be a complicated singular function. When N is large, however, a much simpler and more transparent solution is available if we seek only a lowest-order treatment of the quantum fluctuations.

6.3.5 Fokker–Planck Equation: First-Order Treatment of Quantum Fluctuations

On the basis of the arguments offered in Sect. 6.2.4, in the large N limit we expect to be able to replace the strictly singular distribution representing the density operator by a nonsingular distribution. We hope, therefore, to obtain an adequate treatment of quantum fluctuations to first order in $1/N$ by replacing (6.156) and (6.157) by a Fokker–Planck equation. The procedure that formally takes us to such a description is van Kampen's system size expansion (Sect. 5.1.3). Corresponding to the scaled operators $\bar{J}_-$, $\bar{J}_+$, and $\bar{J}_z$ introduced in (6.142), we define scaled phase-space variables $\bar{v}$, $\bar{v}^*$, and $\bar{m}$, with

$$v = N\bar{v}, \qquad v^* = N\bar{v}^*, \qquad m = N\bar{m}. \quad (6.163)$$

To obtain a systematic expansion of the phase-space equation of motion in inverse powers of the system size, we write

$$\bar{v} = \langle \bar{J}_-(t) \rangle + N^{-1/2}\nu, \quad (6.164\text{a})$$

$$\bar{v}^* = \langle \bar{J}_+(t) \rangle + N^{-1/2}\nu^*, \quad (6.164\text{b})$$

$$\bar{m} = \langle \bar{J}_z(t) \rangle + N^{-1/2}\mu. \quad (6.164\text{c})$$

For the present problem we already have the time-dependent solutions for $\langle \bar{J}_- \rangle$, $\langle \bar{J}_+ \rangle$, and $\langle \bar{J}_z \rangle$ [Eqs. (6.149)], and also the corrections due to quantum fluctuations (for quadratic operator averages) [Eqs. (6.150)]. Our objective is to use the system size expansion for the phase-space equation of motion to reproduce these results. We will obtain a macroscopic law governing the motion of $\langle \bar{J}_- \rangle$, $\langle \bar{J}_+ \rangle$, and $\langle \bar{J}_z \rangle$, and a linear Fokker–Planck equation for the distribution

$$\bar{P}(\nu, \nu^*, \mu, t) \equiv N^{3/2} P\big(N\langle \bar{J}_-(t) \rangle + N^{1/2}\nu, N\langle \bar{J}_+(t) \rangle + N^{1/2}\nu^*,$$
$$N\langle \bar{J}_z(t) \rangle + N^{1/2}\mu, t\big); \tag{6.165}$$

the macroscopic law plus the Gaussian solution to the Fokker–Planck equation should reproduce (6.149) and (6.150).

From (6.165) we write

$$\frac{\partial \bar{P}}{\partial t} = N^{3/2}\bigg(\frac{\partial P}{\partial \langle \bar{J}_-(t) \rangle} \frac{d\langle \bar{J}_-(t) \rangle}{dt} + \frac{\partial P}{\partial \langle \bar{J}_+(t) \rangle} \frac{d\langle \bar{J}_+(t) \rangle}{dt}$$

$$+ \frac{\partial P}{\partial \langle \bar{J}_z(t) \rangle} \frac{d\langle \bar{J}_z(t) \rangle}{dt} + \frac{\partial P}{\partial t} \bigg)$$

$$= N^{1/2}\bigg(\frac{\partial \bar{P}}{\partial \nu} \frac{d\langle \bar{J}_-(t) \rangle}{dt} + \frac{\partial \bar{P}}{\partial \nu^*} \frac{d\langle \bar{J}_+(t) \rangle}{dt} + \frac{\partial \bar{P}}{\partial \mu} \frac{d\langle \bar{J}_z(t) \rangle}{dt} \bigg)$$

$$+ \frac{\partial}{\partial t}(N^{3/2}P).$$

Then, substituting the phase-space equation (6.156) for $\partial P/\partial t$, and using the scaling relations (6.163) and (6.164), we find

$$\frac{\partial \bar{P}}{\partial t} = N^{1/2}\bigg\{ \frac{\partial \bar{P}}{\partial \nu} \frac{d\langle \bar{J}_-(t) \rangle}{dt} + \frac{\partial \bar{P}}{\partial \nu^*} \frac{d\langle \bar{J}_+(t) \rangle}{dt} + \frac{\partial \bar{P}}{\partial \mu} \frac{d\langle \bar{J}_z(t) \rangle}{dt}$$

$$+ \frac{\partial}{\partial \nu} \Big[\frac{\gamma}{2}(2\bar{n}+1) + i\omega_A \Big] \big[\langle \bar{J}_-(t) \rangle + N^{-1/2}\nu \big]$$

$$+ \frac{\partial}{\partial \nu^*} \Big[\frac{\gamma}{2}(2\bar{n}+1) - i\omega_A \Big] \big[\langle \bar{J}_+(t) \rangle + N^{-1/2}\nu^* \big]$$

$$+ \frac{\partial}{\partial \mu} \gamma(2\bar{n}+1) \Big[(2\bar{n}+1)^{-1} + \langle \bar{J}_z(t) \rangle + N^{-1/2}\mu \Big]$$

$$+ \frac{\partial^2}{\partial \nu \partial \nu^*} \gamma\bar{n} + \frac{\partial^2}{\partial \mu^2} \gamma\big[2\bar{n} + 1 + \langle \bar{J}_z(t) \rangle \big]$$

$$- \frac{\partial^2}{\partial \nu \partial \mu} 2\gamma\bar{n}\langle \bar{J}_-(t) \rangle - \frac{\partial^2}{\partial \nu^* \partial \mu} 2\gamma\bar{n}\langle \bar{J}_+(t) \rangle \bigg\} \bar{P} + O\big(N^{-1/2}\big).$$

Collecting terms of order $N^{1/2}$ and N^0, we have

$$\frac{\partial \bar{P}}{\partial t} = N^{1/2} \left\{ \frac{\partial \bar{P}}{\partial \nu} \left(\frac{d\langle \bar{J}_-(t)\rangle}{dt} + \left[\frac{\gamma}{2}(2\bar{n}+1) + i\omega_A \right] \langle \bar{J}_-(t)\rangle \right) \right.$$

$$+ \frac{\partial \bar{P}}{\partial \nu^*} \left(\frac{d\langle \bar{J}_+(t)\rangle}{dt} + \left[\frac{\gamma}{2}(2\bar{n}+1) - i\omega_A \right] \langle \bar{J}_+(t)\rangle \right)$$

$$+ \frac{\partial \bar{P}}{\partial \mu} \left(\frac{d\langle \bar{J}_z(t)\rangle}{dt} + \gamma(2\bar{n}+1)\left[\langle \bar{J}_z(t)\rangle + (2\bar{n}+1)^{-1} \right] \right) \Big\}$$

$$+ \left\{ \left[\frac{\gamma}{2}(2\bar{n}+1) + i\omega_A \right] \frac{\partial}{\partial \nu}\nu + \left[\frac{\gamma}{2}(2\bar{n}+1) - i\omega_A \right] \frac{\partial}{\partial \nu^*}\nu^* \right.$$

$$+ \gamma(2\bar{n}+1)\frac{\partial}{\partial \mu}\mu$$

$$+ \gamma\bar{n}\frac{\partial^2}{\partial \nu \partial \nu^*} + \gamma\left[(2\bar{n}+1) + \langle \bar{J}_z(t)\rangle \right]\frac{\partial^2}{\partial \mu^2}$$

$$\left. -2\gamma\bar{n}\langle \bar{J}_-(t)\rangle \frac{\partial^2}{\partial \nu \partial \mu} - 2\gamma\bar{n}\langle \bar{J}_+(t)\rangle \frac{\partial^2}{\partial \nu^* \partial \mu} \right\} \bar{P} + O(N^{-1/2}).$$

$$(6.166)$$

In the large N limit the terms of order $N^{1/2}$ vanish if $\langle \bar{J}_-\rangle$, $\langle \bar{J}_+\rangle$, and $\langle \bar{J}_z\rangle$ obey the macroscopic law

$$\frac{d\langle \bar{J}_-\rangle}{dt} = -\left[\frac{\gamma}{2}(2\bar{n}+1) + i\omega_A \right]\langle \bar{J}_-\rangle, \qquad (6.167a)$$

$$\frac{d\langle \bar{J}_+\rangle}{dt} = -\left[\frac{\gamma}{2}(2\bar{n}+1) - i\omega_A \right]\langle \bar{J}_+\rangle, \qquad (6.167b)$$

$$\frac{d\langle \bar{J}_z\rangle}{dt} = -\gamma(2\bar{n}+1)\left[\langle \bar{J}_z(t)\rangle + (2\bar{n}+1)^{-1} \right]. \qquad (6.167c)$$

Quantum fluctuations about this deterministic motion are described by the *Fokker–Planck equation for a radiatively damped two-level medium*:

$$\frac{\partial \bar{P}}{\partial t} = \left\{ \left[\frac{\gamma}{2}(2\bar{n}+1) + i\omega_A \right] \frac{\partial}{\partial \nu}\nu + \left[\frac{\gamma}{2}(2\bar{n}+1) - i\omega_A \right] \frac{\partial}{\partial \nu^*}\nu^* \right.$$

$$+ \gamma(2\bar{n}+1)\frac{\partial}{\partial \mu}\mu$$

$$+ \gamma\bar{n}\frac{\partial^2}{\partial \nu \partial \nu^*} + \gamma\left[(2\bar{n}+1) + \langle \bar{J}_z(t)\rangle \right]\frac{\partial^2}{\partial \mu^2}$$

$$\left. -2\gamma\bar{n}\langle \bar{J}_-(t)\rangle \frac{\partial^2}{\partial \nu \partial \mu} - 2\gamma\bar{n}\langle \bar{J}_+(t)\rangle \frac{\partial^2}{\partial \nu^* \partial \mu} \right\} \bar{P}. \qquad (6.168) \cdot$$

Note 6.7 When the dephasing term (6.158) is included, the damping rate γ_p is added inside the square bracket in (6.167a) and (6.167b), and inside the first two square brackets on the right-hand side of (6.168). Also, the term $\gamma_p[1 + \langle \bar{J}_z(t)\rangle]\partial^2/\partial \nu \partial \nu^*$ is added to the diffusion in (6.168).

Equations (6.167) are precisely the single-atom equations of motion (2.37), and reproduce the deterministic dynamics defined by (6.148) and (6.149). It remains to show that the quantum fluctuations described by (6.168) satisfy (6.150); from the definition (6.78) of the phase-space average, and the scaling (6.163) and (6.164), we must show that

$$\left(\overline{\nu^{*2}}\right)_P = N\left(\langle \bar{J}_+^2 \rangle - \langle \bar{J}_+ \rangle^2\right) = -\langle \bar{J}_+ \rangle^2, \tag{6.169a}$$

$$\left(\overline{\nu^2}\right)_P = N\left(\langle \bar{J}_-^2 \rangle - \langle \bar{J}_- \rangle^2\right) = -\langle \bar{J}_- \rangle^2, \tag{6.169b}$$

$$\left(\overline{\mu^2}\right)_P = N\left(\langle \bar{J}_z^2 \rangle - \langle \bar{J}_z \rangle^2\right) = 1 - \langle \bar{J}_z \rangle^2, \tag{6.169c}$$

$$\left(\overline{\nu^*\nu}\right)_P = N\left(\langle \bar{J}_+\bar{J}_- \rangle - \langle \bar{J}_+ \rangle\langle \bar{J}_- \rangle\right) = \tfrac{1}{2}\left(1 + \langle \bar{J}_z \rangle\right) - \langle \bar{J}_+ \rangle\langle \bar{J}_- \rangle, \tag{6.169d}$$

$$\left(\overline{\nu^*\mu}\right)_P = N\left(\langle \bar{J}_+\bar{J}_z \rangle - \langle \bar{J}_+ \rangle\langle \bar{J}_z \rangle\right) = -\langle \bar{J}_+ \rangle\left(1 + \langle \bar{J}_z \rangle\right), \tag{6.169e}$$

$$\left(\overline{\nu\mu}\right)_P = N\left(\langle \bar{J}_z\bar{J}_- \rangle - \langle \bar{J}_z \rangle\langle \bar{J}_- \rangle\right) = -\langle \bar{J}_- \rangle\left(1 + \langle \bar{J}_z \rangle\right). \tag{6.169f}$$

Since the Fokker–Planck equation (6.168) has a time-dependent diffusion, we cannot derive the covariance matrix directly from the results of Sect. 5.2.4. Nevertheless, it is not difficult to solve for the moments on the left-hand sides of (6.169); the calculation is made relatively easy by the diagonal drift. The details are left as an exercise.

Exercise 6.11 Use the Fokker–Planck equation (6.168) to show that the elements of the covariance matrix obey the equations of motion

$$\frac{d}{dt}\left(\overline{\nu^{*2}}\right)_P = -2\left[\frac{\gamma}{2}(2\bar{n}+1) - i\omega_A\right]\left(\overline{\nu^{*2}}\right)_P, \tag{6.170a}$$

$$\frac{d}{dt}\left(\overline{\nu^2}\right)_P = -2\left[\frac{\gamma}{2}(2\bar{n}+1) + i\omega_A\right]\left(\overline{\nu^2}\right)_P, \tag{6.170b}$$

$$\frac{d}{dt}\left(\overline{\mu^2}\right)_P = -2\gamma(2\bar{n}+1)\left(\overline{\mu^2}\right)_P + 2\gamma\left[(2\bar{n}+1) + \langle \bar{J}_z(t) \rangle\right], \tag{6.170c}$$

$$\frac{d}{dt}\left(\overline{\nu^*\nu}\right)_P = -\gamma(2\bar{n}+1)\left(\overline{\nu^*\nu}\right)_P + \gamma\bar{n}, \tag{6.170d}$$

$$\frac{d}{dt}\left(\overline{\nu^*\mu}\right)_P = -\left[\frac{3\gamma}{2}(2\bar{n}+1) - i\omega_A\right]\left(\overline{\nu^*\mu}\right)_P - 2\gamma\bar{n}\langle \bar{J}_+(t) \rangle, \tag{6.170e}$$

$$\frac{d}{dt}\left(\overline{\nu\mu}\right)_P = -\left[\frac{3\gamma}{2}(2\bar{n}+1) + i\omega_A\right]\left(\overline{\nu\mu}\right)_P - 2\gamma\bar{n}\langle \bar{J}_-(t) \rangle. \tag{6.170f}$$

Solve (6.167a)–(6.167c) to define the noise sources, then solve (6.170a)–(6.170f) and show that (6.169a)–(6.169f) are satisfied for all times.

6.3.6 Steady-State Distribution of Inversion

The solution to the Fokker–Planck equation (6.168) is a multidimensional Gaussian distribution with time-dependent means and covariance matrix given by the solutions to (6.167) and (6.170). Moments of all orders can be constructed from the means and covariance matrix using the Gaussian moment theorem. We noted previously [below (6.151)] that first-order corrections to deterministic dynamics, based on a large N limit, are only expected to be accurate when the order of the moments considered is much less than N. With this qualification, all normal-ordered operator averages constructed from (6.167) and (6.170) via the Gaussian moment theorem will agree, up to terms of order N^{-1}, with the exact results. In our present example, where all of the atoms are statistically independent, this is just a consequence of the central limit theorem. As a final illustration let us see how the exact and approximate distributions for the inversion compare in the steady state.

In the asymptotic limit $t \to \infty$, solutions to (6.167) approach the steady state

$$\langle \bar{J}_{\pm} \rangle_{\mathrm{ss}} = 0, \qquad \langle \bar{J}_z \rangle_{\mathrm{ss}} = -(2\bar{n}+1)^{-1}. \tag{6.171}$$

The steady-state statistics are described by the Fokker–Planck equation

$$\frac{\partial \bar{P}}{\partial t} = \left\{ \left[\frac{\gamma}{2}(2\bar{n}+1) + i\omega_A \right] \frac{\partial}{\partial \nu} \nu + \left[\frac{\gamma}{2}(2\bar{n}+1) - i\omega_A \right] \frac{\partial}{\partial \nu^*} \nu^* \right.$$
$$+ \gamma(2\bar{n}+1)\frac{\partial}{\partial \mu}\mu$$
$$\left. + \gamma\bar{n}\frac{\partial^2}{\partial \nu \partial \nu^*} + 4\gamma\bar{n}\frac{\bar{n}+1}{2\bar{n}+1}\frac{\partial^2}{\partial \mu^2} \right\} \bar{P}. \tag{6.172}$$

The steady-state distribution $\bar{P}_{\mathrm{ss}}(\nu, \mu^*, \mu)$ is constructed using (5.80):

$$\bar{P}_{\mathrm{ss}}(\nu, \nu^*, \mu) = \bar{\mathcal{N}}_{\mathrm{ss}}(\nu, \nu^*)\bar{\mathcal{M}}_{\mathrm{ss}}(\mu), \tag{6.173}$$

with

$$\bar{\mathcal{N}}_{\mathrm{ss}}(\nu, \nu^*) = \frac{2\bar{n}+1}{\pi\bar{n}} \exp\left(-\frac{2\bar{n}+1}{\bar{n}}|\nu|^2\right), \tag{6.174a}$$

$$\bar{\mathcal{M}}_{\mathrm{ss}}(\mu) = \frac{1}{\sqrt{2\pi}} \frac{2\bar{n}+1}{\sqrt{4\bar{n}(\bar{n}+1)}} \exp\left[-\frac{1}{2}\frac{(2\bar{n}+1)^2}{4\bar{n}(\bar{n}+1)}\mu^2\right]. \tag{6.174b}$$

We will focus on the distribution over inversion states $m = -N, -N + 2, \ldots, N$, or $\bar{m} = -1, -1+2/N, \ldots, 1$; from the scaling (6.164c) and (6.174b), the steady-state distribution for $\bar{m}$ is given by

$$\bar{M}_{\mathrm{ss}}(\bar{m}) = \frac{1}{\sqrt{2\pi}} \sqrt{N\frac{(2\bar{n}+1)^2}{4\bar{n}(\bar{n}+1)}} \exp\left[-\frac{1}{2}N\frac{(2\bar{n}+1)^2}{4\bar{n}(\bar{n}+1)}\left(\bar{m} + \frac{1}{2\bar{n}+1}\right)^2\right],$$

$$\tag{6.175}$$

where we have defined

$$\bar{M}_{\rm ss}(\bar{m}) \equiv N^{1/2} \bar{\mathcal{M}}_{\rm ss}\left(N^{1/2}(\bar{m} - \langle \bar{J}_z \rangle_{\rm ss})\right). \tag{6.176}$$

The exact distribution over inversion states can be found from the statistical independence of the atoms and the single-atom solution

$$\langle \sigma_{jz} \rangle_{\rm ss} = -(2\bar{n} + 1)^{-1}. \tag{6.177}$$

Each atom is found in its upper state with the probability $(\rho_j)_{22} = \frac{1}{2}(1 + \langle \sigma_{jz} \rangle_{\rm ss}) = \bar{n}/(2\bar{n} + 1)$, and in its lower state with probability $(\rho_j)_{11} = \frac{1}{2}(1 - \langle \sigma_{jz} \rangle_{\rm ss}) = (\bar{n} + 1)/(2\bar{n} + 1)$. The probability for k atoms to be in their upper states and $N - k$ atoms to be in their lower states is then given by

$$p_k = \frac{N!}{k!(N - k)!} \left(\frac{\bar{n}}{2\bar{n} + 1}\right)^k \left(\frac{\bar{n} + 1}{2\bar{n} + 1}\right)^{N-k}; \tag{6.178}$$

the inversion (per atom) in this state, with k excitations, is

$$\bar{m}_k = [k - (N - k)]/N = 2k/N - 1. \tag{6.179}$$

Thus, the medium excitation obeys a binomial distribution, to which the Gaussian (6.175) is an approximation. The moments $\langle k^n \rangle$ can be calculated using the generating function $\langle e^{kx} \rangle$:

$$
\begin{aligned}
\langle k^n \rangle &= \frac{d^n}{dx^n} \langle e^{kx} \rangle \bigg|_{x=0} \\
&= \frac{d^n}{dx^n} \sum_{k=0}^{N} \frac{N!}{k!(N - k)!} \left(e^x \frac{\bar{n}}{2\bar{n} + 1}\right)^k \left(\frac{\bar{n} + 1}{2\bar{n} + 1}\right)^{N-k} \bigg|_{x=0} \\
&= \frac{d^n}{dx^n} \left(\frac{\bar{n} + 1}{2\bar{n} + 1} + e^x \frac{\bar{n}}{2\bar{n} + 1}\right)^N \bigg|_{x=0}.
\end{aligned}
\tag{6.180}
$$

The mean and the variance for the inversion readily follow. We have

$$
\begin{aligned}
\langle k \rangle &= N \left(\frac{\bar{n} + 1}{2\bar{n} + 1} + e^x \frac{\bar{n}}{2\bar{n} + 1}\right)^{N-1} e^x \frac{\bar{n}}{2\bar{n} + 1} \bigg|_{x=0} \\
&= N \frac{\bar{n}}{2\bar{n} + 1},
\end{aligned}
\tag{6.181a}
$$

and

$$
\begin{aligned}
\langle k^2 \rangle &= N(N - 1) \left(\frac{\bar{n} + 1}{2\bar{n} + 1} + e^x \frac{\bar{n}}{2\bar{n} + 1}\right)^{N-2} \left(e^x \frac{\bar{n}}{2\bar{n} + 1}\right)^2 \bigg|_{x=0} + \langle k \rangle \\
&= N(N - 1) \left(\frac{\bar{n}}{2\bar{n} + 1}\right)^2 + N \frac{\bar{n}}{2\bar{n} + 1} \\
&= N^2 \left(\frac{\bar{n}}{2\bar{n} + 1}\right)^2 + N \frac{\bar{n}(\bar{n} + 1)}{(2\bar{n} + 1)^2}.
\end{aligned}
\tag{6.181b}
$$

Then

$$\langle \bar{m}_k \rangle = 2\langle k \rangle / N - 1 = -(2\bar{n}+1)^{-1}, \qquad (6.182a)$$

and

$$\langle (\bar{m}_k)^2 \rangle - \langle \bar{m}_k \rangle^2$$

$$= \frac{4}{N^2}\langle k^2 \rangle - \frac{4}{N}\langle k \rangle + 1 - \left(\frac{1}{2\bar{n}+1}\right)^2$$

$$= 4\left(\frac{\bar{n}}{2\bar{n}+1}\right)^2 + \frac{4}{N}\frac{\bar{n}(\bar{n}+1)}{(2\bar{n}+1)^2} - 4\frac{\bar{n}}{2\bar{n}+1} + 1 - \left(\frac{1}{2\bar{n}+1}\right)^2$$

$$= \frac{4}{N}\frac{\bar{n}(\bar{n}+1)}{(2\bar{n}+1)^2}. \qquad (6.182b)$$

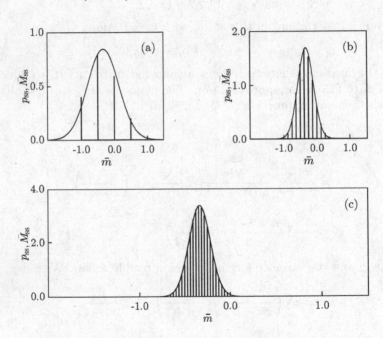

Fig. 6.4 Comparison of the exact (discrete) distribution over inversion states $p_{ss}(\bar{m}_k) \equiv (N/2)p_k$ [Eq. (6.178)] and the Gaussian approximation $\bar{M}_{ss}(\bar{m})$ [Eq. (6.175)] for $\bar{n} = 1$: (a) $N = 4$, (b) $N = 16$, (c) $N = 64$.

We see that the exact mean and variance are correctly given by the approximate Gaussian distribution (6.175); indeed, this is so even if N is not large. The distributions, however, will approach a close functional similarity, in the sense of Fig. 6.2, only as N becomes large. The explicit comparison between $p_{ss}(\bar{m}_k) \equiv (N/2)p_k$ and $\bar{M}_{ss}(\bar{m})$ is illustrated in Fig. 6.4. The difference between these distributions for finite N will show up in the higher moments as corrections beyond the order N^{-1}. (When unlike-atom correlations are present, the mean and variance will not be obtained exactly for finite N from the Gaussian approximation as they are here.)

Exercise 6.12 Show that (6.175) and (6.178) give the following results for the third moment of the steady-state inversion:

$$\left(\overline{m^3}\right)_M - \left(\overline{\overline{m}}\right)_M^3 = -N^{-1}\frac{12\bar{n}(\bar{n}+1)}{(2\bar{n}+1)^3}, \tag{6.183}$$

$$\langle (\bar{m}_k)^3 \rangle - \langle \bar{m}_k \rangle^3 = -N^{-1}\frac{4\bar{n}(\bar{n}+1)}{(2\bar{n}+1)^3}(3-2/N). \tag{6.184}$$

7. The Single-Mode Homogeneously Broadened Laser I: Preliminaries

We have now developed the bulk of the formalism we need and can turn our attention to rather more ambitious applications than the damped harmonic oscillator and the damped two-level atom. We restrict our attention in this book to the single-mode laser. In Volume 2 we consider the degenerate parametric oscillator and cavity QED. As can be judged from a quick look at Haken's book on laser theory [7.1], the first of these examples can easily fill a book on its own. We will therefore have to be rather selective in what we cover in two chapters. Our main objective is to illustrate the things we have learned in a practical application: the derivation of a master equation and associated phase-space equation of motion, the reduction of the phase-space equation to a manageable form using van Kampen's system size expansion, and the extraction of useful results from the resulting stochastic model. The topics that we address are covered in sections V and VI of Haken's book. The treatment will be similar to the one found there; although, we do not follow Haken's notation, and we will fill in the details in some of his calculations. The laser Fokker–Planck equation is derived using somewhat different methods by Louisell [7.2]. Laser theory can also be built around density matrix equations, following the approach of Scully and Lamb [7.3]. For a comparison with the phase-space method, the Scully-Lamb theory can be studied in the text by Sargent, Scully and Lamb [7.4].

We are going to look into a small, and some might say dark corner of laser physics. We want to understand the fundamental statistical character of laser light resulting from the probabilistic nature of quantum mechanics. In the real world, the noise in lasers has more to do with practical engineering concerns, such as mechanical stability, hydrodynamic stability in a dye flow, and so on. The noise we are interested in – the quantum noise – is what remains after all of this is gone.

The basic physics underlying laser action is simple. There is much beyond the basics; but this is all design and engineering – to achieve a different operating wavelength, more power, a different pulse width. The physics behind the quantum fluctuations is also simple. One would hardly think so, however, after wading through master equations, phase-space representations, and system size expansions to arrive at an answer. We will therefore begin by deriving the laser Fokker–Planck equation using rate equations and a little intuition. Hopefully, we can then appreciate the simple physics before it gets lost in the mathematical details of the full quantum-statistical theory.

7.1 Laser Theory from Einstein Rate Equations

A conventional gas discharge generates light by spontaneous emission from an excited medium. The central idea behind laser action is to extract energy by stimulated emission. Two essential conditions must be met: The excitation (pumping) of the medium must produce population inversion on the lasing transition so that stimulated emission will dominate absorption, and the energy density in the laser mode must be raised sufficiently for the stimulated emission rate to exceed the total loss rate, including the removal of energy in the output beam. The means of achieving population inversion are as diverse as the types of available lasers. There is often much physics involved in the details of an inversion mechanism. For our purpose, however, this is a practical concern; certainly central to the design of a real laser, but not important for understanding generic quantum-statistical properties of the laser field. We will model the pumping process by a simple idealized scheme involving two or three levels.

Control over the energy density of the laser mode is provided by an optical cavity. The output beam represents loss from this cavity, and therefore the laser mode must be modeled as a damped oscillator, driven by the inverted medium. Energy is injected by stimulated emission, and the rate at which energy is deposited in the field depends, nonlinearly (gain saturation), on the amount of energy already present. The dynamical paradigm is that of a *driven damped nonlinear oscillator*. Our goal is to find the oscillator equation in quantized form.

7.1.1 Rate Equations and Laser Threshold

Lamb's semiclassical laser theory derives a classical oscillator equation from Maxwell's equations driven by a nonlinear polarization [7.5, 7.6]. Our first attempt at a quantum theory we will follow an equally simple, but different approach. This approach has the advantage that it is formulated from the outset in quantum-mechanical language, and naturally includes the source of quantum fluctuations. With the help of a little intuition it will lead us to the laser Fokker–Planck equation with considerably less labor than is required by the rigorous quantum theory. We focus on the energy exchange between the laser mode and the lasing medium, and accept quantum ideas at the level of Einstein's theory: the energy quantum (photon) is the basic unit of excitation for the laser oscillator, and the exchange of quanta by spontaneous emission, stimulated emission, and absorption [7.7] describes the interaction of the laser light with the lasing atoms.

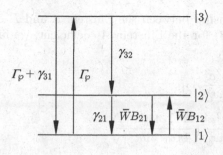

Fig. 7.1 Three-level model for the laser medium. The transition $|2\rangle \to |1\rangle$ is the laser transition.

A collection of N three-level atoms provides the simplest model for the laser gain medium capable of producing population inversion by incoherent pumping. For the full quantum-statistical theory there is advantage in the simplicity of a two-level model. But at least one more level is needed for the pumping process. We therefore start with a three-level model and see how this reduces to a two-level description later on. We adopt the level scheme illustrated in Fig. 7.1. The laser transition, $|2\rangle \to |1\rangle$, interacts on resonance with a single cavity mode with frequency ω_C; γ_{31}, γ_{32}, and γ_{21} are decay rates for the various atomic levels, $\Gamma_\wp$ is the incoherent pump rate, and $\overline{W}B_{21}$ is the stimulated emission and absorption rate for the laser transition, where $\overline{W}$ is the energy density per unit atomic linewidth in the laser mode, and $B_{21} = B_{12}$ is the Einstein B coefficient.

If N_1, N_2, and N_3 denote the atomic state populations ($N_1 + N_2 + N_3 = N$), the rate equations for the atoms are

$$\dot{N}_1 = -\Gamma_\wp N_1 + \gamma_{21}N_2 + (\Gamma_\wp + \gamma_{31})N_3 + \overline{W}B_{21}(N_2 - N_1), \quad (7.1a)$$

$$\dot{N}_2 = -\gamma_{21}N_2 + \gamma_{32}N_3 - \overline{W}B_{21}(N_2 - N_1), \quad (7.1b)$$

$$\dot{N}_3 = -(\Gamma_\wp + \gamma_{31} + \gamma_{32})N_3 + \Gamma_\wp N_1. \quad (7.1c)$$

The number of quanta in the laser mode is determined by the balance between stimulated emission and loss to the output beam:

$$\dot{n} = -2\kappa n + \overline{W}B_{21}(N_2 - N_1), \quad (7.2)$$

where 2κ is the photon loss rate from the cavity, and for the present we neglect spontaneous emission into the laser mode.

The energy density in the laser mode is given in terms of the photon number by

$$\overline{W} = n\frac{\hbar\omega_C}{V_Q\pi(\gamma_h/2)}, \quad (7.3)$$

where V_Q is the mode volume and

$$\gamma_h/2 = (\Gamma_\wp + \gamma_{21})/2 + (\gamma_p) \quad (7.4)$$

is the homogeneous width (half-width at half-maximum) for the laser transition. The term γ_p is added when phase destroying collisions are important.

To calculate B_{21} we use the relationship between the Einstein A and B coefficients [7.9]. From the result (2.33) for the Einstein A coefficient, we find ($\omega_A = \omega_C$)

$$
\begin{aligned}
B_{21} &= \left(\frac{3d^2}{d_{12}^2}\right)\frac{\pi^2 c^3}{\hbar\omega_C^3}A_{21} \\
&= \left(\frac{3d^2}{d_{12}^2}\right)\frac{\pi^2 c^3}{\hbar\omega_C^3}\left(\frac{1}{4\pi\epsilon_0}\frac{4\omega_C^3 d_{12}^2}{3\hbar c^3}\right) \\
&= \frac{2V_Q\pi}{\hbar\omega_C}g^2,
\end{aligned}
\tag{7.5}
$$

where

$$
g = \sqrt{\frac{\omega_C d^2}{2\hbar\epsilon_0 V_Q}}
\tag{7.6}
$$

is the dipole coupling constant (assumed real) between the laser field and the laser transition [Eq. (2.16)]. We are considering a single laser mode with a particular polarization $\hat{e}$; thus, the factor $3d^2/d_{12}^2$, $d = \hat{e}\cdot\mathbf{d}_{12}$, appears on the right-hand side of (7.5) to remove the dipole orientation average $\langle d^2\rangle = d_{12}^2/3$ from the expression for the Einstein A coefficient. Equations (7.3) and (7.5) now give the *stimulated emission rate into the laser mode*:

$$
\overline{W}B_{21}(N_2 - N_1) = n\frac{4g^2}{\gamma_h}(N_2 - N_1).
\tag{7.7}
$$

Note 7.1 Einstein theory only provides an approximate treatment of the interaction between light and atoms. It is normally used in situations involving broadband excitation, which corresponds to the conditions in the blackbody problem Einstein considered. But the interaction between the laser mode and laser transition is not broadband. Nonetheless, we can still use Einstein rate equations. Rate equations are valid for narrowband excitation so long as the homogeneous width is much broader than the natural width [7.8]. These are the conditions that justify our rate equation model for the laser. Under these conditions the energy density $\overline{W}$ that enters the stimulated emission rate is an average of $\overline{W}(\omega)$ for the exciting field over the atomic absorption line. It is for this reason that we find the homogeneous width $\gamma_h/2$, rather than the laser linewidth, in the denominator of (7.3). The required energy density is

$$
\begin{aligned}
\overline{W} &= \int_0^\infty d\omega\,\overline{W}(\omega)\frac{\gamma_h/2\pi}{(\gamma_h/2)^2 + (\omega - \omega_A)^2} \\
&= \int_0^\infty d\omega\left[n\frac{\hbar\omega}{V_Q}\frac{\gamma_L/2\pi}{(\gamma_L/2)^2 + (\omega - \omega_L)^2}\right]\frac{\gamma_h/2\pi}{(\gamma_h/2)^2 + (\omega - \omega_A)^2} \\
&= n\frac{\hbar\omega_L}{V_Q\pi(\gamma_h/2)}\frac{1}{1 + \Delta^2},
\end{aligned}
\tag{7.8}
$$

where $\Delta \equiv 2(\omega_A - \omega_L)/\gamma_h$ is a dimensionless detuning, and ω_L and $\gamma_L/2$ are the laser frequency and linewidth. To evaluate the integral we have assumed that $\gamma_L \ll \gamma_h$. With $\omega_L = \omega_A = \omega_C$, (7.8) reduces to (7.3).

Note 7.2 The homogeneous width γ_h [Eq. (7.4)] is obtained by summing contributions from all decay and pumping rates *out of* states $|1\rangle$ and $|2\rangle$. This general rule can be derived from a master equation treatment of the incoherent transitions. For transitions $|i\rangle \to |j\rangle$, at rates γ_{ij}, between an arbitrary set of atomic states, we obtain the master equation

$$\left(\dot{\rho}\right)_{incoh} = \sum_{ij} \frac{\gamma_{ij}}{2} \left(2|j\rangle\langle i|\rho|i\rangle\langle j| - |i\rangle\langle i|\rho - \rho|i\rangle\langle i|\right). \tag{7.9}$$

This is the obvious generalization of (2.26), which describes two incoherent transitions: $|2\rangle \to |1\rangle$, at the rate $\gamma_{21} = \gamma(\bar{n}+1)$ (with $|j\rangle\langle i| \equiv |1\rangle\langle 2| = \sigma_-$), and $|1\rangle \to |2\rangle$, at the rate $\gamma_{12} = \gamma\bar{n}$ (with $|j\rangle\langle i| \equiv |2\rangle\langle 1| = \sigma_+$). Now from (7.9), the equations of motion for the atomic coherences ρ_{kl}, $k \neq l$, acquire damping terms

$$\left(\dot{\rho}_{kl}\right)_{incoh} = -\left(\frac{1}{2}\sum_{j}(\gamma_{kj} + \gamma_{lj})\right)\rho_{kl}. \tag{7.10}$$

The homogeneous width for the $|k\rangle \to |l\rangle$ transition is therefore $\frac{1}{2}\sum_{j}(\gamma_{kj} + \gamma_{lj})$, which is the sum over all rates *out of* the states $|k\rangle$ and $|l\rangle$.

The stimulated emission term in the rate equations (7.1) and (7.2) introduces the nonlinearity that causes the laser threshold behavior. This term also couples the atomic populations to the photon number. When there are no photons in the laser mode the atomic populations settle into an equilibrium state balancing decay and incoherent pumping. Solving (7.1a)–(7.1c) with $n = 0$ and $\dot{N}_1 = \dot{N}_2 = \dot{N}_3 = 0$ gives the *unsaturated steady-state inversion*

$$N_2^0 - N_1^0 = N\frac{\gamma_{32}(\Gamma_\wp - \gamma_{21}) - \gamma_{21}(\Gamma_\wp + \gamma_{31})}{\gamma_{32}(\Gamma_\wp + \gamma_{21}) + \gamma_{21}(2\Gamma_\wp + \gamma_{31})}. \tag{7.11}$$

Once photons appear in the laser mode the atomic populations change as the laser transition begins to saturate. If $n(t)$ changes slowly compared with atomic decay and pumping rates, the atomic populations will follow adiabatically, maintaining an equilibrium with the instantaneous photon number. Assuming that these conditions hold (this is true in the vicinity of threshold) we make an *adiabatic elimination of atomic populations*; the algebra is left as an exercise:

Exercise 7.1 Solve (7.1a)–(7.1c) with $\dot{N}_1 = \dot{N}_2 = \dot{N}_3 = 0$ and show that the *saturated steady-state inversion* is given by

$$N_2 - N_1 = (N_2^0 - N_1^0)\frac{1}{1 + n/n_{\text{sat}}}, \tag{7.12}$$

where the *saturation photon number* n_{sat} is

$$n_{\text{sat}} = \frac{\gamma_h\left[\gamma_{32}(\Gamma_\wp + \gamma_{21}) + \gamma_{21}(2\Gamma_\wp + \gamma_{31})\right]}{8g^2(\gamma_{32} + \gamma_{31} + 3\Gamma_\wp/2)}. \tag{7.13}$$

Note the relationship between (7.12) and the steady-state inversion for the driven two-level atom given by (2.120b). Show that the saturation photon number for the two-level atom is $\gamma^2/8g^2$.

Substituting (7.7) and (7.12) into (7.2), we obtain the photon number rate equation

$$\dot{n} = -2\kappa n\left(1 - \frac{\wp}{1 + n/n_{\text{sat}}}\right), \tag{7.14}$$

where the *pump parameter* $\wp$ is defined by

$$\wp \equiv \frac{2g^2}{\gamma_h\kappa}(N_2^0 - N_1^0). \tag{7.15}$$

The steady-state photon number n_{ss} satisfies the quadratic equation

$$n_{\text{ss}}(1 - \wp + n_{\text{ss}}/n_{\text{sat}}) = 0. \tag{7.16}$$

Solutions to this equation show the threshold behavior illustrated in Fig. 7.2.

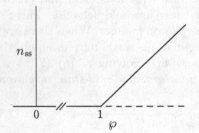

Fig. 7.2 Laser threshold behavior. The steady-state shown by the broken line is unstable.

Below Threshold – $\wp < 1$: Below threshold the photon loss rate exceeds the stimulated emission rate. Since n_{ss} must be positive (7.16) has only one acceptable solution,

$$n_< \equiv n_{\text{ss}}^< = 0. \tag{7.17}$$

From (7.12) and (7.15) the below threshold inversion is

$$\left(N_2 - N_1\right)_< = N_2^0 - N_1^0 = \frac{\gamma_h\kappa}{2g^2}\wp. \tag{7.18}$$

At Threshold $-\wp = 1$: At threshold the photon loss rate is equal to the stimulated emission rate. Equation (7.16) has the doubly degenerate solution

$$n_{\text{thr}} \equiv n_{\text{ss}}^{\text{thr}} = 0, \tag{7.19}$$

and the threshold inversion is

$$\left(N_2 - N_1\right)_{\text{thr}} = N_2^0 - N_1^0 = \frac{\gamma_h \kappa}{2g^2}. \tag{7.20}$$

Above Threshold $-\wp > 1$: Above threshold the stimulated emission rate determined by the unsaturated inversion exceeds the photon loss rate. There are two acceptable solutions for n_{ss}:

$$n_{\text{ss}} = 0, \qquad \text{(unstable)}$$
$$n_{\text{ss}} = n_{\text{sat}}(\wp - 1). \qquad \text{(stable)}$$

The $n_{\text{ss}} = 0$ state is unstable; linearizing (7.14) about this state gives $\dot{n} = 2\kappa(\wp-1)n > 0$, and therefore any small nonzero photon number is amplified. Amplification stops when the saturation term $(1 + n/n_{\text{sat}})^{-1}$ brings the gain minus loss back to zero, which happens when

$$n_> \equiv n_{\text{ss}}^> = n_{\text{sat}}(\wp - 1). \tag{7.21}$$

The saturated inversion above threshold is

$$\begin{aligned}
\left(N_2 - N_1\right)_> &= (N_2^0 - N_1^0)\frac{1}{1 + n/n_{\text{sat}}} \\
&= (N_2^0 - N_1^0)\frac{1}{\wp} \\
&= \frac{\gamma_h \kappa}{2g^2} \\
&= \left(N_2 - N_1\right)_{\text{thr}}.
\end{aligned} \tag{7.22}$$

The saturated inversion is held at the threshold value for all $\wp > 1$, a phenomenon known as *inversion clamping* (or *inversion pinning*).

7.1.2 Spontaneous Emission and Thermal Photons

Equation (7.14) describes the amplification by stimulated emission that leads to the lasing state above threshold. What, however, is to be amplified? Where does the first photon come from? According to (7.14) the laser will not turn on as the pump parameter is raised if the laser mode does not initially contain at least one photon.

The answer, of course, is that we have omitted two photon sources. We have omitted the thermal source that brings the laser mode to thermal equilibrium when the pumping is turned off. This is corrected by adding the term

$2\kappa\bar{n}$ to (7.14), as in the equation of motion for the mean photon number of the damped harmonic oscillator [Eq. (1.79)]. We have also omitted spontaneous emission. Spontaneously emitted photons are included by adding a term $\gamma_{\rm spon}N_2$ to (7.14); $\gamma_{\rm spon}$ is the spontaneous emission rate into the laser mode. With the addition of these two terms the photon number rate equation becomes

$$\dot{n} = -2\kappa n\left(1 - \frac{\wp}{1+n/n_{\rm sat}}\right) + 2\kappa\bar{n} + \gamma_{\rm spon}N_2. \tag{7.23}$$

At optical frequencies $\bar{n}$ is negligible; therefore the first photon must be supplied by spontaneous emission. At microwave frequencies $\bar{n}$ is not negligible; even at low temperatures ($T \sim 4$ K) a few thermal photons are present.

The rate $\gamma_{\rm spon}$ is not equal to the Einstein A coefficient for the $|2\rangle \to |1\rangle$ transition. $\gamma_{\rm spon}$ only accounts for spontaneous photons emitted into the laser mode, while A is the emission rate to all modes of the vacuum electromagnetic field. We can calculate $\gamma_{\rm spon}$, however, using the standard method for relating Einstein A and B coefficients. We write the stimulated emission term in (7.23) in the form (7.7), and then

$$\dot{n} = -2\kappa n\left[1 - \frac{2g^2}{\gamma_h \kappa}(N_2 - N_1)\right] + 2\kappa\bar{n} + \gamma_{\rm spon}N_2. \tag{7.24}$$

Now if the atomic populations are maintained in thermal equilibrium, with

$$\frac{N_1}{N_2} = e^{(E_2-E_1)/k_BT} = e^{\hbar\omega_C/k_BT}, \tag{7.25}$$

(7.24) must bring the cavity photons into equilibrium with the atoms. Thus, this equation must have the steady-state solution $n_{\rm ss} = \bar{n} = (e^{\hbar\omega_C/k_BT}-1)^{-1}$ [Eq. (1.52)]. This requires

$$\gamma_{\rm spon} = \frac{4g^2}{\gamma_h}\bar{n}\left(\frac{N_1}{N_2}-1\right) = \frac{4g^2}{\gamma_h}.$$

After we substitute for $\bar{n}$ and N_1/N_2, the *spontaneous emission rate into the laser mode* is

$$\gamma_{\rm spon}N_2 = \frac{4g^2}{\gamma_h}N_2. \tag{7.26}$$

With the help of (7.12) and (7.15), we write this in the form

$$\gamma_{\rm spon}N_2 = \frac{2g^2}{\gamma_h}(N_2 + N_1) + \frac{2g^2}{\gamma_h}(N_2 - N_1)$$
$$= 2\kappa\left(C\frac{N_{21}}{N_{21}^0} + \frac{1}{2}\frac{\wp}{1+n/n_{\rm sat}}\right), \tag{7.27}$$

where

$$C \equiv \frac{N_{21}^0 g^2}{\gamma_h \kappa}, \tag{7.28}$$

and

$$N_{21} \equiv N_2 + N_1 \tag{7.29}$$

is the total number of atoms distributed between the two levels of the laser transition.

Exercise 7.2 For the three-level model illustrated in Fig. 7.1, show that

$$N_{21} = N_{21}^0 \frac{1 + n/n'_{\text{sat}}}{1 + n/n_{\text{sat}}}, \tag{7.30}$$

where

$$N_{21}^0 \equiv N \frac{\gamma_{32}(\Gamma_\wp + \gamma_{21}) + \gamma_{21}(\Gamma_\wp + \gamma_{31})}{\gamma_{32}(\Gamma_\wp + \gamma_{21}) + \gamma_{21}(2\Gamma_\wp + \gamma_{31})}, \tag{7.31}$$

and

$$n'_{\text{sat}} \equiv \frac{\gamma_h [\gamma_{32}(\Gamma_\wp + \gamma_{21}) + \gamma_{21}(\Gamma_\wp + \gamma_{31})]}{8g^2(\gamma_{32} + \gamma_{31} + \Gamma_\wp)}. \tag{7.32}$$

We can now write down the complete photon number rate equation. Substituting (7.27) into (7.23) and using (7.30), we have

$$(2\kappa)^{-1}\dot{n} = -n\left(1 - \frac{\wp}{1 + n/n_{\text{sat}}}\right) + \left(\bar{n} + C\frac{1 + n/n'_{\text{sat}}}{1 + n/n_{\text{sat}}} + \frac{1}{2}\frac{\wp}{1 + n/n_{\text{sat}}}\right). \tag{7.33}$$

This equation can be simplified considerably if the laser is not operated too far above threshold. Typically n_{sat} and n'_{sat} are large numbers, and under normal operating conditions $n/n_{\text{sat}} \ll 1$, and $n/n'_{\text{sat}} \ll 1$. Then we may neglect the saturation terms in the second bracket on the right-hand side of (7.33). We will discuss how to estimate n_{sat} shortly (Sect. 7.1.4); for a He-Ne laser $n_{\text{sat}} \sim 10^8$. Of course, we cannot neglect the saturation of the stimulated emission term appearing in the first bracket. It is this gain saturation that prevents the photon number from growing without bound above threshold. However, for $n/n_{\text{sat}} \ll 1$, we only need to include the gain saturation to first order. Then the final form of the *rate equation for photon number in the laser mode*, including thermal and spontaneous emission sources, is

$$(2\kappa)^{-1}\dot{n} = -n(1 - \wp + \wp n/n_{\text{sat}}) + \bar{n} + n_{\text{spon}}. \tag{7.34}$$

We have written (7.27) as

$$\gamma_{\text{spon}}N_2 \approx \gamma_{\text{spon}}N_2^0 = 2\kappa\left(C + \tfrac{1}{2}\wp\right) = 2\kappa n_{\text{spon}}; \tag{7.35}$$

$n_{\text{spon}} = C + \tfrac{1}{2}\wp$ is the *spontaneous emission photon number* – the number of photons in the laser mode due to spontaneous emission well below threshold.

The laser can now turn on above threshold using a thermal or spontaneous photon to start the amplification process. The steady-state photon number satisfies the quadratic equation

$$n_{ss}(1 - \wp + \wp n_{ss}/n_{sat}) - (\bar{n} + n_{spon}) = 0, \tag{7.36}$$

which has a *single* (positive) solution for all values of the pump parameter:

$$n_{ss} = -\frac{1}{2\wp}n_{sat}(1 - \wp) + \frac{1}{2\wp}n_{sat}\sqrt{(1 - \wp)^2 + 4\wp(\bar{n} + n_{spon})/n_{sat}}. \tag{7.37}$$

Note 7.3 We must not confuse the pump parameter $\wp$ with the pump rate $\Gamma_\wp$. In particular, for $\Gamma_\wp = 0$, all of the atoms are in the lower state of the laser transition (if $\bar{n} = 0$), and for $\wp = 0$ ($\Gamma_\wp = \gamma_{21}$), the populations in the two states of the laser transition are equal. When $\Gamma_\wp < \gamma_{21}$ the pump parameter takes negative values and the atoms act as an absorber rather than as a gain medium. The solution (7.37) is valid for both positive and negative values of $\wp$.

In contrast to Fig. 7.2, the plot of (7.37) in Fig. 7.3 shows a smooth transition through the threshold region. The sharpness of the transition is determined by the ratio $(\bar{n} + n_{spon})/n_{sat}$. This ratio determines the range of the pump parameter over which the second term in the square root on the right-hand side of (7.37) is important. We define the *laser threshold region* by

$$|1 - \wp| < |1 - \wp|_{thr} \equiv 2\sqrt{\frac{\bar{n} + n_{spon}}{n_{sat}}}. \tag{7.38}$$

This should be a small number. If it is not, even spontaneous emission into the laser mode is sufficient to saturate the laser transition, which would negate the whole aim of building up a field by stimulated emission. The range of the threshold region in a He-Ne laser operated on the $0.63\,\mu m$ line is $|1 - \wp|_{thr} \sim 10^{-4}$ [7.10]. Since $\bar{n}$ is negligible at optical frequencies, if $n_{sat} \sim 10^8$ and $|1 - \wp|_{thr} \sim 10^{-4}$, then $n_{spon} \sim 1$. We can estimate the change in photon number over the threshold region by using (7.37) to write

$$n_{thr}^{\pm} = \pm\tfrac{1}{2}n_{sat}|1 - \wp|_{thr} + \tfrac{1}{2}n_{sat}|1 - \wp|_{thr}\sqrt{2}$$
$$= \tfrac{1}{2}n_{sat}|1 - \wp|_{thr}(\sqrt{2} \pm 1), \tag{7.39}$$

where n_{thr}^{+} and n_{thr}^{-} are the steady-state photon numbers at the upper and lower boundaries of the threshold region, respectively. Then

$$\frac{n_{thr}^{+}}{n_{thr}^{-}} = \frac{\sqrt{2} + 1}{\sqrt{2} - 1} \approx 6. \tag{7.40}$$

Thus, the photon number changes through the threshold region by approximately one order of magnitude.

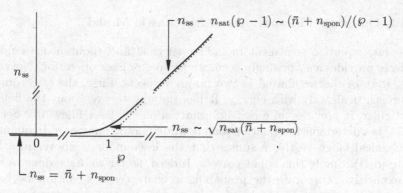

Fig. 7.3 Passage from the nonlasing to the lasing state with thermal and spontaneous emission sources included.

Equation (7.37) gives a number of results that are in rather remarkable agreement with those obtained from the full quantum statistical theory.

Below Threshold $- 1 - \wp \gg |1 - \wp|_{\mathrm{thr}}$: With the inclusion of thermal and spontaneous photons the steady-state photon number below threshold is no longer zero. Expanding the square root in (7.37) to first order, we find

$$n_< \equiv n_{\mathrm{ss}}^< = \frac{\bar{n} + n_{\mathrm{spon}}}{1 - \wp} = \frac{\bar{n} + C + \frac{1}{2}\wp}{1 - \wp}. \tag{7.41}$$

This agrees exactly with the result obtained from the full quantum-statistical theory (Sects. 8.1.3 and 8.1.4).

At Threshold $- \wp = 1$: From (7.37) the threshold photon number is given by

$$n_{\mathrm{thr}} \equiv n_{\mathrm{ss}}^{\mathrm{thr}} = \sqrt{n_{\mathrm{sat}}(\bar{n} + n_{\mathrm{spon}})} = \sqrt{n_{\mathrm{sat}}(\bar{n} + C + \frac{1}{2}\wp)}. \tag{7.42}$$

This result is larger, by the factor $\sqrt{\pi/2} \approx 1.25$, than the result obtained from the full quantum-statistical theory (Sect. 8.2.2). (For $n_{\mathrm{sat}} \sim 10^8$, $n_{\mathrm{spon}} \sim 1$, and $\bar{n}$ negligible, $n_{\mathrm{thr}} \sim 10^4$.)

Above Threshold $- 1 \gg \wp - 1 \gg |1 - \wp|_{\mathrm{thr}}$: We restrict our attention to the region not too far above threshold. The requirement $\wp - 1 \ll 1$ ensures that $n_{\mathrm{ss}}/n_{\mathrm{sat}} \ll 1$. Expanding the square root in (7.37) to first order gives the correction to the solution $n_> = n_{\mathrm{sat}}(\wp - 1)$ [Eq. (7.21)] due to the thermal and spontaneous emission sources:

$$n_> - n_{\mathrm{sat}}(\wp - 1) \equiv n_{\mathrm{ss}}^> - n_{\mathrm{sat}}(\wp - 1) = \frac{\bar{n} + n_{\mathrm{spon}}}{\wp - 1} = \frac{\bar{n} + C + \frac{1}{2}\wp}{\wp - 1}. \tag{7.43}$$

This result is a factor of four larger than that derived from the full quantum-statistical theory (Sect. 8.3.3).

7.1.3 Quantum Fluctuations: A Stochastic Model

The rate equation treatment including thermal and spontaneous emission sources provides a surprisingly accurate picture of laser operation. Nevertheless, it is, of course, limited in two major respects. First, the rate equation approach deals only with energy. It has nothing to say about the field. In particular, it provides no direct information on the laser linewidth. Second, (7.34) is a deterministic equation for a definite photon number n. In quantum-mechanical language this assumes that the laser mode is always in a Fock state $|n(t)\rangle$. Surely this is not correct. Indeed, so far we have done nothing to explicitly incorporate the probabilistic character of quantum-mechanics into our theory, and we are therefore in no position to speculate on quantum states. Equation (7.34) should be interpreted as an equation for the mean photon number.

One way to build a probabilistic theory around (7.34) is to "invent" an underlying birth-death equation for the probabilities, P_n, that there are n photons in the laser mode. Such an equation should produce (7.34), at least in some approximation, as the equation of motion for $\langle n\rangle \equiv \sum_{n=0}^{\infty} nP_n$. This approach leads to a connection with Scully-Lamb theory. Let us consider this connection briefly before turning to our main interest, a stochastic description in terms of the laser field.

There are two approaches that we might take when inventing the underlying birth-death equation. The first makes a mathematical extrapolation from (7.34) with little additional physical input. The approach is direct and simple; although, taken on its own it is somewhat unconvincing since the mathematical extrapolation is not unique. The more convincing approach builds upon well-defined physical arguments to construct a unique birth-death equation. We will look at both approaches. It as well to build some confidence in the mathematical extrapolation since this is the path we must follow to construct a stochastic laser model.

In the mathematical approach we invent the underlying birth-death equation by first writing (7.34) as an equation for the mean photon number:

$$(2\kappa)^{-1}\langle \dot{n}\rangle = -\langle n\rangle\big(1 - \wp + \wp\langle n\rangle/n_{\text{sat}}\big) + \bar{n} + n_{\text{spon}}. \tag{7.44}$$

Something must be done about the nonlinear term $\langle n\rangle^2$ appearing on the right-hand side, since such a term cannot appear in the exact equation for $\langle n\rangle$. In a statistical theory we expect the equation for the mean photon number to couple to equations for higher moments of n; therefore $\langle n\rangle^2$ must be a factorized approximation for $\langle n^2\rangle$. The correct equation for the mean photon number must be

$$(2\kappa)^{-1}\langle \dot{n}\rangle = -\langle n\rangle(1 - \wp) - \wp\langle n^2\rangle/n_{\text{sat}} + \bar{n} + n_{\text{spon}}. \tag{7.45}$$

Each term in (7.45) now suggests a corresponding term in the underlying birth-death equation. We write $F(n) \longrightarrow \langle f(n)\rangle$ to mean $\sum_{n=0}^{\infty} nF(n) = \langle f(n)\rangle$. Then the following correspondences hold:

$$\langle \dot{n} \rangle \longleftarrow \dot{P}_n, \tag{7.46a}$$

$$\langle n \rangle \longleftarrow nP_n - (n+1)P_{n+1}, \tag{7.46b}$$

$$\langle n^2 \rangle \longleftarrow n^2 P_n - (n+1)^2 P_{n+1}, \tag{7.46c}$$

$$1 \longleftarrow nP_{n-1} - (2n+1)P_n + (n+1)P_{n+1}. \tag{7.46d}$$

The choice of terms on the right-hand sides is not entirely "black magic." Familiarity with the damped harmonic oscillator leads us to (7.46b) and (7.46d); these can be deduced from the birth-death equation obtained by taking diagonal matrix elements of the master equation (1.73). The nonlinear term in (7.46c) can be found with a little trial and error. Note that each of the right-hand sides sums to zero, guaranteeing the conservation of total probability ($\sum_{n=0}^{\infty} \dot{P}_n = 0$). This removes some obvious ambiguities—the possibility of replacing (7.46b) by $\langle n \rangle \longleftarrow P_n$, for example. Some ambiguities remain, nonetheless; based solely on the mathematics, the extrapolation we have made is not unique. For example, we can replace (7.46c) by $\langle n^2 \rangle \longleftarrow (n-1)^2 P_{n-1} - n^2 P_n$, and (7.46d) by $1 \longleftarrow P_{n-1} - P_n$ (which might seem quite reasonable for the spontaneous emission term if not for the source of thermal photons). We will return to this issue shortly. Continuing for the moment with what we have, after putting the pieces together we arrive at a probabilistic model in the form of the *birth-death equation for photon number in the laser mode*:

$$(2\kappa)^{-1}\dot{P}_n = -(1-\wp)[nP_n - (n+1)P_{n+1}]$$
$$- (\wp/n_{\mathrm{sat}})[n^2 P_n - (n+1)^2 P_{n+1}]$$
$$+ (\bar{n} + n_{\mathrm{spon}})[nP_{n-1} - (2n+1)P_n + (n+1)P_{n+1}]. \tag{7.47}$$

We can now make a connection with Scully-Lamb theory. To do this we must rewrite (7.47) in a slightly different form. The details are left as an exercise:

Exercise 7.3 Verify that (7.45) follows from (7.47). Then set $N_1 = N_1^0 = 0$ in (7.15) and (7.29), and show that if $\bar{n}$ is negligible, (7.47) can be written in the form

$$\dot{P}_n = A'[nP_{n-1} - (n+1)P_n] - B'[n^2 P_n - (n+1)^2 P_{n+1}]$$
$$- C'[nP_n - (n+1)P_{n+1}], \tag{7.48}$$

with $A' \equiv 2\kappa\wp = 4g^2 N_2^0/\gamma_h$, $B' \equiv 2\kappa\wp/n_{\mathrm{sat}} = 4g^2 N_2^0/\gamma_h n_{\mathrm{sat}}$, and $C' \equiv 2\kappa$.

Equation (7.48) agrees with the first-order expansion – for $nB'/A' = n/n_{\mathrm{sat}} \ll 1$ – of the Scully-Lamb laser equation (Eq. (4) of Ref. [7.3a]). Most of the parameters in the Scully-Lamb definitions of A', B', and C' (Eq. (5) of Ref. [7.3a]) may be identified in a one-to-one correspondence with parameters in our own theory. We have $N_2^0 \leftrightarrow r_a/\gamma_a$, $\gamma_h/2 \leftrightarrow \gamma_{ab}$, $n_{\mathrm{sat}} \leftrightarrow 4g^2/\gamma_a\gamma_b$, and $2\kappa \leftrightarrow \nu/Q$. Because, however, of the different pumping model used by Scully and Lamb, there is no direct correspondence between the atomic decay rates appearing in their expression for n_{sat} ($n_{\text{sat}} = 4g^2/\gamma_a\gamma_b$) and those appearing in ours [Eq. (7.13)].

Note 7.4 As it appears in the original papers [7.3], the Scully-Lamb laser equation assumes there is no population in the lower laser level. For this reason we set $N_1 = N_1^0 = 0$ in order to make a connection with Scully-Lamb theory. These conditions can be achieved with the appropriate pumping scheme [see the discussion below (7.75)] and in our notation give $n_{\text{spon}} = \wp$; at threshold $n_{\text{spon}} = 1$. Scully-Lamb theory may readily be extended to allow for a nonzero population in the lower laser level [7.11].

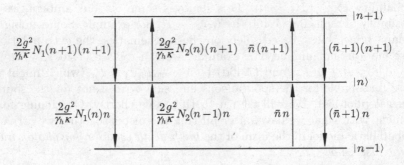

Fig. 7.4 A simple birth-death laser model. All transition rates are in units of 2κ.

We noted that there is some ambiguity in the correspondence (7.46). The ambiguity does not arise if we construct the birth-death equation starting from a physical picture of the elementary absorption and emission processes it must describe. These are the same absorption and emission processes accounted for in the Einstein rate equations formulated in Sects. 7.1.1 and 7.1.2. Now, however, they are to be expressed in terms of transition rates governing the flow of probability between states of photon number n. The various transition rates are shown in Fig. 7.4. Reading from the left they describe: the loss of photons through the cavity mirrors, the absorption of photons through the cavity mirrors, stimulated and spontaneous emission into the laser mode, and the absorption of photons by the laser medium. $N_2(n)$ and $N_1(n)$ denote the numbers of atoms occupying states $|2\rangle$ and $|1\rangle$, respectively, given the photon number n; when the atomic populations may be adiabatically eliminated we write [Eqs. (7.12) and (7.30)]

$$N_2(n) = N_2^0 \frac{1}{1 + n/n_{\text{sat}}} + \frac{1}{2}N_{21}^0 \frac{n/n'_{\text{sat}}}{1 + n/n_{\text{sat}}}, \qquad (7.49a)$$

$$N_1(n) = N_1^0 \frac{1}{1 + n/n_{\text{sat}}} + \frac{1}{2}N_{21}^0 \frac{n/n'_{\text{sat}}}{1 + n/n_{\text{sat}}}. \qquad (7.49b)$$

The birth-death equation corresponding to Fig. 7.4 takes the form

$$\dot{P}_n = -2\kappa(\bar{n}+1)[nP_n - (n+1)P_{n+1}]$$
$$+ 2\kappa\bar{n}[nP_{n-1} - (n+1)P_n]$$
$$+ \frac{4g^2}{\gamma_h}[N_2(n-1)nP_{n-1} - N_2(n)(n+1)P_n]$$
$$+ \frac{4g^2}{\gamma_h}[N_1(n+1)(n+1)P_{n+1} - N_1(n)nP_n]. \tag{7.50}$$

In order to compare this equation with (7.47) we must expand the saturation terms to lowest order in the manner described below (7.33). We first separate the saturation terms in the expressions for the atomic populations, writing (7.49a) and (7.49b) as

$$N_2(n) = N_2^0 - \frac{n/n_{\text{sat}}}{1 + n/n_{\text{sat}}}\left(N_2^0 - \tfrac{1}{2}N_{21}^0 n_{\text{sat}}/n'_{\text{sat}}\right), \tag{7.51a}$$

$$N_1(n) = N_1^0 - \frac{n/n_{\text{sat}}}{1 + n/n_{\text{sat}}}\left(N_1^0 - \tfrac{1}{2}N_{21}^0 n_{\text{sat}}/n'_{\text{sat}}\right). \tag{7.51b}$$

Then, with the help of (7.15), (7.26), (7.28), and (7.35), we write

$$\frac{2g^2}{\gamma_h\kappa}N_2^0 = n_{\text{spon}}, \qquad \frac{2g^2}{\gamma_h\kappa}N_1^0 = n_{\text{spon}} - \wp, \qquad \frac{2g^2}{\gamma_h\kappa}\frac{1}{2}N_{21}^0 = C;$$

thus,

$$\frac{2g^2}{\gamma_h\kappa}N_2(n) = n_{\text{spon}} - S(n)(n_{\text{spon}} - Cn_{\text{sat}}/n'_{\text{sat}}), \tag{7.52a}$$

$$\frac{2g^2}{\gamma_h\kappa}N_1(n) = (n_{\text{spon}} - \wp) - S(n)(n_{\text{spon}} - \wp - Cn_{\text{sat}}/n'_{\text{sat}}), \tag{7.52b}$$

where $S(n)$ is the saturation factor given by

$$S(n) \equiv \frac{n/n_{\text{sat}}}{1 + n/n_{\text{sat}}}. \tag{7.53}$$

Now the birth-death equation may be written as

$$(2\kappa)^{-1}\dot{P}_n = -(1 - \wp)[nP_n - (n+1)P_{n+1}]$$
$$- \wp[S(n)nP_n - S(n+1)(n+1)P_{n+1}]$$
$$+ (\bar{n} + n_{\text{spon}})[nP_{n-1} - (2n+1)P_n + (n+1)P_{n+1}]$$
$$+ (n_{\text{spon}} - Cn_{\text{sat}}/n'_{\text{sat}})[S(n-1)nP_{n-1}$$
$$- S(n)(2n+1)P_n + S(n+1)(n+1)P_{n+1}]. \tag{7.54}$$

The second term on the right-hand side of (7.54) is a nonlinear correction – due to gain saturation – to the first; the fourth term is a nonlinear correction to the third. We must keep the correction to the first term since this term vanishes when $\wp = 1$. For $n/n_{\text{sat}} \ll 1$, we include the correction to lowest order, setting $S(n) = n/n_{\text{sat}}$. The nonlinear correction to the third term may be neglected entirely. In this approximation (7.54) reduces to the birth-death equation (7.47).

The steady-state probabilities, P_n^{ss}, defined by setting $\dot{P}_n = 0$ in the birth-death equation, can be found by referring to Fig. 7.4. The steady state is maintained by the balancing of transitions between neighboring photon states:

$$[\bar{n} + 1 + (2g^2/\gamma_h\kappa)N_1(n)]P_n^{\text{ss}} = [\bar{n} + (2g^2/\gamma_h\kappa)N_2(n-1)]P_{n-1}^{\text{ss}}. \quad (7.55)$$

This is what is known as *detailed balance*. Equation (7.55) has the solution

$$P_n^{\text{ss}} = P_0^{\text{ss}} \prod_{k=1}^{n} \frac{\bar{n} + (2g^2/\gamma_h\kappa)N_2(k-1)}{\bar{n} + 1 + (2g^2/\gamma_h\kappa)N_1(k)}; \quad (7.56)$$

P_0^{ss} is determined by normalization. We can convert this solution into a simpler form which satisfies (7.47) using (7.52a) and (7.52b) to write

$$
\begin{aligned}
&\frac{\bar{n} + (2g^2/\gamma_h\kappa)N_2(k-1)}{\bar{n} + 1 + (2g^2/\gamma_h\kappa)N_1(k)} \\
&\quad = \frac{\bar{n} + n_{\text{spon}} - S(k-1)(n_{\text{spon}} - Cn_{\text{sat}}/n_{\text{sat}}')}{\bar{n} + n_{\text{spon}} + 1 - \wp - S(k)(n_{\text{spon}} - \wp - Cn_{\text{sat}}/n_{\text{sat}}')} \\
&\quad = \frac{\bar{n} + n_{\text{spon}} - \wp S(k) + \sum_{j=1}^{\infty} \delta_j(k)}{\bar{n} + n_{\text{spon}} + 1 - \wp},
\end{aligned} \quad (7.57)
$$

where the $\delta_j(k)$ are saturation terms that follow by making an expansion in powers of $S(k)$; for example,

$$
\begin{aligned}
\delta_1(k) &= \wp S(k) - S(k-1)(n_{\text{spon}} - Cn_{\text{sat}}/n_{\text{sat}}') \\
&\quad + S(k)\frac{\bar{n} + n_{\text{spon}}}{\bar{n} + n_{\text{spon}} + 1 - \wp}(n_{\text{spon}} - \wp - Cn_{\text{sat}}/n_{\text{sat}}') \\
&= [S(k) - S(k-1)](n_{\text{spon}} - Cn_{\text{sat}}/n_{\text{sat}}') \\
&\quad - S(k)(1 - \wp)\frac{n_{\text{spon}} - \wp - Cn_{\text{sat}}/n_{\text{sat}}'}{\bar{n} + n_{\text{spon}} + 1 - \wp}.
\end{aligned} \quad (7.58)
$$

When $k/n_{\text{sat}} \ll 1$, all of the $\delta_j(k)$ may be neglected. $\delta_1(k)$ is the dominant term, and noting that this term is negligible is sufficient. The contribution to $\delta_1(k)$ proportional to $[S(k) - S(k-1)]$ is of order $1/n_{\text{sat}}$ and clearly negligible. The contribution proportional to $S(k)(1 - \wp)$ is negligible since all saturation terms are unimportant below threshold, while at, and not too far above threshold, the $S(k)(1 - \wp)$ in $\delta_1(k)$ is very much smaller than the $\wp S(k)$ separated out explicitly in the numerator of (7.57). Thus, dropping the sum and setting $S(k) = k/n_{\text{sat}}$ in (7.57), the steady-state probabilities (7.56) become

$$P_n^{\text{ss}} = P_0^{\text{ss}} \prod_{k=1}^{n} \frac{\bar{n} + n_{\text{spon}} - \wp k/n_{\text{sat}}}{\bar{n} + n_{\text{spon}} + 1 - \wp}; \tag{7.59}$$

this is the solution for the *distribution of photon number in the laser mode* in steady state which we obtain by applying the detailed balance condition to (7.47).

Figure 7.5 illustrates the way in which the photon number distribution changes through the threshold region. Various approximate expressions for P_n^{ss} capture this evolution well. Their derivation from (7.59) is left as an exercise:

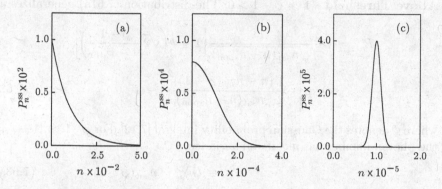

Fig. 7.5 Laser photon number distribution for $\bar{n} + n_{\text{spon}} = 1$, $n_{\text{sat}} = 10^8$ ($|1 - \wp|_{\text{thr}} \sim 10^{-4}$): (a) below threshold, $1 - \wp = 10^{-2}$; (b) at threshold, $\wp = 1$; (c) above threshold, $\wp - 1 = 10^{-3}$.

Exercise 7.4 Prove the following results from (7.59):

Below Threshold $- 1 \gg 1 - \wp \gg |1 - \wp|_{\text{thr}}$: The photon number is distributed according to the "thermal" distribution

$$P_n^{\text{ss}} = \frac{(\langle n \rangle_<)^n}{(1 + \langle n \rangle_<)^{n+1}}, \tag{7.60a}$$

where

$$\langle n \rangle_< \equiv \langle n \rangle_{\text{ss}}^< = \frac{\bar{n} + n_{\text{spon}}}{1 - \wp} = \frac{\bar{n} + C + \frac{1}{2}\wp}{1 - \wp}. \tag{7.60b}$$

Note the agreement between (7.60b) and (7.41) (also see Sects. 8.1.3 and 8.1.4).

At Threshold $- \wp = 1$: The photon number is essentially distributed continuously, with

$$P_n^{ss} = \sqrt{\frac{2}{\pi}} \frac{1}{\sqrt{n_{sat}(\bar{n} + n_{spon})}} \exp\left[-\frac{1}{2} \frac{n^2}{n_{sat}(\bar{n} + n_{spon})}\right], \qquad (7.61a)$$

with mean threshold photon number

$$\langle n \rangle_{thr} \equiv \langle n \rangle_{ss}^{thr} = \sqrt{\frac{2}{\pi}} \sqrt{n_{sat}(\bar{n} + n_{spon})}$$

$$= \sqrt{\frac{2}{\pi}} \sqrt{n_{sat}(\bar{n} + C + \tfrac{1}{2})}. \qquad (7.61b)$$

Note the agreement within a factor of $\sqrt{2/\pi}$ between (7.61b) and (7.42) (also see Sect. 8.2.2).

Above Threshold $-1 \gg \wp - 1 > 0$: The distribution (7.61a) generalizes as

$$P_n^{ss} = \sqrt{\frac{2}{\pi}} \frac{1}{\sqrt{n_{sat}(\bar{n} + n_{spon})}} \left[1 + \Phi\left(\sqrt{2}\frac{\wp - 1}{|1 - \wp|_{thr}}\right)\right]^{-1}$$

$$\times \exp\left[-\frac{1}{2} \frac{(n - n_{sat}(\wp - 1))^2}{n_{sat}(\bar{n} + n_{spon})}\right], \qquad (7.62)$$

where Φ denotes the Gaussian probability integral [7.12]. For $\wp - 1 \gg |1 - \wp|_{thr}$ the photon number has mean and variance

$$\langle n \rangle_> \equiv \langle n \rangle_{ss}^> = n_{sat}(\wp - 1), \qquad (7.63a)$$

$$\langle n^2 \rangle_> - (\langle n \rangle_>)^2 \equiv \langle n^2 \rangle_{ss}^> - (\langle n \rangle_{ss}^>)^2 = n_{sat}(\bar{n} + n_{spon})$$

$$= n_{sat}(\bar{n} + C + \tfrac{1}{2}\wp). \qquad (7.63b)$$

Note the agreement (to dominant order) between (7.63a) and (7.43). Also, (7.63a) and (7.63b) indicate that the photon number distribution above threshold is significantly broader than a Poisson distribution (also see Sect. 8.3.3).

Equation (7.47) provides a simple probabilistic theory. But it is still limited to a description in terms of energy states. A second way to build a probabilistic theory from the rate equation (7.44) is to "invent" an underlying stochastic model for the laser field amplitude $\tilde{\alpha} = e^{i\omega_c t}\alpha$. Such a model can answer questions about the field – questions like: what is the laser linewidth? A stochastic model provides a connection with the phase-space version of the full quantum-statistical theory which we will discuss shortly.

To "invent" a stochastic model we must construct a Fokker–Planck equation for the probability density $\tilde{P}(\tilde{\alpha}, \tilde{\alpha}^*)$ that reproduces (7.44), with

$$\langle n \rangle \equiv \langle |\tilde{\alpha}|^2 \rangle \equiv \int d^2\tilde{\alpha} \, |\tilde{\alpha}|^2 \tilde{P}(\tilde{\alpha}, \tilde{\alpha}^*); \qquad (7.64)$$

more precisely, our Fokker–Planck equation must give the mean value equation

$$(2\kappa)^{-1}\langle|\dot{\tilde{\alpha}}|^2\rangle = -\langle|\tilde{\alpha}|^2\rangle(1 - \wp) - \wp\langle|\tilde{\alpha}|^4\rangle/n_{\text{sat}} + \bar{n} + n_{\text{spon}}, \qquad (7.65)$$

which reduces to (7.44) after the factorization $\langle|\tilde{\alpha}|^4\rangle \approx (\langle|\tilde{\alpha}|^2\rangle)^2$. It is not difficult to come up with the appropriate Fokker–Planck equation. The first two terms on the right-hand side of (7.65) describe cavity loss and saturable stimulated emission gain. These are *coherent* processes. They remove and add energy by coherently decreasing and increasing the field amplitude, and should be contributed by deterministic (drift) terms in the Fokker–Planck equation. In contrast, the third term on the right-hand side of (7.65) describes the *incoherent* thermal and spontaneous emission energy sources. This term should be contributed by a noise (diffusion) term in the Fokker–Planck equation. Writing $F(\tilde{\alpha}, \tilde{\alpha}^*) \longrightarrow \langle f(\tilde{\alpha}, \tilde{\alpha}^*)\rangle$ to mean $\int d^2\tilde{\alpha}|\tilde{\alpha}|^2 F(\tilde{\alpha}, \tilde{\alpha}^*) = \langle f(\tilde{\alpha}, \tilde{\alpha}^*)\rangle$, the term by term correspondence is

$$\langle|\dot{\tilde{\alpha}}|^2\rangle \longleftarrow \dot{\tilde{P}}(\tilde{\alpha}, \tilde{\alpha}^*), \qquad (7.66a)$$

$$\langle|\tilde{\alpha}|^2\rangle \longleftarrow -\frac{1}{2}\left(\frac{\partial}{\partial\tilde{\alpha}}\tilde{\alpha} + \frac{\partial}{\partial\tilde{\alpha}^*}\tilde{\alpha}^*\right)\tilde{P}(\tilde{\alpha}, \tilde{\alpha}^*), \qquad (7.66b)$$

$$\langle|\tilde{\alpha}|^4\rangle \longleftarrow -\frac{1}{2}\left(\frac{\partial}{\partial\tilde{\alpha}}\tilde{\alpha} + \frac{\partial}{\partial\tilde{\alpha}^*}\tilde{\alpha}^*\right)|\tilde{\alpha}|^2\tilde{P}(\tilde{\alpha}, \tilde{\alpha}^*), \qquad (7.66c)$$

$$1 \longleftarrow \frac{\partial^2}{\partial\tilde{\alpha}\partial\tilde{\alpha}^*}\tilde{P}(\tilde{\alpha}, \tilde{\alpha}^*). \qquad (7.66d)$$

Again, familiarity with the damped harmonic oscillator helps us to select the right-hand sides; (7.66b) and (7.66d) are deduced from the terms on the right-hand side of the harmonic oscillator Fokker–Planck equation (3.52).

Exercise 7.5 Show that the mean value equation (7.65) follows from the Fokker–Planck equation

$$\kappa^{-1}\frac{\partial\tilde{P}}{\partial t} = \left[\left(\frac{\partial}{\partial\tilde{\alpha}}\tilde{\alpha} + \frac{\partial}{\partial\tilde{\alpha}^*}\tilde{\alpha}^*\right)(1 - \wp + \wp|\tilde{\alpha}|^2/n_{\text{sat}})\right.$$
$$\left. + 2(\bar{n} + n_{\text{spon}})\frac{\partial^2}{\partial\tilde{\alpha}\partial\tilde{\alpha}^*}\right]\tilde{P}, \qquad (7.67a)$$

with corresponding stochastic differential equation

$$d\tilde{\alpha} = -\tilde{\alpha}\left(1 - \wp + \wp|\tilde{\alpha}|^2/n_{\text{sat}}\right)(\kappa dt) + \sqrt{\kappa(\bar{n} + n_{\text{spon}})}(dW_1 + idW_2); \quad (7.67b)$$

W_1 and W_2 are independent Wiener processes.

Equation (7.67a) is the *laser Fokker–Planck equation*. From this equation results (7.60)–(7.63) can be recovered, together with new results, such as the linewidth of the laser field. In fact, from this equation we can derive all of the results we will obtain from the full quantum-statistical theory. The role of the full theory is really only to show how (7.67a) can be rigorously derived; and to show in some detail what approximations are needed to arrive at this stochastic model. It is important to realize that what we have here is essentially a *classical* stochastic description for the laser field. Although, one might choose to couch the theory in quantum-mechanical language, the fluctuations of the laser field are the fluctuations of a classical stochastic complex field amplitude. Generally, quantum probability cannot be accommodated within a classical stochastic description, and therefore some "reduction" of the quantum mechanics must take place as we pass from a fully quantum-mechanical microscopic formulation to the macroscopic laser Fokker–Planck equation. We will see how this reduction is made in the next few sections. This is not to say that we cannot label the noise on the laser field, or part of it, as *quantum* noise. Quantum mechanics leaves its mark in the diffusion coefficient n_{spon}. The size of this coefficient, specifically its nonzero value, is a statement from quantum mechanics. This statement is found in the relationship between (7.7) and (7.26): *stimulated emission gain is necessarily accompanied by spontaneous emission noise*. This is something that Einstein theory tells us. But Einstein theory is *ad hoc* – it is not integrated with the mathematical formulations of mechanics and electromagnetism. Our job now is to derive the stochastic model defined by (7.67) from the theory of quantum electrodynamics.

7.1.4 Two-Level Model and Laser Parameters

First let us look again at the model for the laser medium. It is cumbersome to carry a detailed description of the pumping process through into the quantum-statistical theory. Moreover, the model shown in Fig. 7.1 is, on the one hand, an idealization of any real pumping process, and on the other, not the only reasonable idealized model that might be chosen (Ref. [7.13] describes a commonly used four-level model). We therefore build our microscopic theory around a two-level model of the laser gain medium, excluding all additional states needed to achieve population inversion. This model includes everything essential to our purpose. The model is illustrated in Fig. 7.6. It may be derived as a limiting case of our three-level model by taking $\gamma_{32} \gg \Gamma_\wp + \gamma_{31}, \gamma_{21}$. In this limit, negligible population resides in state $|3\rangle$, and the pumping rate to the upper level of the laser transition is determined by $\Gamma_\wp$. We set

$$\gamma_\uparrow \equiv \Gamma_\wp, \qquad \gamma_\downarrow \equiv \gamma_{21}, \tag{7.68}$$

and (7.4), (7.11), (7.13), (7.30) and (7.31), and (7.32), now read

$$\gamma_h = \gamma_\uparrow + \gamma_\downarrow \, (+\gamma_p), \tag{7.69a}$$

$$N_2^0 - N_1^0 = N\frac{\gamma_\uparrow - \gamma_\downarrow}{\gamma_\uparrow + \gamma_\downarrow}, \tag{7.69b}$$

$$n_{\text{sat}} = \frac{\gamma_h(\gamma_\uparrow + \gamma_\downarrow)}{8g^2}, \tag{7.69c}$$

$$N_{21} = N_{21}^0 = N, \tag{7.69d}$$

$$n'_{\text{sat}} = n_{\text{sat}} = \frac{\gamma_h(\gamma_\uparrow + \gamma_\downarrow)}{8g^2}. \tag{7.69e}$$

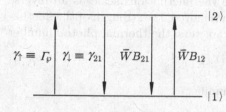

Fig. 7.6 Two-level model for the laser medium. The three-level model of Fig. 7.1 reduces to this model when $\gamma_{32} \gg \Gamma_\wp + \gamma_{31}, \gamma_{21}$.

It is useful to summarize the parameters controlling the behavior of the stochastic model (7.67). In developing the full quantum-statistical theory, we will use van Kampen's system size expansion (Sect. 5.1.3) to provide a description of fluctuations around a deterministic macroscopic state – a rigorous version of the addition of thermal and spontaneous emission noise to the noiseless laser behavior illustrated in Fig. 7.2. As we look at each parameter, we might take this opportunity to identify the role it plays in the context of the system size expansion. To this end, we first introduce a scaled time $\bar{t}$ and scaled field amplitude $\tilde{\bar{\alpha}}$, defining

$$\bar{t} \equiv \kappa t, \qquad \tilde{\bar{\alpha}} \equiv n_{\text{sat}}^{-1/2}\tilde{\alpha}. \tag{7.70}$$

Then (7.67a) and (7.67b) become

$$\frac{\partial \tilde{\bar{P}}}{\partial \bar{t}} = \left[\left(\frac{\partial}{\partial \tilde{\bar{\alpha}}}\tilde{\bar{\alpha}} + \frac{\partial}{\partial \tilde{\bar{\alpha}}^*}\tilde{\bar{\alpha}}^*\right)(1 - \wp + \wp|\tilde{\bar{\alpha}}|^2) + 2n_{\text{sat}}^{-1}(\bar{n} + n_{\text{spon}})\frac{\partial^2}{\partial \tilde{\bar{\alpha}}\partial \tilde{\bar{\alpha}}^*}\right]\tilde{\bar{P}}, \tag{7.71a}$$

and

$$d\tilde{\bar{\alpha}} = -\tilde{\bar{\alpha}}(1 - \wp + \wp|\tilde{\bar{\alpha}}|^2)d\bar{t} + \sqrt{(\bar{n} + n_{\text{spon}})/n_{\text{sat}}}\,(d\bar{W}_1 + id\bar{W}_2); \tag{7.71b}$$

the Wiener processes $\bar{W}_1$ and $\bar{W}_2$ have variances $\bar{t}$ (the processes W_1 and W_2 have variances t).

Equations (7.71) depend on four parameters: $\wp$, $\bar{n}$, n_{spon}, and n_{sat}. The saturation photon number plays the role of the parameter Ω characterizing the system size in Sect. 5.1.3. The macroscopic limit – the limit of zero fluctuations – is reached for $n_{\text{sat}} \to \infty$. In this limit the behavior of the laser is controlled by a single intensive parameter, the pump parameter

$$\wp \equiv \frac{2g^2}{\gamma_h \kappa}(N_2^0 - N_1^0) = 2C\frac{\gamma_\uparrow - \gamma_\downarrow}{\gamma_\uparrow + \gamma_\downarrow}, \tag{7.72}$$

with

$$C \equiv \frac{Ng^2}{\gamma_h \kappa}. \tag{7.73}$$

In a system of finite size there are fluctuations about the macroscopic state. We will see shortly that the rigorous system size expansion needed to describe these fluctuations is different for operation below threshold, at threshold, and above threshold. However, in each case the fluctuations scale as an inverse power of n_{sat}. After this scaling has been removed, the strength of the fluctuations is determined by two intensive parameters: the thermal photon number

$$\bar{n} = \left(e^{\hbar\omega_C/k_B T} - 1\right)^{-1}, \tag{7.74}$$

and the spontaneous emission photon number

$$n_{\text{spon}} = C + \tfrac{1}{2}\wp = \frac{2Ng^2}{\gamma_h \kappa}\frac{\gamma_\uparrow}{\gamma_\uparrow + \gamma_\downarrow}. \tag{7.75}$$

The two-level model for the laser medium suppresses the details of the pumping mechanism. It does, however, allow the pump parameter to be changed in two different ways. In the first, $\gamma_\uparrow$ provides the control. This applies when the lower level of the lasing transition is populated at thermal equilibrium, with $N_1^0 = N$ for $\gamma_\uparrow = 0$. When $\gamma_\uparrow = \gamma_\downarrow$, the populations N_1^0 and N_2^0 are equalized, and gain is available once $\gamma_\uparrow$ exceeds $\gamma_\downarrow$. Generally the laser operates with $\gamma_\uparrow - \gamma_\downarrow \ll \gamma_\uparrow + \gamma_\downarrow$. Then, from (7.72), $2C$ must be large so that the small difference between $\gamma_\uparrow$ and $\gamma_\downarrow$ translates into sufficient gain to reach threshold. Equation (7.75) gives $n_{\text{spon}} = C + \tfrac{1}{2}\wp \gg 1$.

The second mode of operation applies in situations where the lower laser level is an excited state that is not normally populated at thermal equilibrium. If decay out of this state is fast enough, it may be assumed that this level remains unpopulated while the laser is in operation. This requires that we take $\gamma_\uparrow \gg \gamma_\downarrow$ in the two-level model. The pumping excites atoms from an energy state below the lower level of the laser transition into the upper laser level. Thus, control over the pump parameter $\wp$ is provided by changing $N_2^0 = N$ (consequently changing C). A four-level scheme is needed to give the simplest complete description of this process [7.13]. Since $N_1^0 = 0$ ($\gamma_\uparrow \gg \gamma_\downarrow$), we find $n_{\text{spon}} = \wp$; at threshold, $n_{\text{spon}} = 1$. This is the situation in Scully-Lamb theory [7.3].

Since the saturation photon number determines the importance of fluctuations, let us spend a little time to give it special consideration. If we use (2.33) to express d_{12}^2 in terms of the radiative decay rate $\gamma_\downarrow$, (7.69c) and (7.6) give

$$n_{\text{sat}} = \frac{\gamma_h(\gamma_\uparrow + \gamma_\downarrow)}{8} \frac{2\hbar\epsilon_0 V_Q}{\omega_C} \left(\frac{d_{12}}{\hat{e}\cdot\boldsymbol{d}_{12}}\right)^2 \left(\frac{1}{4\pi\epsilon_0} \frac{4\omega_C^3}{3\hbar c^3} \frac{1}{\gamma_\downarrow}\right)$$

$$= \frac{1}{2}\left(\frac{\pi w_0^2}{\sigma_{12}}\right)\left(\frac{L/c}{1/\gamma_\downarrow}\right)\frac{\gamma_h(\gamma_\uparrow + \gamma_\downarrow)}{\gamma_\downarrow^2}, \tag{7.76}$$

where we have written $V_Q = \pi w_0^2 L$, where πw_0^2 is the cross-sectional area of the laser mode and L is the cavity length;

$$\sigma_{12} = \frac{(\hat{e}\cdot\boldsymbol{d}_{12})^2}{d_{12}^2/3}(\lambda^2/2\pi) \tag{7.77}$$

is the weak-field absorption cross-section for the laser transition.

Note 7.5 The cross-section σ_{12} can be calculated from the radiated power in weak-field resonance fluorescence. Using (2.127), (2.120b), (2.112), and (2.91), we have ($Y^2 \ll 1$)

$$P = \gamma\hbar\omega_A\langle 2|\rho_{\text{ss}}|2\rangle = \gamma\hbar\omega_A \frac{1}{2}Y^2 = \frac{4\omega_A d^2}{\hbar\gamma}E^2,$$

where E is the electric field amplitude for the coherent excitation. The power density in W/m^2 for the exciting field is $2\epsilon_0 c E^2$; thus, σ_{12} is obtained from the statement of conservation of energy, $2\epsilon_0 c E^2 \sigma_{12} = P$:

$$\sigma_{12} = \frac{2\omega_A d^2}{\epsilon_0 c\hbar\gamma} = \frac{2\omega_A d^2}{\epsilon_0 c\hbar}\left(4\pi\epsilon_0 \frac{3\hbar c^3}{4\omega_A^3 d_{12}^2}\right) = \frac{(\hat{e}\cdot\boldsymbol{d}_{12})^2}{d_{12}^2/3}(\lambda^2/2\pi).$$

The physical content of (7.76) can be appreciated by setting $\gamma_\uparrow = 0$, so that the laser medium acts as a two-level absorber. The ratio $(1/\gamma_\downarrow) \div (L/c)$ is the number of times a photon inside the cavity revisits each atom during the atomic lifetime. (Photons remain in the cavity for many atomic lifetimes when $\kappa \ll \gamma_\downarrow$.) Equation (7.76) can now be written as

$$n/n_{\text{sat}} = 2n\left(\frac{\sigma_{12}}{\pi w_0^2}\right)\frac{1/\gamma_\downarrow}{L/c}$$

$$= 2 \times \left(\begin{array}{c}\text{number of photons presented to each atom within}\\ \text{an absorption cross−section per atomic lifetime}\end{array}\right). \tag{7.78}$$

Thus, the condition $n = n_{\text{sat}}$ means that half a photon is presented to the atom within an absorption cross-section per atomic lifetime. [Note that the greatest rate at which a two-level atom can scatter photons is *half* a photon per lifetime, corresponding to full saturation ($n/n_{\text{sat}} \gg 1$), for which the probability for the atom to be in its excited state is one half.]

The saturation photon number must be large if the fluctuations are to be small. In this regard there is an important qualitative difference between the contributions to the diffusion constant in (7.71) coming from $\bar{n}$ and n_{spon}. The thermal photon number may be extremely small. At optical frequencies $\bar{n} \sim 10^{-33}$ at $T = 300K$. Thus, thermal fluctuations are not intrinsic to the quantum dynamics of the laser and may be negligible quite independent of the size of n_{sat}. On the other hand, the fluctuations represented by n_{spon} *are* intrinsic to the quantized dynamics; they cannot be reduced (beyond a certain limit) other than by increasing n_{sat}. To see this, note that at threshold $n_{\text{spon}} = C + \frac{1}{2}$, and, from (7.72), C has a minimum value $C = \frac{1}{2}$. Therefore, if the laser is to lase, it is necessary that $n_{\text{spon}} \geq 1$; the minimum $n_{\text{spon}} = 1$ occurs when the laser operates with $N_1^0 = 0$.

The laser is therefore intrinsically aware of its "size," as measured by n_{sat}. This is because the laser is a *nonlinear* device whose energy is *quantized* in "lumps" of finite size. There is an interplay between nonlinearity and the size of the lumps. The nonlinearity derives from the saturation of the laser transition. The degree of saturation depends on the energy density per unit atomic linewidth at the site of each laser atom. A characteristic energy density $\overline{W}_{\text{sat}}$ is needed to "turn on" the nonlinearity. The number of photons (lumps) that must be present in the cavity to provide this energy density scales proportional to the mode volume. The same energy density may be achieved in a cavity with a large mode volume or a cavity with a small mode volume. The smaller mode volume provides the same $\overline{W}_{\text{sat}}$ with fewer photons. In a small volume the fluctuations associated with the coming and going of the lumps are large relative to the mean energy needed to turn on the nonlinearity.

The role of n_{sat} as a measure of system size is displayed explicitly in (7.76). This expression is proportional to the mode volume $V_Q = \pi w_0^2 L$. Choosing $w_0 = 1\,\text{mm}$, $L = 15\,\text{cm}$, $\lambda = 0.6\mu\text{m}$, $\gamma_\perp \sim 10^7\,\text{s}^{-1}$, and $\gamma_\uparrow \sim \gamma_h \approx 3 \times 10^8\,\text{s}^{-1}$ – numbers appropriate for a He-Ne laser – we obtain $n_{\text{sat}} \sim 10^8$.

7.2 Phase-Space Formulation in the Normal-Ordered Representation

7.2.1 Model and Hamiltonian

The microscopic model for the single-mode homogeneously broadened laser is built on our descriptions of the damped harmonic oscillator (Sect. 1.4) and the damped two-level medium (Sect. 6.3). The model is shown schematically in Fig. 7.7. We use a ring cavity rather than a standing-wave cavity to avoid the awkward problem of spatial effects (Note 6.1).

Following the approach of Chap. 1, we formulate the mathematical description in terms of a system S interacting with various reservoirs R. Our main interest is in the system S, which is comprised of a single ring-cavity

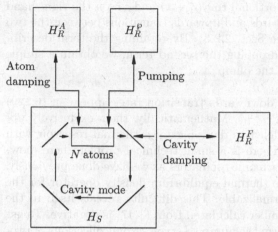

Fig. 7.7 Schematic diagram of the microscopic laser model.

mode for the laser field, coupled to N identical two-level atoms representing the laser medium. In the dipole and rotating-wave approximations, the Hamiltonian for this coupled atom-field system is given by

$$H_S = H_A + H_F + H_{AF}$$
$$\equiv \tfrac{1}{2}\hbar\omega_C J_z + \hbar\omega_C a^\dagger a + i\hbar g(a^\dagger J_- - aJ_+), \qquad (7.79)$$

where J_-, J_+, and J_z are the collective atomic operators defined by (6.44); the phases ϕ_j in the definition of $J_\pm$ are $\phi_j = \boldsymbol{k}_C \cdot \boldsymbol{r}_j$, where $\boldsymbol{r}_j$ is the position of the jth atom; for simplicity we assume exact resonance between the atoms and the field.

Hamiltonian (7.79) has been studied extensively. Jaynes and Cummings where the first to study its single-atom version [7.14] and the many-atom Hamiltonian was first studied by Tavis and Cummings [7.15]. In our laser model H_S is only part of the story. The laser is intrinsically a *dissipative* system, and we must add various interactions between S and the environment to account for the flows of energy into and out of the system. First, the ring-cavity mode has three perfectly reflecting mirrors and one partially reflecting output coupler. We model the loss of energy through the partially reflecting mirror by a weak interaction with a reservoir of electromagnetic field modes outside the cavity (see Sect. 7.2.5), and then use the formalism for the damped harmonic oscillator from Chap. 1. Second, each atom loses energy by spontaneous emission (fluorescence) out the sides of the cavity. This energy loss is described by coupling each atom to the many modes of the radiation field, as in Sect. 6.3.1. Third, we need a model for the pumping that injects energy into the system. In Sect. 7.1.4 we saw how the pumping mechanism might be reduced to a simple rate $\gamma_\uparrow$ for the transfer of population from the lower to the upper state of the laser transition. From (6.129) we see that this may also be modeled by the reservoir interaction used to treat atomic decay. The terms proportional to $\gamma(\bar{n}+1)$ and $\gamma\bar{n}$ on the right-hand side of (6.129) describe downwards and upwards transitions between the two atomic levels, respectively (see Sect. 2.2.3). By retaining the term describing upwards transitions, and dropping the second term, we obtain a simple quantum statistical model for the pump.

Note 7.6 The upwards and downwards transition rates appear in (6.129) in the ratio $\bar{n}/(\bar{n}+1) = e^{-\hbar\omega_A/k_B T}$. Mathematically, then, exclusively upwards transitions can be modeled as damping by a thermal reservoir with a low *negative* temperature. There is a small technical problem here, however, if we use a reservoir of harmonic oscillators as we have done previously. For negative temperatures the thermal equilibrium density operator for the harmonic oscillator is not normalizable. This difficulty reveals itself in the fact that the mean photon number calculated from (1.47) is negative. To get around this problem we can form the pump reservoirs from collections of two-level systems. The master equation for the damped two-level atom may be derived following an almost identical calculation to that of Sect. 2.2.1 using a reservoir of two-level systems rather than harmonic oscillators. In place of (2.15b) and (2.15c) we have

$$H_R \equiv \tfrac{1}{2}\sum_k \hbar\omega_k \Sigma_{kz}, \tag{7.80a}$$

$$H_{SR} \equiv \sum_k \hbar(\kappa_k \sigma_- \Sigma_{k+} + \kappa_k^* \sigma_+ \Sigma_{k-}), \tag{7.80b}$$

where Σ_{k-}, Σ_{k+}, and Σ_{kz} are pseudo-spin operators for the kth two-level system of the reservoir, with frequency ω_k and coupling constant κ_k. The thermal equilibrium density operator for the reservoir is

$$R_0 = \prod_k \frac{(|1\rangle\langle 1|)_k + e^{-\hbar\omega_k/k_B T}(|2\rangle\langle 2|)_k}{e^{-\hbar\omega_k/k_B T}+1}. \tag{7.81}$$

The only changes to the master equation (2.26) are the replacements

$$\bar{n} \longrightarrow \mathrm{tr}_R\big(R_0 \Sigma_{k+}\Sigma_{k-}\big)_{\omega_k=\omega_A} = \frac{e^{-\hbar\omega_A/k_B T}}{e^{-\hbar\omega_A/k_B T}+1}, \tag{7.82a}$$

$$\bar{n}+1 \longrightarrow \mathrm{tr}_R\big(R_0 \Sigma_{k-}\Sigma_{k+}\big)_{\omega_k=\omega_A} = \frac{1}{e^{-\hbar\omega_A/k_B T}+1}, \tag{7.82b}$$

and, of course, γ is not to be read as the Einstein A coefficient, but as $\Gamma_\wp$. For low negative temperatures, (7.82a) and (7.82b) approach the limits $\bar{n} \to 1$ and $\bar{n}+1 \to 0$; in (2.26), only the term proportional to $\bar{n}$, describing upwards transitions, survives.

To complete our microscopic laser model we add the environmental interactions, coupling the system S described by the Hamiltonian (7.79) to a reservoir R with Hamiltonian

$$H_R = H_R^F + H_R^A + H_R^\wp, \tag{7.83}$$

where

$$H_R^F \equiv \sum_k \hbar\omega_k r_k^\dagger r_k, \tag{7.84a}$$

$$H_R^A \equiv \sum_{k,\lambda} \hbar\omega_k r_{k,\lambda}^\dagger r_{k,\lambda}, \tag{7.84b}$$

$$H_R^\wp \equiv \sum_{j=1}^N \left(\sum_k \tfrac{1}{2}\hbar\omega_k^j \Sigma_{kz}^j \right). \tag{7.84c}$$

H_R^F is the Hamiltonian for the electromagnetic field modes that couple to the laser mode through the output mirror; H_R^A is the Hamiltonian for the free-space electromagnetic field that causes the radiative decay of the atoms; and $H_R^\wp$ is the Hamiltonian for N independent two-level pumping reservoirs. The electromagnetic field modes in (7.84a) and (7.84b) are in thermal equilibrium at the ambient temperature T, and each two-level system in (7.84c) is described by an equilibrium density operator of the form (7.81), at some low *negative* effective temperature $T_\wp$. The interaction between S and R is described by the Hamiltonian

$$H_{SR} = H_{SR}^F + H_{SR}^A + H_{SR}^\wp, \tag{7.85}$$

where

$$H_{SR}^F \equiv \hbar(a\Gamma^\dagger + a^\dagger\Gamma), \tag{7.86a}$$

$$H_{SR}^A \equiv \sum_{j=1}^N \hbar(\sigma_{j-}\Gamma_j^\dagger + \sigma_{j+}\Gamma_j), \tag{7.86b}$$

$$H_{SR}^\wp \equiv \sum_{j=1}^N \hbar(\sigma_{j-}\Gamma_{j\wp}^\dagger + \sigma_{j+}\Gamma_{j\wp}). \tag{7.86c}$$

$\Gamma^\dagger$ and Γ are defined in (1.39b); $\Gamma_j^\dagger$ and Γ_j are defined in (6.125b); and

$$\Gamma_{j\wp}^\dagger \equiv \sum_k \kappa_k^{j*}\Sigma_{k+}^j, \qquad \Gamma_{j\wp} \equiv \sum_k \kappa_k^j \Sigma_{k-}^j. \tag{7.87}$$

Equations (7.79) and (7.83)–(7.86) define the complete laser Hamiltonian

$$H = H_S + H_R + H_{SR}. \tag{7.88}$$

7.2.2 Master Equation for the Single-Mode Homogeneously Broadened Laser

The laser master equation is now derived using the formalism of Sects. 1.3.1–1.3.3. We may pass directly to the general non-Markovian equation (1.34), with $4N + 2$ system operators

$$\{s_i\} \equiv \left(a, a^\dagger; \sigma_{1-}, \sigma_{1+}, \ldots, \sigma_{N-}, \sigma_{N+}; \sigma_{1-}, \sigma_{1+}, \ldots, \sigma_{N-}, \sigma_{N+}\right), \quad (7.89a)$$

which couple to $4N + 2$ reservoir operators

$$\{\Gamma_i\} \equiv \left(\Gamma^\dagger, \Gamma; \Gamma_1^\dagger, \Gamma_1, \ldots, \Gamma_N^\dagger, \Gamma_N; \Gamma_{1\wp}^\dagger, \Gamma_{1\wp}, \ldots, \Gamma_{N\wp}^\dagger, \Gamma_{N\wp}\right). \quad (7.89b)$$

Operators in the interaction picture are defined by

$$\{\tilde{s}_i\} \equiv e^{(i/\hbar)H_S t}\{s_i\}e^{-(i/\hbar)H_S t}, \quad (7.90a)$$

$$\{\tilde{\Gamma}_i\} \equiv e^{(i/\hbar)H_R t}\{\Gamma_i\}e^{-(i/\hbar)H_R t}. \quad (7.90b)$$

Two simplifications enable us to write down the master equation directly from results we have already derived. First, we note that the reservoir operators $\{\tilde{\Gamma}_i\}$ are all statistically independent, except for the pairs $(\Gamma^\dagger, \Gamma)$, $(\Gamma_j^\dagger, \Gamma_j)$, and $(\Gamma_{j\wp}^\dagger, \Gamma_{j\wp})$. This follows because the reservoirs for damping the laser mode, damping the atoms, and performing the pumping, are statistically independent; and although the operators $(\Gamma_j^\dagger, \Gamma_j)$ are derived from the same reservoir for all j, the spatial distribution of the atoms ensures the independence of these operators for different j [see the discussion below (6.128)]. We may now write

$$\dot{\rho} = \left(\dot{\rho}\right)_F + \left(\dot{\rho}\right)_A + \left(\dot{\rho}\right)_\wp, \quad (7.91)$$

where each of the three terms on the right-hand side has the form (1.34), with

$$\left(\tilde{s}_1, \tilde{s}_2\right)_F \equiv \left(\tilde{a}, \tilde{a}^\dagger\right), \quad (7.92a)$$

$$\left(\tilde{\Gamma}_1, \tilde{\Gamma}_2\right)_F \equiv \left(\tilde{\Gamma}^\dagger, \tilde{\Gamma}\right), \quad (7.92b)$$

$$\left(\tilde{s}_1, \ldots, \tilde{s}_{2N}\right)_A \equiv \left(\tilde{\sigma}_{1-}, \tilde{\sigma}_{1+}, \ldots, \tilde{\sigma}_{N-}, \tilde{\sigma}_{N+}\right), \quad (7.92c)$$

$$\left(\tilde{\Gamma}_1, \ldots, \tilde{\Gamma}_{2N}\right)_A \equiv \left(\tilde{\Gamma}_1^\dagger, \tilde{\Gamma}_1, \ldots, \tilde{\Gamma}_N^\dagger, \tilde{\Gamma}_N\right), \quad (7.92d)$$

$$\left(\tilde{s}_1, \ldots, \tilde{s}_{2N}\right)_\wp \equiv \left(\tilde{\sigma}_{1-}, \tilde{\sigma}_{1+}, \ldots, \tilde{\sigma}_{N-}, \tilde{\sigma}_{N+}\right), \quad (7.92e)$$

$$\left(\tilde{\Gamma}_1, \ldots, \tilde{\Gamma}_{2N}\right)_\wp \equiv \left(\tilde{\Gamma}_{1\wp}^\dagger, \tilde{\Gamma}_{1\wp}, \ldots, \tilde{\Gamma}_{N\wp}^\dagger, \tilde{\Gamma}_{N\wp}\right). \quad (7.92f)$$

One obstacle remains: the form of the Hamiltonian H_S that is to be substituted into (7.90a). This Hamiltonian includes the interaction H_{AF} between the laser mode and the laser atoms. In the absence of this interaction, using the standard Markov approximation described in Sect. 1.4.1, $(\dot{\rho})_F$ and $(\dot{\rho})_A$ produce the terms on the right-hand sides of (1.73) and (6.129), respectively; as we have noted, the contribution from $(\dot{\rho})_\wp$ can also be deduced from

(6.129). The presence of the interaction term H_{AF} means, however, that the three environmental interactions are not completely independent. Although the reservoir operators are statistically independent, there is a communication from one reservoir interaction to the other through the internal coupling in the system S. We discussed the effects of such coupling in Sect. 2.3.2. The system S interacts with the reservoir R at the eigenfrequencies of the full system Hamiltonian $H_S = H_A + H_F + H_{AF}$, rather than at the frequencies of the decoupled components H_A and H_F. But for reasonable coupling strengths, this change is negligible. This is shown explicitly for Scully-Lamb laser theory by Carmichael and Walls [7.16]. We therefore neglect the effects of H_{AF} on the reservoir interactions by replacing H_S with $H_A + H_F$ in the definitions of $\{\tilde{s}_i\}$ and $\tilde{\rho}$; the interaction term $[H_{AF}, \tilde{\rho}]/i\hbar$ must then be added to the right-hand side of (7.91). With these modifications we may pass directly from (7.91) to the *master equation for the single-mode homogeneously broadened laser*:

$$
\begin{aligned}
\dot{\rho} = {}&-i\tfrac{1}{2}\omega_C[J_z, \rho] - i\omega_C[a^\dagger a, \rho] + g[a^\dagger J_- - aJ_+, \rho] \\
&+ \kappa(2a\rho a^\dagger - a^\dagger a\rho - \rho a^\dagger a) + 2\kappa\bar{n}(a\rho a^\dagger + a^\dagger \rho a - a^\dagger a\rho - \rho aa^\dagger) \\
&+ \frac{\gamma_\downarrow}{2}\left(\sum_{j=1}^{N} 2\sigma_{j-}\rho\sigma_{j+} - \tfrac{1}{2}J_z\rho - \tfrac{1}{2}\rho J_z - N\rho\right) \\
&+ \frac{\gamma_\uparrow}{2}\left(\sum_{j=1}^{N} 2\sigma_{j+}\rho\sigma_{j-} + \tfrac{1}{2}J_z\rho + \tfrac{1}{2}\rho J_z - N\rho\right).
\end{aligned}
\tag{7.93}
$$

The damping terms for the laser mode are taken from (1.73) with $\gamma \to 2\kappa$; the terms describing the atomic damping and pumping are taken from (6.129) with $\gamma(\bar{n}+1) \to \gamma_\downarrow$ and $\gamma\bar{n} \to \gamma_\uparrow$.

Note 7.7 The thermal photon number (determined by the ambient temperature) enters the atomic transition rates as well as appearing explicitly in the source term for the laser mode; the interaction H_{SR}^A between the laser transition and the free-space electromagnetic field generates transitions describing the absorption and emission of thermal photons. The upwards thermal transitions add to the pump transitions generated by H_{SR}^φ. Thus, if γ is the Einstein A coefficient for the laser transition, $\gamma_\uparrow \equiv \gamma\bar{n} + \Gamma_\varphi \approx \Gamma_\varphi$. When the pump reservoirs are taken to be in a *low* negative effective temperature T_φ, downwards transitions are only produced by H_{SR}^A; $\gamma_\downarrow \equiv \gamma_{21} = \gamma(\bar{n}+1)$.

Note 7.8 With nonradiative dephasing processes included, the term

$$
\left(\dot{\rho}\right)_{\text{dephase}} = \frac{\gamma_p}{2}\left(\sum_{j=1}^{N} \sigma_{jz}\rho\sigma_{jz} - N\rho\right)
\tag{7.94}
$$

is added to the master equation (7.93).

7.2.3 The Characteristic Function and Associated Distribution

We wish to convert the operator equation (7.93) into a phase-space equation for the full laser system of atoms plus laser field. For this purpose we introduce a distribution function in the normal-ordered representation using the Glauber–Sudarshan representation for the field, and Haken's representation for N two-level atoms. Combining the definitions of Sects. 3.2.1 and 6.2.3, we define the characteristic function

$$\chi_N(z, z^*, \xi, \xi^*, \eta) \equiv \text{tr}\left(\rho e^{iz^* a^\dagger} e^{iza} e^{i\xi^* J_+} e^{i\eta J_z} e^{i\xi J_-}\right). \tag{7.95}$$

The normal-ordered averages for the field operators and collective atomic operators are given by

$$\langle a^{\dagger p'} a^{q'} J_+^p J_z^r J_-^q \rangle \equiv \text{tr}\left(\rho a^{\dagger p'} a^{q'} J_+^p J_z^r J_-^p\right)$$

$$= \left. \frac{\partial^{p'+q'+p+r+q}}{\partial(iz^*)^{p'}\, \partial(iz)^{q'}\, \partial(i\xi^*)^p \partial(i\eta)^r \partial(i\xi)^q} \chi_N \right|_{\substack{z=z^*=0 \\ \xi=\xi^*=\eta=0}}. \tag{7.96}$$

The distribution $P(\alpha, \alpha^*, v, v^*, m)$ is the five-dimensional Fourier transform of $\chi_N(z, z^*, \xi, \xi^*, \eta)$:

$$P(\alpha, \alpha^*, v, v^*, m)$$

$$\equiv \frac{1}{2\pi^5} \int d^2z \int d^2\xi \int d\eta\, \chi_N(z, z^*, \xi, \xi^*, \eta) e^{-iz^*\alpha^*} e^{-iz\alpha} e^{-i\xi^* v^*} e^{-i\xi v} e^{-i\eta m}$$

$$\equiv \frac{1}{2\pi^5} \int_{-\infty}^{\infty} d\mu \int_{-\infty}^{\infty} d\nu \int_{-\infty}^{\infty} dw \int_{-\infty}^{\infty} ds \int_{-\infty}^{\infty} d\eta \chi_N(\mu + i\nu, \mu - i\nu,$$

$$w + is, w - is, \eta) e^{-2i(\mu x - \nu y)} e^{-2i(w\vartheta - s\varphi)} e^{-i\eta m}, \tag{7.97}$$

with the inverse relationship

$$\chi_N(z, z^*, \xi, \xi^* \eta)$$

$$= \int d^2\alpha \int d^2v \int dm\, P(\alpha, \alpha^*, v, v^*, m) e^{iz^*\alpha^*} e^{iz\alpha} e^{i\xi^* v^*} e^{i\xi v} e^{i\eta m}$$

$$= \int_{-\infty}^{\infty} dx \int_{-\infty}^{\infty} dy \int_{-\infty}^{\infty} d\vartheta \int_{-\infty}^{\infty} d\varphi \int_{-\infty}^{\infty} dm\, P(x + iy, x - iy,$$

$$\vartheta + i\varphi, \vartheta - i\varphi, m) e^{2i(\mu x - \nu y)} e^{2i(w\vartheta - s\varphi)} e^{i\eta m}. \tag{7.98}$$

From (7.96) and (7.98), we have

$$\langle a^{\dagger p'} a^{q'} J_+^p J_z^r J_-^q \rangle$$

$$= \frac{\partial^{p'+q'+p+r+q}}{\partial(iz^*)^{p'}\partial(iz)^{q'}\partial(i\xi^*)^p\partial(i\eta)^r\partial(i\xi)^q} \int d^2\alpha \int d^2v \int dm\, P(\alpha,\alpha^*,v,v^*,m)$$

$$\times e^{iz^*\alpha^*} e^{iz\alpha} e^{i\xi^*v^*} e^{i\xi v} e^{i\eta m} \Bigg|_{\substack{z=z^*=0 \\ \xi=\xi^*=\eta=0}}$$

$$= \left(\overline{\alpha^{*p'} \alpha^{q'} v^{*p} m^r v^q} \right)_P, \tag{7.99a}$$

with

$$\left(\overline{\alpha^{*p'} \alpha^{q'} v^{*p} m^r v^q} \right)_P \equiv \int d^2\alpha \int d^2v \int dm\, \alpha^{*p'} \alpha^{q'} z^{*p} m^r v^q P(\alpha,\alpha^*,v,v^*,m). \tag{7.99b}$$

7.2.4 Phase-Space Equation of Motion for the Single-Mode Homogeneously Broadened Laser

Converting the master equation (7.93) into a phase-space equation of motion is now accomplished by a straightforward application of the techniques we have learned in Sects. 3.2.2, and 6.1.3 and 6.3.4. In fact, we have already derived most of the laser phase-space equation; the interaction term $g[a^\dagger J_- - aJ_+, \rho]$ is the only term in (7.93) that we have not converted to phase-space form in one of our earlier calculations. Drawing on our previous results for the damped harmonic oscillator and the damped two-level medium, we may write the *phase-space equation of motion for the single-mode homogeneously broadened laser* in the form

$$\frac{\partial P}{\partial t} = \left[L_A\left(v, v^*, m, \frac{\partial}{\partial v}, \frac{\partial}{\partial v^*}, \frac{\partial}{\partial m}\right) + L_F\left(\alpha, \alpha^*, \frac{\partial}{\partial \alpha}, \frac{\partial}{\partial \alpha^*}\right) \right.$$

$$\left. + L_{AF}\left(\alpha, \alpha^*, v, v^*, m, \frac{\partial}{\partial \alpha}, \frac{\partial}{\partial \alpha^*}, \frac{\partial}{\partial v}, \frac{\partial}{\partial v^*}, \frac{\partial}{\partial m}\right) \right] P, \tag{7.100}$$

where, from (6.157) [with $\gamma(\bar{n}+1) \to \gamma_\downarrow$, $\gamma\bar{n} \to \gamma_\uparrow$],

$$L_A\left(v, v^*, m, \frac{\partial}{\partial v}, \frac{\partial}{\partial v^*}, \frac{\partial}{\partial m}\right)$$

$$\equiv i\omega_C\left(\frac{\partial}{\partial v}v - \frac{\partial}{\partial v^*}v^*\right)$$

$$+ \frac{\gamma_\downarrow}{2}\left[\left(e^{2\frac{\partial}{\partial m}} - 1\right)(N+m) + \frac{\partial}{\partial v}v + \frac{\partial}{\partial v^*}v^*\right]$$

$$+ \frac{\gamma_\uparrow}{2}\left[\left(e^{-2\frac{\partial}{\partial m}} - 1\right)(N - m) + \frac{\partial^4}{\partial v^2 \partial v^{*2}} e^{2\frac{\partial}{\partial m}}(N + m)\right.$$

$$\left. + 2\left(e^{-2\frac{\partial}{\partial m}} + \frac{\partial^2}{\partial v \partial v^*} - \frac{1}{2}\right)\left(\frac{\partial}{\partial v}v + \frac{\partial}{\partial v^*}v^*\right) + 2N\frac{\partial^2}{\partial v \partial v^*}\right];$$

$$(7.101\mathrm{a})$$

and from (3.47) (with $\gamma \to 2\kappa$),

$$L_F\left(\alpha, \alpha^*, \frac{\partial}{\partial \alpha}, \frac{\partial}{\partial \alpha^*}\right)$$

$$\equiv (\kappa + i\omega_C)\frac{\partial}{\partial \alpha}\alpha + (\kappa - i\omega_C)\frac{\partial}{\partial \alpha^*}\alpha^* + 2\kappa\bar{n}\frac{\partial^2}{\partial \alpha \partial \alpha^*}. \quad (7.101\mathrm{b})$$

The derivation of the one term we have not met previously is left as an exercise:

Exercise 7.6 Show that the interaction term $g[a^\dagger J_- - aJ_+, \rho]$ in the master equation produces the differential operator

$$L_{AF}\left(\alpha, \alpha^*, v, v^*, m, \frac{\partial}{\partial \alpha}, \frac{\partial}{\partial \alpha^*}, \frac{\partial}{\partial v}, \frac{\partial}{\partial v^*}, \frac{\partial}{\partial m}\right)$$

$$\equiv -g\left\{\left[\left(e^{-2\frac{\partial}{\partial m}} - 1\right)v^* + \frac{\partial}{\partial v}m - \frac{\partial^2}{\partial v^2}v\right]\alpha + \frac{\partial}{\partial \alpha}v\right.$$

$$\left. + \left[\left(e^{-2\frac{\partial}{\partial m}} - 1\right)v + \frac{\partial}{\partial v^*}m - \frac{\partial^2}{\partial v^{*2}}v^*\right]\alpha^* + \frac{\partial}{\partial \alpha^*}v^*\right\}$$

$$(7.101\mathrm{c})$$

in the phase-space equation of motion.

Note 7.9 If the nonradiative dephasing term (7.94) is included in the master equation we add the term $L_{\mathrm{dephase}}P$ to (7.100), where L_{dephase} is given by (6.158).

Equation (7.100) is considerably more complicated than the Fokker–Planck equation we invented from the rate equation treatment of the laser [Eq. (7.67a)]. First, it describes not only the laser field, but also the atoms. Second, it is not a Fokker–Planck equation; the atomic variables introduce derivatives to all orders, and we cannot hope to find the exact solution for $P(\alpha, \alpha^*, v, v^*, m)$. In fact, from the discussion in Sects. 6.1.4 and 6.2.4 we know that the exact solution to (7.100) is highly singular, and only approximately represented by a smooth, well-behaved function. Our task now is to introduce the approximations that allow us to extract useful information from this equation. With the use of van Kampen's system size expansion and the adiabatic elimination of atomic variables, we will be able to connect the complicated phase-space equation of motion (7.100) with the Fokker–Planck equation (7.67a). Before we begin this exercise we need to spend a little time on one last detail of our laser model.

7.3 The Laser Output Field

7.3.1 Free Field and Source Field for a Lossy Cavity Mode

The master equation treatment of resonance fluorescence (Sect. 2.3) was built around a description of the *source* of the fluorescence, the driven two-level atom. The master equation (2.96) provided a mathematical description of the atomic dynamics. To obtain information about the fluorescence we needed a relationship between the atomic source and its radiated field; this was provided by the operator version of the dipole radiation formula, given by (2.76) and (2.83).

We have an analogous situation here. The intracavity laser field is described by the driven, damped oscillator, obeying the master equation (7.93). This is not, however, the laser output field. Classically, the field at the output of an optical cavity is obtained from the intracavity field after multiplying by a mirror transmission coefficient. Quantum mechanically, this simple relationship will not do. It asserts that the output field is described by operators $\sqrt{T}e^{i\phi_T}a$ and $\sqrt{T}e^{-i\phi_T}a^\dagger$, where T is the transmission coefficient for the output mirror and ϕ_T is a phase change on transmission through the mirror. But a and $a^\dagger$ obey the commutation relation $[a, a^\dagger] = 1$, and therefore $[\sqrt{T}e^{i\phi_T}a, \sqrt{T}e^{-i\phi_T}a^\dagger] = T < 1$. As we saw at the very beginning of the book (Sect. 1.2), special care must be taken to preserve commutators when dealing with dissipation in quantum mechanics.

What does the transmission of an intracavity field through an output mirror have to do with dissipation? Well, the cavity output field carries the energy dissipated by the laser mode. The energy lost from the cavity is not simply discarded, it is radiated into the many modes of the electromagnetic field outside the cavity. These modes form the reservoir that damps the intracavity field, and it is these modes that carry the useful laser output. We modeled the laser mode losses by the reservoir interaction (7.86a). The master equation describes one end of this interaction; by eliminating the reservoir variables, a simple description is obtained for the system S which retains the dissipative effects of the reservoir R. We must now consider the other end of the interaction – the effect of the system S on the reservoir R. We can construct the laser output field by calculating the source contribution from S to the reservoir mode operators r_k and $r_k^\dagger$, a calculation analogous to that of Sect. 2.3.1.

Figure 7.8 shows the laser cavity with external traveling-wave modes r_k satisfying periodic boundary conditions at $z = -L'/2$ and $z = L'/2$. The field outside the cavity is described by the Heisenberg operator

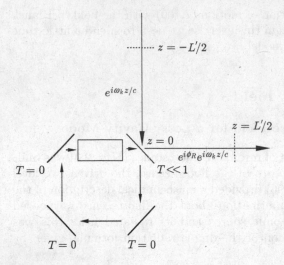

Fig. 7.8 Schematic diagram of the laser cavity and output field modes.

$$\hat{\boldsymbol{E}}(z,t) = \hat{\boldsymbol{E}}^{(+)}(z,t) + \hat{\boldsymbol{E}}^{(-)}(z,t), \tag{7.102a}$$

with

$$\hat{\boldsymbol{E}}^{(+)}(z,t) = i\hat{e}_0 \sum_k \sqrt{\frac{\hbar\omega_k}{2\epsilon_0 AL'}}\, r_k(t) e^{i[(\omega_k/c)z + \phi(z)]}, \tag{7.102b}$$

$$\hat{\boldsymbol{E}}^{(-)}(z,t) = \hat{\boldsymbol{E}}^{(+)}(z,t)^\dagger, \tag{7.102c}$$

where

$$\phi(z) \equiv \begin{cases} \phi_R & z > 0 \\ 0 & z < 0 \end{cases}; \tag{7.103}$$

ϕ_R is the phase change on reflection at the cavity output mirror. Of course, there are also counterpropagating modes, and modes polarized orthogonal to $\hat{e}_0$; but these can be neglected since they do not couple to the laser mode. Using the Hamiltonian (7.88), we obtain Heisenberg equations of motion

$$\dot{r}_k = -i\omega_k r_k - i\kappa_k^* a. \tag{7.104}$$

The term $i\kappa_k^* a$ couples energy from the intracavity field into the modes of the external field; for the present, the coupling constant κ_k^* need not be specified. Integrating (7.104) formally, we have

$$r_k(t) = r_k(0)e^{-i\omega_k t} - i\kappa_k^* e^{-i\omega_C t}\int_0^t dt'\, \tilde{a}(t')e^{i(\omega_k - \omega_C)(t'-t)}, \tag{7.105}$$

where $\tilde{a}(t)$ is the slowly-varying operator

$$\tilde{a}(t) \equiv e^{i\omega_C t} a(t). \tag{7.106}$$

Then the laser output field is given by

$$\hat{E}^{(+)}(z,t) = \hat{E}_f^{(+)}(z,t) + \hat{E}_s^{(+)}(z,t), \tag{7.107}$$

with

$$\hat{E}_f^{(+)}(z,t) = i\hat{e}_0 \sum_k \sqrt{\frac{\hbar\omega_k}{2\epsilon_0 A L'}}\, r_k(0) e^{-i[\omega_k(t-z/c)-\phi(z)]}, \tag{7.108}$$

and

$$\hat{E}_s^{(+)}(z,t) = \hat{e}_0 \sqrt{\frac{\hbar}{2\epsilon_0 A L'}}\, e^{-i[\omega_C(t-z/c)-\phi(z)]}$$

$$\times \sum_k \sqrt{\omega_k}\, \kappa_k^* \int_0^t dt'\, \tilde{a}(t') e^{i(\omega_k-\omega_C)(t'-t+z/c)}. \tag{7.109}$$

This field decomposes into the sum of a freely evolving field $\hat{E}_f^{(+)}(z,t)$ (free field), and a source field $\hat{E}_s^{(+)}(z,t)$, in a manner analogous to the decomposition (2.76)–(2.78) for resonance fluorescence.

To express the source field in manageable form we introduce the density of states $g(\omega) = L'/2\pi c$ for traveling-wave modes in one dimension, and perform the summation over k as an integral:

$$\hat{E}_s^{(+)}(z,t) = \hat{e}_0 \sqrt{\frac{\hbar}{2\epsilon_0 A c}} \sqrt{\frac{L'}{c}}\, e^{-i[\omega_C(t-z/c)-\phi(z)]}$$

$$\times \frac{1}{2\pi} \int_0^\infty d\omega\, \sqrt{\omega}\, \kappa^*(\omega) \int_0^t dt'\, \tilde{a}(t') e^{i(\omega-\omega_C)(t'-t+z/c)}. \tag{7.110}$$

Assuming that $\tilde{a}(t')$ varies slowly compared with the optical period $2\pi/\omega_C$, we can treat the integrals in the manner described below (2.82) – we set $\sqrt{\omega}\kappa^*(\omega) \approx \sqrt{\omega_C}\kappa^*(\omega_C)$ and extend the range of the frequency integral to $-\infty$; after evaluating the frequency integral, we obtain

$$\hat{E}_s^{(+)}(z,t)$$

$$= \hat{e}_0 \sqrt{\frac{\hbar\omega_C}{2\epsilon_0 A c}} \sqrt{\frac{L'}{c}}\, \kappa^*(\omega_C) e^{-i[\omega_C(t-z/c)-\phi(z)]} \int_0^t dt'\, \tilde{a}(t')\delta(t'-t+z/c)$$

$$= \begin{cases} \hat{e}_0 \sqrt{\dfrac{\hbar\omega_C}{2\epsilon_0 A c}} \sqrt{\dfrac{L'}{c}}\, \kappa^*(\omega_C) e^{i\phi_R} a(t-z/c) & ct > z > 0 \\[2ex] 0 & z < 0. \end{cases} \tag{7.111}$$

Thus, for $ct > z > 0$, the source field is proportional to the intracavity field evaluated at the retarded time $t - z/c$.

Note 7.10 The source field radiated by a standing-wave cavity can be found in much the same way. In this case the external reservoir field is expanded in the *standing-wave* modes of a large external cavity of length L'. The time integral in (7.110) is taken over the sum of two terms, one proportional to $\exp[i(\omega - \omega_C)(t' - t + z/c)]$, and the other proportional to $\exp[i(\omega - \omega_C)(t' - t - z/c)]$; the two terms are contributed by the counterpropagating components of the reservoir modes. We expect the source field to propagate in only one direction, despite the presence of counterpropagating terms in the reservoir modes. This follows naturally from the mathematics. Two δ-functions, $\delta(t' - t + z/c)$ and $\delta(t' - t - z/c)$, appear inside the integral leading to (7.111). The range of this integral selects the contribution from the δ-function that gives a retarded field propagating away from the cavity output mirror, and rejects the second contribution. [An analogous situation is illustrated by (2.82) and (2.83).]

It is now time to determine the value of the reservoir coupling constant $\kappa^*(\omega_C)$. If (7.111) is to give the expected relationship, $\langle a \rangle \to \sqrt{T} e^{i\phi_T} \langle a \rangle$, between the mean intracavity field and the mean output field, we must choose

$$-i\sqrt{\frac{L'}{c}}\kappa^*(\omega_C)e^{i\phi_R} = \sqrt{\frac{c}{L}}\sqrt{T}e^{i\phi_T} = \sqrt{2\kappa}e^{i\phi_T}, \qquad (7.112)$$

where $\kappa = Tc/2L$ is the cavity decay rate appearing in the master equation (7.93). [Note that the field inside the laser cavity is expanded like (7.102a), with no sum, and with $r_k \to a, \omega_k \to \omega_C$, and $L' \to L$.] We can also derive this relationship from (1.70a) (without the phase factor). In the present notation, (1.70a) gives

$$2\kappa = 2\pi g(\omega_C)|\kappa(\omega_C)|^2.$$

Substituting $g(\omega_C) = L'/2\pi c$ for the reservoir density of states, we find $\sqrt{L'/c}|\kappa(\omega_C)| = \sqrt{2\kappa}$, which is the modulus of the relationship (7.112). The final form of the source term in the cavity output field is now

$$\hat{E}_s^{(+)}(z,t) = \begin{cases} i\hat{e}_0\sqrt{\dfrac{\hbar\omega_C}{2\epsilon_0 Ac}}\sqrt{2\kappa}e^{i\phi_T}a(t - z/c) & ct > z > 0 \\ 0 & z < 0. \end{cases} \qquad (7.113)$$

In fact, (7.113) is the relationship we would write down directly from the classical result for the transmission of the intracavity field through the cavity output mirror; we could have constructed the complete expression (7.107) for the cavity output field from our understanding of the classical boundary conditions at the output mirror; the free-field term is just the contribution from the reflection of incoming reservoir modes into the cavity output (our theory assumes $R = 1 - T \approx 1$). The only difference between the quantum-mechanical and classical pictures is that $\hat{E}_f^{(+)}(z,t)$ and $\hat{E}_s^{(+)}(z,t)$ are operators in the quantum-mechanical theory, and therefore play an algebraic role that is absent in a classical theory. The source field does not commute with the free field – the operators $\hat{E}_f^{(\pm)}(z,t)$ and $\hat{E}_s^{(\mp)}(z,t)$ do not commute; it is their noncommutation that preserves the commutation relation for the operators, $\hat{E}^{(+)}(z,t)$ and $\hat{E}^{(-)}(z,t)$, of the total field. Thus, the free-field term cannot be dropped from (7.107) even when the reservoir modes are in the vacuum state. On the other hand, when the reservoir modes are in the vacuum state this concern for algebraic integrity in the quantum theory really has little practical consequence, since we are generally interested in normal-ordered, time-ordered operator averages, quantities that are insensitive to vacuum contributions. A discussion of these issues is given in Ref. [7.17].

Note 7.11 Equation (7.113) yields exactly what we would expect for the average photon flux from the laser cavity:

$$\frac{2\epsilon_0 cA}{\hbar\omega_C}\langle \hat{E}_s^{(-)}(z,t)\hat{E}_s^{(+)}(z,t)\rangle = 2\kappa\langle a^\dagger(t-z/c)a(t-z/c)\rangle. \qquad (7.114)$$

The right-hand side is the product of the photon escape probability per unit time and the mean number of photons in the cavity.

Of course, the free-field term does contribute to normal-ordered, time-ordered averages when the reservoir modes are not in the vacuum state. Moreover, there are situations in which non-normal-ordered, or non-time-ordered averages are needed. Then things are not so straightforward; the free field contributes to the output, and to calculate its contribution we generally need nontrivial information about how it is correlated with the source. Now is a good time to see how this information can be obtained.

7.3.2 Coherently Driven Cavities

We start with a simple example. Consider an empty cavity driven by a coherent field. The reservoir mode with frequency $\omega_k = \omega_C$ is in the coherent state $|\beta\rangle$, and all other modes are in the vacuum state. Thus, from (7.107), (7.108), and (7.113), the cavity is driven on resonance by the mean field ($z < 0$)

$$\langle \hat{E}^{(+)}(z,t)\rangle = \langle \hat{E}_f^{(+)}(z,t)\rangle$$

$$= i\hat{e}_0\sqrt{\frac{\hbar\omega_C}{2\epsilon_0 AL'}}\beta e^{-i\omega_C(tz-/c)}, \qquad (7.115)$$

with mean output field ($z > 0$)

$$\langle \hat{E}^{(+)}(z,t)\rangle = \langle \hat{E}_f^{(+)}(z,t)\rangle + \langle \hat{E}_s^{(+)}(z,t)\rangle$$

$$= i\hat{e}_0\sqrt{\frac{\hbar\omega_C}{2\epsilon_0 AL'}}\left(e^{i\phi_R}\beta\right.$$

$$\left.+\sqrt{L'/c}\sqrt{2\kappa}e^{i\phi_T}\langle\tilde{a}(t-z/c)\rangle\right)e^{-i\omega_C(t-z/c)}. \qquad (7.116)$$

The geometry is shown in Fig. (7.8). The first term inside the bracket in (7.116) is the input field, reflected into the output, and the second term is the field radiated by the cavity. Since the cavity has only one partially transmitting mirror, in the steady state the two contributions must interfere to reconstruct the input amplitude, with a possible phase change. To check that this is so we need $\langle \tilde{a} \rangle_{ss}$. This is obtained from the mean-value equation

$$\langle \dot{\tilde{a}} \rangle = -\kappa \langle \tilde{a} \rangle - i\kappa(\omega_C)\beta, \tag{7.117}$$

where $\kappa(\omega_C)$ is the system-reservoir coupling coefficient given by (7.112); the driving term in (7.117) is derived from the interaction Hamiltonian

$$H_{SR}|_{\omega_k = \omega_C} \equiv \hbar(\kappa_k^* a r_k^\dagger + \kappa_k a^\dagger r_k)|_{\omega_k = \omega_C}. \tag{7.118}$$

Substituting the steady-state solution to (7.117), $\langle \tilde{a} \rangle_{ss} = -i\kappa(\omega_C)\beta/\kappa$, into (7.116), we find $(z > 0)$

$$\langle \hat{E}^{(+)}(z, t) \rangle = i\hat{e}_0 \sqrt{\frac{\hbar\omega_C}{2\epsilon_0 AL'}} \left[e^{i\phi_R} \beta \right.$$

$$\left. + \sqrt{L'/c}\sqrt{2\kappa}\, e^{i\phi_T}\left(-\sqrt{\frac{c}{L'}}\frac{\sqrt{2\kappa}\beta}{\kappa}\, e^{i(\phi_R - \phi_T)} \right) \right] e^{-i\omega_C(t - z/c)}$$

$$= -e^{i\phi_R} i\hat{e}_0 \sqrt{\frac{\hbar\omega_C}{2\epsilon_0 AL'}}\, \beta e^{-i\omega_C(t - z/c)}. \tag{7.119}$$

This is the mean driving field amplitude multiplied by the phase factor $-e^{i\phi_R}$. We do, therefore, recover the anticipated result.

Note 7.12 Equation (7.112) gives the phase of the coupling coefficient $\kappa(\omega_C)$ as $\arg[\kappa(\omega_C)] = \phi_R - \phi_T - \pi/2$. This phase was chosen so that it is consistent with the boundary condition that couples the field $\langle a \rangle$ out of the cavity. We might, alternatively, choose the phase of $\kappa(\omega_C)$ so that the driving term $-i\kappa(\omega_C)\beta$ in (7.117) is consistent with the boundary condition that couples the external driving field into the cavity. This requires $\arg[\kappa(\omega_C)] = \phi_T - \phi_R + \pi/2$. The two choices of phase are consistent if $\phi_R - \phi_T = \pi/2$. It can be verified that this relationship between the phase changes on transmission and reflection at a mirror does, indeed, hold. It must hold for energy to be conserved. Consider fields of amplitude A and B incident on the two faces of a mirror such that the outgoing field amplitudes are $C = \sqrt{R}e^{i\phi_R} A + \sqrt{T}e^{i\phi_T} B$ and $D = \sqrt{R}e^{i\phi_R} B + \sqrt{T}e^{i\phi_T} A$. The incoming and outgoing energy fluxes are proportional to $|A|^2 + |B|^2$ and $|C|^2 + |D|^2 = |A|^2 + |B|^2 + 2\sqrt{RT}\cos(\phi_R - \phi_T)$, respectively. Thus, we require $\phi_R - \phi_T = \pi/2$ for the energy to be conserved.

What happens if we allow the cavity in Fig. 7.8 to have two partially transmitting mirrors, with transmission coefficients $T_1 \ll 1$ $(\kappa_1 = cT_1/2L)$ and $T_2 \ll 1$ $(\kappa_2 = cT_2/2L)$? Let T_1 refer to the mirror at which we input the coherent amplitude β, and T_2 refer to a second cavity mirror with a vacuum field input. The mean output field at mirror 1 is given by (7.116) with $\kappa \to \kappa_1$. The steady-state field amplitude is

$$|\hat{e}_0 \cdot \langle \hat{\boldsymbol{E}}_1^{(+)}(z,t)\rangle| = \sqrt{\frac{\hbar \omega_C}{2\epsilon_0 AL'}} \left| e^{i\phi_R}\beta + \sqrt{L'/c}\sqrt{2\kappa_1}e^{i\phi_T}\langle \tilde{a}\rangle_{\mathrm{ss}}\right|. \quad (7.120a)$$

At mirror 2 the steady-state output field amplitude is given by a similar expression:

$$|\hat{e}_0 \cdot \langle \hat{\boldsymbol{E}}_2^{(+)}(z,t)\rangle| = \sqrt{\frac{\hbar \omega_C}{2\epsilon_0 AL'}} \left| \sqrt{L'/c}\sqrt{2\kappa_2}e^{i\phi_T}\langle \tilde{a}\rangle_{\mathrm{ss}}\right|. \quad (7.120b)$$

In the mean-value equation (7.117) we now have $\kappa_1 + \kappa_2$ in place of κ, and $\kappa(\omega_C) = -i\sqrt{c/L'}\sqrt{2\kappa_1}e^{i(\phi_R-\phi_T)}$. Thus,

$$|\hat{e}_0 \cdot \langle \hat{\boldsymbol{E}}_1^{(+)}(z,t)\rangle| = \sqrt{\frac{\hbar \omega_C}{2\epsilon_0 AL'}} \left| \beta - \frac{2\kappa_1}{\kappa_1 + \kappa_2}\beta\right|$$

$$= \sqrt{\frac{\hbar \omega_C}{2\epsilon_0 AL'}} \left| \frac{\kappa_1 - \kappa_2}{\kappa_1 + \kappa_2}\right| |\beta|, \quad (7.121a)$$

and

$$|\hat{e}_0 \cdot \langle \hat{\boldsymbol{E}}_2^{(+)}(z,t)\rangle| = \sqrt{\frac{\hbar \omega_C}{2\epsilon_0 AL'}} \frac{2\sqrt{\kappa_1 \kappa_2}}{\kappa_1 + \kappa_2} |\beta|. \quad (7.121b)$$

When $\kappa_2 = 0$ we recover the result from (7.119) – the full input field amplitude appears in the output at mirror 1, with no output at mirror 2. When $\kappa_1 = \kappa_2$ the free-field and source-field contributions cancel at mirror 1; there is no output at mirror 1, and the full incident field amplitude is transmitted by the cavity through mirror 2. More generally, we obtain partial transmission and partial reflection by the cavity with $|\hat{e}_0\langle \hat{\boldsymbol{E}}_1^{(+)}(z,t)\rangle|^2 + |\hat{e}_0\langle \hat{\boldsymbol{E}}_2^{(+)}(z,t)|^2 = (\hbar\omega_C/2\epsilon_0 AL')|\beta|^2$, as expected from the classical theory of interferometers. It is clear from this example that the free-field term in (7.107) is not always negligible.

7.3.3 Correlations Between the Free Field and Source Field for Thermal Reservoirs

Accounting for free-field contributions is more difficult when this field is not in a coherent state. It is common to encounter thermal reservoirs, as in our laser theory, and reservoirs with different statistical properties are also sometimes of interest – for example, squeezed reservoirs, where the free field is in a broadband squeezed state. We can appreciate the difficulties that arise, as well as the road to their resolution, by considering the first-order correlation function for the full cavity output field $\hat{\boldsymbol{E}}(z,t)$. First, let us simplify the notation in (7.107), (7.108), and (7.113) by scaling the field operators so that the source field appears in units of photon flux. We write ($ct > z > 0$)

$$\hat{\mathcal{E}}(z,t) = \sqrt{c/L'}\, r_f(t - z/c) + \sqrt{2\kappa}\, a(t - z/c), \qquad (7.122)$$

where

$$\hat{\mathcal{E}}(z,t) \equiv -ie^{-i\phi_T}\sqrt{\frac{2\epsilon_0 Ac}{\hbar\omega_C}}\,\hat{e}_0 \cdot \hat{\boldsymbol{E}}^{(+)}(z,t), \qquad (7.123a)$$

$$r_f(t - z/c) \equiv e^{i(\phi_R - \phi_T)}\sum_k \sqrt{\frac{\omega_k}{\omega_C}}\, r_k(0)e^{-i\omega_k(t\ z/c)}. \qquad (7.123b)$$

Then the normalized first-order correlation function for the field $\hat{\boldsymbol{E}}(z,t)$ is given by

$$g_{\text{ss}}^{(1)}(\tau) = \left(\langle\hat{\mathcal{E}}^\dagger\hat{\mathcal{E}}\rangle_{\text{ss}}\right)^{-1}\Big\{(c/L')\langle r_f^\dagger(0)r_f(\tau)\rangle + 2\kappa\Big[\lim_{t\to\infty}\langle a^\dagger(t)a(t+\tau)\rangle\Big]$$
$$+\sqrt{c/L'}\sqrt{2\kappa}\Big[\lim_{t\to\infty}\langle r_f^\dagger(t)a(t+\tau)\rangle + \lim_{t\to\infty}\langle a^\dagger(t)r_f(t+\tau)\rangle\Big]\Big\},$$
$$(7.124)$$

with

$$\langle\hat{\mathcal{E}}^\dagger\hat{\mathcal{E}}\rangle_{\text{ss}} = (c/L')\langle r_f^\dagger r_f\rangle + 2\kappa\langle a^\dagger a\rangle_{\text{ss}} + \sqrt{c/L'}\sqrt{2\kappa}(\langle r_f^\dagger a\rangle_{\text{ss}} + \langle a^\dagger r_f\rangle_{\text{ss}}). \qquad (7.125)$$

We need more than the source-field correlation function $\langle a^\dagger(t)a(t+\tau)\rangle$ if we are going to calculate this quantity. The free-field correlation function $\langle r_f^\dagger(t)r_f(t+\tau)\rangle$ is presumably straightforward to calculate, given the state of the reservoir. But how do we calculate the correlations between the free field and the source field, the correlation functions $\langle r_f^\dagger(t)a(t+\tau)\rangle$ and $\langle a^\dagger(t)r_f(t+\tau)\rangle$?

When the free field is in a coherent state these correlation functions factorize; because they are in normal order, the action of $r_f^\dagger$ and r_f to the left and right, respectively, on the reservoir state, replaces the operators by coherent amplitudes. In general, however, there is no similarly straightforward procedure available. Gardiner and Collett [7.18] provide a method for calculating these correlation functions using an input-output theory built around quantum stochastic differential equations – a Heisenberg picture formulation of reservoir theory. We will follow a different approach which is more closely tied to the Schrödinger picture formulation of reservoir theory we have been using. It is not possible to perform a single calculation that is applicable to all master equations. As an illustration we consider a fairly general form of the master equation, with $\dot{\rho} = \mathcal{L}\rho$, where the action of $\mathcal{L}$ is defined by

$$\mathcal{L}\hat{O} = \frac{1}{i\hbar}[H_S, \hat{O}] + \mathcal{L}_{\text{out}} + \mathcal{L}'\hat{O}, \qquad (7.126)$$

with

$$\mathcal{L}_{\text{out}}\hat{O} \equiv \kappa(2a\hat{O}a^\dagger - a^\dagger a\hat{O} - \hat{O}a^\dagger a)$$

$$+ 2\kappa\bar{n}(a\hat{O}a^\dagger + a^\dagger\hat{O}a - a^\dagger a\hat{O} - \hat{O}aa^\dagger), \qquad (7.127a)$$

$$\mathcal{L}'\hat{O} \equiv \sum_{jk} c_{jk}[\hat{O}_j\hat{O}, \hat{O}_k] + d_{jk}[\hat{O}_j, \hat{O}\hat{O}_k]; \qquad (7.127b)$$

$\hat{O}$ is an arbitrary system operator; $\mathcal{L}_{\text{out}}$ describes the interaction of the cavity mode (source) through the output mirror with a reservoir in thermal equilibrium, and $\mathcal{L}'$ includes reservoir interaction terms involving system operators $\hat{O}_j$ and $\hat{O}_k$ that commute with a and $a^\dagger$. The laser master equation (7.93) has this form, where the operators $\hat{O}_j$ and $\hat{O}_k$ are the Pauli pseudo-spin operators describing the lasing medium.

We must begin our calculation at a level that still includes the reservoir operators explicitly. The master equation is of no direct use since the reservoir operators have been traced out of this equation. We return to the Heisenberg equations of motion. The Heisenberg equation for the mode operators of the reservoir field is given by (7.104). The Heisenberg equation for the lossy cavity mode is

$$\dot{a} = \frac{1}{i\hbar}[a, H_S + H_R + H_{SR}]$$

$$= \frac{1}{i\hbar}[a, H_S + H_{SR}^F]$$

$$= \frac{1}{i\hbar}[a, H_S] - i\sum_k \kappa_k r_k, \qquad (7.128)$$

where we have used $H_{SR}^F \equiv \hbar(a\Gamma^\dagger + a^\dagger\Gamma)$, with $\Gamma^\dagger$ and Γ given by (1.39b). Substituting the solution (7.105) for $r_k(t)$, and treating the mode summation and time integral as we did in passing from (7.110) to (7.111), we have

$$\dot{a} = \frac{1}{i\hbar}[a, H_S] - e^{-i\omega_C t}\sum_k |\kappa_k|^2 \int_0^t dt' \tilde{a}(t')e^{i(\omega_k - \omega_C)(t'-t)}$$

$$- i\sum_k \kappa_k r_k(0)e^{-i\omega_k t}$$

$$= \frac{1}{i\hbar}[a, H_S] - (L'/c)e^{-i\omega_C t}\frac{1}{2\pi}\int_0^\infty d\omega |\kappa(\omega)|^2 \int_0^t dt' \tilde{a}(t')e^{i(\omega - \omega_C)(t'-t)}$$

$$- i\sum_k \kappa_k r_k(0)e^{-i\omega_k t}$$

$$= \frac{1}{i\hbar}[a, H_S] - \frac{1}{2}(L'/c)|\kappa(\omega_C)|^2 a - i\sum_k \kappa_k r_k(0)e^{-i\omega_k t}$$

$$= \frac{1}{i\hbar}[a, H_S] - \kappa a - i\sum_k \kappa_k r_k(0)e^{-i\omega_k t}. \qquad (7.129)$$

The last term on the right-hand side of (7.129) describes the driving of the cavity mode by the freely evolving modes of the reservoir field. The cavity mode will only respond to those free-field modes with frequencies close to ω_C. For these frequencies we may read (7.129) with $\kappa_k = \kappa(\omega_C) = -ie^{i(\phi_R - \phi_T)}\sqrt{c/L'}\sqrt{2\kappa}$, and (7.123b) with $\sqrt{\omega_k/\omega_C} = 1$. Thus, (7.129) may be written in the form

$$\dot{a} = \frac{1}{i\hbar}[a, H_S] - \kappa a - \sqrt{c/L'}\sqrt{2\kappa}\, r_f. \tag{7.130}$$

Equation (7.130) allows us to express the correlations between the free field and the source field in terms of averages involving system operators alone. By multiplying this equation on the left or right by an arbitrary system operator $\hat{O}$, we find

$$\sqrt{c/L'}\sqrt{2\kappa}\langle \hat{O}(t+\tau)r_f(t)\rangle$$
$$= \frac{1}{i\hbar}\langle \hat{O}(t+\tau)[a, H_S](t)\rangle - \kappa\langle \hat{O}(t+\tau)a(t)\rangle - \langle \hat{O}(t+\tau)\dot{a}(t)\rangle, \tag{7.131a}$$

$$\sqrt{c/L'}\sqrt{2\kappa}\langle r_f(t)\hat{O}(t+\tau)\rangle$$
$$= \frac{1}{i\hbar}\langle [a, H_S](t)\hat{O}(t+\tau)\rangle - \kappa\langle a(t)\hat{O}(t+\tau)\rangle - \langle \dot{a}(t)\hat{O}(t+\tau)\rangle, \tag{7.131b}$$

and, for $\tau > 0$,

$$\sqrt{c/L'}\sqrt{2\kappa}\langle \hat{O}(t)r_f(t+\tau)\rangle$$
$$= -\left(\frac{d}{d\tau} + \kappa\right)\langle \hat{O}(t)a(t+\tau)\rangle + \frac{1}{i\hbar}\langle \hat{O}(t)[a, H_S](t+\tau)\rangle, \tag{7.132a}$$

$$\sqrt{c/L'}\sqrt{2\kappa}\langle r_f(t+\tau)\hat{O}(t)\rangle$$
$$= -\left(\frac{d}{d\tau} + \kappa\right)\langle a(t+\tau)\hat{O}(t)\rangle + \frac{1}{i\hbar}\langle [a, H_S](t+\tau)\hat{O}(t)\rangle. \tag{7.132b}$$

We will use these relationships to prove

$$\sqrt{c/L'}\sqrt{2\kappa}\langle \hat{O}(t+\tau)r_f(t)\rangle = \begin{cases} 0 & \tau < 0 \\ \kappa\bar{n}\langle[\hat{O}(t+\tau), a(t)]\rangle & \tau = 0 \\ 2\kappa\bar{n}\langle[\hat{O}(t+\tau), a(t)]\rangle & \tau > 0, \end{cases} \tag{7.133a}$$

and

$$\sqrt{c/L'}\sqrt{2\kappa}\langle r_f(t)\hat{O}(t+\tau)\rangle = \begin{cases} 0 & \tau < 0 \\ \kappa(\bar{n}+1)\langle[\hat{O}(t+\tau),a(t)]\rangle & \tau = 0 \\ 2\kappa(\bar{n}+1)\langle[\hat{O}(t+\tau),a(t)]\rangle & \tau > 0. \end{cases}$$

$$(7.133b)$$

Most of the two-time averages appearing on the right-hand sides of (7.131a) and (7.131b), and (7.132a) and (7.132b), can be evaluated directly using the quantum regression formula in the version (1.97) or (1.98). A little thought is required, however, to evaluate the averages involving $\dot{a}$.

The proof of (7.133a) and (7.133b) draws on the explicit form of $\mathcal{L}_{\text{out}}$, since it is through this operator that information on the state of the reservoir field enters. Specifically, we will need the following results:

Exercise 7.7 For the superoperator $\mathcal{L}_{\text{out}} + \mathcal{L}'$ defined in (7.127), show that

$$\text{tr}[a(\mathcal{L}_{\text{out}} + \mathcal{L}')\hat{O}] = -\kappa\text{tr}(a\hat{O}),\tag{7.134a}$$

$$(\mathcal{L}_{\text{out}} + \mathcal{L}')(a\hat{O}) = a[(\mathcal{L}_{\text{out}} + \mathcal{L}')\hat{O}] + \kappa a\hat{O} + \kappa 2\bar{n}[a,\hat{O}],\tag{7.134b}$$

$$(\mathcal{L}_{\text{out}} + \mathcal{L}')(\hat{O}a) = [(\mathcal{L}_{\text{out}} + \mathcal{L}')\hat{O}]a + \kappa a\hat{O} + \kappa(2\bar{n}+1)[a,\hat{O}],\tag{7.134c}$$

where $\hat{O}$ is an arbitrary system operator.

We will follow the proof of (7.133a) through in detail, and leave the similar proof of (7.133b) as an exercise.

Proof of (7.133a) – $\tau < 0$: The vanishing of correlations between the free field and source field for $\tau < 0$ is expected on physical grounds. Correlations arise through the driving term proportional to r_f in (7.130). But this equation predicts that $a(t)$ will only depend on $r_f(t')$ for $t' \leq t$. If r_f is correlated with itself at later times, correlations between $a(t)$ and $r_f(t')$ for $t' > t$ could still arise. However, r_f appears as a δ-correlated field to the cavity mode. Of course, it is not strictly δ-correlated. But the cavity mode only responds to a narrow band of frequencies around ω_C, in which case (7.123b) leads to δ-correlated free-field fluctuations in the sense of the discussion below (1.52). We therefore expect that $\langle \hat{O}(t+\tau)r_f(t)\rangle = \langle\hat{O}(t+\tau)\rangle\langle r_f(t)\rangle = 0$ for $\tau < 0$.

We prove this result from (7.132a). We must show that the correlation function on the left-hand side vanishes for $\tau > 0$. Notice that if the right-hand side of (7.132a) is set to zero, we obtain the equation of motion for $\langle \hat{O}(t)a(t+\tau)\rangle$ given by the quantum regression formula [in the form (1.107)] from the mean-value equation

$$-\langle\dot{a}\rangle - \kappa\langle a\rangle + \frac{1}{i\hbar}\langle[a,H_S]\rangle = 0.$$

The vanishing of the correlation function on the left-hand side is required, therefore, for the quantum regression formula to hold. More formally, from (1.97), we have ($\tau > 0$)

$$\left(\frac{d}{d\tau} + \kappa\right)\langle \hat{O}(t)a(t+\tau)\rangle = \left(\frac{d}{d\tau} + \kappa\right)\text{tr}\{ae^{\mathcal{L}\tau}(\rho(t)\hat{O})\}$$

$$= \text{tr}\{a(\mathcal{L} + \kappa)e^{\mathcal{L}\tau}(\rho(t)\hat{O})\}$$

$$= \frac{1}{i\hbar}\text{tr}\{a[H_S, e^{\mathcal{L}\tau}(\rho(t)\hat{O})]\}.$$

where we have used (7.134a). Using the cyclic property of the trace, and again using (1.97), we find

$$\left(\frac{d}{d\tau} + \kappa\right)\langle \hat{O}(t)a(t+\tau)\rangle = \frac{1}{i\hbar}\text{tr}\{[a, H_S]e^{\mathcal{L}\tau}(\rho(t)\hat{O})\}$$

$$= \frac{1}{i\hbar}\langle \hat{O}(t)[a, H_S](t+\tau)\rangle. \qquad (7.135)$$

Substituting (7.135) into (7.132a) gives $\langle \hat{O}(t)r_f(t+\tau)\rangle = 0$ for $\tau > 0$. $\quad\square$

Proof of (7.133a) — $\tau = 0$: We prove this result using (7.131a). To calculate the average $\langle \hat{O}(t)\dot{a}(t)\rangle$ we may write, for any two system operators $\hat{O}_1$ and $\hat{O}_2$,

$$\langle \hat{O}_1\dot{\hat{O}}_2\rangle + +\langle \dot{\hat{O}}_1\hat{O}_2\rangle = \frac{d}{dt}\langle \hat{O}_1\hat{O}_2\rangle, \qquad (7.136)$$

and

$$\langle \hat{O}_1\dot{\hat{O}}_2\rangle - \langle \dot{\hat{O}}_2\hat{O}_1\rangle = \text{tr}_{S\otimes R}\left\{\chi(t)\left(\hat{O}_1\frac{1}{i\hbar}[\hat{O}_2, H] - \frac{1}{i\hbar}[\hat{O}_1, H]\hat{O}_2\right)\right\}$$

$$= \text{tr}_{S\otimes R}\left\{\hat{O}_2\frac{1}{i\hbar}[H, \chi(t)\hat{O}_1] - \hat{O}_1\frac{1}{i\hbar}[H, \hat{O}_2\chi(t)]\right\}$$

$$= \text{tr}_S\{\hat{O}_2\mathcal{L}[\rho(t)\hat{O}_1] - \hat{O}_1\mathcal{L}[\hat{O}_2\rho(t)]\}, \qquad (7.137)$$

where H and $\chi(t)$ are the Hamiltonian and density operator, respectively, for $S \otimes R$, and the trace over the reservoir is taken in the same Born–Markov approximation used to derive the master equation (Chap. 1). From (7.136) and (7.137),

$$2\langle \hat{O}_1\dot{\hat{O}}_2\rangle = \frac{d}{dt}\langle \hat{O}_1\hat{O}_2\rangle + \text{tr}\{\hat{O}_2\mathcal{L}[\rho(t)\hat{O}_1] - \hat{O}_1\mathcal{L}[\hat{O}_2\rho(t)]\}. \qquad (7.138)$$

Equations (7.138) and (7.126), and the master equation $\dot{\rho} = \mathcal{L}\rho$, now give

$$2\langle \hat{O}\dot{a}\rangle = \frac{d}{dt}\langle \hat{O}a\rangle + \text{tr}\{a\mathcal{L}[\rho(t)\hat{O}] - \hat{O}\mathcal{L}[a\rho(t)]\}$$

$$= \text{tr}[\hat{O}a\mathcal{L}\rho(t)] + \text{tr}\{a\mathcal{L}[\rho(t)\hat{O}] - \hat{O}\mathcal{L}[a\rho(t)]\}$$

$$= \frac{1}{i\hbar}\text{tr}[\hat{O}aH_S\rho(t) - \hat{O}a\rho(t)H_S + aH_S\rho(t)\hat{O} - a\rho(t)\hat{O}H_S$$

$$- \hat{O}H_Sa\rho(t) + \hat{O}a\rho(t)H_S] + \text{tr}\{\hat{O}a(\mathcal{L}_{\text{out}} + \mathcal{L}')\rho(t)$$

$$+ a(\mathcal{L}_{\text{out}} + \mathcal{L}')[\rho(t)\hat{O}] - \hat{O}(\mathcal{L}_{\text{out}} + \mathcal{L}')[a\rho(t)]\}. \qquad (7.139)$$

We rewrite the second term in the curly bracket using (7.134a), and the third term in the curly bracket using (7.134b). Then, after reordering operator products using the cyclic property of the trace, we find

$$\langle \hat{O}\dot{a} \rangle = \frac{1}{i\hbar} \langle \hat{O}[a, H_S] \rangle - \kappa \langle \hat{O}a \rangle - \kappa\bar{n}\langle [\hat{O}, a] \rangle. \tag{7.140}$$

Substituting (7.140) into (7.131a) completes the proof of (7.133a) for $\tau = 0$.

□

Proof of (7.133a) – $\tau > 0$: The proof again follows from (7.131a). We now need the average $\langle \hat{O}(t + \tau)\dot{a}(t) \rangle$. For any two system operators $\hat{O}_1$ and $\hat{O}_2$, we have ($\tau > 0$)

$$\langle \hat{O}_1(t+\tau)\dot{\hat{O}}_2(t) \rangle$$

$$= \mathrm{tr}_{S\otimes R} \left\{ \chi(t) e^{(i/\hbar)H\tau} \hat{O}_1 e^{-(i/\hbar)H\tau} \frac{1}{i\hbar} [\hat{O}_2, H] \right\}$$

$$= \mathrm{tr}_{S\otimes R} \left\{ \hat{O}_1 e^{-(i/\hbar)H\tau} \frac{1}{i\hbar} \left[\hat{O}_2 H \chi(t) - H\hat{O}_2 \chi(t) \right] e^{(i/\hbar)H\tau} \right\}$$

$$= \mathrm{tr}_{S\otimes R} \left\{ \hat{O}_1 e^{-(i/\hbar)H\tau} \frac{1}{i\hbar} \left[\hat{O}_2 \big(H\chi(t) - \chi(t)H \big) \right. \right.$$
$$\left. \left. - \big(H\hat{O}_2 \chi(t) - \hat{O}_2 \chi(t)H \big) \right] e^{(i/\hbar)H\tau} \right\}$$

$$= \left(\frac{d}{dt} - \frac{d}{d\tau} \right) \mathrm{tr}_{S\otimes R} \left\{ \hat{O}_1 e^{-(i/\hbar)H\tau} \hat{O}_2 \chi(t) e^{(i/\hbar)H\tau} \right\}$$

$$= \left(\frac{d}{dt} - \frac{d}{d\tau} \right) \mathrm{tr}_S \left\{ \hat{O}_1 e^{\mathcal{L}\tau} [\hat{O}_2 \rho(t)] \right\}$$

$$= \mathrm{tr}_S \left\{ \hat{O}_1 e^{\mathcal{L}\tau} \big(\hat{O}_2 [\mathcal{L}\rho(t)] - \mathcal{L}[\hat{O}_2 \rho(t)] \big) \right\}. \tag{7.141}$$

The trace over the reservoir has again been taken in the Born–Markov approximation. We now calculate $\langle \hat{O}(t + \tau)\dot{a}(t) \rangle$, $\tau > 0$, from (7.141):

$$\langle \hat{O}(t + \tau)\dot{a}(t) \rangle$$

$$= \mathrm{tr} \left\{ \hat{O} e^{\mathcal{L}\tau} \big(a[\mathcal{L}\rho(t)] - \mathcal{L}[a\rho(t)] \big) \right\}$$

$$= \frac{1}{i\hbar} \mathrm{tr} \left\{ \hat{O} e^{\mathcal{L}\tau} [a H_S \rho(t) - a\rho(t) H_S - H_S a\rho(t) + a\rho(t) H_S] \right\}$$

$$+ \mathrm{tr} \left\{ \hat{O} e^{\mathcal{L}\tau} \big(a[(\mathcal{L}_{\mathrm{out}} + \mathcal{L}')\rho(t)] - (\mathcal{L}_{\mathrm{out}} + \mathcal{L}')[a\rho(t)] \big) \right\}$$

$$= \mathrm{tr} \left\{ \hat{O} e^{\mathcal{L}\tau} \left(\frac{1}{i\hbar} [a, H_S]\rho(t) - \kappa a\rho(t) - \kappa 2\bar{n}[a, \rho(t)] \right) \right\},$$

where the last line follows from (7.134b). Then, using (1.97) and (1.98), we find

$$\langle \hat{O}(t+\tau)\dot{a}(t)\rangle = \frac{1}{i\hbar}\langle \hat{O}(t+\tau)[a,H_S](t)\rangle - \kappa\langle \hat{O}(t+\tau)a(t)\rangle$$
$$- 2\kappa\bar{n}\langle[\hat{O}(t+\tau),a(t)]\rangle. \tag{7.142}$$

Substituting (7.142) into (7.131a) completes the proof of (7.133a) for $\tau > 0$.
□

Exercise 7.8 Show that $(\tau > 0)$

$$\left(\frac{d}{d\tau}+\kappa\right)\langle a(t+\tau)\hat{O}(t)\rangle = \frac{1}{i\hbar}\langle[a,H_S](t+\tau)\hat{O}(t)\rangle, \tag{7.143a}$$

$$\langle \dot{a}\hat{O}\rangle = \frac{1}{i\hbar}\langle[a,H_S]\hat{O}\rangle - \kappa\langle \hat{O}a\rangle - \kappa\bar{n}\langle[\hat{O},a]\rangle, \tag{7.143b}$$

$$\langle \dot{a}(t)\hat{O}(t+\tau)\rangle = \frac{1}{i\hbar}\langle[a,H_S](t)\hat{O}(t+\tau)\rangle - \kappa\langle a(t)\hat{O}(t+\tau)\rangle$$
$$- 2\kappa(\bar{n}+1)\langle[\hat{O}(t+\tau),a(t)]\rangle, \tag{7.143c}$$

and hence prove (7.133b).

7.3.4 Spectrum of the Free Field plus Source Field for the Laser Below Threshold

We can now evaluate all of the terms in (7.124) and (7.125) for a cavity mode radiating into a thermal reservoir. Using (7.133a) we have

$$g_{ss}^{(1)}(\tau) = \left(\langle\hat{\mathcal{E}}^\dagger\hat{\mathcal{E}}\rangle_{ss}\right)^{-1}\left\{(c/L')\langle r_f^\dagger(0)r_f(\tau)\rangle + 2\kappa\left[\lim_{t\to\infty}\langle a^\dagger(t)a(t+\tau)\rangle\right]\right.$$
$$\left. + 2\kappa\bar{n}\left[\lim_{t\to\infty}\langle[a^\dagger(t),a(t+\tau)]\rangle\right]\right\}, \tag{7.144}$$

with

$$\langle\hat{\mathcal{E}}^\dagger\hat{\mathcal{E}}\rangle_{ss} = (c/L')\langle r_f^\dagger r_f\rangle + 2\kappa\langle a^\dagger a\rangle_{ss} + 2\kappa\bar{n}\langle[a^\dagger,a]\rangle_{ss}$$
$$= (c/L')\langle r_f^\dagger r_f\rangle + 2\kappa(\langle a^\dagger a\rangle_{ss} - \bar{n}). \tag{7.145}$$

To go beyond this point we must specify the details of the source that determines the correlation functions $\langle a^\dagger(t)a(t+\tau)\rangle$ and $\langle a(t+\tau)a^\dagger(t)\rangle$. Before we perform the calculation for the laser, let us consider a simpler problem. The damped harmonic oscillator model of Chap. 1 provides a description of a cavity mode coming to thermal equilibrium with the reservoir field. In steady state the presence of the cavity should be invisible to a measurement made on the total reservoir field; effectively, the cavity mode is simply "absorbed" into the reservoir, becoming part of a slightly larger thermal equilibrium system. If we calculate the spectrum of $\hat{E}(z,t)$ $(z > 0)$ from the source field alone, taking the Fourier transform of the correlation function (1.116), we obtain a Lorentzian line with halfwidth κ. This is not correct since it is not the blackbody spectrum. Equations (7.144) and (7.145) give the correct result. From (1.80) and (1.116) we have

$$\langle a^\dagger a\rangle_{ss} = \bar{n}, \tag{7.146a}$$

and

$$\lim_{t\to\infty} \langle a^\dagger(t)a(t+\tau)\rangle = \bar{n}e^{-i\omega_C\tau}e^{-\kappa|\tau|}, \tag{7.146b}$$

$$\lim_{t\to\infty} \langle [a^\dagger(t), a(t+\tau)]\rangle = -e^{-i\omega_C t}e^{-\kappa|\tau|}; \tag{7.146c}$$

the correlation function $\lim_{t\to\infty}\langle a(t+\tau)a^\dagger(t)\rangle = (\bar{n}+1)e^{-i\omega_C t}e^{-\kappa|\tau|}$ needed to obtain the commutator (7.146c) is calculated in a similar manner to (1.116). When these results are substituted into (7.144) and (7.145), we see that the interference term, $2\kappa\bar{n}\lim_{t\to\infty}\langle [a^\dagger(t), a(t+\tau)]\rangle$, between the free field and the source field cancels the source term $2\kappa\lim_{t\to\infty}\langle a^\dagger(t)a(t+\tau)\rangle$. Thus,

$$\begin{aligned}
g_{ss}^{(1)}(\tau) &= \left(\langle r_f^\dagger r_f\rangle\right)^{-1}\langle r_f^\dagger(0)r_f(\tau)\rangle \\
&= \left[\sum_k \omega_k\bar{n}(\omega_k,T)\right]^{-1}\sum_k \omega_k\bar{n}(\omega_k,T)e^{-i\omega_k\tau} \\
&= \left[\int_0^\infty d\omega\,\omega\bar{n}(\omega,T)\right]^{-1}\int_0^\infty d\omega\,\omega\bar{n}(\omega,T)e^{-i\omega\tau} \\
&= \frac{6}{\pi^2}\psi'(1+i\tau/t_R), \tag{7.147}
\end{aligned}$$

where $t_R = \hbar/k_B T$ is the thermal correlation time, and we have used (1.56) to calculate the normalization $\psi'(1)^{-1} = 6/\pi^2$. This is the reservoir correlation function plotted in Fig. 1.1(a). Its Fourier transform gives the (one-dimensional) blackbody spectrum

$$\frac{1}{2\pi}\int_{-\infty}^\infty d\tau e^{i\omega\tau}g_{ss}^{(1)}(\tau) = (6t_R^2/\pi^2)\omega\bar{n}(\omega,T). \tag{7.148}$$

The laser is a nonequilibrium device. Above threshold the photon flux $2\kappa\langle a^\dagger a\rangle_> = 2\kappa n_{sat}(\wp - 1)$ will dominate any thermal background, since, by design, the laser is to act as a source of coherent radiation. Below threshold, from (7.41) and (7.60b) we have

$$\langle a^\dagger a\rangle_{ss} = \frac{\bar{n} + n_{spon}}{1 - \wp}; \tag{7.149a}$$

the mean-value equation $\langle \dot{a}\rangle = -[i\omega_C + \kappa(1-\wp)]\langle a\rangle$ [from (7.71b); also see (8.61b)] and the quantum regression formula [Eqs. (1.107) and (1.108)] give

$$\lim_{t\to\infty} \langle a^\dagger(t)a(t+\tau)\rangle = \frac{\bar{n}+n_{\mathrm{spon}}}{1-\wp} e^{-i\omega_C\tau} e^{-\kappa(1-\wp)|\tau|}, \quad (7.149\mathrm{b})$$

$$\lim_{t\to\infty} \langle [a^\dagger(t), a(t+\tau)]\rangle = -e^{-i\omega_C t} e^{-\kappa(1-\wp)|\tau|}. \quad (7.149\mathrm{c})$$

We then have

$$g_{\mathrm{ss}}^{(1)}(\tau) = \left(\langle \hat{\mathcal{E}}^\dagger \hat{\mathcal{E}}\rangle_{\mathrm{ss}}\right)^{-1} \left\{ \frac{1}{2\pi} (\omega_C t_R^2)^{-1} \psi'(1+i\tau/t_R)\right.$$

$$\left. +2\kappa\left(\frac{\bar{n}+n_{\mathrm{spon}}}{1-\wp} - \bar{n}\right) e^{-i\omega_C\tau} e^{-\kappa(1-\wp)|\tau|} \right\}, \quad (7.150)$$

with

$$\langle \hat{\mathcal{E}}^\dagger \hat{\mathcal{E}}\rangle_{\mathrm{ss}} = \frac{3}{\pi^3} (\omega_C t_R^2)^{-1} + 2\kappa\left(\frac{\bar{n}+n_{\mathrm{spon}}}{1-\wp} - \bar{n}\right). \quad (7.151)$$

The Fourier transform gives a Lorentzian line with halfwidth $\kappa(1-\wp)$ sitting on the background blackbody spectrum (7.148). The Lorentzian component signifies a departure from thermal equilibrium. It has two pieces. First, a Lorentzian proportional to $n_{\mathrm{spon}}/(1-\wp)$ is added to the background blackbody spectrum due to amplified ($\wp > 0$) or deamplified ($\wp < 0$) spontaneous emission. Second, the blackbody spectrum is reshaped over the cavity bandwidth due to the amplification or deamplification of thermal fluctuations by the laser medium. The second effect is accounted for by the term proportional to $\bar{n}/(1-\wp) - \bar{n} = \bar{n}\wp/(1-\wp)$ in (7.151); for $\wp > 0$ thermal fluctuations are amplified over the cavity bandwidth, while for $\wp < 0$ they are deamplified, or absorbed.

8. The Single-Mode Homogeneously Broadened Laser II: Phase-Space Analysis

We now set about the task of reducing the laser phase-space equation of motion (7.100) to the Fokker–Planck equation (7.71a). There are two steps to be taken. We must eliminate derivatives beyond second order, and we must eliminate the explicit appearance of the variables v, v^*, and m describing the laser medium. Actually, we are not quite going to pass directly from (7.100) to (7.71a). We eliminate derivatives beyond second order using van Kampen's system size expansion. But in Sect. 5.1.3 we discussed the fact that a systematic "small noise" expansion generally leads directly to a *linear* Fokker–Planck equation. The laser Fokker–Planck equation (7.71a) is nonlinear. It is possible to arrive at this nonlinear equation from (7.100) by dropping derivatives and performing the adiabatic elimination of atomic variables. This approach, however, does not treat the fluctuations in a systematic way. Equation (7.71a) retains terms of the same order as terms that are dropped; at least it does so in certain operating regions. A systematic system size expansion leads directly to the linearized version of (7.71). This expansion should tell us if, and when, the linearization breaks down. We will therefore first seek a self-consistent laser theory, including fluctuations, analogous to the theory of the radiatively damped two-level medium developed in Sect. 6.3.5; we seek a set of macroscopic equations like (6.167a)–(6.167c) and a linear Fokker–Planck equation describing fluctuations about the macroscopic state. We will find that this linearized theory holds below threshold.

8.1 Linearization: Laser Fokker–Planck Equation Below Threshold

8.1.1 System Size Expansion Below Threshold

We observed in Sect. 7.1.4 that n_{sat} provides a measure of the system "size"; it is a natural choice for the system size parameter Ω [Eqs. (5.39)]. However, our microscopic laser model includes variables for the laser field and the laser medium on an equal basis. The natural system size parameter for the medium is N, the number of atoms occupying the states of the laser transition. At this stage it is simplest to use just one of these parameters to scale all of the variables. It does not really matter which one we choose. Using (7.69c) and (7.73) we see that they are related by

$$N = n_{\text{sat}}\, 4C\xi, \qquad \xi \equiv \frac{2\kappa}{\gamma_\uparrow + \gamma_\downarrow}. \qquad (8.1)$$

We must take $\xi \ll 1$ to justify the adiabatic elimination of atomic variables, and $4C$ need not be correspondingly large. It is possible then that N and n_{sat} differ by a few orders of magnitude. But we can assume that both are much larger than their ratio, so either one can be chosen for the system size parameter and still be the largest parameter (orders of magnitude larger than $1/\xi$) in the problem. We choose N as the system size parameter, in keeping with the work of Haken [8.1].

The system size expansion begins with the definition of scaled variables. The appropriate scaling can often be determined by a combination of guess work and physical intuition. For example, our heuristic derivation of the Fokker–Planck equation (7.71a) can tell us how to scale α, and the size we can expect the fluctuations in α to be. The idea behind van Kampen's method, however, is that a systematic approach will *tell* us the scaling and the size of the fluctuations. We saw how this works for a one-dimensional example in Sect. 5.1.3. Although the algebra is a little tedious, we will perform the present calculation without making *a priori* assumptions about scaling. We set

$$\alpha = N^{p_1}\bar{\alpha}, \qquad \alpha^* = N^{p_1}\bar{\alpha}^*,$$
$$v = N^{p_2}\bar{v}, \qquad v^* = N^{p_2}\bar{v}^*, \qquad m = N^{p_3}\bar{m}, \qquad (8.2)$$

with

$$\bar{\alpha} = \langle \bar{a}(t) \rangle + N^{-q_1} z, \qquad (8.3a)$$

$$\bar{\alpha}^* = \langle \bar{a}^\dagger(t) \rangle + N^{-q_1} z^*, \qquad (8.3b)$$

$$\bar{v} = \langle \bar{J}_-(t) \rangle + N^{-q_2} \nu, \qquad (8.3c)$$

$$\bar{v}^* = \langle \bar{J}_+(t) \rangle + N^{-q_2} \nu^*, \qquad (8.3d)$$

$$\bar{m} = \langle \bar{J}_z(t) \rangle + N^{-q_3} \mu, \qquad (8.3e)$$

where

$$a = N^{p_1}\bar{a}, \qquad a^\dagger = N^{p_1}\bar{a}^\dagger,$$
$$J_- = N^{p_2}\bar{J}_-, \qquad J_+{}^* = N^{p_2}\bar{J}_+, \qquad J_z = N^{p_3}\bar{J}_z. \qquad (8.4)$$

The laser equations themselves will determine the correct choices for p_1, p_2, p_3, q_1, q_2, and q_3. We define

$$\bar{P}(z, z^*, \nu, \nu^*, \mu, t) \equiv N^{p_1 + p_2 + p_3 - q_1 - q_2 - q_3} P\big(\alpha(z, t), \alpha^*(z^*, t),$$
$$v(\nu, t), v^*(\nu, t), m(\mu, t), t\big), \qquad (8.5)$$

and then the phase-space equation of motion in scaled variables is

$$\frac{\partial \bar{P}}{\partial t} = N^{p_1+p_2+p_3-q_1-q_2-q_3} \left(\frac{\partial P}{\partial \alpha}\frac{\partial \alpha}{\partial t} + \frac{\partial P}{\partial \alpha^*}\frac{\partial \alpha^*}{\partial t} + \frac{\partial P}{\partial v}\frac{\partial v}{\partial t} + \frac{\partial P}{\partial v^*}\frac{\partial v^*}{\partial t} \right.$$

$$\left. + \frac{\partial P}{\partial m}\frac{\partial m}{\partial t} + \frac{\partial P}{\partial t} \right)$$

$$= N^{q_1} \left(\frac{\partial \bar{P}}{\partial z}\frac{d\langle \bar{a}(t) \rangle}{dt} + \text{c.c.} \right) + N^{q_2} \left(\frac{\partial \bar{P}}{\partial v}\frac{d\langle \bar{J}_-(t) \rangle}{dt} + \text{c.c.} \right)$$

$$+ N^{q_3}\frac{\partial \bar{P}}{\partial \mu}\frac{d\langle \bar{J}_z(t) \rangle}{dt} + \frac{\partial}{\partial t}\left(N^{p_1+p_2+p_3-q_1-q_2-q_3} P \right)$$

$$= N^{q_1} \left(\frac{\partial \bar{P}}{\partial z}\frac{d\langle \bar{a}(t) \rangle}{dt} + \text{c.c.} \right) + N^{q_2} \left(\frac{\partial \bar{P}}{\partial v}\frac{d\langle \bar{J}_-(t) \rangle}{dt} + \text{c.c.} \right)$$

$$+ N^{q_3}\frac{\partial \bar{P}}{\partial \mu}\frac{d\langle \bar{J}_z(t) \rangle}{dt} + \left[\bar{L}_A\left(v, v^*, \mu, \frac{\partial}{\partial v}, \frac{\partial}{\partial v^*}, \frac{\partial}{\partial \mu}, t \right) \right.$$

$$+ \bar{L}_F\left(z, z^*, \frac{\partial}{\partial z}, \frac{\partial}{\partial z^*}, t \right)$$

$$\left. + \bar{L}_{AF}\left(z, z^*, v, v^*, \mu, \frac{\partial}{\partial z}, \frac{\partial}{\partial z^*}, \frac{\partial}{\partial v}, \frac{\partial}{\partial v^*}, \frac{\partial}{\partial \mu}, t \right) \right] \bar{P}, \qquad (8.6)$$

where

$$\bar{L}_A\left(v, v^*, \mu, \frac{\partial}{\partial v}, \frac{\partial}{\partial v^*}, \frac{\partial}{\partial \mu}, t \right)$$

$$\equiv L_A\left(v(v,t), v^*(v^*,t), m(\mu,t), N^{q_2-p_2}\frac{\partial}{\partial v}, N^{q_2-p_2}\frac{\partial}{\partial v^*}, N^{q_3-p_3}\frac{\partial}{\partial \mu} \right),$$
$$\qquad (8.7a)$$

$$\bar{L}_F\left(z, z^*, \frac{\partial}{\partial z}, \frac{\partial}{\partial z^*}, t \right) \equiv L_F\left(\alpha(z,t), \alpha^*(z^*,t), N^{q_1-p_1}\frac{\partial}{\partial z}, N^{q_1-p_1}\frac{\partial}{\partial z^*} \right),$$
$$\qquad (8.7b)$$

and

$$\bar{L}_{AF}\left(z, z^*, v, v^*, \mu, \frac{\partial}{\partial z}, \frac{\partial}{\partial z^*}, \frac{\partial}{\partial v}, \frac{\partial}{\partial v^*}, \frac{\partial}{\partial \mu}, t \right)$$

$$\equiv L_{AF}\left(\alpha(z,t), \alpha^*(z^*,t), v(v,t), v^*(v^*,t), m(\mu,t), \right.$$

$$\left. N^{q_1-p_1}\frac{\partial}{\partial z}, N^{q_1-p_1}\frac{\partial}{\partial z^*}, N^{q_2-p_2}\frac{\partial}{\partial v}, N^{q_2-p_2}\frac{\partial}{\partial v^*}, N^{q_3-p_3}\frac{\partial}{\partial \mu} \right).$$
$$\qquad (8.7c)$$

Equation (8.6) includes terms of order N^{q_1}, N^{q_2}, and N^{q_3}. Since q_1, q_2, and q_3 are positive, these terms diverge for large N unless their coefficients vanish identically. The requirement that their coefficients vanish determines the macroscopic law, or the laser equations without fluctuations. It also fixes the values of p_1, p_2, and p_3. We substitute the explicit expressions (7.101a)–(7.101c) into (8.7a)–(8.7c), and collect all first-order derivatives of $\bar{P}$ with constant coefficients (coefficients with no dependence on z, z^*, ν, ν^*, and μ). This gives the equation

$$
\frac{\partial \bar{P}}{\partial t} = N^{q_1} \left\{ \frac{\partial \bar{P}}{\partial z} \left[\frac{d\langle \bar{a}(t) \rangle}{dt} + (\kappa + i\omega_C)\langle \bar{a}(t) \rangle \right. \right.
$$

$$
\left. \left. - N^{-p_1 + p_2 - 1/2} \sqrt{N} g \langle \bar{J}_-(t) \rangle \right] + \text{c.c.} \right\}
$$

$$
+ N^{q_2} \left\{ \frac{\partial \bar{P}}{\partial \nu} \left[\frac{d\langle \bar{J}_-(t) \rangle}{dt} + \left(\frac{\gamma_\uparrow + \gamma_\downarrow}{2} + i\omega_C \right) \langle \bar{J}_-(t) \rangle \right. \right.
$$

$$
\left. \left. - N^{p_1 - p_2 + p_3 - 1/2} \sqrt{N} g \langle \bar{J}_z(t) \rangle \langle \bar{a}(t) \rangle \right] + \text{c.c.} \right\}
$$

$$
+ N^{q_3} \frac{\partial \bar{P}}{\partial \mu} \left[\frac{d\langle \bar{J}_z(t) \rangle}{dt} + (\gamma_\uparrow + \gamma_\downarrow)\left(\langle \bar{J}_z(t) \rangle - N^{1-p_3} \frac{\gamma_\uparrow - \gamma_\downarrow}{\gamma_\uparrow + \gamma_\downarrow} \right) \right.
$$

$$
\left. + N^{p_1 + p_2 - p_3 - 1/2} 2\sqrt{N} g \left(\langle \bar{J}_+(t) \rangle \langle \bar{a}(t) \rangle + \text{c.c.} \right) \right]
$$

$$
+ \left(\begin{array}{c} \text{first–order derivatives} \\ \text{with nonconstant coefficients} \end{array} \right) + \left(\begin{array}{c} \text{higher–order} \\ \text{derivatives} \end{array} \right). \tag{8.8}
$$

If the coefficients of N^{q_1}, N^{q_2}, and N^{q_3} are to vanish, the individual terms inside each square bracket must first be of the same order in N. To determine these orders we must recognize that $\sqrt{N} g$ is to be treated as a term of order N^0. This follows from (7.73), which gives

$$
\frac{N g^2}{\kappa^2} = 2C \frac{\gamma_h}{2\kappa} = 2C \frac{\gamma_\uparrow + \gamma_\downarrow + (2\gamma_p)}{2\kappa}. \tag{8.9}
$$

We have divided by κ^2 so that we can compare dimensionless quantities; the square root of the ratio on the left-hand side of (8.9) is the quantity that appears in (8.8) when time is scaled by κ^{-1}. The adiabatic elimination of atomic variables will require $\gamma_h / 2\kappa \gg 1$, and C may also be large [see the discussion below (7.75)]. Therefore $\sqrt{N} g / \kappa$ may be much larger than unity. When we assert that it is of order N^0, however, we claim only that it is much smaller than the lowest nonzero power of N appearing in the system size expansion. Assuming, then, that this is so, the requirement that all terms inside the square brackets in (8.8) are of order N^0 leads to the equations

$$
-p_1 + p_2 - 1/2 = 0, \qquad p_1 - p_2 + p_3 - 1/2 = 0,
$$

$$
1 - p_3 = 0, \qquad p_1 + p_2 - p_3 - 1/2 = 0,
$$

with solutions

$$
p_1 = 1/2, \qquad p_2 = p_3 = 1. \tag{8.10}
$$

Then the requirement that the coefficients of N^{q_1}, N^{q_2}, and N^{q_3} vanish gives the macroscopic law (equations)

$$\frac{d\langle \bar{a} \rangle}{dt} = -(\kappa + i\omega_C)\langle \bar{a} \rangle + \sqrt{N}g\langle \bar{J}_- \rangle, \tag{8.11a}$$

$$\frac{d\langle \bar{a}^\dagger \rangle}{dt} = -(\kappa - i\omega_C)\langle \bar{a}^\dagger \rangle + \sqrt{N}g\langle \bar{J}_+ \rangle, \tag{8.11b}$$

$$\frac{d\langle \bar{J}_- \rangle}{dt} = -\left(\frac{\gamma_\uparrow + \gamma_\downarrow}{2} + i\omega_C\right)\langle \bar{J}_- \rangle + \sqrt{N}g\langle \bar{J}_z \rangle\langle \bar{a} \rangle, \tag{8.11c}$$

$$\frac{d\langle \bar{J}_+ \rangle}{dt} = -\left(\frac{\gamma_\uparrow + \gamma_\downarrow}{2} - i\omega_C\right)\langle \bar{J}_+ \rangle + \sqrt{N}g\langle \bar{J}_z \rangle\langle \bar{a}^\dagger \rangle, \tag{8.11d}$$

$$\frac{d\langle \bar{J}_z \rangle}{dt} = -(\gamma_\uparrow + \gamma_\downarrow)\left(\langle \bar{J}_z \rangle - \frac{\gamma_\uparrow - \gamma_\downarrow}{\gamma_\uparrow + \gamma_\downarrow}\right) - 2\sqrt{N}g(\langle \bar{J}_+ \rangle\langle \bar{a} \rangle + \langle \bar{J}_- \rangle\langle \bar{a}^\dagger \rangle). \tag{8.11e}$$

The first two equations describe the damped field amplitude driven by the polarized laser medium. The last three equations are the optical Bloch equations for the medium [compare Eqs. (2.97)], driven, self-consistently, by the field.

The powers, q_1, q_2, and q_3, that govern the size of the fluctuations, remain to be determined. To do this we must look at the explicit form of the terms in (8.8) designated as "first-order derivatives with nonconstant coefficients" and "higher-order derivatives". Expanding these terms using (8.7) and (7.101), and substituting the known values of p_1, p_2, and p_3, we find

$$\begin{aligned}
\frac{\partial \bar{P}}{\partial t} = &\left\{ \frac{\partial}{\partial z}\left[(\kappa + i\omega_C)z - N^{q_1-q_2}\sqrt{N}g\nu\right] + \text{c.c.} \right. \\
&+ \frac{\partial}{\partial \nu}\left[\left(\frac{\gamma_\uparrow + \gamma_\downarrow}{2} + i\omega_C\right)\nu \right. \\
&\left. -N^{-q_1+q_2}\sqrt{N}g\langle \bar{J}_z(t) \rangle z - N^{q_2-q_3}\sqrt{N}g\langle \bar{a}(t) \rangle\mu\right] + \text{c.c.} \\
&+ \frac{\partial}{\partial \mu}\left[(\gamma_\uparrow + \gamma_\downarrow)\mu + N^{q_3-q_1}2\sqrt{N}g(\langle \bar{J}_+(t) \rangle z + \text{c.c.})\right. \\
&\left. + N^{-q_2+q_3}2\sqrt{N}g(\langle \bar{a}(t) \rangle\nu^* + \text{c.c.})\right] \\
&+ N^{2q_1-1}2\kappa\bar{n}\frac{\partial^2}{\partial z\partial z^*} \\
&+ N^{2q_2-1}\left[\gamma_\uparrow\frac{\partial^2}{\partial \nu\partial \nu^*} + \sqrt{N}g\left(\langle \bar{J}_-(t) \rangle\langle \bar{a}(t) \rangle\frac{\partial^2}{\partial \nu^2} + \text{c.c.}\right)\right] \\
&\left. - N^{q_2+q_3-1}2\gamma_\uparrow\left(\langle \bar{J}_-(t) \rangle\frac{\partial^2}{\partial \nu\partial \mu} + \text{c.c.}\right)\right.
\end{aligned}$$

$$+ N^{2q_3-1} \left[(\gamma_\uparrow + \gamma_\downarrow) \left(1 - \langle \bar{J}_z(t) \rangle \frac{\gamma_\uparrow - \gamma_\downarrow}{\gamma_\uparrow + \gamma_\downarrow} \right) \right.$$

$$\left. - 2\sqrt{N} g (\langle \bar{J}_+(t) \rangle \langle \bar{a}(t) \rangle + \text{c.c.}) \right] \frac{\partial^2}{\partial \mu^2}$$

$$+ \left(\begin{array}{c} \text{first-order derivatives} \\ \text{with nonlinear coefficients} \end{array} \right) + \left(\begin{array}{c} \text{second-order derivatives} \\ \text{with nonconstant coefficients} \end{array} \right)$$

$$\left. + \left(\begin{array}{c} \text{higher-order} \\ \text{derivatives} \end{array} \right) \right\} \bar{P}; \tag{8.12}$$

we have explicitly displayed the terms found in a linear Fokker–Planck equation – first-order derivatives with linear coefficients, and second-order derivatives with constant coefficients; the remaining terms are

$$\left(\begin{array}{c} \text{first-order derivatives} \\ \text{with nonlinear coefficients} \end{array} \right)$$

$$= -N^{q_2-q_3-q_1} \sqrt{N} g \left(\frac{\partial}{\partial \nu} \mu z + \text{c.c.} \right) + N^{q_3-q_1-q_2} 2\sqrt{N} g \frac{\partial}{\partial \mu} (\nu^* z + \text{c.c.}), \tag{8.13a}$$

$$\left(\begin{array}{c} \text{second-order derivatives} \\ \text{with nonconstant coefficients} \end{array} \right)$$

$$= N^{q_3-1} \left[2\gamma_\uparrow \frac{\partial}{\partial \mu} \left(\frac{\partial}{\partial \nu} \nu + \text{c.c.} \right) - (\gamma_\uparrow - \gamma_\downarrow) \frac{\partial^2}{\partial \mu^2} \mu \right]$$

$$+ \sqrt{N} g \left(N^{-q_1+2q_2-1} \langle \bar{J}_-(t) \rangle \frac{\partial^2}{\partial \nu^2} z + N^{q_2-1} \langle \bar{a}(t) \rangle \frac{\partial^2}{\partial \nu^2} \nu \right.$$

$$- N^{2q_3-q_1-1} 2 \langle \bar{J}_+(t) \rangle \frac{\partial^2}{\partial \mu^2} z - N^{-q_2+2q_3-1} 2 \langle \bar{a}(t) \rangle \frac{\partial^2}{\partial \mu^2} \nu^*$$

$$\left. + N^{-q_1+q_2-1} \frac{\partial^2}{\partial \nu^2} \nu z - N^{2q_3-q_1-q_2-1} 2 \frac{\partial^2}{\partial \mu^2} \nu^* z + \text{c.c.} \right), \tag{8.13b}$$

and

$$\left(\begin{array}{c} \text{higher-order} \\ \text{derivatives} \end{array} \right)$$

$$= \gamma_\uparrow \left[N^{3q_2-2} \frac{\partial^2}{\partial \nu \partial \nu^*} \left(\langle \bar{J}_-(t) \rangle \frac{\partial}{\partial \nu} + \text{c.c.} \right) + N^{2q_2-2} \frac{\partial^2}{\partial \nu \partial \nu^*} \left(\frac{\partial}{\partial \nu} + \text{c.c.} \right) \right.$$

$$\left. + N^{4q_2-3} \frac{1}{2} (\langle \bar{J}_z(t) \rangle + 1) \frac{\partial^4}{\partial \nu^2 \partial \nu^{*2}} + N^{4q_2-q_3-3} \frac{1}{2} \frac{\partial^4}{\partial \nu^2 \partial \nu^{*2}} \mu \right]$$

$$+ \left(\begin{array}{c} \text{derivatives higher} \\ \text{than second-order in } \mu \end{array} \right). \tag{8.13c}$$

We hope to choose q_1, q_2, and q_3 so that the terms appearing explicitly in (8.12) are of order N^0, and every term in (8.13a)–(8.13c) vanishes, for large

N, as some negative power of N. All powers of N appearing explicitly in (8.12) are zero if

$$q_1 - q_2 = q_2 - q_3 = q_3 - q_1 = 0, \qquad 2q_1 - 1 = 2q_2 - 1 = 2q_3 - 1 = 0,$$

$$q_2 + q_3 - 1 = 0.$$

These equations are satisfied with

$$q_1 = q_2 = q_3 = \tfrac{1}{2}. \tag{8.14}$$

Then, each term in (8.13a)–(8.13c) does, indeed, vanish as some negative power of N; these terms follow an expansion in powers of $N^{-1/2}$. The terms in (8.13c) designated as "derivatives higher than second-order in μ" come from third- and higher-order derivatives in the expansion of the shift operators $e^{\pm 2 \frac{\partial}{\partial m}}$ in (7.101a) and (7.101c). We can be sure that they also vanish for large N since they are of higher-order in $N^{-1/2}$ than the second-order derivatives in μ that are explicitly displayed. We have therefore found a self-consistent system size expansion. Fluctuations about the macroscopic state described by (8.11) obey the Fokker–Planck equation

$$
\begin{aligned}
\frac{\partial \bar{P}}{\partial t} = \Bigg\{ &\frac{\partial}{\partial z}\Big[(\kappa + i\omega_C)z - \sqrt{N}g\nu\Big] + \frac{\partial}{\partial z^*}\Big[(\kappa - i\omega_C)z^* - \sqrt{N}g\nu^*\Big] \\
&+ \frac{\partial}{\partial \nu}\left[\left(\frac{\gamma_\uparrow + \gamma_\downarrow}{2} + i\omega_C\right)\nu - \sqrt{N}g\big(\langle \bar{J}_z(t)\rangle z + \langle \bar{a}(t)\rangle \mu\big)\right] \\
&+ \frac{\partial}{\partial \nu^*}\left[\left(\frac{\gamma_\uparrow + \gamma_\downarrow}{2} - i\omega_C\right)\nu^* - \sqrt{N}g\big(\langle \bar{J}_z(t)\rangle z^* + \langle \bar{a}^\dagger(t)\rangle \mu\big)\right] \\
&+ \frac{\partial}{\partial \mu}\Big[(\gamma_\uparrow + \gamma_\downarrow)\mu \\
&\quad + 2\sqrt{N}g\big(\langle \bar{J}_+(t)\rangle z + \langle \bar{a}(t)\rangle \nu^* + \langle \bar{J}_-(t)\rangle z^* + \langle \bar{a}^\dagger(t)\rangle \nu\big)\Big] \\
&+ 2\kappa\bar{n}\frac{\partial^2}{\partial z \partial z^*} + \gamma_\uparrow \frac{\partial^2}{\partial \nu \partial \nu^*} + \sqrt{N}g\left(\langle \bar{J}_-(t)\rangle\langle \bar{a}(t)\rangle \frac{\partial^2}{\partial \nu^2}\right. \\
&\left. + \langle \bar{J}_+(t)\rangle\langle \bar{a}^\dagger(t)\rangle \frac{\partial^2}{\partial \nu^{*2}}\right) - \gamma_\uparrow\left(\langle \bar{J}_-(t)\rangle \frac{\partial^2}{\partial \nu \partial \mu} + \langle \bar{J}_+(t)\rangle \frac{\partial^2}{\partial \nu^* \partial \mu}\right) \\
&+ \left[(\gamma_\uparrow + \gamma_\downarrow)\left(1 - \langle \bar{J}_z(t)\rangle \frac{\gamma_\uparrow - \gamma_\downarrow}{\gamma_\uparrow + \gamma_\downarrow}\right)\right. \\
&\left. - 2\sqrt{N}g\big(\langle \bar{J}_+(t)\rangle\langle \bar{a}(t)\rangle + \langle \bar{J}_-(t)\rangle\langle \bar{a}^\dagger(t)\rangle\big)\right]\frac{\partial^2}{\partial \mu^2} \Bigg\} \bar{P}.
\end{aligned}
\tag{8.15}
$$

Note 8.1 When the dephasing term $L_{\text{dephase}}P$ [Eq. (6.158)] is included in the phase-space equation of motion, two changes are required in the linearized theory: In (8.11) and (8.15),

$$\left(\frac{\gamma_\uparrow + \gamma_\downarrow}{2} \pm i\omega_C\right) \to \left(\frac{\gamma_\uparrow + \gamma_\downarrow + 2\gamma_p}{2} \pm i\omega_C\right) = \left(\frac{\gamma_h}{2} \pm i\omega_C\right), \qquad (8.16a)$$

and in (8.15),

$$\gamma_\uparrow \frac{\partial^2}{\partial\nu\partial\nu^*} \to \left[\gamma_\uparrow + \gamma_p(1 + \langle\bar{J}_z(t)\rangle)\right]\frac{\partial^2}{\partial\nu\partial\nu^*}. \qquad (8.16b)$$

8.1.2 Laser Equations Without Fluctuations

Now that we have identified the scaling law for the variables α, α^*, v, v^*, and m, in terms of powers of N, it is convenient to fine tune the scaling relations to simplify the final equations of the linearized theory. It is natural to scale field variables in terms of n_{sat} rather than N, and this choice leads to a simpler Fokker–Planck equation if atomic fluctuations are scaled in the same way. Also, judicious insertion of $2\sqrt{2}C$ in the definition of $\bar{v}$ $(\bar{J}_-)$ and $\bar{v}^*$ $(\bar{J}_+)$, and $2C$ in the definition of $\bar{m}$ $(\bar{J}_z)$, helps simplify the equations. Thus, using (8.10) and (8.14), for the field variables we write

$$\alpha = n_{\text{sat}}^{1/2}\bar{\alpha}, \qquad \alpha^* = n_{\text{sat}}^{1/2}\bar{\alpha}^*, \qquad (8.17)$$

with

$$\bar{\alpha} = \langle\bar{a}(t)\rangle + n_{\text{sat}}^{-1/2}z, \qquad (8.18a)$$

$$\bar{\alpha}^* = \langle\bar{a}^\dagger(t)\rangle + n_{\text{sat}}^{-1/2}z^*, \qquad (8.18b)$$

where

$$a = n_{\text{sat}}^{1/2}\bar{a}, \qquad a^\dagger = n_{\text{sat}}^{1/2}\bar{a}^\dagger; \qquad (8.19)$$

for the atomic variables we write

$$2\sqrt{2}Cv = N\bar{v}, \qquad 2\sqrt{2}Cv^* = N\bar{v}^*, \qquad 2Cm = N\bar{m}, \qquad (8.20)$$

with

$$\bar{v} = \langle\bar{J}_-(t)\rangle + n_{\text{sat}}^{-1/2}\nu, \qquad (8.21a)$$

$$\bar{v}^* = \langle\bar{J}_+(t)\rangle + n_{\text{sat}}^{-1/2}\nu^*, \qquad (8.21b)$$

$$\bar{m} = \langle\bar{J}_z(t)\rangle + n_{\text{sat}}^{-1/2}\mu, \qquad (8.21c)$$

where

$$2\sqrt{2}CJ_- = N\bar{J}_-, \qquad 2\sqrt{2}CJ_+ = N\bar{J}_+, \qquad 2CJ_z = N\bar{J}_z. \qquad (8.22)$$

We must now read the macroscopic equations (8.11) and the Fokker–Planck equation (8.15) with

$$\bar{a} \to \sqrt{\frac{n_{\text{sat}}}{N}}\,\bar{a}, \qquad \bar{a}^\dagger \to \sqrt{\frac{n_{\text{sat}}}{N}}\,\bar{a}^\dagger, \tag{8.23}$$

and

$$\bar{J}_- \to \bar{J}_-/2\sqrt{2}C, \qquad \bar{J}_+ \to \bar{J}_+/2\sqrt{2}C, \qquad \bar{J}_z \to \bar{J}_z/2C, \tag{8.24a}$$

$$\nu \to \sqrt{\frac{N}{n_{\text{sat}}}}\,\nu/2\sqrt{2}C, \quad \nu^* \to \sqrt{\frac{N}{n_{\text{sat}}}}\,\nu^*/2\sqrt{2}C, \quad \mu \to \sqrt{\frac{N}{n_{\text{sat}}}}\,\mu/2C. \tag{8.24b}$$

Now, in a frame rotating at the frequency ω_C of the laser mode, the *laser equations without fluctuations* [Eqs. (8.11)] become

$$\kappa^{-1}\frac{d\langle\tilde{\bar{a}}\rangle}{dt} = -\langle\tilde{\bar{a}}\rangle + \langle\tilde{\bar{J}}_-\rangle, \tag{8.25a}$$

$$\kappa^{-1}\frac{d\langle\tilde{\bar{a}}^\dagger\rangle}{dt} = -\langle\tilde{\bar{a}}^\dagger\rangle + \langle\tilde{\bar{J}}_+\rangle, \tag{8.25b}$$

$$\left(\frac{\gamma_\uparrow + \gamma_\downarrow}{2}\right)^{-1}\frac{d\langle\tilde{\bar{J}}_-\rangle}{dt} = -\langle\tilde{\bar{J}}_-\rangle + \langle\bar{J}_z\rangle\langle\tilde{\bar{a}}\rangle, \tag{8.25c}$$

$$\left(\frac{\gamma_\uparrow + \gamma_\downarrow}{2}\right)^{-1}\frac{d\langle\tilde{\bar{J}}_+\rangle}{dt} = -\langle\tilde{\bar{J}}_+\rangle + \langle\bar{J}_z\rangle\langle\tilde{\bar{a}}^\dagger\rangle, \tag{8.25d}$$

$$(\gamma_\uparrow + \gamma_\downarrow)^{-1}\frac{d\langle\bar{J}_z\rangle}{dt} = -(\langle\bar{J}_z\rangle - \wp) - \tfrac{1}{2}(\langle\tilde{\bar{J}}_+\rangle\langle\tilde{\bar{a}}\rangle + \langle\tilde{\bar{J}}_-\rangle\langle\tilde{\bar{a}}^\dagger\rangle), \tag{8.25e}$$

where

$$\tilde{\bar{a}} \equiv e^{i\omega_C t}\bar{a}, \qquad \tilde{\bar{a}}^\dagger \equiv e^{-i\omega_C t}\bar{a}^\dagger, \tag{8.26a}$$

$$\tilde{\bar{J}}_- \equiv e^{i\omega_C t}\bar{J}_-, \qquad \tilde{\bar{J}}_+ \equiv e^{-i\omega_C t}\bar{J}_+. \tag{8.26b}$$

We have used (8.1) and (7.72), and (8.9) with $\gamma_h = \gamma_\uparrow + \gamma_\downarrow$.

We noted in Sect. 7.1.4 that the laser equations without fluctuations depend on a single intensive parameter, the pump parameter $\wp$. This is the case in (8.25a)–(8.25e). These equations take the place of the rate equations (7.1) and (7.2) in our previous theory. They are more complete than the rate equations since they describe the laser field and the polarization of the laser medium rather than just the photon number and atomic populations. We are interested in steady-state operation. The steady-state solutions are

$$\langle\tilde{\bar{a}}\rangle_{\text{ss}} = |\langle\bar{a}\rangle|_{\text{ss}}\,e^{i\phi}, \tag{8.27a}$$

$$\langle\tilde{\bar{a}}^\dagger\rangle_{\text{ss}} = |\langle\bar{a}\rangle|_{\text{ss}}\,e^{-i\phi}, \tag{8.27b}$$

and

$$\langle \tilde{\bar{J}}_- \rangle_{\text{ss}} = \frac{\wp}{1 + |\langle \bar{a} \rangle|^2_{\text{ss}}} |\langle \bar{a} \rangle|_{\text{ss}} e^{i\phi}, \tag{8.27c}$$

$$\langle \tilde{\bar{J}}_+ \rangle_{\text{ss}} = \frac{\wp}{1 + |\langle \bar{a} \rangle|^2_{\text{ss}}} |\langle \bar{a} \rangle|_{\text{ss}} e^{-i\phi}, \tag{8.27d}$$

$$\langle \bar{J}_z \rangle_{\text{ss}} = \frac{\wp}{1 + |\langle \bar{a} \rangle|^2_{\text{ss}}}, \tag{8.27e}$$

where the mean field amplitude obeys the quadratic equation

$$|\langle \bar{a} \rangle|_{\text{ss}} \left(1 - \wp + |\langle \bar{a} \rangle|^2_{\text{ss}} \right) = 0, \tag{8.28}$$

and the phase ϕ is arbitrary. Solutions to (8.19) reproduce the laser threshold behavior illustrated in Fig. 7.2.

Below Threshold – $\wp < 1$: Since $|\langle \bar{a} \rangle|^2_{\text{ss}}$ must be positive, (8.19) has the single solution

$$|\langle \bar{a} \rangle|_< \equiv |\langle \bar{a} \rangle|^<_{\text{ss}} = 0; \tag{8.29}$$

we then have

$$\langle \tilde{\bar{a}} \rangle_< \equiv \langle \tilde{\bar{a}} \rangle^<_{\text{ss}} = 0, \tag{8.30a}$$

$$\langle \tilde{\bar{a}}^\dagger \rangle_< \equiv \langle \tilde{\bar{a}}^\dagger \rangle^<_{\text{ss}} = 0, \tag{8.30b}$$

$$\langle \tilde{\bar{J}}_- \rangle_< \equiv \langle \tilde{\bar{J}}_- \rangle^<_{\text{ss}} = 0, \tag{8.30c}$$

$$\langle \tilde{\bar{J}}_+ \rangle_< \equiv \langle \tilde{\bar{J}}_+ \rangle^<_{\text{ss}} = 0, \tag{8.30d}$$

$$\langle \bar{J}_z \rangle_< \equiv \langle \bar{J}_z \rangle^<_{\text{ss}} = \wp. \tag{8.30e}$$

At Threshold – $\wp = 1$: Equation (8.19) still has the single solution

$$|\langle \bar{a} \rangle|_{\text{thr}} \equiv |\langle \bar{a} \rangle|^{\text{thr}}_{\text{ss}} = 0, \tag{8.31}$$

and

$$\langle \tilde{\bar{a}} \rangle_{\text{thr}} \equiv \langle \tilde{\bar{a}} \rangle^{\text{thr}}_{\text{ss}} = 0, \tag{8.32a}$$

$$\langle \tilde{\bar{a}}^\dagger \rangle_{\text{thr}} \equiv \langle \tilde{\bar{a}}^\dagger \rangle^{\text{thr}}_{\text{ss}} = 0, \tag{8.32b}$$

$$\langle \tilde{\bar{J}}_- \rangle_{\text{thr}} \equiv \langle \tilde{\bar{J}}_- \rangle^{\text{thr}}_{\text{ss}} = 0, \tag{8.32c}$$

$$\langle \tilde{\bar{J}}_+ \rangle_{\text{thr}} \equiv \langle \tilde{\bar{J}}_+ \rangle^{\text{thr}}_{\text{ss}} = 0, \tag{8.32d}$$

$$\langle \bar{J}_z \rangle_{\text{thr}} \equiv \langle \bar{J}_z \rangle^{\text{thr}}_{\text{ss}} = 1. \tag{8.32e}$$

Above Threshold – $\wp > 1$: Equation (8.19) has two solutions. The solution $|\langle \bar{a} \rangle|_{\text{ss}} = 0$ is unstable and for stable operation above threshold

$$|\langle \bar{a} \rangle|_> \equiv |\langle \bar{a} \rangle|^>_{\text{ss}} = \sqrt{\wp - 1}. \tag{8.33}$$

We then have

$$\langle \tilde{\bar{a}} \rangle_> \equiv \langle \tilde{\bar{a}} \rangle_{ss}^> = \sqrt{\wp - 1}\, e^{i\phi}, \tag{8.34a}$$

$$\langle \tilde{\bar{a}}^\dagger \rangle_> \equiv \langle \tilde{\bar{a}}^\dagger \rangle_{ss}^> = \sqrt{\wp - 1}\, e^{-i\phi}, \tag{8.34b}$$

$$\langle \tilde{\bar{J}}_- \rangle_> \equiv \langle \tilde{\bar{J}}_- \rangle_{ss}^> = \sqrt{\wp - 1}\, e^{i\phi}, \tag{8.34c}$$

$$\langle \tilde{\bar{J}}_+ \rangle_> \equiv \langle \tilde{\bar{J}}_+ \rangle_{ss}^> = \sqrt{\wp - 1}\, e^{-i\phi}, \tag{8.34d}$$

$$\langle \bar{J}_z \rangle_> \equiv \langle \bar{J}_z \rangle_{ss}^> = 1. \tag{8.34e}$$

Note 8.2 Actually, the steady-state solution (8.34) is not always stable. Under certain conditions a *second laser threshold* is reached beyond the threshold at $\wp = 1$. Above the second laser threshold the solution (8.34) is unstable and the laser settles into either a periodic oscillatory state or a chaotic state. This behavior is readily appreciated by noting the relationship between the laser equations (8.25) and the Lorenz equations [8.2]

$$\dot{X} = -\sigma(X - Y), \tag{8.35a}$$

$$\dot{Y} = -Y + rX - XZ, \tag{8.35b}$$

$$\dot{Z} = -bZ + XY. \tag{8.35c}$$

The Lorenz equations have been extensively studied for their interesting non-linear dynamics, in particular, for the chaotic solutions they exhibit [8.3]. If we assume that the phases of the laser field and the medium polarization are equal and constant in time, (8.25a)–(8.25e) are mapped into (8.35a)–(8.35c) by writing

$$X \equiv |\langle \bar{a} \rangle|, \qquad Y \equiv |\langle \bar{J}_- \rangle|, \qquad Z \equiv \wp - \langle \bar{J}_z \rangle, \tag{8.36}$$

and

$$\sigma \equiv \frac{2\kappa}{\gamma_\uparrow + \gamma_\downarrow}, \qquad b \equiv 2, \qquad r \equiv \wp. \tag{8.37}$$

Haken found this mapping connecting the single-mode laser equations with the Lorenz equations [8.4]. In recent years there has been a considerable amount of research in the area of laser instabilities. The field is reviewed by Abraham, Mandel, and Narducci [8.5], and Narducci and Abraham [8.6].

Note 8.3 We have derived the laser equations (8.25) with $\gamma_h = \gamma_\uparrow + \gamma_\downarrow$. When nonradiative dephasing processes are included, if $\bar{J}_-$ and $\bar{J}_+$ are defined by

$$\sqrt{\frac{\gamma_h}{\gamma_\uparrow + \gamma_\downarrow}} 2\sqrt{2} C J_- = N\bar{J}_-, \qquad \sqrt{\frac{\gamma_h}{\gamma_\uparrow + \gamma_\downarrow}} 2\sqrt{2} C J_+ = N\bar{J}_+, \tag{8.38}$$

[in place of (8.22)] the laser equations without fluctuations take the same form as (8.25), with the replacement $\gamma_\uparrow + \gamma_\downarrow \to \gamma_h = \gamma_\uparrow + \gamma_\downarrow + 2\gamma_p$ in (8.25c) and

(8.25d). The identification with the Lorenz equations then requires $\sigma = 2\kappa/\gamma_h$ and $b = 2(\gamma_\uparrow + \gamma_\downarrow)/\gamma_h$.

Exercise 8.1 In Note 7.1 we noted that rate equations are valid when the homogeneous width is much broader than the natural width ($\gamma_h = \gamma_\uparrow + \gamma_\downarrow + \gamma_p \gg \gamma_\uparrow + \gamma_\downarrow$). When this condition is satisfied the polarization variables may be adiabatically eliminated. Show that adiabatic elimination of the polarization from (8.25a)–(8.25e) gives the rate equations

$$(\gamma_\uparrow + \gamma_\downarrow)^{-1}\frac{d\langle \bar{J}_z \rangle}{dt} = -\langle \bar{J}_z \rangle \left(1 + |\langle \bar{a} \rangle|^2\right) + \wp, \qquad (8.39a)$$

$$(2\kappa)^{-1}\frac{d|\langle \bar{a} \rangle|^2}{dt} = -|\langle \bar{a} \rangle|^2 \left(1 - \langle \bar{J}_z \rangle\right). \qquad (8.39b)$$

Show that these equations are equivalent to (7.1) and (7.2) when the three-level model of Sect. 7.1.1 is reduced to our current two-level model for the laser medium (for $\gamma_{32} \gg \Gamma_\wp + \gamma_{31}$, γ_{21}). Equations (8.39) always predict stable steady-state operation above threshold. The comparison between this prediction and the behavior of the Lorenz equations illustrates the limitations of a rate-equation description of laser dynamics.

8.1.3 Linearized Treatment of Quantum Fluctuations Below Threshold

We now use the Fokker–Planck equation (8.15) to describe the fluctuations about the steady-state (8.30). We saw in Sect. 7.1.2 that, in a rate equation theory, the inclusion of thermal and spontaneous photons gives a nonzero photon number in the laser mode below threshold. The laser equations without fluctuations give $|\langle \bar{a} \rangle|_<^2 = 0$; thus, in the present treatment, all of the energy in the laser mode below threshold is carried by the fluctuations. Our first task is to reproduce the rate equation result for the mean photon number [Eq. (7.41)]. Actually, we will derive a more general result which applies for arbitrary values of $\xi = 2\kappa/(\gamma_\uparrow + \gamma_\downarrow)$. Equation (7.41) is recovered in the limit $\xi \ll 1$, the limit that justifies the adiabatic elimination of atomic variables. After making this elimination we will derive the laser linewidth below threshold, something the rate equation treatment was not able to give to us.

Before we begin, we must rewrite the Fokker–Planck equation (8.15) to reflect the new scaling adopted in the last section. For the scaling defined by (8.17)–(8.22) the distribution $\bar{P}$ is given by

$$\bar{P}(z, z^*, \nu, \nu^*, \mu, t)$$
$$= N^3 n_{\text{sat}}^{-3/2}\frac{1}{16C^3}P\big(\alpha(z, t), \alpha^*(z^*, t), v(\nu, t), v^*(\nu^*, t), m(\mu, t), t\big). \qquad (8.40)$$

Then, applying the transformations (8.23) and (8.24), and transforming to a frame rotating at the frequency ω_C, (8.15) becomes

$$\frac{\partial \tilde{\bar{P}}}{\partial t} = \left\{ \kappa \frac{\partial}{\partial \tilde{z}}(\tilde{z} - \tilde{\nu}) + \kappa \frac{\partial}{\partial \tilde{z}^*}(\tilde{z}^* - \tilde{\nu}^*) \right.$$

$$+ \frac{\gamma_\uparrow + \gamma_\downarrow}{2} \frac{\partial}{\partial \tilde{\nu}}\left(\tilde{\nu} - \langle \bar{J}_z(t)\rangle \tilde{z} - \langle \tilde{\bar{a}}(t)\rangle \mu\right)$$

$$+ \frac{\gamma_\uparrow + \gamma_\downarrow}{2} \frac{\partial}{\partial \tilde{\nu}^*}\left(\tilde{\nu}^* - \langle \bar{J}_z(t)\rangle \tilde{z}^* - \langle \tilde{\bar{a}}^\dagger(t)\rangle \mu\right)$$

$$+ (\gamma_\uparrow + \gamma_\downarrow) \frac{\partial}{\partial \mu}\left[\mu + \tfrac{1}{2}\left(\langle \tilde{\bar{J}}_+(t)\rangle \tilde{z} + \langle \tilde{\bar{a}}(t)\rangle \tilde{\nu}^*\right.\right.$$

$$\left.\left. + \langle \tilde{\bar{J}}_-(t)\rangle \tilde{z}^* + \langle \tilde{\bar{a}}^\dagger(t)\rangle \tilde{\nu}\right)\right]$$

$$+ 2\kappa \bar{n} \frac{\partial^2}{\partial \tilde{z}\partial \tilde{z}^*} + \xi^{-1} 2C\gamma_\uparrow \frac{\partial^2}{\partial \tilde{\nu}\partial \tilde{\nu}^*}$$

$$+ \xi^{-1}\frac{\gamma_\uparrow + \gamma_\downarrow}{4}\left(\langle \tilde{\bar{J}}_-(t)\rangle\langle \tilde{\bar{a}}(t)\rangle \frac{\partial^2}{\partial \tilde{\nu}^2} + \langle \tilde{\bar{J}}_+(t)\rangle\langle \tilde{\bar{a}}^\dagger(t)\rangle \frac{\partial^2}{\partial \tilde{\nu}^{*2}}\right)$$

$$- \xi^{-1}\gamma_\uparrow\left(\langle \tilde{\bar{J}}_-(t)\rangle \frac{\partial^2}{\partial \tilde{\nu}\partial \mu} + \langle \tilde{\bar{J}}_+(t)\rangle \frac{\partial^2}{\partial \tilde{\nu}^*\partial \mu}\right)$$

$$+ \xi^{-1}C(\gamma_\uparrow + \gamma_\downarrow)\left[\left(1 - \frac{\wp}{4C^2}\langle \bar{J}_z(t)\rangle\right)\right.$$

$$\left.\left. - \frac{1}{4C}\left(\langle \tilde{\bar{J}}_+(t)\rangle\langle \tilde{\bar{a}}(t)\rangle + \langle \tilde{\bar{J}}_-(t)\rangle\langle \tilde{\bar{a}}^\dagger(t)\rangle\right)\right]\frac{\partial^2}{\partial \mu^2}\right\}\tilde{\bar{P}}, \tag{8.41}$$

where

$$\tilde{\bar{P}}(\tilde{z}, \tilde{z}^*, \tilde{\nu}, \tilde{\nu}^*, \mu, t) \equiv \bar{P}\left(z(\tilde{z}, t), z^*(\tilde{z}^*, t), \nu(\tilde{\nu}, t), \nu^*(\tilde{\nu}^*, t), \mu, t\right), \tag{8.42}$$

with

$$z = e^{-i\omega_C t}\tilde{z}, \qquad z^* = e^{i\omega_C t}\tilde{z}^*, \tag{8.43a}$$

$$\nu = e^{-i\omega_C t}\tilde{\nu}, \qquad \nu^* = e^{i\omega_C t}\tilde{\nu}^*. \tag{8.43b}$$

We have used (8.1) and (7.72), and (8.9) with $\gamma_h = \gamma_\uparrow + \gamma_\downarrow$.

Equation (8.41) is more general than the equation we require here. It provides a linearized description of fluctuations about transient solutions to the macroscopic equations. After substituting the steady-state solutions (8.30), we obtain the much simpler equation

$$\frac{\partial \tilde{\bar{P}}}{\partial t} = \left[\kappa \frac{\partial}{\partial \tilde{z}}(\tilde{z} - \tilde{\nu}) + \kappa \frac{\partial}{\partial \tilde{z}^*}(\tilde{z}^* - \tilde{\nu}^*) + \frac{\gamma_\uparrow + \gamma_\downarrow}{2}\frac{\partial}{\partial \tilde{\nu}}(\tilde{\nu} - \wp\tilde{z})\right.$$

$$+ \frac{\gamma_\uparrow + \gamma_\downarrow}{2}\frac{\partial}{\partial \tilde{\nu}^*}(\tilde{\nu}^* - \wp\tilde{z}^*) + (\gamma_\uparrow + \gamma_\downarrow)\frac{\partial}{\partial \mu}\mu$$

$$+ 2\kappa \bar{n}\frac{\partial^2}{\partial \tilde{z}\partial \tilde{z}^*} + \xi^{-1}2C\gamma_\uparrow \frac{\partial^2}{\partial \tilde{\nu}\partial \tilde{\nu}^*}$$

$$\left. + \xi^{-1}C(\gamma_\uparrow + \gamma_\downarrow)\left(1 - \frac{\wp^2}{4C^2}\right)\frac{\partial^2}{\partial \mu^2}\right]\tilde{\bar{P}}. \tag{8.44}$$

This is the *laser Fokker–Planck equation below threshold without adiabatic elimination of the medium polarization.*

Equation (8.44) can be solved by the separation of variables. We write

$$\tilde{\bar{P}}(\tilde{z}, \tilde{z}^*, \tilde{\nu}, \tilde{\nu}^*, \mu, t) = \tilde{\bar{X}}(\tilde{z}_1, \tilde{\nu}_1, t)\tilde{\bar{Y}}(\tilde{z}_2, \tilde{\nu}_2, t)\bar{\mathcal{M}}(\mu, t), \qquad (8.45)$$

with

$$\tilde{z} = \tilde{z}_1 + i\tilde{z}_2, \qquad \tilde{z}^* = \tilde{z}_1 - i\tilde{z}_2, \qquad (8.46a)$$

$$\tilde{\nu} = \tilde{\nu}_1 + i\tilde{\nu}_2, \qquad \tilde{\nu}^* = \tilde{\nu}_1 - i\tilde{\nu}_2. \qquad (8.46b)$$

Then (8.44) separates into three equations. There are two equations describing real and imaginary parts of the coupled fluctuations in the laser field and the medium polarization:

$$\frac{\partial \tilde{\bar{X}}}{\partial t} = \left[\kappa \frac{\partial}{\partial \tilde{z}_1}(\tilde{z}_1 - \tilde{\nu}_1) + \frac{\gamma_\uparrow + \gamma_\downarrow}{2} \frac{\partial}{\partial \tilde{\nu}_1}(\tilde{\nu}_1 - \wp\tilde{z}_1) + \frac{1}{2}\kappa\bar{n}\frac{\partial^2}{\partial \tilde{z}_1^2} \right.$$
$$\left. + \frac{1}{2}\xi^{-1}C\gamma_\uparrow \frac{\partial^2}{\partial \tilde{\nu}_1^2} \right]\tilde{\bar{X}}, \qquad (8.47a)$$

$$\frac{\partial \tilde{\bar{Y}}}{\partial t} = \left[\kappa \frac{\partial}{\partial \tilde{z}_2}(\tilde{z}_2 - \tilde{\nu}_2) + \frac{\gamma_\uparrow + \gamma_\downarrow}{2} \frac{\partial}{\partial \tilde{\nu}_2}(\tilde{\nu}_2 - \wp\tilde{z}_2) + \frac{1}{2}\kappa\bar{n}\frac{\partial^2}{\partial \tilde{z}_2^2} \right.$$
$$\left. + \frac{1}{2}\xi^{-1}C\gamma_\uparrow \frac{\partial^2}{\partial \tilde{\nu}_2^2} \right]\tilde{\bar{Y}}, \qquad (8.47b)$$

and a third equation describing fluctuations in the atomic inversion:

$$\frac{\partial \bar{\mathcal{M}}}{\partial t} = (\gamma_\uparrow + \gamma_\downarrow)\left[\frac{\partial}{\partial \mu}\mu + \xi^{-1}C\left(1 - \frac{\wp^2}{4C^2}\right)\frac{\partial^2}{\partial \mu^2} \right]\bar{\mathcal{M}}. \qquad (8.47c)$$

We are primarily interested in (8.47a) and (8.47b), since these equations contain the information about the laser field. It is useful, however, to make one observation about (8.47c). This equation is equivalent to the Fokker–Planck equation governing inversion fluctuations in our treatment of the radiatively damped two-level medium (Sects. 6.3.5 and 6.3.6). Specifically, (8.47c) corresponds to the Fokker–Planck equation derived from the third and fifth terms on the right-hand side of (6.172). To see this correspondence we use (8.1) and (7.72) to rewrite the diffusion term in (8.47c) as

$$(\gamma_\uparrow + \gamma_\downarrow)\xi^{-1}C\left(1 - \frac{\wp^2}{4C^2}\right)\frac{\partial^2}{\partial \mu^2}$$
$$= (\gamma_\uparrow + \gamma_\downarrow)\left[1 - \left(\frac{\gamma_\uparrow - \gamma_\downarrow}{\gamma_\uparrow + \gamma_\downarrow}\right)^2\right]\left(4C^2\frac{n_{sat}}{N}\frac{\partial^2}{\partial \mu^2}\right)$$
$$= \frac{4\gamma_\uparrow\gamma_\downarrow}{\gamma_\uparrow + \gamma_\downarrow}\left(4C^2\frac{n_{sat}}{N}\frac{\partial^2}{\partial \mu^2}\right). \qquad (8.48)$$

When we allow for the different definitions of μ [Eq. (8.24b)], (8.48) reproduces the diffusion term in (6.172) with the identifications $\gamma_\uparrow \to \gamma\bar{n}$ and $\gamma_\downarrow \to \gamma(\bar{n}+1)$. So far as the inversion is concerned then, below threshold the laser medium behaves in the same way as the collection of *statistically independent* atoms in our treatment of the radiatively damped two-level medium; of course, with the negative temperature pumping reservoirs needed to produce a positive steady-state inversion. We will see shortly that this connection can often also be made for the polarization, despite the fact that the individual atomic dipoles are coupled by their interaction with the common laser field.

Equations (8.47a) and (8.47b) can be analyzed using results from the treatment of linear Fokker–Planck equations in Sect. 5.2. Fluctuations in the real and imaginary parts of the laser field and medium polarization are statistically independent, and have zero steady-state mean. Therefore, from the phase-space expression for operator averages [Eq. (7.99)] and the scaling (8.17)–(8.19), the average steady-state photon number is given by

$$\langle a^\dagger a \rangle_< \equiv \langle a^\dagger a \rangle_{\text{ss}}^< = \left(\overline{\tilde{z}_1 \tilde{z}_1}\right)_{\tilde{X}_{\text{ss}}} + \left(\overline{\tilde{z}_2 \tilde{z}_2}\right)_{\tilde{Y}_{\text{ss}}}. \tag{8.49}$$

Of course, the fluctuations in the laser field are phase symmetric, and therefore the two contributions to $\langle a^\dagger a \rangle_<$ are equal. Explicit solutions for $\tilde{X}_{\text{ss}}$ and $\tilde{Y}_{\text{ss}}$ are given by (5.80). We do not need the distributions themselves, however, to solve for $\langle a^\dagger a \rangle_<$. The variances appearing on the right-hand side of (8.49) can be found by solving directly for the steady-state covariance matrix

$$\boldsymbol{C}_{\text{ss}} \equiv \begin{pmatrix} \left(\overline{\tilde{z}_1 \tilde{z}_1}\right)_{\tilde{X}_{\text{ss}}} & \left(\overline{\tilde{z}_1 \tilde{\nu}_1}\right)_{\tilde{X}_{\text{ss}}} \\ \left(\overline{\tilde{z}_1 \tilde{\nu}_1}\right)_{\tilde{X}_{\text{ss}}} & \left(\overline{\tilde{\nu}_1 \tilde{\nu}_1}\right)_{\tilde{X}_{\text{ss}}} \end{pmatrix} = \begin{pmatrix} \left(\overline{\tilde{z}_2 \tilde{z}_2}\right)_{\tilde{Y}_{\text{ss}}} & \left(\overline{\tilde{z}_2 \tilde{\nu}_2}\right)_{\tilde{Y}_{\text{ss}}} \\ \left(\overline{\tilde{z}_2 \tilde{\nu}_2}\right)_{\tilde{Y}_{\text{ss}}} & \left(\overline{\tilde{\nu}_2 \tilde{\nu}_2}\right)_{\tilde{Y}_{\text{ss}}} \end{pmatrix}. \tag{8.50}$$

From (5.102a), $\boldsymbol{C}_{\text{ss}}$ satisfies the matrix equation

$$\begin{pmatrix} -2\kappa & 2\kappa \\ (\gamma_\uparrow + \gamma_\downarrow)\wp & -(\gamma_\uparrow + \gamma_\downarrow) \end{pmatrix} \boldsymbol{C}_{\text{ss}} + \boldsymbol{C}_{\text{ss}} \begin{pmatrix} -2\kappa & (\gamma_\uparrow + \gamma_\downarrow)\wp \\ 2\kappa & -(\gamma_\uparrow + \gamma_\downarrow) \end{pmatrix}$$
$$= -2 \begin{pmatrix} \kappa\bar{n} & 0 \\ 0 & \xi^{-1}C\gamma_\uparrow \end{pmatrix}. \tag{8.51}$$

This provides us with a set of three simultaneous equations for the variance of fluctuations in the laser field, the variance of the fluctuations in the medium polarization, and the correlations between the field and polarization. The solution of these equations is left as an exercise:

Exercise 8.2 Solve the matrix equation (8.51) to obtain

$$\left(\overline{\tilde{z}_1 \tilde{z}_1}\right)_{\tilde{X}_{\text{ss}}} = \left(\overline{\tilde{z}_2 \tilde{z}_2}\right)_{\tilde{Y}_{\text{ss}}} = \frac{1}{2} \frac{\bar{n}[2\kappa(1-\wp) + \gamma_\uparrow + \gamma_\downarrow] + 2C\gamma_\uparrow}{(1-\wp)(2\kappa + \gamma_\uparrow + \gamma_\downarrow)}, \tag{8.52a}$$

$$\left(\overline{\tilde{\nu}_1 \tilde{\nu}_1}\right)_{\tilde{X}_{ss}} = \left(\overline{\tilde{\nu}_2 \tilde{\nu}_2}\right)_{\tilde{Y}_{ss}} = \frac{1}{2}\left[\wp \frac{\bar{n}\wp(\gamma_\uparrow + \gamma_\downarrow) + 2C\gamma_\uparrow}{(1-\wp)(2\kappa + \gamma_\uparrow + \gamma_\downarrow)} + \frac{\gamma_\uparrow}{\kappa}C\right], \quad (8.52b)$$

$$\left(\overline{\tilde{z}_1 \tilde{\nu}_1}\right)_{\tilde{X}_{ss}} = \left(\overline{\tilde{z}_2 \tilde{\nu}_2}\right)_{\tilde{Y}_{ss}} = \frac{1}{2}\wp \frac{\bar{n}\wp(\gamma_\uparrow + \gamma_\downarrow) + 2C\gamma_\uparrow}{(1-\wp)(2\kappa + \gamma_\uparrow + \gamma_\downarrow)}. \quad (8.52c)$$

Equations (8.49) and (8.52a) give us the *average photon number in the laser mode below threshold without adiabatic elimination of the medium polarization*:

$$\langle a^\dagger a \rangle_< = \frac{\bar{n}[2\kappa(1-\wp) + \gamma_\uparrow + \gamma_\downarrow] + 2C\gamma_\uparrow}{(1-\wp)(2\kappa + \gamma_\uparrow + \gamma_\downarrow)}. \quad (8.53)$$

8.1.4 Adiabatic Elimination of the Polarization and Laser Linewidth

When the laser medium relaxes much faster than the field [for $\xi \equiv 2\kappa/(\gamma_\uparrow + \gamma_\downarrow) \ll 1$] (8.53) reduces to the expression

$$\begin{aligned}
\langle a^\dagger a \rangle_< &= \frac{\bar{n}(\gamma_\uparrow + \gamma_\downarrow) + C(\gamma_\uparrow + \gamma_\downarrow) + C(\gamma_\uparrow - \gamma_\downarrow)}{(1-\wp)(\gamma_\uparrow + \gamma_\downarrow)} \\
&= \frac{\bar{n} + C + \frac{1}{2}\wp}{1-\wp} \\
&= \frac{\bar{n} + n_{\text{spon}}}{1-\wp}.
\end{aligned} \quad (8.54)$$

This is the result (7.41) obtained from the rate equation theory. The reason for this exact agreement can be appreciated more readily after we adiabatically eliminate the polarization to obtain a stochastic model involving the laser field alone. The adiabatic elimination is made using the Ito stochastic differential equations equivalent to the Fokker–Planck equations (8.47a) and (8.47b). The equivalence between Fokker–Planck equations and Ito stochastic differential equations is defined by (5.149); in the present case it gives

$$d\tilde{z}_i = -\kappa(\tilde{z}_i - \tilde{\nu}_i)dt + \sqrt{\kappa\bar{n}}\, dW_z^i, \quad (8.55a)$$

$$d\tilde{\nu}_i = -\frac{\gamma_\uparrow + \gamma_\downarrow}{2}(\tilde{\nu}_i - \wp\tilde{z}_i)dt + \sqrt{\xi^{-1}C\gamma_\uparrow}\, dW_\nu^i, \quad (8.55b)$$

where W_z^i and W_ν^i are independent Wiener processes, and $i = 1, 2$. For $\xi \ll 1$, we set $d\tilde{\nu}_i = 0$ on the left-hand side of (8.55b), and write

$$\tilde{\nu}_i dt = \wp\tilde{z}_i dt + \left(\frac{\gamma_\uparrow + \gamma_\downarrow}{2}\right)^{-1}\sqrt{\xi^{-1}C\gamma_\uparrow}\, dW_\nu^i. \quad (8.56)$$

Substituting this result into (8.55a), we have

$$d\tilde{z}_i = -\kappa(1 - \wp)\tilde{z}_i dt + \sqrt{\kappa\bar{n}}\, dW_z^i + \xi\sqrt{\xi^{-1}C\gamma_\uparrow}\, dW_\nu^i. \tag{8.57}$$

The last two terms on the right-hand side of (8.57) describe the two sources of fluctuations that drive the laser field: one from the thermal reservoir that damps the laser mode and the other from the laser medium. These fluctuations are statistically independent, and therefore the sum of the two Wiener processes may be replaced by a single process whose variance is the sum of the individual variances. We write

$$\kappa\bar{n} + \xi C\gamma_\uparrow = \kappa\left[\bar{n} + \frac{C(\gamma_\uparrow + \gamma_\downarrow) + C(\gamma_\uparrow - \gamma_\downarrow)}{\gamma_\uparrow + \gamma_\downarrow}\right]$$

$$= \kappa(\bar{n} + C + \tfrac{1}{2}\wp)$$

$$= \kappa(\bar{n} + n_{\text{spon}}). \tag{8.58}$$

The stochastic differential equations for the real and imaginary parts of the laser field amplitude are now

$$d\tilde{z}_i = -\kappa(1 - \wp)\tilde{z}_i dt + \sqrt{\kappa(\bar{n} + n_{\text{spon}})}\, dW_i. \tag{8.59}$$

The Fokker–Planck equations corresponding to (8.59) are

$$\kappa^{-1}\frac{\partial\tilde{X}}{\partial t} = \left[(1 - \wp)\frac{\partial}{\partial\tilde{z}_1}\tilde{z}_1 + \frac{1}{2}(\bar{n} + n_{\text{spon}})\frac{\partial^2}{\partial\tilde{z}_1^2}\right]\tilde{X}, \tag{8.60a}$$

$$\kappa^{-1}\frac{\partial\tilde{Y}}{\partial t} = \left[(1 - \wp)\frac{\partial}{\partial\tilde{z}_2}\tilde{z}_2 + \frac{1}{2}(\bar{n} + n_{\text{spon}})\frac{\partial^2}{\partial\tilde{z}_2^2}\right]\tilde{Y}. \tag{8.60b}$$

Written in complex notation, the *laser Fokker–Planck equation below threshold* is given by

$$\frac{\partial\tilde{P}}{\partial\bar{t}} = \left[(1 - \wp)\left(\frac{\partial}{\partial\tilde{z}}\tilde{z} + \frac{\partial}{\partial\tilde{z}^*}\tilde{z}^*\right) + 2(\bar{n} + n_{\text{spon}})\frac{\partial^2}{\partial\tilde{z}\partial\tilde{z}^*}\right]\tilde{P}, \tag{8.61a}$$

with corresponding stochastic differential equation

$$d\tilde{z} = -(1 - \wp)\tilde{z}d\bar{t} + \sqrt{(\bar{n} + n_{\text{spon}})}(d\bar{W}_1 + id\bar{W}_2), \tag{8.61b}$$

where

$$\bar{t} \equiv \kappa t. \tag{8.62}$$

Equations (8.61a) and (8.61b) are the same as the equations obtained by linearizing the stochastic model (7.71) about the steady state $\langle\tilde{a}\rangle_< = 0$. In contrast to the heuristic derivation offered on the basis of rate equations, the self-consistent treatment we have just followed leads directly to a linear Fokker–Planck equation. This derivation also reveals the role of the adiabatic elimination of the polarization, and the relationship between polarization fluctuations and spontaneous emission into the laser mode. Let us look at these issues briefly before proceeding with the derivation of the laser linewidth.

Polarization fluctuations are fed into the stochastic differential equation for the laser field by the expression (8.56) for $\tilde{\nu}_i dt$. The first thing to note, is that in the adiabatic limit ($\xi \ll 1$), the polarization fluctuations are produced by *statistically independent* atoms. To confirm this assertion we calculate the strength of the polarization fluctuations assuming statistical independence and compare it with (8.52b). For statistically independent atoms, with $\langle J_- \rangle_< = \langle J_+ \rangle_< = 0$, we can write [Eq. (6.150d)]

$$\langle J_+ J_- \rangle_< = \tfrac{1}{2}\big(N + \langle J_z \rangle_< \big),$$

and using the scaling (8.20)–(8.22), and $\langle \bar{J}_z \rangle_< = \wp$, we find

$$
\begin{aligned}
(\overline{\tilde{\nu}_1 \tilde{\nu}_1})_{\tilde{X}} + (\overline{\tilde{\nu}_2 \tilde{\nu}_2})_{\tilde{Y}} &= \langle \bar{J}_+ \bar{J}_- \rangle_< \\
&= 8C^2 \frac{n_{\text{sat}}}{N^2} \frac{N}{2}\Big(1 + \frac{\wp}{2C}\Big) \\
&= \frac{\gamma_\uparrow}{\kappa} C;
\end{aligned}
\tag{8.63}
$$

we have also used (8.1) and (7.72). This result agrees with that given by (8.52b) for $\xi \ll 1$. The second thing to observe is that the source of spontaneous photons in the rate equation theory is represented in the present theory, after adiabatic elimination, by the fluctuations from the medium polarization. The variance of these fluctuations – from the last term in (8.57) – is

$$
\xi C \gamma_\uparrow = \frac{2g^2}{\gamma_h}\Big(N \frac{\gamma_\uparrow}{\gamma_\uparrow + \gamma_\downarrow}\Big).
\tag{8.64}
$$

This is just one half of the spontaneous emission rate $\gamma_{\text{spon}} N_2$ [Eq. (7.26)] into the laser mode; the sum of the variances for the real and imaginary parts of the field give the full spontaneous emission rate. This relationship between polarization fluctuations and spontaneous emission illustrates the ambiguity, or flexibility, of interpretation that often arises in quantum mechanics. We can trace the source of the polarization fluctuations all the way back to the last term, $\gamma_\uparrow \partial^2 / \partial v \partial v^*$, on the right-hand side of (7.101a). This origin suggests that these fluctuations are introduced by the pumping process. Indeed, they are the analog from the pump reservoir of the thermal reservoir fluctuations that drive the damped cavity mode through the term $2\kappa \bar{n} \partial^2 / \partial \alpha \partial \alpha^*$ in (7.101b). Is the source of fluctuations, then, the medium pump fluctuations, or spontaneous emission? Well, for the model we are studying these are the same thing. The polarization fluctuations are certainly derived from the pump interaction and depend directly on the pump rate $\gamma_\uparrow$. But this rate also determines the excited state population $N_2 = N_2^0 = N \gamma_\uparrow / (\gamma_\uparrow + \gamma_\downarrow)$, which determines the rate of spontaneous emission. Thus, the strength of polarization fluctuations driven by the pump process is tied in a self-consistent way to the inversion achieved, and therefore to the spontaneous emission rate into the laser mode. Pump fluctuations and spontaneous emission are two different views of the same thing.

The interpretation in terms of spontaneous emission should, however, be reserved for those conditions that justify adiabatic elimination of the polarization. It is really only under these conditions that the Einstein description of the quantum dynamics in terms of spontaneous emission and stimulated emission is well defined. We can appreciate from (8.52b) that a qualitative change takes place when adiabatic elimination of the polarization is not justified. For $\xi = 2\kappa/(\gamma_\uparrow + \gamma_\downarrow) \sim 1$ the polarization fluctuations no longer take the form (8.63) derived for statistically independent atoms – even if we neglect thermal fluctuations. The atoms communicate with each other through their interaction with the laser field, and the dynamics are truly those of coupled field and polarization oscillators. We will return to the issue of atom-atom correlations in Volume 2 (Sect. 14.1.4).

Exercise 8.3   When nonradiative dephasing processes are included the changes (8.16) are made in the Fokker–Planck equation (8.15). Show that if $\bar{J}_+$ and $\bar{J}_-$ are defined by (8.38), rather than (8.22), and $\bar{v}$ and $\bar{v}^*$ are defined by

$$\sqrt{\frac{\gamma_h}{\gamma_\uparrow + \gamma_\downarrow}} 2\sqrt{2}Cv = N\bar{v}, \qquad \sqrt{\frac{\gamma_h}{\gamma_\uparrow + \gamma_\downarrow}} 2\sqrt{2}Cv^* = N\bar{v}^*, \qquad (8.65)$$

rather than (8.20), the Fokker–Planck equation (8.41) holds with the changes

$$\frac{\gamma_\uparrow + \gamma_\downarrow}{2} \to \frac{\gamma_\uparrow + \gamma_\downarrow + 2\gamma_p}{2} = \frac{\gamma_h}{2}, \qquad (8.66a)$$

$$\xi^{-1} 2C\gamma_\uparrow \frac{\partial^2}{\partial\tilde{v}\partial\tilde{v}^*} \to \xi^{-1} 2C \frac{\gamma_h}{\gamma_\uparrow + \gamma_\downarrow}\left[\gamma_\uparrow + \gamma_p\left(1 + \frac{\langle \bar{J}_z(t) \rangle}{2C}\right)\right] \frac{\partial^2}{\partial\tilde{v}\partial\tilde{v}^*}. \qquad (8.66b)$$

The stochastic differential equation for the polarization [Eq. (8.55b)] is now

$$d\tilde{v}_i = -\frac{\gamma_h}{2}(\tilde{v}_i - \wp z_i)dt + \sqrt{\xi^{-1}C\frac{\gamma_h}{\gamma_\uparrow + \gamma_\downarrow}\left[\gamma_\uparrow + \gamma_p\left(1 + \frac{\wp}{2C}\right)\right]}\,dW_\nu^i. \qquad (8.67)$$

Show that adiabatic elimination of the polarization still gives (8.59) with $n_{\text{spon}} = C + \frac{1}{2}\wp$.

To complete our discussion of fluctuations below threshold we calculate the laser linewidth. The spectrum of the output field is given by the Fourier transform of the normalized first-order correlation function

$$g_<^{(1)}(\tau)$$
$$\equiv \left(\langle a^\dagger a\rangle_<\right)^{-1}\left[\lim_{t\to\infty}\langle a^\dagger(t)a(t+\tau)\rangle\right]$$
$$= \left(\langle a^\dagger a\rangle_<\right)^{-1}e^{-i\omega_C\tau}\left[\lim_{t\to\infty}\left(\overline{\tilde{z}_1(t)\tilde{z}_1(t+\tau)}\right)_{\tilde{X}} + \lim_{t\to\infty}\left(\overline{\tilde{z}_2(t)\tilde{z}_2(t+\tau)}\right)_{\tilde{Y}}\right].$$
$$(8.68)$$

Recall that the laser output field is related to the intracavity source field by (7.107) and (7.113); we can drop the free-field term from (7.107) when evaluating averages that are normal ordered and time ordered (see Sect. 7.3). The correlation functions that appear on the right-hand side of (8.68) are calculated from the Fokker–Planck equations (8.60) using a trivial application of (5.93). We find

$$g_<^{(1)}(\tau) = e^{-i\omega_C\tau}e^{-\kappa(1-\wp)|\tau|}. \qquad (8.69)$$

The Fourier transform is a Lorentzian and the *laser linewidth below threshold* (half-width at half-maximum) is

$$\left(\Delta\omega\right)_< = \kappa(1-\wp) = \kappa\frac{\bar{n}+n_{\text{spon}}}{\langle a^\dagger a\rangle_<} = 2\kappa^2\hbar\omega_C\frac{\bar{n}+n_{\text{spon}}}{P_<}, \qquad (8.70)$$

where we have used (7.114) to express $\langle a^\dagger a\rangle_<$ in terms of the output power $P_<$.

Exercise 8.4 Green function solutions, $\tilde{\bar{X}}(\tilde{z}_1,t|\tilde{z}_1^0,0)$ and $\tilde{\bar{Y}}(\tilde{z}_2,t|\tilde{z}_2^0,0)$, to the Fokker–Planck equations (8.60a) and (8.60b), can be written down from (5.18). Use these to show by direct integration that

$$g_<^{(2)}(\tau)$$
$$\equiv \left(\langle a^\dagger a\rangle_<\right)^{-2}\left[\lim_{t\to\infty}\langle a^\dagger(t)a^\dagger(t+\tau)a(t+\tau)a(t)\rangle\right]$$
$$= \tfrac{1}{2} + \left(\langle a^\dagger a\rangle_<\right)^{-2}\left[\lim_{t\to\infty}\left(\overline{\tilde{z}_1^2(t)\tilde{z}_1^2(t+\tau)}\right)_{\tilde{X}} + \lim_{t\to\infty}\left(\overline{\tilde{z}_2^2(t)\tilde{z}_2^2(t+\tau)}\right)_{\tilde{Y}}\right]$$
$$= 1 + e^{-2\kappa(1-\wp)|\tau|}. \qquad (8.71)$$

This result shows the photon bunching of a "thermal" field [Eq. (1.122)]. Of course, this is expected since (8.61a) has the same form as the Fokker–Planck equation for an oscillator damped by a thermal reservoir; also, we saw that the photon number distribution below threshold is that of a "thermal" field [Eq. (7.60a)].

8.2 Laser Fokker–Planck Equation at Threshold

The linearized treatment of fluctuations breaks down at the laser threshold. For $\wp = 1$ the drift term in the Fokker–Planck equation (8.61a) vanishes, and then there is no restoring force to prevent the fluctuations from growing without bound. This breakdown is apparent from the result (8.52) for the steady-state covariance matrix. Fluctuations in both $\tilde{z}$ and $\tilde{\nu}$ diverge for $\wp = 1$. This problem arises at any critical point, or, more generally, at any bifurcation point where one of the eigenvalues of the linearized deterministic dynamics vanishes. In Sect. 5.1.4 we discussed the resolution of this problem for a one-dimensional example. If the linear coefficient of the first-order derivative term in the system size expansion vanishes, we are not justified in dropping the lowest-order nonlinear coefficient. With the nonlinear term included a new scaling can be found which gives a self-consistent description of fluctuations in terms of a nonlinear Fokker–Planck equation.

Things are not quite so straightforward in a multidimensional problem. For example, none of the first-order derivative terms in the Fokker–Planck equation (8.44) vanish when $\wp = 1$; although, the problem of the divergence shown by (8.52) must be buried in there somewhere. To find it let us calculate the eigenvalues λ of the deterministic equations

$$\dot{\tilde{z}}_i = -\kappa(\tilde{z}_i - \tilde{\nu}_i), \tag{8.72a}$$

$$\dot{\tilde{\nu}}_i = -\frac{\gamma_\uparrow + \gamma_\downarrow}{2}(\tilde{\nu}_i - \wp\tilde{z}_i), \tag{8.72b}$$

derived from the drift terms in (8.44). The eigenvalues satisfy the characteristic equation

$$\lambda^2 + \lambda\left(\kappa + \frac{\gamma_\uparrow + \gamma_\downarrow}{2}\right) + \kappa\frac{\gamma_\uparrow + \gamma_\downarrow}{2}(1 - \wp) = 0. \tag{8.73}$$

One of the eigenvalues vanishes for $\wp = 1$. If the Fokker–Planck equation (8.44) is written in terms of new variables determined by the eigenvectors of its drift matrix, the coefficients of the first-order derivatives are just the eigenvalues λ (see Sect. 5.2.1); then some of the first-order derivative terms do vanish for $\wp = 1$, and nonlinear contributions to the coefficients of these derivatives must be retained in the system size expansion.

The general approach in the multidimensional case starts, therefore, with the diagonalization of the linear drift to determine the *"slow" variables* – the eigenvectors whose eigenvalues vanish at the bifurcation point. In the laser example, however, we need not perform the diagonalization if we plan to adiabatically eliminate the atomic variables. It is clear that after the atomic variables have been eliminated the "slow" variables are the real and imaginary parts of the laser field; all of the *"fast" variables* have already been removed. There is still a small catch. We need to be aware that by taking this approach the system size expansion and adiabatic elimination of atomic variables will not separate as independent calculations. We will have to wait until after the adiabatic elimination of atomic variables has been performed to fix the scaling in the system size expansion.

8.2.1 System Size Expansion and Adiabatic Elimination of Atomic Variables

We refer back to the expansion in Sect. 8.1.1. Two observations allow us to simplify the exact phase-space equation of motion [Eq. (8.12)] and obtain a Fokker–Planck equation with just one undetermined scaling parameter. First, the linear drift terms should keep the form they have in the linearized theory. This requires us to take

$$q_1 - q_2 = q_2 - q_3 = q_3 - q_1 = 0.$$

We define a single scaling parameter

$$q = q_1 = q_2 = q_3. \tag{8.74}$$

We also know that the threshold fluctuations are larger, not smaller than the fluctuations below threshold. Indeed, the linear theory says they become infinite. This cannot be correct, but certainly the exponent q must satisfy the constraint $0 < q < \frac{1}{2}$, since the linearization gave $q = \frac{1}{2}$. This second observation ensures that the "second-order derivatives with nonlinear coefficients" given in (8.13b), and "higher-order derivatives" given in (8.13c), are negligible for large N compared with the second-order derivatives that appear explicitly in (8.12).

We can now write down a Fokker–Planck equation that includes the first-order and second-order derivatives which appear explicitly in (8.12), and the "second-order derivatives with nonlinear coefficients" given in (8.13a). If we scale in powers of n_{sat}, the scaled variables are defined by

$$\bar{\alpha} = e^{-i\omega_C t} \langle \tilde{\bar{a}} \rangle_{\text{thr}} + n_{\text{sat}}^{-q} z, \tag{8.75a}$$

$$\bar{\alpha}^* = e^{i\omega_C t} \langle \tilde{\bar{a}}^\dagger \rangle_{\text{thr}} + n_{\text{sat}}^{-q} z^*, \tag{8.75b}$$

and

$$\bar{v} = e^{-i\omega_C t} \langle \tilde{\bar{J}}_- \rangle_{\text{thr}} + n_{\text{sat}}^{-q} \nu, \tag{8.76a}$$

$$\bar{v}^* = e^{i\omega_C t} \langle \tilde{\bar{J}}_+ \rangle_{\text{thr}} + n_{\text{sat}}^{-q} \nu^*, \tag{8.76b}$$

$$\bar{m} = \langle \bar{J}_z \rangle_{\text{thr}} + n_{\text{sat}}^{-q} \mu. \tag{8.76c}$$

The parameter q remains to be determined. The scaling equations (8.17), (8.19), (8.20), and (8.22) are unchanged. The distribution in scaled variables is defined by

$$\bar{P}(z, z^*, \nu, \nu^*, \mu, t)$$

$$= N^3 n_{\text{sat}}^{1-5q} \frac{1}{16C^3} P(\alpha(z), \alpha^*(z^*), v(\nu), v^*(\nu^*), m(\mu), t), \quad (8.77)$$

and (8.12) and (8.13a) are now to be read with

$$z \to \left(\frac{N}{n_{\text{sat}}}\right)^{q-1/2} z, \qquad z^* \to \left(\frac{N}{n_{\text{sat}}}\right)^{q-1/2} z^*, \qquad (8.78a)$$

and

$$\nu \to \left(\frac{N}{n_{\text{sat}}}\right)^q \nu/2\sqrt{2}C, \quad \nu^* \to \left(\frac{N}{n_{\text{sat}}}\right)^q \nu^*/2\sqrt{2}C, \quad \mu \to \left(\frac{N}{n_{\text{sat}}}\right)^q \mu/2C. \tag{8.78b}$$

Substituting the steady-state solution (8.32), and transforming to a frame rotating at the frequency ω_C, we find

$$\frac{\partial \tilde{\tilde{P}}}{\partial t} = \left\{ \kappa \frac{\partial}{\partial \tilde{z}} (\tilde{z} - \tilde{\nu}) + \kappa \frac{\partial}{\partial \tilde{z}^*} (\tilde{z}^* - \tilde{\nu}^*) \right.$$

$$+ \frac{\gamma_\uparrow + \gamma_\downarrow}{2} \frac{\partial}{\partial \tilde{\nu}} (\tilde{\nu} - \tilde{z} - n_{\text{sat}}^{-q} \mu \tilde{z})$$

$$+ \frac{\gamma_\uparrow + \gamma_\downarrow}{2} \frac{\partial}{\partial \tilde{\nu}^*} (\tilde{\nu}^* - \tilde{z}^* - n_{\text{sat}}^{-q} \mu \tilde{z}^*)$$

$$+ (\gamma_\uparrow + \gamma_\downarrow) \frac{\partial}{\partial \mu} \left[\mu + n_{\text{sat}}^{-q} \tfrac{1}{2} (\tilde{\nu}^* \tilde{z} + \tilde{\nu} \tilde{z}^*) \right]$$

$$+ n_{\text{sat}}^{2q-1} \left[2\kappa \bar{n} \frac{\partial^2}{\partial \tilde{z} \partial \tilde{z}^*} + \xi^{-1} 2C \gamma_\uparrow \frac{\partial^2}{\partial \tilde{\nu} \partial \tilde{\nu}^*} \right.$$

$$\left. \left. + \xi^{-1} C (\gamma_\uparrow + \gamma_\downarrow) \left(1 - \frac{1}{4C^2}\right) \frac{\partial^2}{\partial \mu^2} \right] \right\} \tilde{\tilde{P}}, \qquad (8.79)$$

where $\tilde{\tilde{P}}$ is defined by (8.42). We have used (8.1) and (7.72), and (8.9) with $\gamma_h = \gamma_\uparrow + \gamma_\downarrow$.

Before we can determine the value of q we must perform the adiabatic elimination of atomic variables. In complex notation the Ito stochastic differential equations equivalent to (8.79) are

$$d\tilde{z} = -\kappa(\tilde{z} - \tilde{\nu})dt + n_{\text{sat}}^{q-1/2} \sqrt{\kappa \bar{n}} (dW_z^1 + idW_z^2), \qquad (8.80a)$$

$$d\tilde{\nu} = -\frac{\gamma_\uparrow + \gamma_\downarrow}{2} (\tilde{\nu} - \tilde{z} - n_{\text{sat}}^{-q} \mu \tilde{z})dt + n_{\text{sat}}^{q-1/2} \sqrt{\xi^{-1} C \gamma_\uparrow} (dW_\nu^1 + idW_\nu^2), \tag{8.80b}$$

and

$$d\mu = -(\gamma_\uparrow + \gamma_\downarrow)\left[\mu + n_{\text{sat}}^{-q}\tfrac{1}{2}(\tilde{\nu}^*\tilde{z} + \tilde{\nu}\tilde{z}^*)\right]dt$$
$$+ n_{\text{sat}}^{q-1/2}\sqrt{\xi^{-1}2C(\gamma_\uparrow + \gamma_\downarrow)(1 - 1/4C^2)}\,dW_\mu. \tag{8.80c}$$

To eliminate the atomic variables, we set $d\tilde{\nu} = 0$ and $d\mu = 0$, and write

$$\tilde{\nu}dt = \tilde{z}\big(1 + n_{\text{sat}}^{-q}\mu\big)dt + \left(\frac{\gamma_\uparrow + \gamma_\downarrow}{2}\right)^{-1}n_{\text{sat}}^{q-1/2}\sqrt{\xi^{-1}C\gamma_\uparrow}\,(dW_\nu^1 + idW_\nu^2), \tag{8.81a}$$

$$\mu dt = -n_{\text{sat}}^{-q}\tfrac{1}{2}(\tilde{\nu}^*\tilde{z} + \tilde{\nu}\tilde{z}^*)dt$$
$$+ (\gamma_\uparrow + \gamma_\downarrow)^{-1}n_{\text{sat}}^{q-1/2}\sqrt{\xi^{-1}2C(\gamma_\uparrow + \gamma_\downarrow)(1 - 1/4C^2)}\,dW_\mu. \tag{8.81b}$$

Our objective is to find a solution for $\tilde{\nu}dt$ in terms of the field variables and the Wiener increments dW_ν^1, dW_ν^2, and dW_μ alone. We must therefore eliminate μdt from (8.81a). We follow an approximate procedure that includes the fluctuations to dominant order: Since μdt appears in (8.81a) multiplied by n_{sat}^{-q}, and each of the Wiener increments is multiplied by $n_{\text{sat}}^{q-1/2}$, we may write

$$\mu dt = -\frac{n_{\text{sat}}^{-q}|\tilde{z}|^2}{1 + n_{\text{sat}}^{-2q}|\tilde{z}|^2}dt. \tag{8.82}$$

This is the solution obtained by setting $dW_\nu^1 = dW_\nu^2 = dW_\mu = 0$ in (8.81a) and (8.81b). When we substitute this solution into (8.81a) we are only neglecting fluctuation terms of order $n_{\text{sat}}^{-1/2} \ll n_{\text{sat}}^{q-1/2}$; we obtain

$$\tilde{\nu}dt = \frac{\tilde{z}}{1 + n_{\text{sat}}^{-2q}|\tilde{z}|^2}dt + \left(\frac{\gamma_\uparrow + \gamma_\downarrow}{2}\right)^{-1}n_{\text{sat}}^{q-1/2}\sqrt{\xi^{-1}C\gamma_\uparrow}\,(dW_\nu^1 + idW_\nu^2). \tag{8.83}$$

The stochastic differential equation (8.80a) for the laser field becomes

$$d\tilde{z} = -\kappa n_{\text{sat}}^{-2q}|\tilde{z}|^2\tilde{z}dt + n_{\text{sat}}^{q-1/2}\sqrt{\kappa(\bar{n} + n_{\text{spon}})}\,(dW_1 + idW_2), \tag{8.84a}$$

where the nonlinearity is kept to lowest order in n_{sat}^{-2q}. The corresponding Fokker–Planck equation is

$$\kappa^{-1}\frac{\partial \tilde{\tilde{P}}}{\partial t} = \left[n_{\text{sat}}^{-2q}\left(\frac{\partial}{\partial \tilde{z}}\tilde{z} + \frac{\partial}{\partial \tilde{z}^*}\tilde{z}^*\right)|\tilde{z}|^2 + n_{\text{sat}}^{2q-1}2(\bar{n} + n_{\text{spon}})\frac{\partial^2}{\partial \tilde{z}\partial \tilde{z}^*}\right]\tilde{\tilde{P}}. \tag{8.84b}$$

We can now determine q. For a self-consistent treatment of the fluctuations the drift and diffusion terms in (8.84b) must be of the same order in n_{sat}. Thus,

$$q = \tfrac{1}{4}, \tag{8.85}$$

and the *laser Fokker–Planck equation at threshold* is

$$\frac{\partial \tilde{\bar{P}}}{\partial \bar{t}} = \left[\left(\frac{\partial}{\partial \tilde{z}} \tilde{z} + \frac{\partial}{\partial \tilde{z}^*} \tilde{z}^* \right) |\tilde{z}|^2 + 2(\bar{n} + n_{\text{spon}}) \frac{\partial^2}{\partial \tilde{z} \partial \tilde{z}^*} \right] \tilde{\bar{P}}, \tag{8.86a}$$

with corresponding stochastic differential equation

$$d\tilde{z} = -\tilde{z}|\tilde{z}|^2 d\bar{t} + \sqrt{\bar{n} + n_{\text{spon}}}\,(d\bar{W}_1 + i d\bar{W}_2), \tag{8.86b}$$

where time has the nontrivial scaling

$$\bar{t} \equiv n_{\text{sat}}^{-1/2} \kappa t. \tag{8.87}$$

The fluctuations at threshold scale as $n_{\text{sat}}^{-1/4}$, rather than $n_{\text{sat}}^{-1/2}$, and are there-fore much larger than the fluctuations below threshold. Equations (8.86a) and (8.86b) are the same as the equations derived from the stochastic model (7.71) by setting $\wp = 1$ (taking into account the change of scaling).

Note 8.4 When nonradiative dephasing processes are included, the changes (8.65) and (8.66) are made in the Fokker Planck equation (8.79). After adi-abatic elimination of the atomic variables the laser Fokker–Planck equation at threshold is obtained in the same form [Eq. (8.86a)].

8.2.2 Steady-State Solution and Threshold Photon Number

Because of the nonlinearity, it is quite difficult to find time-dependent so-lutions to the Fokker–Planck equation (8.86a). The methods used to obtain such quantities as $g_{\text{thr}}^{(1)}(\tau)$ and $g_{\text{thr}}^{(2)}(\tau)$ are reviewed by Haken [8.7]. The book on Fokker–Planck equations by Risken [8.8] is also a good source of infor-mation on this subject. The *steady-state* solution to (8.86a) is, on the other hand, easily obtained. From this we can calculate the average photon number at threshold,

$$\langle a^\dagger a \rangle_{\text{thr}} \equiv \langle a^\dagger a \rangle_{\text{ss}}^{\text{thr}} = n_{\text{sat}}^{1/2} (\overline{\tilde{z}^* \tilde{z}})_{\tilde{\bar{P}}_{\text{ss}}}, \tag{8.88}$$

for comparison with the result (7.42) given by rate equations.

The Fokker–Planck equation (8.86a) is phase independent. Therefore, to solve for $\tilde{\bar{P}}_{\text{ss}}$, we first transform to amplitude and phase variables, writing

$$\tilde{z} = r e^{i\psi}, \qquad\qquad \tilde{z}^* = r e^{-i\psi}, \tag{8.89a}$$

$$\frac{\partial}{\partial \tilde{z}} = e^{-i\psi} \frac{1}{2} \left(\frac{\partial}{\partial r} - i \frac{1}{r} \frac{\partial}{\partial \psi} \right), \qquad \frac{\partial}{\partial \tilde{z}^*} = e^{i\psi} \frac{1}{2} \left(\frac{\partial}{\partial r} + i \frac{1}{r} \frac{\partial}{\partial \psi} \right). \tag{8.89b}$$

We define

$$\tilde{\mathcal{P}}(r, \psi, t) \equiv r \tilde{\bar{P}}(r e^{i\psi}, r e^{-i\psi}, t), \tag{8.90}$$

and after some algebra (8.86a) gives

$$\frac{\partial \tilde{\bar{P}}}{\partial \bar{t}} = \left\{ \frac{\partial}{\partial r}\left[r^3 - (\bar{n} + n_{\mathrm{spon}})\frac{1}{2r} \right] + \frac{1}{2}(\bar{n} + n_{\mathrm{spon}})\left(\frac{\partial^2}{\partial r^2} + \frac{1}{r^2}\frac{\partial^2}{\partial \psi^2} \right) \right\} \tilde{\bar{P}}.$$

(8.91)

The steady-state solution has the form

$$\tilde{\bar{P}}_{\mathrm{ss}}(r, \psi) = \frac{1}{2\pi}\bar{R}_{\mathrm{ss}}(r),$$

(8.92)

where

$$\frac{d}{dr}\left[r^3 - (\bar{n} + n_{\mathrm{spon}})\frac{1}{2r} + \frac{1}{2}(\bar{n} + n_{\mathrm{spon}})\frac{d}{dr} \right] \bar{R}_{\mathrm{ss}} = 0.$$

(8.93)

The solution to (8.93) can now be constructed from the general steady-state solution (5.30) for a one-dimensional Fokker–Planck equation:

$$\bar{R}_{\mathrm{ss}}(r) = \frac{1}{\mathcal{N}}\frac{1}{\bar{n} + n_{\mathrm{spon}}}\exp\left[\int dr\left(\frac{1}{r} - \frac{2r^3}{\bar{n} + n_{\mathrm{spon}}} \right) \right]$$

$$= \frac{1}{\mathcal{N}'}r\exp\left(-\frac{1}{2}\frac{r^4}{\bar{n} + n_{\mathrm{spon}}} \right);$$

(8.94a)

the normalization constant is

$$\mathcal{N}' = \int_0^\infty dr\, r\exp\left(-\frac{1}{2}\frac{r^4}{\bar{n} + n_{\mathrm{spon}}} \right)$$

$$= \frac{1}{2}\int_0^\infty dy\exp\left(-\frac{1}{2}\frac{y^2}{\bar{n} + n_{\mathrm{spon}}} \right)$$

$$= \frac{1}{2}\sqrt{\frac{\pi}{2}}\sqrt{\bar{n} + n_{\mathrm{spon}}}.$$

(8.94b)

For the *average photon number in the laser mode at threshold* we find

$$\langle a^\dagger a \rangle_{\mathrm{thr}} = n_{\mathrm{sat}}^{1/2}\left(\overline{r^2} \right)_{\bar{R}_{\mathrm{ss}}}$$

$$= n_{\mathrm{sat}}^{1/2}\sqrt{\frac{2}{\pi}}\frac{2}{\sqrt{\bar{n} + n_{\mathrm{sat}}}}\int_0^\infty dr\, r^3\exp\left(-\frac{1}{2}\frac{r^4}{\bar{n} + n_{\mathrm{spon}}} \right)$$

$$= \sqrt{\frac{2}{\pi}}\sqrt{n_{\mathrm{sat}}(\bar{n} + n_{\mathrm{spon}})}.$$

(8.95)

This result differs from that given by the rate equation theory [Eq.(7.42)] by the factor $\sqrt{2/\pi}$. It agrees with the expression (7.61b) obtained from the birth-death equation constructed in Sect. 7.1.3. In Sect. 7.1.3 we saw that the connection between the photon number rate equation and the probabilistic birth-death, or stochastic models, was the factorization $\langle n^2 \rangle \rightarrow \langle n \rangle^2$, or $\langle |\tilde{\alpha}|^4 \rangle \rightarrow (\langle |\tilde{\alpha}|^2 \rangle)^2$. This factorization is unimportant below threshold where the nonlinearity is negligible; we therefore obtain the average photon number below threshold exactly using rate equations. At threshold, however, the linear term in the photon number rate equation vanishes, and the nonlinear term is then dominant. The disagreement between (8.95) and (7.42) is a result of the factorization assumed by the photon number rate equation.

Note 8.5 We can now assess the range of validity of the linearized treatment of fluctuations. Linearization is valid below threshold so long as the linear drift term in the Fokker–Planck equation (8.61a) is much larger than the nonlinear drift term in the Fokker–Planck equation (8.86a). Taking the different scaling of the two Fokker–Planck equations into account, this requires

$$1 - \wp \gg n_{\text{sat}}^{-1/2} |\tilde{z}|_{\text{thr}}^2 \equiv n_{\text{sat}}^{-1/2} (\overline{\tilde{z}^* \tilde{z}})_{\tilde{P}_{ss}} = n_{\text{sat}}^{-1/2} \langle a^\dagger a \rangle_{\text{thr}}$$

$$= \sqrt{\frac{2}{\pi}} \sqrt{\frac{\bar{n} + n_{\text{spon}}}{n_{\text{sat}}}}. \qquad (8.96)$$

This condition is consistent with our definition of the laser threshold region in (7.38). For the range of $\wp$ that matches the linearized theory below threshold to the nonlinear theory at threshold, both linear and nonlinear drift terms can be included, as in (7.71). (Of course, this Fokker–Planck equation is not systematic to the order of the smaller of the two drift terms at either end of the matching range.)

Exercise 8.5 Show that, for $|1 - \wp| \ll 1$, the Fokker–Planck equation (7.71) has the steady-state solution

$$\tilde{\bar{P}}_{ss}(\tilde{\alpha}) = \frac{1}{2\pi} \sqrt{\frac{2}{\pi}} \sqrt{\frac{n_{\text{sat}}}{\bar{n} + n_{\text{spon}}}} 2 \left[1 + \Phi \left(\sqrt{2} \frac{\wp - 1}{|1 - \wp|_{\text{thr}}} \right) \right]^{-1}$$

$$\times \exp \left[-\frac{1}{2} \frac{\left(|\tilde{\alpha}|^2 - (\wp - 1) \right)^2}{(\bar{n} + n_{\text{spon}})/n_{\text{sat}}} \right]. \qquad (8.97)$$

8.3 Quasi-Linearization: Laser Fokker–Planck Equation Above Threshold

Sufficiently far above threshold the nonlinear drift will again be negligible with respect to the linear drift, as it was below threshold. We might then expect to return to the linearized treatment of fluctuations described in Sects. 8.1.2 and 8.1.3. Unfortunately, things are not this simple. Equations (8.18) and (8.21) expand the fluctuations about macroscopic field and polarization states with defined amplitude *and* phase. Above threshold the macroscopic equations (8.25) fix the amplitudes of the laser field and medium polarization in the steady state. They also require the phases of the field and polarization to be locked. But the common phase ϕ for the field and polarization is left undetermined. This means that fluctuations in ϕ are free to grow without bound to produce a phase-symmetric steady-state distribution. Figure 8.1 illustrates the distribution $\bar{P}_{ss}(\tilde{\alpha})$, plotted from (8.97), close to threshold. Above threshold a phase-symmetric state with nonzero amplitude develops [Fig. 8.1(c)]. Clearly the scaling defined by (8.18) and (8.21) is inadequate to treat such phase-symmetric fluctuations. The steady-state (8.34) gives $|\langle \bar{a} \rangle|_> = |\langle \bar{J}_- \rangle|_> = \sqrt{\wp - 1}$, and fluctuations $z \sim n_{\text{sat}}^{1/2}\sqrt{\wp - 1}$ and $\nu \sim n_{\text{sat}}^{1/2}\sqrt{\wp - 1}$ are needed to distribute the phases of the field and polarization over the range $-\pi < \phi \leq \pi$. If $n_{\text{sat}}^{1/2}\sqrt{\wp - 1} \gg 1$, this requires z and ν to be large, contrary to the assumption that the scaled fluctuations are of order unity. To deal with this difficulty we must base our treatment of quantum fluctuations above threshold on a system size expansion in amplitude and phase variables.

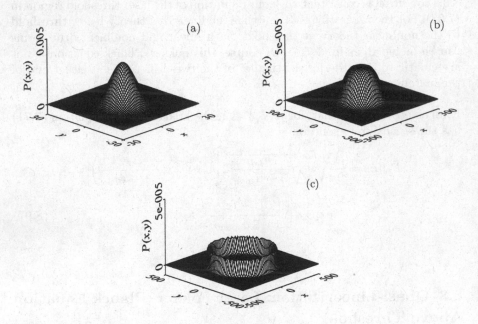

Fig. 8.1 Laser phase-space distribution for $\bar{n} + n_{\text{spon}} = 1$, $n_{\text{sat}} = 10^8$: (a) below threshold, $1 - \wp = 10^{-2}$; (b) at threshold, $\wp = 1$; (c) above threshold, $\wp - 1 = 10^{-3}$. (The parameter values are the same as those in Fig. 7.5.)

8.3.1 System Size Expansion Above Threshold

The system size expansion in amplitude and phase variables is, in principle, no more complicated than the expansion made in Sect. 8.1.1. The algebra can get a little confusing, however, because the change of variables involves a nonlinear transformation; we must systematically expand the nonlinearities arising from the change of variables along with the nonlinearities and higher-order derivatives we have already met in the phase-space equation of motion. Also, we must be particularly careful in our treatment of the phases. Without fluctuations, the phases of the laser field and the medium polarization are locked. In the presence of fluctuations, small differences between the phases of the field and polarization can arise. These small fluctuations in the phase difference must be retained to give a correct treatment of the much larger fluctuations that develop in the phase sum. Fluctuations in the phase sum are responsible for the laser linewidth above threshold. To try and separate the different aspects of the problem let us first make an expansion of the field and polarization amplitudes alone. Then we will address the question of phase fluctuations separately.

It is a laborious task to begin again from scratch, with arbitrary scaling parameters p_1, p_2, p_3, q_1, q_2, and q_3, as in (8.2)–(8.4). In fact, this is not necessary. It is reasonable to assume that the fluctuations in field and polarization amplitudes scale as they did below threshold, as $n_{\text{sat}}^{-1/2}$. If this choice is incorrect, we will certainly discover our mistake in the course of the calculations, since the expansion will not be self-consistent. Thus, in place of (8.18) and (8.21) we write

$$\bar{\alpha} = e^{i\phi}\big(\bar{\mathcal{A}}(t) + n_{\text{sat}}^{-1/2}z\big), \tag{8.98a}$$

$$\bar{\alpha}^* = e^{-i\phi}\big(\bar{\mathcal{A}}(t) + n_{\text{sat}}^{-1/2}z\big), \tag{8.98b}$$

and

$$\bar{v} = e^{i\theta}\big(\bar{\mathcal{J}}(t) + n_{\text{sat}}^{-1/2}\nu\big), \tag{8.99a}$$

$$\bar{v}^* = e^{-i\theta}\big(\bar{\mathcal{J}}(t) + n_{\text{sat}}^{-1/2}\nu\big), \tag{8.99b}$$

$$\bar{m} = \langle \bar{J}_z(t)\rangle + n_{\text{sat}}^{-1/2}\mu. \tag{8.99c}$$

In these expressions z and ν have a new meaning; they now represent *real* amplitude fluctuations rather than complex amplitude fluctuations. These fluctuations can be both positive and negative, but must fall within the ranges $-n_{\text{sat}}^{1/2}\bar{\mathcal{A}}(t) \leq z < \infty$ and $-n_{\text{sat}}^{1/2}\bar{\mathcal{J}}(t) \leq \nu < \infty$, bounded below, since $\big(\bar{\mathcal{A}}(t) + n_{\text{sat}}^{-1/2}\big)$ and $\big(\bar{\mathcal{J}}(t) + n_{\text{sat}}^{-1/2}\nu\big)$ must be positive. The phase-space distribution in scaled variables, normalized with respect to the integration measure $dz\,d\phi\,d\nu\,d\theta\,d\mu$, is defined by

$$P'(z, \phi, \nu, \theta, \mu, t)$$

$$\equiv N^3 n_{\text{sat}}^{-3/2} \frac{1}{16C^3} \left(\bar{\mathcal{A}}(t) + n_{\text{sat}}^{-1/2} z\right)\left(\bar{\mathcal{J}}(t) + n_{\text{sat}}^{-1/2} \nu\right)$$

$$\times P\left(\alpha(z, \phi, t), \alpha^*(z, \phi, t), v(\nu, \theta, t), v^*(\nu, \theta, t), m(\mu, t), t\right). \quad (8.100)$$

Note 8.6 In (8.18) and (8.21), the field and polarization fluctuations are expanded about operator averages $\langle \bar{a}(t) \rangle$, $\langle \bar{a}^\dagger(t) \rangle$, $\langle \bar{J}_-(t) \rangle$, and $\langle \bar{J}_+(t) \rangle$ (by definition the fluctuations z and ν have zero mean). A little thought shows that it is not generally possible to give simple expressions for $\bar{\mathcal{A}}(t)$ and $\bar{\mathcal{J}}(t)$ in terms of operator averages. We can certainly relate these quantities to averages of the stochastic variables $\bar{\alpha}$ and $\bar{v}$. Taking the mean of z and ν to be zero, we have $\bar{\mathcal{A}}(t) = \left(|\bar{\alpha}|(t)\right)_P$, and $\bar{\mathcal{J}}(t) = \left(|\bar{v}|(t)\right)_{P'}$. The difficulty arises when we try to relate these stochastic averages to operator averages. The relationship (7.99) only applies for operators that can be written as normal-ordered power series. Amplitude and phase operators are not of this type. Consider the field amplitude. The definition of amplitude and phase operators for the field is not unique, but a self-consistent definition is possible [8.9]; for the sake of argument let us say that the scaled amplitude operator is $\sqrt{\bar{a}^\dagger \bar{a}}$, whose action on the Fock state basis is given by $\sqrt{\bar{a}^\dagger \bar{a}}|n\rangle = n_{\text{sat}}^{-1/2} \sqrt{n}|n\rangle$. It is easy to see that $\langle \sqrt{\bar{a}^\dagger \bar{a}} \rangle$ is not given by $(|\bar{\alpha}|)_{\bar{P}}$ [we write $\bar{P}(\bar{\alpha}, \bar{\alpha}^*) = n_{\text{sat}} P\left(n_{\text{sat}}^{1/2} \bar{\alpha}, n_{\text{sat}}^{1/2} \bar{\alpha}^*\right)$.] To illustrate this point we take the field to be in the coherent state $|e^{-i\omega_C t} n_{\text{sat}}^{1/2} \bar{\alpha}_0\rangle$. Then $\bar{P}(\bar{\alpha}, \bar{\alpha}^*) = \delta(\bar{\alpha} - \bar{\alpha}_0)$ and $(|\bar{\alpha}|)_{\bar{P}} = |\bar{\alpha}_0|$. However,

$$\langle \sqrt{\bar{a}^\dagger \bar{a}} \rangle = \langle e^{-i\omega_C t} n_{\text{sat}}^{1/2} \bar{\alpha}_0 | \sqrt{\bar{a}^\dagger \bar{a}} | e^{-i\omega_C t} n_{\text{sat}}^{1/2} \bar{\alpha}_0 \rangle$$

$$= n_{\text{sat}}^{-1/2} \sum_{n=0}^\infty \sqrt{n} \frac{\left(n_{\text{sat}} |\bar{\alpha}_0|^2\right)^n}{n!} \exp\left(- n_{\text{sat}} |\bar{\alpha}_0|^2\right).$$

This is a Poisson average of $\sqrt{n}$ which is not generally equal to $|\bar{\alpha}_0|$. For arbitrary field states we can use (3.15) to write

$$\langle \sqrt{\bar{a}^\dagger \bar{a}} \rangle = n_{\text{sat}}^{-1/2} \int d^2\bar{\alpha} \left(\sum_{n=0}^\infty \sqrt{n} \frac{\left(n_{\text{sat}} |\bar{\alpha}|^2\right)^n}{n!} \exp\left(- n_{\text{sat}} |\bar{\alpha}|^2\right) \right) \bar{P}(\bar{\alpha}, \bar{\alpha}^*);$$

$$(8.101)$$

this is not generally equal to $(|\bar{\alpha}|)_{\bar{P}}$. Nevertheless, having said all this, from (8.201) we see that in the limit of large n_{sat} these subtleties are rather unimportant. When $\bar{P}(\bar{\alpha}, \bar{\alpha}^*)$ is peaked about some $|\bar{\alpha}| \approx \bar{\mathcal{A}}$, and $n_{\text{sat}} \bar{\mathcal{A}}^2 \gg 1$, the Poisson average in the integrand of (8.201) is approximately equal to $n_{\text{sat}}^{1/2} \bar{\mathcal{A}}$ over the dominant range of the integral, and to dominant order we can write $\bar{\mathcal{A}} = \langle \sqrt{\bar{a}^\dagger \bar{a}} \rangle$.

To expand the phase-space equation of motion we must transform the derivatives with respect to $\bar{\alpha}$, $\bar{\alpha}^*$, $\bar{v}$, and $\bar{v}^*$ into derivatives with respect to z, ϕ, ν, and θ. The proof of the basic transformation formulas is left as an exercise:

Exercise 8.6 Show that

$$(\bar{A}(t) + n_{\text{sat}}^{-1/2}z)\frac{\partial}{\partial\bar{\alpha}}$$

$$= \frac{1}{2}\left(n_{\text{sat}}^{1/2}\frac{\partial}{\partial z} - i\frac{1}{\bar{A}(t) + n_{\text{sat}}^{-1/2}z}\frac{\partial}{\partial\phi}\right)e^{-i\phi}(\bar{A}(t) + n_{\text{sat}}^{-1/2}z),$$

(8.102a)

$$(\bar{A}(t) + n_{\text{sat}}^{-1/2}z)\frac{\partial}{\partial\bar{\alpha}^*}$$

$$= \frac{1}{2}\left(n_{\text{sat}}^{1/2}\frac{\partial}{\partial z} + i\frac{1}{\bar{A}(t) + n_{\text{sat}}^{-1/2}z}\frac{\partial}{\partial\phi}\right)e^{i\phi}(\bar{A}(t) + n_{\text{sat}}^{-1/2}z),$$

(8.102b)

and

$$(\bar{J}(t) + n_{\text{sat}}^{-1/2}\nu)\frac{\partial}{\partial\bar{v}}$$

$$= \frac{1}{2}\left(n_{\text{sat}}^{1/2}\frac{\partial}{\partial\nu} - i\frac{1}{\bar{J}(t) + n_{\text{sat}}^{-1/2}\nu}\frac{\partial}{\partial\theta}\right)e^{-i\theta}(\bar{J}(t) + n_{\text{sat}}^{-1/2}\nu),$$

(8.103a)

$$(\bar{J}(t) + n_{\text{sat}}^{-1/2}\nu)\frac{\partial}{\partial\bar{v}^*}$$

$$= \frac{1}{2}\left(n_{\text{sat}}^{1/2}\frac{\partial}{\partial\nu} + i\frac{1}{\bar{J}(t) + n_{\text{sat}}^{-1/2}\nu}\frac{\partial}{\partial\theta}\right)e^{i\theta}(\bar{J}(t) + n_{\text{sat}}^{-1/2}\nu).$$

(8.103b)

Now, from the scaling transformation (8.98)–(8.100), and (8.102) and (8.103), we find

$$\frac{\partial P'}{\partial t} = N^3 n_{\text{sat}}^{-3/2}\frac{1}{16C^3}(\bar{A}(t) + n_{\text{sat}}^{-1/2}z)(\bar{J}(t) + n_{\text{sat}}^{-1/2}\nu)$$

$$\times\left(\frac{\partial P}{\partial\alpha}\frac{\partial\alpha}{\partial t} + \frac{\partial P}{\partial\alpha^*}\frac{\partial\alpha^*}{\partial t} + \frac{\partial P}{\partial v}\frac{\partial v}{\partial t} + \frac{\partial P}{\partial v^*}\frac{\partial v^*}{\partial t} + \frac{\partial P}{\partial m}\frac{\partial m}{\partial t} + \frac{\partial P}{\partial t}\right)$$

$$+ \frac{d\bar{A}(t)}{dt}\frac{1}{\bar{A}(t) + n_{\text{sat}}^{-1/2}z}P' + \frac{d\bar{J}(t)}{dt}\frac{1}{\bar{J}(t) + n_{\text{sat}}^{-1/2}\nu}P'$$

$$
= n_{\text{sat}}^{1/2} \left(\frac{\partial P'}{\partial z} \frac{d\bar{A}(t)}{dt} + \frac{\partial P'}{\partial \nu} \frac{d\bar{\mathcal{J}}(t)}{dt} + \frac{\partial P'}{\partial \mu} \frac{d\langle \bar{J}_z(t) \rangle}{dt} \right)
$$

$$
+ \left(\bar{A}(t) + n_{\text{sat}}^{-1/2} z \right) \left(\bar{\mathcal{J}}(t) + n_{\text{sat}}^{-1/2} \nu \right) \frac{\partial}{\partial t} \left(N^3 n_{\text{sat}}^{-3/2} \frac{1}{16C^3} P \right).
$$

$$(8.104)$$

The explicit form of the last term on the right-hand side is given by (7.100). There is rather a lot of algebra involved in writing this out in terms of amplitude and phase variables. However, we only need to find the explicit form for terms involving derivatives up to second-order. We can convince ourselves that all higher-order derivatives vanish as some power of $n_{\text{sat}}^{-1/2}$, just as the terms (8.13b) and (8.13c) vanished in the expansion in Sect. 8.1.1. Nonlinear terms corresponding to those given in (8.13a) vanish in the same way. After completing the algebra we can write

$$
\frac{\partial P'}{\partial t} = n_{\text{sat}}^{1/2} \left\{ \frac{\partial P'}{\partial z} \left[\frac{d\bar{A}(t)}{dt} + \kappa \left(\bar{A}(t) - \bar{\mathcal{J}}(t) \cos(\phi - \theta) \right) \right] \right.
$$

$$
+ \frac{\partial P'}{\partial \nu} \left[\frac{d\bar{\mathcal{J}}(t)}{dt} + \frac{\gamma_\uparrow + \gamma_\downarrow}{2} \left(\bar{\mathcal{J}}(t) - \langle \bar{J}_z(t) \rangle \bar{A}(t) \cos(\phi - \theta) \right) \right]
$$

$$
\left. + \frac{\partial P'}{\partial \mu} \left[\frac{d\langle \bar{J}_z(t) \rangle}{dt} + (\gamma_\uparrow + \gamma_\downarrow) \left(\langle \bar{J}_z(t) \rangle - \wp + \bar{\mathcal{J}}(t) \bar{A}(t) \cos(\phi - \theta) \right) \right] \right\}
$$

$$
+ \left\{ \kappa \frac{\partial}{\partial z} \left[z - \nu \cos(\phi - \theta) - n_{\text{sat}}^{-1/2} \frac{1}{2} \bar{n} \frac{1}{\bar{A}(t) + n_{\text{sat}}^{-1/2} z} \right] \right.
$$

$$
+ \frac{\partial}{\partial \phi} \left[\omega_C + \kappa \bar{\mathcal{J}}(t) \frac{1}{\bar{A}(t) + n_{\text{sat}}^{-1/2} z} \sin(\phi - \theta) \right]
$$

$$
+ \frac{\gamma_\uparrow + \gamma_\downarrow}{2} \frac{\partial}{\partial \nu} \left[\nu - \left(\langle \bar{J}_z(t) \rangle z + \bar{A}(t) \mu \right) \cos(\phi - \theta) \right.
$$

$$
\left. - n_{\text{sat}}^{-1/2} \xi^{-1} \frac{1}{2} \left(n_{\text{spon}} + \tfrac{1}{2} \bar{\mathcal{J}}(t) \bar{A}(t) \cos(\phi - \theta) \right) \frac{1}{\bar{\mathcal{J}}(t) + n_{\text{sat}}^{-1/2} \nu} \right]
$$

$$
+ \frac{\partial}{\partial \theta} \left[\omega_C - \frac{\gamma_\uparrow + \gamma_\downarrow}{2} \langle \bar{J}_z(t) \rangle \bar{A}(t) \frac{1}{\bar{\mathcal{J}}(t) + n_{\text{sat}}^{-1/2} \nu} \sin(\phi - \theta) \right]
$$

$$
+ (\gamma_\uparrow + \gamma_\downarrow) \frac{\partial}{\partial \mu} \left[\mu + \left(\bar{\mathcal{J}}(t) z + \bar{A}(t) \nu \right) \cos(\phi - \theta) \right]
$$

$$
+ \frac{1}{2} \kappa \bar{n} \left[\frac{\partial^2}{\partial z^2} + n_{\text{sat}}^{-1} \frac{1}{\left(\bar{A}(t) + n_{\text{sat}}^{-1/2} z \right)^2} \frac{\partial^2}{\partial \phi^2} \right]
$$

$$
+ \frac{1}{2} \xi^{-1} \left[C \gamma_\uparrow + \frac{\gamma_\uparrow + \gamma_\downarrow}{4} \bar{\mathcal{J}}(t) \bar{A}(t) \cos(\phi - \theta) \right]
$$

$$\times \left[\frac{\partial^2}{\partial \nu^2} + n_{sat}^{-1} \frac{1}{\left(\bar{\mathcal{J}}(t) + n_{sat}^{-1/2} \nu \right)^2} \frac{\partial^2}{\partial \theta^2} \right] - \xi^{-1} \gamma_\uparrow \bar{\mathcal{J}}(t) \frac{\partial^2}{\partial \nu \partial \mu}$$

$$+ \xi^{-1} C (\gamma_\uparrow + \gamma_\downarrow) \left[1 - \frac{\wp}{4C^2} \langle \bar{J}_z(t) \rangle - \frac{1}{2C} \bar{\mathcal{J}}(t) \bar{A}(t) \cos(\phi - \theta) \right] \frac{\partial^2}{\partial \mu^2}$$

$$+ \left(\begin{array}{c} \text{first-order and higher-} \\ \text{order terms in } n_{sat}^{-1/2} \end{array} \right) \Bigg\} P'. \tag{8.105}$$

Equation (8.105) displays the terms of order $n_{sat}^{-1/2}$ that result from the change of variables explicitly. These terms enter in two places: in the factors $\left(\bar{A}(t) + n_{sat}^{-1/2} z \right)^{-1}$ and $\left(\bar{\mathcal{J}}(t) + n_{sat}^{-1/2} \nu \right)^{-1}$, and in contributions to the drift in field and polarization amplitudes (appearing on the fourth and seventh lines). The contributions of order $n_{sat}^{-1/2}$ in the field and polarization amplitude drift, arise, mathematically, from passing the factor $\left(\bar{A}(t) + n_{sat}^{-1/2} z \right) \left(\bar{\mathcal{J}}(t) + n_{sat}^{-1/2} \nu \right)$ multiplying the last term in (8.103) through the second derivatives with respect to z and ν. Such terms are sometimes retained as corrections to the drift. They should not be, however. For a self-consistent expansion these terms must be dropped along with the nonlinear terms and higher-order derivatives collected together in the last term on the right-hand side of (8.105). Thus, we will drop all terms of order $n_{sat}^{-1/2}$ arising from the change of variables. We can then quickly complete the system size expansion once we decide what to do with the phase fluctuations.

There are two points to notice about how the phases enter (8.105). The first is that phase variables only appear (aside from the derivatives) in the combination $\phi - \theta$. The natural variables to use for treating phase fluctuations are therefore the phase difference $\phi - \theta$ and the phase sum $\phi + \theta$. Secondly, fluctuations in the phase difference are driven by the diffusion terms in ϕ and θ, which are of order n_{sat}^{-1}; these fluctuations should therefore scale as $n_{sat}^{-1/2}$. It is convenient to scale the phase sum in the same manner; although, we will find that the fluctuations in $\phi + \theta$ are undamped, and over sufficiently long times can grow arbitrarily large; scaling the fluctuations in $\phi + \theta$ simply keeps the notation symmetric. On the basis of these observations, we write

$$\phi + \theta = \Psi(t) + n_{sat}^{-1/2} \psi, \tag{8.106a}$$

$$\phi - \theta = \Delta(t) + n_{sat}^{-1/2} \delta. \tag{8.106b}$$

Note 8.7 The comment below (8.100) also applies here. There is generally no simple relationship between $\Psi(t)$ and $\Delta(t)$ and operator averages. However, to dominant order in $n_{sat}^{-1/2}$, we have $\Psi(t) = \arg(\langle \bar{a}(t) \rangle) + \arg(\langle \bar{J}_-(t) \rangle)$ and $\Delta(t) = \arg(\langle \bar{a}(t) \rangle) - \arg(\langle \bar{J}_-(t) \rangle)$.

We now expand the phase-dependent terms in (8.105) using

$$\cos(\phi - \theta) = \cos(\Delta(t)) + O(n_{\text{sat}}^{-1/2}), \qquad (8.107a)$$

and

$$\frac{\partial}{\partial \phi} \sin(\phi - \theta)$$

$$= \left(\frac{\partial}{\partial \psi} + \frac{\partial}{\partial \delta}\right)\left[n_{\text{sat}}^{1/2} \sin(\Delta(t)) + \delta \cos(\Delta(t)) + O(n_{\text{sat}}^{-1/2})\right], \quad (8.107b)$$

$$\frac{\partial}{\partial \theta} \sin(\phi - \theta)$$

$$= \left(\frac{\partial}{\partial \psi} - \frac{\partial}{\partial \delta}\right)\left[n_{\text{sat}}^{1/2} \sin(\Delta(t)) + \delta \cos(\Delta(t)) + O(n_{\text{sat}}^{-1/2})\right]. \quad (8.107c)$$

It is important to note that these expansions only require fluctuations in the phase difference to be small; the phase sum can be arbitrarily large. The distribution in scaled amplitude and phase variables is now defined by

$$\bar{P}(z, \nu, \psi, \delta, t) \equiv n_{\text{sat}}^{-1} \frac{1}{2} P'(z, \phi(\psi, \delta, t), \nu, \theta(\psi, \delta, t), \mu, t), \qquad (8.108)$$

and obeys the phase-space equation of motion

$$\frac{\partial \bar{P}}{\partial t} = \left(n_{\text{sat}}^{-1} \frac{1}{2}\right)\left(\frac{\partial P'}{\partial \phi}\frac{\partial \phi}{\partial t} + \frac{\partial P'}{\partial \theta}\frac{\partial \theta}{\partial t} + \frac{\partial P'}{\partial t}\right)$$

$$= n_{\text{sat}}^{1/2}\left(\frac{\partial \bar{P}}{\partial \psi}\frac{d\Psi(t)}{dt} + \frac{\partial \bar{P}}{\partial \delta}\frac{d\Delta(t)}{dt}\right) + \frac{\partial}{\partial t}\left(n_{\text{sat}}^{-1} \frac{1}{2} P'\right). \qquad (8.109)$$

After substituting from (8.105) and dropping terms of order $n_{\text{sat}}^{-1/2}$, the remaining terms of order $n_{\text{sat}}^{1/2}$ and n_{sat}^0 determine the macroscopic equations and the Fokker–Planck equation describing the fluctuations about the macroscopic state. The requirement that terms of order $n_{\text{sat}}^{1/2}$ vanish identically gives the macroscopic equations, the *laser equations without fluctuations in amplitude and phase variables*:

$$\kappa^{-1}\frac{d\bar{A}}{dt} = -\bar{A} + \bar{J}\cos\Delta, \qquad (8.110a)$$

$$\left(\frac{\gamma_\uparrow + \gamma_\downarrow}{2}\right)^{-1}\frac{d\bar{J}}{dt} = -\bar{J} + \langle\bar{J}_z\rangle\bar{A}\cos\Delta, \qquad (8.110b)$$

$$(\gamma_\uparrow + \gamma_\downarrow)^{-1}\frac{d\langle\bar{J}_z\rangle}{dt} = -\langle\bar{J}_z\rangle + \wp - \bar{J}\bar{A}\cos\Delta, \qquad (8.110c)$$

$$\frac{d\Psi}{dt} = -2\omega_C - \left(\kappa\frac{\bar{J}}{\bar{A}} - \frac{\gamma_\uparrow + \gamma_\downarrow}{2}\langle\bar{J}_z\rangle\frac{\bar{A}}{\bar{J}}\right)\sin\Delta, \qquad (8.110d)$$

$$\frac{d\Delta}{dt} = -\left(\kappa\frac{\bar{J}}{\bar{A}} + \frac{\gamma_\uparrow + \gamma_\downarrow}{2}\langle\bar{J}_z\rangle\frac{\bar{A}}{\bar{J}}\right)\sin\Delta. \qquad (8.110e)$$

Fluctuations about the macroscopic state obey the Fokker–Planck equation

$$
\begin{aligned}
\frac{\partial \bar{P}}{\partial t} = \Bigg\{ &\kappa \frac{\partial}{\partial z}\left[z - \nu \cos\left(\Delta(t)\right)\right] \\
&+ \frac{\gamma_\uparrow + \gamma_\downarrow}{2} \frac{\partial}{\partial \nu}\left[\nu - \left(\langle \bar{J}_z(t)\rangle z + \bar{A}(t)\mu\right)\cos\left(\Delta(t)\right)\right] \\
&+ (\gamma_\uparrow + \gamma_\downarrow)\frac{\partial}{\partial \mu}\left[\mu + \left(\bar{J}(t)z + \bar{A}(t)\nu\right)\cos\left(\Delta(t)\right)\right] \\
&+ \left(\kappa \frac{\bar{J}(t)}{\bar{A}(t)} - \frac{\gamma_\uparrow + \gamma_\downarrow}{2}\langle \bar{J}_z(t)\rangle \frac{\bar{A}(t)}{\bar{J}(t)}\right)\cos\left(\Delta(t)\right)\frac{\partial}{\partial \psi}\delta \\
&+ \left(\kappa \frac{\bar{J}(t)}{\bar{A}(t)} + \frac{\gamma_\uparrow + \gamma_\downarrow}{2}\langle \bar{J}_z(t)\rangle \frac{\bar{A}(t)}{\bar{J}(t)}\right)\cos\left(\Delta(t)\right)\frac{\partial}{\partial \delta}\delta \\
&+ \frac{1}{2}\kappa \bar{n}\frac{\partial^2}{\partial z^2} + \frac{1}{2}\xi^{-1}\left[C\gamma_\uparrow + \frac{\gamma_\uparrow + \gamma_\downarrow}{4}\bar{J}(t)\bar{A}(t)\cos\left(\Delta(t)\right)\right]\frac{\partial^2}{\partial \nu^2} \\
&- \xi^{-1}\gamma_\uparrow \bar{J}(t)\frac{\partial^2}{\partial \nu \partial \mu} \\
&+ \xi^{-1}C(\gamma_\uparrow + \gamma_\downarrow)\left[1 - \frac{\wp}{4C^2}\langle \bar{J}_z(t)\rangle - \frac{1}{2C}\bar{J}(t)\bar{A}(t)\cos\left(\Delta(t)\right)\right]\frac{\partial^2}{\partial \mu^2} \\
&+ \frac{1}{2}\kappa \bar{n}\frac{1}{\bar{A}(t)^2}\left(\frac{\partial}{\partial \psi} + \frac{\partial}{\partial \delta}\right)^2 \\
&+ \frac{1}{2}\xi^{-1}\left[C\gamma_\uparrow \frac{1}{\bar{J}(t)^2} + \frac{\gamma_\uparrow + \gamma_\downarrow}{4}\frac{\bar{A}(t)}{\bar{J}(t)}\cos\left(\Delta(t)\right)\right]\left(\frac{\partial}{\partial \psi} - \frac{\partial}{\partial \delta}\right)^2 \Bigg\}\bar{P}.
\end{aligned}
$$

$$(8.111)$$

Note 8.8 Equations (8.110) are obtained from our earlier version of the laser equations [Eqs. (8.25a)–(8.25e)] by writing

$$
\langle \bar{a}\rangle = \bar{A}\exp\left[i\tfrac{1}{2}(\Psi + \Delta)\right], \qquad \langle \bar{a}^\dagger\rangle = \bar{A}\exp\left[-i\tfrac{1}{2}(\Psi + \Delta)\right],
$$
$$
\langle \bar{J}_-\rangle = \bar{J}\exp\left[i\tfrac{1}{2}(\Psi - \Delta)\right], \qquad \langle \bar{J}_+\rangle = \bar{J}\exp\left[-i\tfrac{1}{2}(\Psi - \Delta)\right].
$$

In general, however, we should not identify $\bar{A}$ and $\bar{J}$ with $|\langle \bar{a}\rangle|$ and $|\langle \bar{J}_-\rangle|$. This identification is possible while the fluctuations, ψ, in the phase sum remain small. But a phase-symmetric state like that illustrated in Fig. 8.1(c) has $|\langle \bar{a}\rangle| = |\langle \bar{J}_-\rangle| = 0$, while $\bar{A}$ and $\bar{J}$ are certainly not zero.

Note 8.9 When nonradiative dephasing processes are included, (8.110a)–(8.110e) and (8.111) hold with the minor modifications resulting from (8.16) described previously: now $\bar{v}$ and $\bar{v}^*$ (and hence $\bar{J}$ and ν) are defined by (8.65); we make the replacement $(\gamma_\uparrow + \gamma_\downarrow)/2 \to (\gamma_\uparrow + \gamma_\downarrow + 2\gamma_p)/2 = \gamma_h/2$; and on the sixth and tenth lines of (8.111),

$$C\gamma_\uparrow \to C\frac{\gamma_h}{\gamma_\uparrow + \gamma_\downarrow}\left[\gamma_\uparrow + \gamma_p\left(1 + \frac{\langle \bar{J}_z(t)\rangle}{2C}\right)\right].$$

8.3.2 Adiabatic Elimination

Above threshold the steady-state solutions to (8.110a)–(8.110e) are

$$\bar{\mathcal{A}}_> = \bar{\mathcal{A}}^>_{ss} = \sqrt{\wp - 1}, \tag{8.112a}$$

$$\bar{\mathcal{J}}_> \equiv \bar{\mathcal{J}}^>_{ss} = \sqrt{\wp - 1}, \tag{8.112b}$$

$$\langle \bar{J}_z\rangle_> \equiv \langle \bar{J}_z\rangle^>_{ss} = 1, \tag{8.112c}$$

$$\Psi_> \equiv \Psi^>_{ss} = 2\omega_C t + \tilde{\Psi}, \tag{8.112d}$$

$$\Delta_> \equiv \Delta^>_{ss} = 0; \tag{8.112e}$$

the phase $\tilde{\Psi}$ is arbitrary. We will restrict our treatment of the fluctuations about this steady state to the region not too far above threshold, where $\sqrt{\wp - 1} \ll 1$. With this restriction we are able to neglect diffusion terms proportional to $\bar{\mathcal{J}}_> = \sqrt{\wp - 1}$ and $\bar{\mathcal{J}}_>\bar{\mathcal{A}}_> = \wp - 1$. Then, from (8.111), the *laser Fokker–Planck equation above threshold without adiabatic elimination* is

$$\frac{\partial \bar{P}}{\partial t} = \left[\kappa\frac{\partial}{\partial z}(z - \nu) + \frac{\gamma_\uparrow + \gamma_\downarrow}{2}\frac{\partial}{\partial \nu}(\nu - z - \sqrt{\wp - 1}\mu)\right.$$

$$+ (\gamma_\uparrow + \gamma_\downarrow)\frac{\partial}{\partial \mu}(\mu - \sqrt{\wp - 1}z - \sqrt{\wp - 1}\nu)$$

$$+ \left(\kappa - \frac{\gamma_\uparrow + \gamma_\downarrow}{2}\right)\frac{\partial}{\partial \psi}\delta + \left(\kappa + \frac{\gamma_\uparrow + \gamma_\downarrow}{2}\right)\frac{\partial}{\partial \delta}\delta$$

$$+ \frac{1}{2}\kappa\bar{n}\frac{\partial^2}{\partial z^2} + \frac{1}{2}\xi^{-1}C\gamma_\uparrow\frac{\partial^2}{\partial \nu^2} + \xi^{-1}C(\gamma_\uparrow + \gamma_\downarrow)\left(1 - \frac{1}{4C^2}\right)\frac{\partial^2}{\partial \mu^2}$$

$$\left.+ \frac{1}{2}\kappa\bar{n}\frac{1}{\wp - 1}\left(\frac{\partial}{\partial \psi} + \frac{\partial}{\partial \delta}\right)^2 + \frac{1}{2}\xi^{-1}C\gamma_\uparrow\frac{1}{\wp - 1}\left(\frac{\partial}{\partial \psi} - \frac{\partial}{\partial \delta}\right)^2\right]\bar{P}. \tag{8.113}$$

Equation (8.113) is separable. It may be separated into an equation describing fluctuations in the field amplitude, the polarization amplitude, and the inversion, and an equation describing phase fluctuations. Our task is to adiabatically eliminate the polarization amplitude and inversion from the former to obtain a stochastic description of amplitude fluctuations in the laser field, and to adiabatically eliminate the phase difference from the latter to obtain a description of the fluctuations in the phase sum, or equivalently, the common phase of the field and polarization. We write the distribution $\bar{P}$ as the product

$$\bar{P}(z,\nu,\psi,\delta,\mu,t) = \bar{A}(z,\nu,\mu,t)\bar{\Phi}(\psi,\delta,t); \qquad (8.114)$$

the distribution $\bar{A}$ satisfies the Fokker–Planck equation

$$\frac{\partial \bar{A}}{\partial t} = \left[\kappa \frac{\partial}{\partial z}(z-\nu) + \frac{\gamma_\uparrow + \gamma_\downarrow}{2}\frac{\partial}{\partial \nu}(\nu - z - \sqrt{\wp-1}\,\mu) \right.$$

$$+ (\gamma_\uparrow + \gamma_\downarrow)\frac{\partial}{\partial \mu}(\mu + \sqrt{\wp-1}\,z + \sqrt{\wp-1}\,\nu)$$

$$\left. + \frac{1}{2}\kappa\bar{n}\frac{\partial^2}{\partial z^2} + \frac{1}{2}\xi^{-1}C\gamma_\uparrow\frac{\partial^2}{\partial \nu^2} + \xi^{-1}C(\gamma_\uparrow + \gamma_\downarrow)\left(1 - \frac{1}{4C^2}\right)\frac{\partial^2}{\partial \mu^2} \right]\bar{A},$$

$$(8.115a)$$

and $\bar{\Phi}$ satisfies the Fokker–Planck equation

$$\frac{\partial \bar{\Phi}}{\partial t} = \left[\left(\kappa - \frac{\gamma_\uparrow + \gamma_\downarrow}{2} \right)\frac{\partial}{\partial \psi}\delta + \left(\kappa + \frac{\gamma_\uparrow + \gamma_\downarrow}{2} \right)\frac{\partial}{\partial \delta}\delta \right.$$

$$\left. + \frac{1}{2}\kappa\bar{n}\frac{1}{\wp-1}\left(\frac{\partial}{\partial \psi} + \frac{\partial}{\partial \delta}\right)^2 + \frac{1}{2}\xi^{-1}C\gamma_\uparrow\frac{1}{\wp-1}\left(\frac{\partial}{\partial \psi} - \frac{\partial}{\partial \delta}\right)^2 \right]\bar{\Phi}.$$

$$(8.115b)$$

Let us first consider the adiabatic elimination of the polarization amplitude and inversion from (8.115a). We wish to adiabatically eliminate the variables ν and μ. The Ito stochastic differential equations equivalent to this Fokker–Planck equation are

$$dz = -\kappa(z-\nu)dt + \sqrt{\kappa\bar{n}}\,dW_z, \qquad (8.116a)$$

$$d\nu = -\frac{\gamma_\uparrow + \gamma_\downarrow}{2}(\nu - z - \sqrt{\wp-1}\,\mu)dt + \sqrt{\xi^{-1}C\gamma_\uparrow}\,dW_\nu, \quad (8.116b)$$

and

$$d\mu = -(\gamma_\uparrow + \gamma_\downarrow)(\mu + \sqrt{\wp-1}\,z + \sqrt{\wp-1}\,\nu)dt$$

$$+ \sqrt{\xi^{-1}2C(\gamma_\uparrow + \gamma_\downarrow)(1 - 1/4C^2)}\,dW_\mu, \qquad (8.116c)$$

where dW_z, dW_ν, and dW_μ are independent Wiener processes. Equations (8.116a) and (8.116b) are similar to the equations that describe fluctuations in the coupled field and polarization amplitudes below threshold [Eqs. (8.55a) and (8.55b)]. The main difference is that above threshold fluctuations in the polarization couple to fluctuations in the inversion; below threshold these fluctuations are separable. This difference arises because the laser field and medium polarization both acquire a nonzero mean amplitude above threshold. But this feature brings little change to the calculations, or to the results, when the laser is not too far above threshold ($\sqrt{\wp-1} \ll 1$). We set $d\nu = d\mu = 0$ on the left-hand sides of (8.116b) and (8.116c), and write

$$\nu dt = \left(z + \sqrt{\wp - 1}\,\mu\right)dt + \left(\frac{\gamma_\uparrow + \gamma_\downarrow}{2}\right)^{-1}\sqrt{\xi^{-1}C\gamma_\uparrow}\,dW_\nu, \quad (8.117\text{a})$$

$$\mu dt = -\sqrt{\wp - 1}\,(z + \nu)dt$$
$$+ (\gamma_\uparrow + \gamma_\downarrow)^{-1}\sqrt{\xi^{-1}2C(\gamma_\uparrow + \gamma_\downarrow)(1 - 1/4C^2)}\,dW_\mu. \quad (8.117\text{b})$$

Substituting (8,117a) into (8.117b), we find

$$\wp\mu dt = -2\sqrt{\wp - 1}\,z dt - \sqrt{\wp - 1}\left(\frac{\gamma_\uparrow + \gamma_\downarrow}{2}\right)^{-1}\sqrt{\xi^{-1}C\gamma_\uparrow}\,dW_\nu$$
$$+ (\gamma_\uparrow + \gamma_\downarrow)^{-1}\sqrt{\xi^{-1}2C(\gamma_\uparrow + \gamma_\downarrow)(1 - 1/4C^2)}\,dW_\mu. \quad (8.118)$$

Then substituting the result back into (8.117a), we have

$$\nu dt = [1 - 2(\wp - 1)/\wp]z dt + [1 - (\wp - 1)/\wp]\left(\frac{\gamma_\uparrow + \gamma_\downarrow}{2}\right)^{-1}\sqrt{\xi^{-1}C\gamma_\uparrow}\,dW_\nu$$
$$+ (\sqrt{\wp - 1}/\wp)(\gamma_\uparrow + \gamma_\downarrow)^{-1}\sqrt{\xi^{-1}2C(\gamma_\uparrow + \gamma_\downarrow)(1 - 1/4C^2)}\,dW_\mu.$$
$$(8.119)$$

We have already neglected diffusion terms of order $\sqrt{\wp - 1}$ and $\wp - 1$ in passing from (8.111) to (8.113). To be consistent we should therefore drop these terms in the coefficients of the Wiener processes appearing in (8.119). As a result, the surviving fluctuations from the laser medium are just the polarization fluctuations we met below threshold [Eq. (8.56)]. After substituting for νdt in (8.116a), we find

$$dz = -\kappa 2(\wp - 1)z dt + \sqrt{\kappa(\bar{n} + n_{\text{spon}})}\,dW. \quad (8.120)$$

Thus, the *laser Fokker–Planck equation for field amplitude fluctuations above threshold* is given by

$$\frac{\partial \bar{A}}{\partial \bar{t}} = \left[2(\wp - 1)\frac{\partial}{\partial z}z + \frac{1}{2}(\bar{n} + n_{\text{spon}})\frac{\partial^2}{\partial z^2}\right]\bar{A}, \quad (8.121\text{a})$$

with corresponding stochastic differential equation

$$dz = -2(\wp - 1)z d\bar{t} + \sqrt{(\bar{n} + n_{\text{spon}})}\,d\bar{W}, \quad (8.121\text{b})$$

where $\bar{t}$ is given by (8.62).

Note 8.10 Close to threshold the adiabatic elimination of atomic variables can be justified even when the condition $\xi \ll 1$ is not satisfied. This is because the fluctuations in the laser field "slow down" near threshold. In (8.121), and also in (8.61), the decay rate for the field is determined by $\kappa|\wp - 1|$ rather than by the empty cavity rate κ.

We now return to (8.115b). We wish to adiabatically eliminate the phase difference δ from this equation. The diffusion matrix in (8.115b) is not diagonal, and therefore to write the equivalent Ito stochastic differential equations we must first factorize the diffusion matrix in the form BB^T. We may write

$$\frac{1}{2}\kappa\bar{n}\frac{1}{\wp-1}\left(\frac{\partial}{\partial\psi}+\frac{\partial}{\partial\delta}\right)^2 + \frac{1}{2}\xi^{-1}C\gamma_\uparrow\frac{1}{\wp-1}\left(\frac{\partial}{\partial\psi}-\frac{\partial}{\partial\delta}\right)^2$$

$$= \frac{1}{2}\left(\frac{\partial}{\partial\psi}\,\frac{\partial}{\partial\delta}\right)BB^T\left(\frac{\partial}{\partial\psi}\,\frac{\partial}{\partial\delta}\right)^T, \quad (8.122)$$

with

$$B = \frac{1}{\sqrt{\wp-1}}\left(\begin{array}{cc}\sqrt{\kappa\bar{n}} & \sqrt{\xi^{-1}C\gamma_\uparrow} \\ \sqrt{\kappa\bar{n}} & -\sqrt{\xi^{-1}C\gamma_\uparrow}\end{array}\right). \quad (8.123)$$

Then, using (5.149), the Fokker–Planck equation (8.115b) is equivalent to the Ito stochastic differential equations

$$d\psi = -\left(\kappa - \frac{\gamma_\uparrow+\gamma_\downarrow}{2}\right)\delta dt + \frac{1}{\sqrt{\wp-1}}\left(\sqrt{\kappa\bar{n}}\,dW_1 + \sqrt{\xi^{-1}C\gamma_\uparrow}\,dW_2\right),$$
$$(8.124a)$$

$$d\delta = -\left(\kappa + \frac{\gamma_\uparrow+\gamma_\downarrow}{2}\right)\delta dt + \frac{1}{\sqrt{\wp-1}}\left(\sqrt{\kappa\bar{n}}\,dW_1 - \sqrt{\xi^{-1}C\gamma_\uparrow}\,dW_2\right);$$
$$(8.124b)$$

dW_1 and dW_2 are independent Wiener processes. The phase difference δ is damped at the rate $\kappa + (\gamma_\uparrow+\gamma_\downarrow)/2$. The phase sum ψ is not damped, and it is driven by the Wiener processes dW_1 and dW_2, both directly, and also indirectly by its coupling to the damped fluctuations in the phase difference. We may adiabatically eliminate the phase difference under the assumption that the relaxation rate $\kappa + (\gamma_\uparrow+\gamma_\downarrow)/2$ is much faster than the rate at which fluctuations in ψ grow. To accomplish the adiabatic elimination we set $d\delta \to 0$ on the left-hand side of (8.124b), and write

$$\delta dt = \left(\kappa + \frac{\gamma_\uparrow+\gamma_\downarrow}{2}\right)^{-1}\frac{1}{\sqrt{\wp-1}}\left(\sqrt{\kappa\bar{n}}\,dW_1 - \sqrt{\xi^{-1}C\gamma_\uparrow}\,dW_2\right). \quad (8.125)$$

Substituting this result into (8.124a), we have

$$d\psi = \frac{2}{\sqrt{\wp-1}}\left(\frac{\gamma_\uparrow+\gamma_\downarrow}{2\kappa+\gamma_\uparrow+\gamma_\downarrow}\sqrt{\kappa\bar{n}}\,dW_1 + \frac{2\kappa}{2\kappa+\gamma_\uparrow+\gamma_\downarrow}\sqrt{\xi^{-1}C\gamma_\uparrow}\,dW_2\right). \quad (8.126)$$

Equation (8.126) is equivalent to the Brownian motion equation

$$d\psi = BdW, \quad (8.127a)$$

where

$$B = \frac{2}{\sqrt{\wp - 1}}(2\kappa + \gamma_\uparrow + \gamma_\downarrow)^{-1}\sqrt{(\gamma_\uparrow + \gamma_\downarrow)^2 \kappa \bar{n} + (2\kappa)^2 \frac{\gamma_\uparrow + \gamma_\downarrow}{2\kappa} C \gamma_\uparrow}$$

$$= \frac{2}{\sqrt{\wp - 1}}\left(1 + \frac{2\kappa}{\gamma_\uparrow + \gamma_\downarrow}\right)^{-1}\sqrt{\kappa(\bar{n} + n_{\text{spon}})}. \tag{8.127b}$$

The corresponding Fokker–Planck equation reads

$$\kappa^{-1}\frac{\partial\bar{\Phi}}{\partial t} = \frac{1}{2}\left(1 + \frac{2\kappa}{\gamma_\uparrow + \gamma_\downarrow}\right)^{-2} 4\frac{(\bar{n} + n_{\text{spon}})}{\wp - 1}\frac{\partial^2}{\partial\psi^2}\bar{\Phi}. \tag{8.128}$$

Now that we have eliminated the phase sum it is convenient to write (8.128) directly in terms of the phase ϕ of the laser field. In the steady-state, from (8.106a) and (8.106b), we may write

$$\phi = \tfrac{1}{2}(\Psi_> + \Delta_>) + n_{\text{sat}}^{-1/2}\tfrac{1}{2}(\psi + \delta)$$

$$= \omega_C t + n_{\text{sat}}^{-1/2}\tfrac{1}{2}\psi. \tag{8.129}$$

We have neglected δ compared to ψ because the damped fluctuations in δ remain finite, while the fluctuations in ψ grow as $\sqrt{t}$. We define

$$\Phi(\phi, t) \equiv \tilde{\Phi}(\tilde{\phi}, t) \equiv n_{\text{sat}}^{-1/2}\tfrac{1}{2}\bar{\Phi}(\psi, t), \tag{8.130a}$$

with

$$\tilde{\phi} \equiv \phi - \omega_C t = n_{\text{sat}}^{-1/2}\tfrac{1}{2}\psi. \tag{8.130b}$$

Then the *laser Fokker–Planck equation for phase fluctuations above threshold* is given by

$$\frac{\partial\tilde{\Phi}}{\partial\tilde{t}} = \frac{1}{2}\left(1 + \frac{2\kappa}{\gamma_\uparrow + \gamma_\downarrow}\right)^{-2}\frac{\bar{n} + n_{\text{spon}}}{n_{\text{sat}}(\wp - 1)}\frac{\partial^2}{\partial\tilde{\phi}^2}\tilde{\Phi}, \tag{8.131a}$$

with corresponding stochastic differential equation

$$d\tilde{\phi} = \left(1 + \frac{2\kappa}{\gamma_\uparrow + \gamma_\downarrow}\right)^{-1}\sqrt{\frac{\bar{n} + n_{\text{spon}}}{n_{\text{sat}}(\wp - 1)}}\, d\bar{W}, \tag{8.131b}$$

where $\bar{t}$ is given by (8.62).

Note 8.11 When nonradiative dephasing processes are included and the changes described in the note below (8.111) are carried over into (8.115a) and (8.115b), the results of the adiabatic elimination are the same, apart from the replacement of $\gamma_\uparrow + \gamma_\downarrow$ by γ_h in (8.131a) and (8.131b).

8.3.3 Quantum Fluctuations Above Threshold

Equations (8.121a) and (8.131a) are linear Fokker–Planck equations in one dimension. The Green function solution to (8.121a) can be written down from (5.18):

$$\bar{A}(z,t|z_0,0) = \frac{1}{\sqrt{2\pi}} \frac{1}{\sqrt{[(\bar{n}+n_{\text{spon}})/4(\wp-1)][1-e^{-4\kappa(\wp-1)t}]}}$$

$$\times \exp\left\{-\frac{1}{2} \frac{[z-z_0 e^{-2\kappa(\wp-1)t}]^2}{[(\bar{n}+n_{\text{spon}})/4(\wp-1)][1-e^{-4\kappa(\wp-1)t}]}\right\}.$$

(8.132)

The solution for $\tilde{\Phi}(\tilde{\phi},t|\tilde{\phi}_0,0)$ is slightly different from that given by (5.18) since we must account for the different boundary condition that applies to the phase variable. We wish to solve (8.131a) with $\tilde{\phi}$ distributed in the range $0 \le \tilde{\phi} < 2\pi$. We write

$$\tilde{\Phi}(\tilde{\phi},t|\tilde{\phi}_0,0) = \sum_{m=-\infty}^{\infty} C_m(t)e^{im\tilde{\phi}}, \qquad (8.133a)$$

and to find the Green function solution take

$$C_m(0) = \frac{1}{2\pi}e^{-im\tilde{\phi}_0}; \qquad (8.133b)$$

this initial condition gives a periodic δ-function at $\tilde{\phi} = \phi_0 + k2\pi$, where k is an integer. For simplicity let us assume $2\kappa \ll \gamma_\uparrow + \gamma_\downarrow$. Then, substituting (8.133a) into (8.131a), the C_m obey the equations

$$\dot{C}_m = -\frac{1}{2}\frac{\bar{n}+n_{\text{spon}}}{n_{\text{sat}}(\wp-1)}m^2 C_m, \qquad (8.134)$$

and hence

$$\tilde{\Phi}(\tilde{\phi},t|\tilde{\phi}_0,0) = \frac{1}{2\pi}\sum_{m=-\infty}^{\infty}\exp\left[im(\tilde{\phi}-\tilde{\phi}_0) - \frac{1}{2}\frac{\bar{n}+n_{\text{spon}}}{n_{\text{sat}}(\wp-1)}m^2 t\right]. \qquad (8.135)$$

The Green function (8.132) gives us the variance of the steady-state amplitude fluctuations, and hence the correction to the average photon number in the laser mode due to spontaneous emission and thermal photon fluctuations:

$$\langle a^\dagger a\rangle_> - n_{\text{sat}}(\wp-1) = \left(\overline{z^2}\right)_{\bar{A}}$$

$$= \frac{\bar{n}+n_{\text{spon}}}{4(\wp-1)}. \qquad (8.136)$$

This result is different from that obtained using the rate equation theory [Eq. (7.43)] by the factor of four in the denominator on the right-hand side. The difference arises from the factorization used to pass from the exact mean energy equation (7.65) to a rate equation description. From the Green function solution (8.135) for phase fluctuations we calculate the laser linewidth above threshold. The normalized first-order correlation function for the laser output is

$$g_>^{(1)}(\tau) \equiv \left(\langle a^\dagger a \rangle_> \right)^{-1} \left[\lim_{t \to \infty} \langle a^\dagger(t) a(t + \tau) \rangle \right]$$

$$= e^{-i\omega_C t} \lim_{t \to \infty} \left(\overline{e^{-i\tilde{\phi}(t)} e^{i\tilde{\phi}(t+\tau)}} \right)_{\tilde{\Phi}}$$

$$= e^{-i\omega_C t} \sum_{m=-\infty}^{\infty} \exp\left[-\kappa \frac{1}{2} \frac{\bar{n} + n_{\text{spon}}}{n_{\text{sat}}(\wp - 1)} m^2 |\tau| \right]$$

$$\times \left(\frac{1}{2\pi} \int_0^{2\pi} d\tilde{\phi}_0 \, e^{-i(m+1)\tilde{\phi}_0} \right) \left(\frac{1}{2\pi} \int_0^{2\pi} d\tilde{\phi} \, e^{i(m+1)\tilde{\phi}} \right)$$

$$= e^{-i\omega_C t} \exp\left[-\kappa \frac{1}{2} \frac{\bar{n} + n_{\text{spon}}}{n_{\text{sat}}(\wp - 1)} |\tau| \right]. \tag{8.137}$$

The Fourier transform gives a Lorentzian line, with the *laser linewidth above threshold* (half-width at half-maximum) given by

$$(\Delta\omega)_> = \kappa \frac{1}{2} \frac{\bar{n} + n_{\text{spon}}}{n_{\text{sat}}(\wp - 1)} = \kappa \frac{1}{2} \frac{\bar{n} + n_{\text{spon}}}{\langle a^\dagger a \rangle_>} = \kappa^2 \hbar \omega_C \frac{\bar{n} + n_{\text{spon}}}{P_>}. \tag{8.138}$$

It is interesting to note, that when written in terms of the output power $P_>$, this expression only differs from the expression (8.70), which holds below threshold, by a factor of two. Of course, the actual linewidth varies a great deal. The output power, or, alternatively, the mean intracavity photon number, increases by many orders of magnitude from below threshold to above threshold. Thus, a linewidth of the order of the cavity width κ below threshold is replaced by a very much narrower line above threshold. Using $\bar{n} + n_{\text{spon}} = 1$ and $n_{\text{sat}} = 10^8$, we find $(\Delta\omega)_>/\kappa \sim 10^{-6}$ when $\wp - 1 = 10^{-2}$ (one percent above threshold).

Exercise 8.7 Show that above threshold

$$g_>^{(2)}(\tau) = 1 + \frac{\bar{n} + n_{\text{spon}}}{n_{\text{sat}}(\wp - 1)^2} e^{-2\kappa(\wp-1)|\tau|}$$

$$= 1 + \frac{\pi}{2} \left(\frac{\langle a^\dagger a \rangle_{\text{thr}}}{\langle a^\dagger a \rangle_>} \right)^2 e^{-2\kappa(\wp-1)|\tau|}. \tag{8.139}$$

Compare this with the "thermal" result (8.71) obtained below threshold. For $\bar{n} + n_{\text{spon}} = 1$, $n_{\text{sat}} = 10^8$, and $\wp - 1 = 10^{-2}$ (one percent above threshold), the difference between (8.139) and the correlation function $g^{(2)}(\tau) = 1$ for coherent light is $\sim 10^{-4}$.

Note 8.12 The quasi-linearization used to arrive at the Fokker–Planck equation (8.113) is not valid too close to threshold. When threshold is approached from above, in place of the deterministic equations (8.72) we have

$$\dot{z} = -\kappa(z - \nu), \qquad (8.140a)$$

$$\dot{\nu} = -\frac{\gamma_\uparrow + \gamma_\downarrow}{2}(\nu - z - \sqrt{\wp - 1}\mu), \qquad (8.140b)$$

$$\dot{\mu} = -(\gamma_\uparrow + \gamma_\downarrow)[\mu - \sqrt{\wp - 1}z - \sqrt{\wp - 1}\nu). \qquad (8.140c)$$

The characteristic equation determining the eigenvalues is

$$\lambda(\lambda + \gamma_\uparrow + \gamma_\downarrow)\left(\lambda + \kappa + \frac{\gamma_\uparrow + \gamma_\downarrow}{2}\right)(\wp - 1)(\gamma_\uparrow + \gamma_\downarrow)\frac{\gamma_\uparrow + \gamma_\downarrow}{2}(\lambda + 2\kappa) = 0. \qquad (8.141)$$

One of the eigenvalues vanishes for $\wp = 1$; as described below (8.73), this means that nonlinear terms must be retained in the system size expansion close to threshold. In addition to the neglect of these nonlinearities, the quasilinearization in amplitude and phase variables assumes $z \ll n_{\text{sat}}^{1/2}\bar{\mathcal{A}}_> = n_{\text{sat}}^{1/2}\sqrt{\wp - 1}$, in order to remove the nonlinearity arising from the change of variables. Using (8.136) to estimate the magnitude of z, we see now, that this requires

$$\wp - 1 \gg \frac{1}{2}\sqrt{\frac{\bar{n} + n_{\text{spon}}}{n_{\text{sat}}}}, \qquad (8.142)$$

a condition that is consistent with our definition of the threshold region in (7.38). [The factor of four – relating (8.136) and (7.43) – shows up again in the comparison between (8.142) and (7.38).]

Exercise 8.8 The Fokker Planck equations (8.121a) and (8.131a) are valid when the laser is operated not too far above threshold, with $\wp - 1 \ll 1$. We may lift this restriction without adding too much complication if we consider the four-level model for the laser gain medium mentioned in the second paragraph below (7.75) ($\gamma_\uparrow \gg \gamma_\downarrow$, $2C = \wp$) and add a strong polarization dephasing process ($\gamma_p \gg \gamma_\uparrow \gg \gamma_\downarrow$) in the manner described in Note 8.9. Introduce these changes and repeat the adiabatic elimination, starting from (8.111), without assuming $\wp - 1 \ll 1$. Show that (8.121a) is replaced by the Fokker–Planck equation

$$\frac{\partial\bar{A}}{\partial\bar{t}} = \left[2\frac{\wp - 1}{\wp}\frac{\partial}{\partial z}z + \frac{1}{2}\left(\bar{n} + \frac{\wp + 1}{2\wp^2}\right)\frac{\partial^2}{\partial z^2}\right]\bar{A}, \qquad (8.143)$$

and (8.131a) is replaced by the Fokker–Planck equation

$$\frac{\partial \tilde{\Phi}}{\partial \bar{t}} = \frac{1}{2}\left(1 + \frac{2\kappa}{\gamma_h}\right)^{-2} \frac{\bar{n} + (\wp + 1)/2}{n_{\mathrm{sat}}(\wp - 1)} \frac{\partial^2}{\partial \tilde{\phi}^2}\tilde{\Phi}; \qquad (8.144)$$

$\bar{t}$ is given by (8.62).

References

Chapter 1

1.1 I. R. Senitzky: Phys. Rev. **119**, 670 (1960); **124**, 642 (1961)
1.2 J. R. Ray: Lett. Nuovo Cim. **25**, 47 (1979)
1.3 A. O. Caldeira and A. J. Leggett: Ann. Phys. **149**, 374 (1983)
1.4 W. H. Louisell: *Quantum Statistical Properties of Radiation* (Wiley, New York, 1973) pp. 331–347
1.5 H. Haken: *Handbuch der Physik*, Vol. XXV/2c, ed. by L. Genzel (Springer-Verlag, Berlin, 1970) pp. 51–56
1.6 M. Sargent III, M. O. Scully, and W. E. Lamb, Jr.: *Laser Physics* (Addison-Wesley, Reading, Massachusetts, 1974) pp. 257–267
1.7 F. Haake: Z. Phys. **223**, 353 (1969); **223**, 364 (1969)
1.8 F. Haake: "Statistical Treatment of Open Systems by Generalized Master Equations", in *Springer Tracts in Modern Physics*, Vol. 66 (Springer-Verlag, Berlin, 1973) pp. 98–168
1.9 W. C. Schieve and J. W. Middleton: International J. Quant. Chem., Quantum Chemistry Symposium 11, 625 (1977)
1.10 M. Abramowitz and I. A. Stegun: *Handbook of Mathematical Functions* (Dover, New York, 1965) pp. 259–260
1.11 E. T. Whittaker and G. N. Watson: *A Course of Modern Analysis*, 4th ed. (Cambridge University Press, London, 1935) p. 75
1.12 G. Lindblad: Commun. Math. Phys. **48**, 119 (1976)
1.13 Reference [1.4] pp. 324, 336; Reference [1.5] pp. 29–30, and references therein
1.14 E. B. Davies: *Quantum Theory of Open Systems* (Academic Press, New York, 1976)
1.15 M. D. Srinivas and E. B. Davies: Optica Acta **28**, 981 (1981)
1.16 G. S. Agarwal: Phys. Rev. A **4**, 1778 (1971)
1.17 G. S. Agarwal: Phys. Rev. A **7**, 1195 (1973)
1.18 K. Lindenberg and B. West: Phys. Rev. A **30**, 568 (1984)
1.19 H. Grabert, P. Schramm, and G.-L. Ingold: Physics Reports **168**, 115 (1988)
1.20 M. Lax: Phys. Rev. **129**, 2342 (1963)
1.21 M. Lax: Phys. Rev. **157**, 213 (1967)
1.22 B. R. Mollow: Phys. Rev. **188**, 1969 (1969) Footnote 7
1.23 L. Onsager: Phys. Rev. **37**, 405 (1931); **38**, 2265 (1931)
1.24 G. W. Ford and R. F. O'Connell, Phys. Rev. Lett. **77**, 798 (1996); Ann. Phys. **276**, 144 (1999); Optics Commun. **179**, 451 (2000)
1.25 A. Einstein: Ann. Phys. (Leipz.) **22**, 180 (1907)
1.26 G. W. Ford, J. T. Lewis, and R. F. O'Connell, Ann. Phys. **252**, 362 (1996)
1.27 G. W. Ford and R. F. O'Connell, Ann. Phys. **269**, 51 (1998)
1.28 M. Lax, Optics Commun. **179**, 463 (2000)
1.29 I. Prigogine, C. George, F. Henin, and L. Rosenfeld: Chem. Scripta **4**, 5 (1973)

1.30 R. Hanbury-Brown and R. Q. Twiss: Nature **177**, 27 (1956); **178**, 1046 (1956);
Proc. R. Soc. Lond. A **242**, 300 (1957); **243**, 291 (1957)

Chapter 2

2.1 W. H. Louisell: *Quantum Statistical Properties of Radiation* (Wiley, New York,
1973) pp. 347–357

2.2 H. Haken: *Handbuch der Physik*, Vol. XXV/2c, ed. by L. Genzel (Springer-
Verlag, Berlin, 1970) pp. 57–58

2.3 M. Sargent III, M. O. Scully, and W. E. Lamb, Jr.: *Laser Physics* (Addison-
Wesley, Reading, Massachusetts, 1974) pp. 273–278

2.4 L. Allen and J. H. Eberly: *Optical Resonance and Two-Level Atoms* (Wiley,
New York, 1975) pp. 28–40

2.5 Reference [2.1] pp. 122–127; Reference [2.3] pp. 9–12

2.6 Reference [2.2] pp. 27–30; Reference [2.3] pp. 14–16, 230–233

2.7 There are subtleties in the derivation of the Hamiltonian for the atom-field
interaction which have given rise to a long-standing debate. For a recent contribution to the debate and a review of the literature, see E. A. Power and T.
Thirunamachandran: J. Opt. Soc. Am. B **2**, 1100 (1985)

2.8 T. F. Gallagher and W. E. Cook: Phys. Rev. Lett. **42**, 835 (1979)

2.9 J. W. Farley and W. H. Wing: Phys. Rev. A **23**, 5 (1981)

2.10 L. Hollberg and J. L. Hall: Phys. Rev. Lett. **53**, 230 (1984)

2.11 G. S. Agarwal: Phys. Rev A **7**, 1195 (1973)

2.12 Reference [2.1] pp. 250–251

2.13 V. G. Weisskopf and E. Wigner: Z. Phys. **63**, 54 (1930)

2.14 Reference [2.1] pp. 281–283; Reference [2.3] pp. 20–23

2.15 R. J. Glauber: "Optical Coherence and Photon Statistics," in *Quantum Optics
and Electronics*, ed. by C. DeWitt, A. Blandin, and C. Cohen-Tannoudji (Gordon and Breach, London, 1965) pp. 78–84; in particular, consider Eq. (4.11)
with a sharply peaked (δ-function) sensitivity function $s(\omega)$

2.16 J. Herschel: Phil. Trans. R. Soc. Lond., 143 (1845)

2.17 D. Brewster: Trans. of Edin., part II, 3 (1846)

2.18 H. A. Lorentz: *The Theory of Electrons* (Dover, New York, 1952)

2.19 W. Heitler: *The Quantum Theory of Radiation* (Oxford, London, 1954) Chapter 1

2.20 Reference [2.3] Chapter III; Reference [2.4] Chapter 1

2.21 Reference [2.19] pp. 196–204

2.22 R. J. Ballagh: Ph. D. Thesis, University of Colorado, Boulder, U.S.A. (1978)

2.23 B. R. Mollow: Phys. Rev. **188**, 1969 (1969)

2.24 F. Y. Schuda, C. R. Stroud, Jr., and M. Hercher: J. Phys. B **7**, L198 (1974)

2.25 F. Y. Wu, R. E. Grove, and S. Ezekiel: Phys. Rev. Lett. **35**, 1426 (1975)

2.26 W. Hartig, W. Rasmussen, R. Schieder, and H. Walther: Z. Phys. **A278**, 205
(1976)

2.27 R. Hanbury-Brown and R. Q. Twiss: Nature **177**, 27 (1956); **178**, 1046 (1956);
Proc. Roy. Soc. Lond. A **242**, 300 (1957); **243**, 291 (1957)

2.28 D. F. Walls: Nature **280**, 451 (1979)

2.29 R. Loudon: Rep. Prog. Phys. **43**, 913 (1980)

2.30 H. Paul: Rev. Mod. Phys. **54**, 1061 (1982)

2.31 B. R. Mollow: Phys. Rev. A **12**, 1919 (1975); the relevant comments appear
below equation (4.15)

2.32 H. J. Carmichael and D. F. Walls: J. Phys. B **9**, L43 (1976); **9**, 1199 (1976)

2.33 H. J. Kimble, M. Dagenais, and L. Mandel: Phys. Rev. Lett. **39**, 691 (1977)

2.34 J. D. Cresser, J. Häger, G. Leuchs, M. Rateike, and H. Walther: "Resonance Fluorescence of Atoms in Strong Monochromatic Laser Fields," in *Dissipative Systems in Quantum Optics*, ed. by R. Bonifacio (Springer-Verlag, Berlin, 1982) pp. 21–59

2.35 I. I. Rabi: Phys. Rev. **51**, 652 (1937)

2.36 H. J. Carmichael and D. F. Walls: J. Phys. A **6**, 1552 (1973)

2.37 H. J. Carmichael and D. F. Walls: Phys. Rev. A **9**, 2686 (1974)

2.38 M. Lewenstein, T. W. Mossberg, and R. J. Glauber: Phys. Rev. Lett. **59**, 775 (1987)

2.39 M. Lewenstein and T. W. Mossberg: Phys. Rev. A **37**, 2048 (1988)

2.40 F. Bloch: Phys. Rev. **70**, 460 (1946)

2.41 C. Cohen-Tannoudji and S. Reynaud: J. Phys. B **10**, 345 (1977)

2.42 H. Sambe: Phys. Rev. A **7**, 2203 (1973)

2.43 J. M. Okuniewicz: J. Math. Phys. **5**, 1587 (1974)

2.44 R. K. Eisenschitz: *Matrix Algebra for Physicists* (Plenum, New York, 1966) Chapter 7

2.45 R. J. Glauber: Phys. Rev. Lett. **10**, 84 (1963)

2.46 R. J. Glauber: Phys. Rev. **130**, 2529 (1963)

2.47 H. J. Kimble and L. Mandel: Phys. Rev. A **13**, 2123 (1976)

2.48 M. Dagenais and L. Mandel: Phys. Rev. A **18**, 2217 (1978)

2.49 R. Short and L. Mandel: Phys. Rev. Lett. **51**, 384 (1983)

2.50 H. J. Carmichael, S. Singh, R. Vyas, and P. R. Rice: Phys. Rev. A **39**, 1200 (1989)

2.51 P. Zoller, M. Marte, and D. F. Walls: Phys. Rev. A **35**, 198 (1987)

2.52 H. J. Carmichael, "Theory of Quantum Fluctuations in Optical Bistability," in *Frontiers in Quantum Optics*, ed. by E. R. Pike and S. Sarkar (Adam Hilger, Bristol, 1986) pp. 120–203 [see Fig. 11(c)]

2.53 P. R. Rice and H. J. Carmichael: IEEE J. Quantum Electron. **QE 24**, 1351 (1988) (see Fig. 3)

2.54 A wide selection of early articles in this field can be found in the following volumes: *Frontiers in Quantum Optics*, ed. by E. R. Pike and S. Sarkar (Adam Hilger, Bristol, 1986); *Quantum Optics IV*, ed. by J. D. Harvey and D. F. Walls (Springer-Verlag, Berlin, 1986); J. Mod. Opt. **34**, Special Issue on "Squeezed Light," June 1987; J. Opt. Soc. Am. B **4**, Feature Issue on "Squeezed States of the Electromagnetic Field," October, 1987

2.55 D. F. Walls and P. Zoller: Phys. Rev. Lett. **47**, 709 (1981)

2.56 D. F. Walls: Nature **306**, 141 (1983)

2.57 H. J. Carmichael: Phys. Rev. Lett. **55**, 2790 (1985)

2.58 L. Mandel: Phys. Rev. Lett. **49**, 136 (1982)

Chapter 3

3.1 H. Risken: *The Fokker Planck Equation* (Springer-Verlag, Berlin, 1984)

3.2 E. P. Wigner: Phys. Rev. **40**, 749 (1932)

3.3 R. J. Glauber: Phys. Rev. **131**, 2766 (1963)

3.4 E. C. G. Sudarshan: Phys. Rev. Lett. **10**, 277 (1963)

3.5 R. J. Glauber: Phys. Rev. Lett. **10**, 84 (1963)

3.6 R. J. Glauber: Phys. Rev. **130**, 2529 (1963)

3.7 W. H. Louisell: *Quantum Statistical Properties of Radiation* (Wiley, New York, 1973) pp. 104–109

3.8 M. Sargent III, M. O. Scully, and W. E. Lamb Jr.: *Laser Physics* (Addison-Wesley, Reading, Massachusetts, 1974) Chapter 15

3.9 G. Temple: J. London Math. Soc. **28**, 134 (1953)

3.10 G. Temple: Proc. Roy. Soc. A **228**, 175 (1955)
3.11 M. J. Lighthill: *Fourier Analysis and Generalized Functions* (Cambridge University Press, Cambridge, 1960)
3.12 L. Schwartz: *Théorie des Distributions*, Vol. I/II (Hermann, Paris, 1950/51; 2nd edition 1957/1959)
3.13 H. Bremermann: *Distributions, Complex Variables, and Fourier Transforms* (Addison-Wesley, Reading, Massachusetts, 1965)
3.14 J. R. Klauder and E. C. G. Sudarshan: *Fundamentals of Quantum Optics* (Benjamin, New York, 1968) pp. 178–201
3.15 H. M. Nussenzveig: *Introduction to Quantum Optics*, (Gordon and Breach, London, 1973) pp. 54–68
3.16 D. Zwillinger: *Handbook of Differential Equations* (Academic Press, Boston, 1989) pp. 325–330
3.17 W. Feller: *An Introduction to Probability Theory and its Applications*, Vol. II (Wiley, New York, 1966; 2nd edition 1971) Chapter XV

Chapter 4

4.1 K. E. Cahill and R. J. Glauber: Phys. Rev. **177**, 1857 (1969); **177** 1882 (1969)
4.2 G. S. Agarwal and E. Wolf: Phys. Rev. D **2**, 2161 (1970); **2**, 2187 (1970); **2**, 2206 (1970)
4.3 P. D. Drummond and C. W. Gardiner: J. Phys. A **13**, 2353 (1980)
4.4 W. H. Louisell: *Quantum Statistical Properties of Radiation* (Wiley, New York, 1973) pp. 138–150,168–176
4.5 H. Haken: *Handbuch der Physik*, Vol. XXV/2c, ed. by L. Genzel (Springer-Verlag, Berlin, 1970 pp. 61–64
4.6 M. Hillery, R. F. O'Connell, M. O. Scully, E. P. Wigner: Phys. Rep. **106**, 121 (1984)
4.7 J. R. Klauder and E. C. G. Sudarshan: *Fundamentals of Quantum Optics* (Benjamin, New York, 1968) pp. 128,129,178–195
4.8 H. M. Nussenzveig: *Introduction to Quantum Optics* (Gordon and Breach, London, 1973) pp. 53–54
4.9 J. M. Normand: *A Lie Group: Rotations in Quantum Mechanics* (North Holland, Amsterdam, 1980) Appendix D, Sect. D.2.2
4.10 H. Weyl: *The Theory of Groups and Quantum Mechanics* (Dover, New York, 1950) pp. 272–276
4.11 R. J. Glauber: Phys. Rev. Lett. **10**, 84 (1963); Phys. Rev. **130**, 2529 (1963); **131**, 2766 (1963)
4.12 P. L. Kelly and W. H. Kleiner: Phys. Rev. **136**, 316 (1964)

Chapter 5

5.1 A. D. Fokker: Ann. Phys. (Leipzig) **43**, 310 (1915)
5.2 M. Planck: Sitzungsber. Preuss. Akad. Wiss. Phys. Math. Kl. **325** (1917)
5.3 C. W. Gardiner: *Handbook of Stochastic Methods for Physics, Chemistry and the Natural Sciences* (Springer-Verlag, Berlin, 1983) pp. 47–53
5.4 N. G. van Kampen: *Stochastic Processes in Physics and Chemistry* (North-Holland, Amsterdam, 1981)
5.5 H. Risken: *The Fokker-Planck Equation* (Springer-Verlag, Berlin, 1984)
5.6 Reference [5.3] pp. 146–147; Reference [5.5] pp. 133–134
5.7 W. Horsthemke and R. Lefever: *Noise-Induced Transitions. Theory and Applications in Physics, Chemistry and Biology* (Springer-Verlag, Berlin, 1984)

5.8 N. G. van Kampen: Can. J. Phys. **39**, 551 (1961); for further tutorial discussion see Reference [5.3] pp. 250–257, and Reference [5.4], Chapter IX – Chapter XI

5.9 H. A. Kramers: Physica **7**, 284 (1940)

5.10 J. E. Moyal: J. R. Stat. Soc. **11**, 151 (1949)

5.11 A. E. R. Woodcock and T. Poston: *A Geometrical Study of the Elementary Catastrophes* (Springer-Verlag, Berlin, 1974)

5.12 R. Gilmore: *Catastrophe Theory for Scientists and Engineers* (Wiley, New York, 1981) Chapter 6

5.13 See, for example, Reference [5.3] Chapter 9; Reference [5.4] pp. 304–311

5.14 R. K. Eisenschitz: *Matrix Algebra for Physicists* (Plenum, New York, 1966) Chapter 7

5.15 M. M. Wang and G. E. Uhlenbeck: Rev. Mod. Phys. **17**, 323 (1945)

5.16 H. T. H. Piaggio: *Differential Equations* (Bell, London, 1965) Chapter XII

5.17 Reference [5.3] pp. 36–37; Reference [5.5] pp. 23–24

5.18 Z. Schuss: *Theory and Applications of Stochastic Differential Equations* (Wiley, New York, 1980)

5.19 T. T. Soong: *Random Differential Equations in Science and Engineering* (Academic Press, New York, 1973)

5.20 See the discussions of Markoff processes in References [5.3-5.5, 5.18, 5.19]

5.21 Reference [5.3] pp. 70–73; Reference [5.4] pp. 17–18

5.22 K. Ito: *Lectures on Stochastic Processes*, Lecture Notes, Tata Inst. of Fundamental Res., Bombay, India, 1961

5.23 R. L. Stratonovich: SIAM J. Control **4**, 369 (1966)

5.24 Reference [5.3] pp. 83–101

5.25 R. E. Mortensen: J. Stat. Phys. **1**, 271 (1969)

5.26 N. G. van Kampen: J. Stat. Phys. **24**, 175 (1981)

5.27 Reference [5.3] pp. 96–97; Reference [5.18] Chapter 5; Reference [5.19] pp. 183–190

Chapter 6

6.1 H. Haken, H. Risken, and W. Weidlich: Z. Physik **206**, 355 (1967)

6.2 H. Haken: *Handbuch der Physik*, Vol. XXV/2c, ed. by L. Genzel (Springer-Verlag, Berlin, 1970) pp. 64–65

6.3 W. H. Louisell: *Quantum Statistical Properties of Radiation* (Wiley, New York, 1973) pp. 375–390

6.4 C. W. Gardiner: *Handbook of Stochastic Methods for Physics, Chemistry and the Natural Sciences* (Springer-Verlag, Berlin, 1983) pp. 78,79,402

6.5 A. Einstein: Phys. Z. **18**, 121 (1917)

6.6 M. Sargent III, M. O. Scully, and W. E. Lamb, Jr.: *Laser Physics* (Addison-Wesley, Reading, Massachusetts, 1974) pp. 20–23

6.7 Min Xiao, H. J. Kimble, and H. J. Carmichael: Phys. Rev. A **35**, 3832 (1987)

6.8 Min Xiao, H. J. Kimble, and H. J. Carmichael: J. Opt. Soc. Am. B **4**, 1546 (1987)

6.9 R. R. Puri and S. V. Lawande: Phys. Lett. **72A**, 200 (1979)

6.10 G. S. Agarwal: "Quantum Statistical Theories of Spontaneous Emission and their Relation to Other Approaches," *Springer Tracts in Modern Physics*, Vol. 70 (Springer-Verlag, Berlin, 1974) pp. 73–83

6.11 H. J. Carmichael: J. Phys. B **13**, 3551 (1980); Phys. Rev. Lett. **43**, 1106 (1979)

6.12 S. Sarkar and J. S. Satchell: Europhys. Lett. **3**, 797 (1987)

6.13 R. H. Dicke: Phys. Rev. **93**, 99 (1954)

6.14 J. M. Radcliffe: J. Phys. A **4**, 313 (1971)

6.15 F. T. Arecchi, E. Courtens, R. Gilmore, and H. Thomas: Phys. Rev. A **6**, 2211 (1972)
6.16 R. Bonifacio, P. Schwendimann, and F. Haake: Phys. Rev. A **4**, 854 (1971); R. Bonifacio and L. A. Lugiato: Phys. Rev. A **12**, 587 (1975)
6.17 F. Haake and R. J. Glauber: Phys. Rev. A **5**, 1457 (1972); Phys. Rev. A **13**, 357 (1976)
6.18 L. M. Narducci, C. A. Coulter, and C. M. Bowden: Phys. Rev. A **9**, 829 (1974)
6.19 J. P. Gordon: Phys. Rev. **161**, 367 (1967)
6.20 M. Gronchi and L. A. Lugiato: Lett. Nuovo Cimento **23**, 593 (1978)
6.21 R. H. Lehmberg: Phys. Rev. A **2**, 883 (1970)
6.22 Reference [6.10] pp. 25–38
6.23 E. Merzbacher: *Quantum Mechanics* (Wiley, New York, 1961) pp. 421–426

Chapter 7

7.1 H. Haken: *Handbuch der Physik*, Vol. XXV/2c, ed. by L. Genzel (Springer-Verlag, Berlin, 1970)
7.2 W. H. Louisell: *Quantum Statistical Properties of Radiation* (Wiley, New York, 1973) Chapter 9
7.3 M. O. Scully and W. E. Lamb Jr.: Phys. Rev. Lett. **16**, 853 (1966); Phys. Rev. **159**, 208 (1967); **166**, 246 (1968)
7.4 M. Sargent III, M. O. Scully, and W. E. Lamb Jr.: *Laser Physics* (Addison-Wesley, Reading, Massachusetts, 1974) Chapter XVII
7.5 W. E. Lamb Jr.: Phys. Rev. **134**, A1429 (1964)
7.6 Reference [7.2] Chapter 8; Reference [7.4] Chapter VIII
7.7 Reference [7.4] pp. 20–23
7.8 Reference [7.4] pp. 104,203
7.9 Reference [7.2] p. 288; Reference [7.4] p. 22
7.10 M. R. Young and S. Singh: Phys. Rev. A **35**, 1453 (1987)
7.11 Reference [7.4] problems 17.13 and 17.14; P. Meystre and M. Sargent III: *Elements of Quantum Optics* (Springer-Verlag, Berlin, 1991) pp. 469–483
7.12 I. S. Gradshteyn and I. M. Ryzhik: *Tables of Integrals Series and Products* (Academic Press, New York, 1965) p. 930
7.13 A. Yariv: *Introduction to Optical Electronics* (Holt, Rinehart and Wilson, New York, 1976) pp. 118–121
7.14 E. T. Jaynes and F. W. Cummings: Proc. IEEE **51**, 89 (1963)
7.15 M. Tavis and F. W. Cummings: Phys. Rev. **170**, 379 (1968); **188**, 692 (1969)
7.16 H. J. Carmichael and D. F. Walls: Phys. Rev. A **9**, 2686 (1974)
7.17 H. J. Carmichael: J. Opt. Soc. Am. B **4**, 1588 (1987)
7.18 C. W. Gardiner and M. J. Collett: Phys. Rev. A **31**, 3761 (1985)

Chapter 8

8.1 H. Haken: *Handbuch der Physik*, Vol. XXV/2c, ed. by L. Genzel (Springer-Verlag, Berlin, 1970) pp. 154–156
8.2 E. N. Lorenz: J. Atmos. Sci. **20**, 130 (1963)
8.3 Sparrow: *The Lorenz Equations: Bifurcations, Chaos, and Strange Attractors* (Springer-Verlag, New York, 1982)
8.4 H. Haken: Phys. Lett. **53A**, 77 (1975)
8.5 N. B. Abraham, P. Mandel, and L. M. Narducci: "Dynamical Instabilities and Pulsations in Lasers," in *Progress in Optics*, Vol. XXV, ed. by E. Wolf (North-Holland, Amsterdam, 1988) pp. 1–190

8.6 L. M. Narducci and N. B. Abraham: *Laser Physics and Laser Instabilties* (World Scientific, Singapore, 1988)

8.7 Reference [8.1] pp. 159–168

8.8 H. Risken: *The Fokker-Planck Equation* (Springer-Verlag, Berlin, 1984)

8.9 One possible definition for amplitude and phase operators is described by Loudon: *The Quantum Theory of Light*, second edition (Oxford University Press, Oxford, 1983) pp. 141–145. There has recently been much discussion of phase in quantum optics; some of the recent literature is reviewed by J. H. Shapiro and S. R. Shepard: Phys. Rev. A **43**, 3795 (1991)

Index

Texts and Monographs in Physics

Series Editors: R. Balian   W. Beiglböck   H. Grosse   E. H. Lieb
N. Reshetikhin   H. Spohn   W. Thirring

Essential Relativity Special, General, and Cosmological   Revised 2nd edition
By W. Rindler

The Elements of Mechanics
By G. Gallavotti

Generalized Coherent States and Their Applications
By A. Perelomov

Quantum Mechanics II
By A. Galindo and P. Pascual

Geometry of the Standard Model of Elementary Particles
By. A. Derdzinski

From Electrostatics to Optics
A Concise Electrodynamics Course
By G. Scharf

Finite Quantum Electrodynamics
The Causal Approach   2nd edition
By G. Scharf

Path Integral Approach to Quantum Physics An Introduction
2nd printing   By G. Roepstorff

Supersymmetric Methods in Quantum and Statistical Physics By G. Junker

Relativistic Quantum Mechanics and Introduction to Field Theory
By F. J. Yndurái n

Local Quantum Physics Fields, Particles, Algebras   2nd revised and enlarged edition
By R. Haag

The Mechanics and Thermodynamics of Continuous Media By M. Šilhavý

Quantum Relativity A Synthesis of the Ideas of Einstein and Heisenberg
By D. R. Finkelstein

Scattering Theory of Classical and Quantum N-Particle Systems
By. J. Derezinski and C. Gérard

Effective Lagrangians for the Standard Model By A. Dobado, A. Gómez-Nicola, A. L. Maroto and J. R. Peláez

Quantum The Quantum Theory of Particles, Fields, and Cosmology   By E. Elbaz

Quantum Groups and Their Representations
By A. Klimyk and K. Schmüdgen

Multi-Hamiltonian Theory of Dynamical Systems By M. Błaszak

Renormalization An Introduction
By M. Salmhofer

Fields, Symmetries, and Quarks
2nd, revised and enlarged edition   By U. Mosel

Statistical Mechanics of Lattice Systems
Volume 1: Closed-Form and Exact Solutions
2nd, revised and enlarged edition
By D. A. Lavis and G. M. Bell

Statistical Mechanics of Lattice Systems
Volume 2: Exact, Series and Renormalization Group Methods
By D. A. Lavis and G. M. Bell

Conformal Invariance and Critical Phenomena By M. Henkel

The Theory of Quark and Gluon Interactions
3rd revised and enlarged edition
By F. J. Yndurái n

Quantum Field Theory in Condensed Matter Physics By N. Nagaosa

Quantum Field Theory in Strongly Correlated Electronic Systems
By N. Nagaosa

Information Theory and Quantum Physics
Physical Foundations for Understanding the Conscious Process   By H.S. Green

Magnetism and Superconductivity
By L.-P. Lévy

The Nuclear Many-Body Problem
By P. Ring and P. Schuck

Perturbative Quantum Electrodynamics and Axiomatic Field Theory By O. Steinmann

Quantum Non-linear Sigma Models
From Quantum Field Theory to Supersymmetry, Conformal Field Theory, Black Holes and Strings   By S. V. Ketov

Series homepage – http://www.springer.de/phys/books/tmp

Texts and Monographs in Physics

Series Editors: R. Balian   W. Beiglböck   H. Grosse   E. H. Lieb
N. Reshetikhin   H. Spohn   W. Thirring

**The Statistical Mechanics
of Financial Markets** By J. Voit

Statistical Mechanics A Short Treatise
By G. Gallavotti

Statistical Physics of Fluids
Basic Concepts and Applications
By V. I. Kalikmanov

**Many-Body Problems and Quantum Field
Theory** An Introduction
By Ph. A. Martin and F. Rothen

Foundations of Fluid Dynamics
By G. Gallavotti

High-Energy Particle Diffraction
By E. Barone and V. Predazzi

Physics of Neutrinos
and Applications to Astrophysics
By M. Fukugita and T. Yanagida

Relativistic Quantum Mechanics
By H. M. Pilkuhn

Series homepage – http://www.springer.de/phys/books/tmp